AIRCRAFT **System Maintenance**

REVISED EDITION

AVOTEK®
INFORMATION RESOURCES

Production Staff

Designer/Photographer Dustin Blyer
Production Manager Holly Bonos
Designer Roberta Byerly
Designer/Lead Illustrator Amy Siever

International Standard Book Number 0-9708109-4-6
ISBN 13: 978-0-9708109-4-6
Order # T-AFSYS-0201

For Sale by: Avotek
A Select Aerospace Industries, Inc. company

Mail to:
P.O. Box 219
Weyers Cave, Virginia 24486
USA

Ship to:
200 Packaging Drive
Weyers Cave, Virginia 24486
USA

Toll Free: 1-800-828-6835
Telephone: 1-540-234-9090
Fax: 1-540-234-9399

Revised Edition
Fourth Printing
Printed in the USA

www.avotek.com

Preface

This textbook, the third in a series of four, was written for the Aviation Maintenance Technician student of today. It is based on the real-world requirements of today's aviation industry. At the same time, it does not eliminate the traditional subject areas taught since the first A&E schools were certificated.

This series of textbooks has evolved through careful study and gathering of information offered by the Federal Aviation Administration, the Blue Ribbon Panel, the Joint Task Analysis report, industry involvement and AMT schools nationwide.

The series is designed to fulfill both current and future requirements for a course of study in Aviation Maintenance Technology.

Textbooks, by their very nature, must be general in their overall coverage of a subject area. As always, the aircraft manufacturer is the sole source of operation, maintenance, repair and overhaul information. Their manuals are approved by the FAA and must always be followed. You may not use any material presented in this or any other textbook as a manual for actual operation, maintenance or repairs.

The writers, individuals and companies which have contributed to the production of this textbook have done so in the spirit of cooperation for the good of the industry. To the best of their abilities, they have tried to provide accuracy, honesty and pertinence in the presentation of the material. However, as with all human endeavors, errors and omissions can show up in the most unexpected places. If any exist, they are unintentional. Please bring them to our attention. ➻

Email us at comments@avotek.com for comments or suggestions.

Avotek® Aircraft Maintenance Series
Introduction to Aircraft Maintenance
Aircraft Structural Maintenance
Aircraft System Maintenance
Aircraft Powerplant Maintenance

Avotek® Aircraft Avionics Series
Avionics: Fundamentals of Aircraft Electronics
Avionics: Beyond the AET
Avionics: Systems and Troubleshooting

Other Books by Avotek®
Aircraft Corrosion Control Guide
Aircraft Structural Technician
Aircraft Turbine Engines
Aircraft Wiring & Electrical Installation
AMT Reference Handbook
Avotek Aeronautical Dictionary
Fundamentals of Modern Aviation
Light Sport Aircraft Inspection Procedures
Structural Composites: Advanced Composites in Aviation

Acknowledgements

Academy of Infrared Training, Inc.

Air Methods (Rocky Mountain Helicopters)

Jim Akovenko — JAARS

Alan Bandes — UE Systems, Inc.

Bob Blouin — National Business Aircraft Association

Boeing Commercial Airplane Co.

Tom Brotz — U.S. Industrial Tools

Greg Campbell, , Sherman Showalter, Stacey Smith — Shenandoah Valley Regional Airport

Champlin Fighter Aircraft Museum

Pat Colgan, Dennis Burnett — Colgan Air

Cal Crowder

De-Ice Systems International

Al Dibble — Snap-On Tools

Dynamic Solution Systems, Inc.

Eaton Aerospace Fluid Systems (Aeroquip)

Rob Gamble — Continental Airlines

Paul Geist, J.R. Dodson — Dodson International Parts

Raymond Goldsby — AIA

Mary Ellen Gubanic — Alcoa

Steve Hanson, Jeff Ellis, Tim Travis — Hawker Beechcraft

Lee Helm — RAM Aircraft Corporation

Ken Hyde, Weldon Britton — The Wright Experience

Lori Johnson, Larry Bartlett — Duncan Aviation

David Jones — Aviation Institute of Maintenance

Debbie Jones — MD Helicopters, Inc.

Robert Kafales — Frontier Airlines

Richard Kiser, Steve Bradley — Classic Aviation Services

Jack Knox — Evergreen Air Center

Lilbern Design

Lufthansa

Chris McGee — USAF Museum

Mckee Foods Flight Department

Jessica Meyers — Cessna Aircraft Company

Micro-Mesh

Richard Milburn — Northrop Grumman Corporation

Harry Moyer, Virgil Gottfried — Samaritan's Purse

Michelle Moyer

Phoenix Composites

Pilgrim's Pride Aviation Department

David Posavec — Barry Controls

Precision Airmotive Corp.

Precision Instruments

Vern Raburn, Andrew Broom — Eclipse Aviation

Jim Schmidt — Cincinnati State Technical

Scott Aviation Enterprises

Select Aerospace Industries, Inc

Select Airparts

Stern Technologies

Brian Stoltzfus — Priority Air Charter

Karl Stoltzfus, Sr., Michael Stoltzfus, Aaron Lorson & staff — Dynamic Aviation Group, Inc.

Ken Stoltzfus, Jr. — Preferred Airparts

Susan Timmons — JRA Executive Air

Virginia State Police Med Flight III

Vought Aircraft Industries, Inc.

Jean Watson — FAA

Rich Welsch — Dynamic Solutions Systems, Inc.

Andy Wilson — B/E Aerospace

Charlie Witman — Avotek

Fred Workley — Workley Aviation Mx

Contents

Preface ———————————————————————————————————— iii

Acknowledgements ————————————————————————————— iv

1 Hydraulic and Pneumatic Systems

1-1	Principles of Hydraulics	1-36	Hydraulic Contamination
1-7	Aircraft Hydraulic Systems	1-46	Typical Corporate Aircraft
1-26	Seals and Gaskets		Hydraulic Systems
1-30	Dampening and Absorbing Units	1-56	Aircraft Pneumatic Systems

2 Cabin Environmental Systems

2-1	Cabin Environmental Systems	2-24	Cabin Heaters
2-5	Pressurization Systems	2-31	Vapor-Cycle Air Conditioning System (Freon)
2-12	Air-Cycle Air Conditioning Systems	2-39	Oxygen Systems

3 Landing Gear Systems

3-1	Aircraft Landing Gear	3-44	Aircraft Brake Systems
3-27	Aircraft Tires and Tubes	3-70	Anti-skid System
3-40	Aircraft Wheels		

4 Aircraft Electrical Systems

4-2	Electrical Theory Overview	4-74	Electrical Components
4-11	Generators and Alternators	4-78	Electrical Power Sources and Monitoring
4-35	Aircraft Wiring		
4-69	Wiring Diagrams		

5 Aircraft Instrument Systems

5-1	Introduction	5-54	Standby and Other Flight Instruments
5-4	Classification of Instruments		
5-14	Flight Instruments	5-59	Engine Instruments
5-35	Gyroscopic Instruments	5-74	Maintenance of Instruments
5-50	Electronic Flight Information Systems	5-80	Autopilot Systems

6 Navigation and Communication Systems

6-2	Basic Radio Principles	6-26	Installation of Communication and Navigation Systems
6-6	Communication Systems		
6-10	Airborne Navigation Equipment	6-31	Advanced Integrated Navigation Instruments

7 Lighting, Warning and Utility Systems

7-1	Aircraft Lighting Systems		7-11	Warning Systems
7-8	Data Recorders		7-15	Interphone Systems
7-10	Traffic Alert and Collision Avoidance System		7-17	Advanced Cabin Entertainment Service System

8 Fire Protection Systems

8-1	Fire Protection Basics		8-20	Fire Prevention and Protection
8-2	Fire Detection Systems			
8-13	Fire Extinguishing Systems			

9 Aircraft Fuel Systems

9-1	Aircraft Fuel Basics		9-22	Inspection, Maintenance and Repair of Fuel Systems
9-6	Fuel System Types and Components			

10 Ice and Rain Control Systems

10-1	Ice Protection		10-11	Rain Protection

11 Aircraft Inspections

11-2	The Annual Inspection		11-15	Conformity Inspections
11-9	Progressive Inspections			

Index — I-1

1

Hydraulic and Pneumatic Systems

Section 1

Principles of Hydraulics

Hydraulics has proven to be the most efficient and economical system adaptable to aviation. First used by the ancient Greeks as a means of elevating the stages of their amphitheaters, the principles of hydraulics were explained scientifically by the seventeenth-century scholars Blaise Pascal and Robert Boyle. The laws discovered by these two men regarding the effects of pressure and temperature on fluids and gases in confined areas form the basis of the principle of mechanical advantage; in other words, the *why* and *how* of hydraulics.

In aviation, hydraulics is the use of fluids under pressure to transmit force developed in one location on an aircraft or other related equipment to some other point on the same aircraft or equipment. Hydraulics also includes the principles underlying hydraulic action and the methods, fluids and equipment used in implementing those principles.

Examples can be found in the everyday use of hydraulics in connection with familiar items such as automobile jacks and brakes. As a further example, the phrase *hydraulic freight elevator* refers to an elevator ascending and descending on a column of liquid instead of using cables and a drum.

On the other hand, the word hydraulics is the generic name of a subject. According to the dictionary, *hydraulics* is defined as a branch of science that deals with practical applications (such as the transmission of energy or the effects of flow) of a liquid in motion.

Left. The most visible parts of any aircraft hydraulic system are the landing gear and brakes. Sizes range from something you can hold in your hand to the behemoth shown in this photo. The smaller units can be overhauled by an individual technician, while the complex assemblies require a certified repair facility.

On fixed-wing aircraft, hydraulics is used to operate retractable landing gear and wheel brakes and to control wing flaps and propeller pitch. In conjunction with gases, hydraulics is used in the operation of rotor and wheel brakes, shock struts, shimmy dampeners, flight control systems, loading ramps and folding stairs.

Characteristics of Hydraulic Systems

Hydraulic systems have many desirable features. However, one disadvantage is the original high cost of the various components. This is more than offset by the many advantages that make hydraulic systems the most economical means of power transmission.

Efficiency. Discounting any losses that can occur in its mechanical linkage, practically all the energy transmitted through a hydraulic system is received at the output end — where the work is performed. The electrical system, its closest competitor, is 15-30 percent lower in efficiency. The best straight mechanical systems are generally 30-70 percent less efficient than comparable hydraulic systems because of high inertia factors and frictional losses. Inertia is the resistance to motion, action or change.

Dependability. The hydraulic system is consistently reliable. Unlike the other systems mentioned, it is not subject to changes in performance or to sudden, unexpected failure.

Control sensitivity. The confined liquid of a hydraulic system operates like a bar of steel in transmitting force. However, the moving parts are lightweight and can be, almost instantaneously, put into motion or stopped. The valves within the system can start or stop the flow of pressurized fluids almost instantly and require very little effort to manipulate. The entire system is very responsive to operator control.

Flexibility of installation. Hydraulic lines can be run almost anywhere. Unlike mechanical systems that must follow straight paths, the lines of a hydraulic system can be led around obstructions. The major components of hydraulic systems, with the exception of power-driven pumps that must be located near the power source, can be installed in a variety of places. The advantages of this feature are readily recognized when you study the many locations of hydraulic components on various types of aircraft.

Low space requirements. The functional parts of a hydraulic system are small in comparison to those of other systems; therefore, the total space requirement is comparatively low.

Lines of any length or contour can readily connect these components. They can be separated and installed in small, unused and out-of-the-way spaces. Large, unoccupied areas for the hydraulic system are unnecessary; in short, special space requirements are reduced to a minimum.

Low weight. The hydraulic system weighs remarkably little in comparison to the amount of work it does. A mechanical or electrical system capable of doing the same job weighs considerably more. Since payload weight is an important factor on aircraft, the hydraulic system is ideal for aviation use.

Self-lubricating. The majority of the parts of a hydraulic system operate in a bath of oil. Thus, hydraulic systems are practically self-lubricating. The few components that do require periodic lubrication are the mechanical linkages of the system.

Low maintenance requirements. Maintenance records have consistently shown that adjustments and emergency repairs to the parts of hydraulic systems are seldom necessary. The aircraft time-change schedules specify the replacement of components on the basis of hours flown or days elapsed and require relatively infrequent change of hydraulic components.

Definition of Hydraulic Terms

Force. The word *force*, used in a mechanical sense, means a push or pull. Force, because it is a push or pull, tends to cause the object on which it is exerted to move. In certain instances, when the force acting on an object is not sufficient to overcome its resistance or drag, no movement will take place. In such cases, force is still considered to be present.

Direction of force. Force can be exerted in any direction. It may act downward: as when gravity acts on a body, pulling it toward the earth. A force may act across: as when the wind pushes a boat across the water. A force can be applied upwards: as when an athlete throws (pushes) a ball into the air. Also, a force can act in all directions at once: as when a balloon bursts.

Magnitude of force. The extent, or magnitude, of a given force is expressed by means of a single measurement. In the United States, the *pound* is the unit of measurement of force. For example, it took 7.5 million lbs. of thrust (force) to lift the Apollo moonship off its launch pad. Hydraulic force is measured in the amount of pounds required to displace an object within a specified area, such as in a square inch.

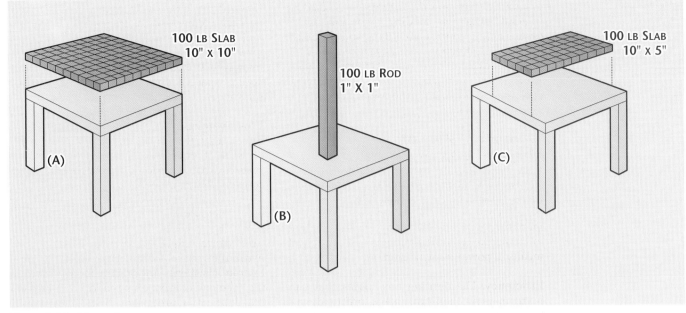

Figure 1-1-1. Measuring pressure

Pressure. The word *pressure*, when used in conjunction with mechanical and hydro-mechanical systems, has two different uses. One is technical, the other, non-technical. These two uses can be easily distinguished from each other by the presence or absence of a number. In technical use, a number always accompanies the word "pressure." In non-technical use, no number is present.

Technical. The number accompanying pressure conveys specific information about the significant strength of the force being applied. The strength of this applied force is expressed as a rate at which the force is distributed over the area on which it is acting. Thus, pounds per square inch (p.s.i.) expresses a rate of pressure just as miles per hour (m.p.h.) expresses a rate of speed.

Non-technical. The word pressure, when used in the non-technical sense, simply indicates that an unspecified amount of force is being applied to an object. Frequently, adjectives such as light, medium or heavy are used to remove some of the vagueness concerning the strength of the applied force.

Pressure Measurement

When used in the technical sense, pressure is defined as *the amount of force per unit area*. To have universal, consistent and definite meaning, standard units of measurement are used to express pressure. In the United States, the *pound* is the unit of measurement used for force, and the *square inch* is the unit for area.

A pressure measurement is always expressed in terms of both units of measurement just explained: *amount of force* and *unit area*. However, only one of these units, the amount of force, is variable. The square inch is used only in the singular — never more or less than one square inch.

A given pressure measurement can be stated in three different ways and still mean the same thing. Therefore, 50 p.s.i. pressure, 50 lbs. pressure and 50 p.s.i. all have identical meanings.

Examples of pressure measurement. A table with a 10-inch by 10-inch flat top contains 100 square inches of surface. If a 100-pound slab of exactly the same dimensions is placed on the tabletop, one pound per square inch pressure is exerted over the entire table surface.

Now, think of the same table (100 square inches) with a 100-pound block instead of the slab resting on its top. Assume this block has a face of only 50 square inches contacting the table. Because the area of contact has been cut in half and the weight of the block remains the same, the pressure exerted on the table doubles to 2 p.s.i.

As a final example, suppose a long rod weighing 100 lbs. with a face of one square inch is balanced upright on the tabletop. The pressure now being exerted on the table is increased to 100 p.s.i., since the entire load is being supported on a single square inch of the table surface. These examples are illustrated in Figure 1-1-1.

Force-area-pressure formulas. The formula to find the pressure acting on a surface is *pressure equals force divided by area*. If *P* is the symbol for pressure, *A* the symbol for area, and *F* the symbol for force, the formula can be expressed as follows:

$$P = \frac{F}{A}$$

By transposing the symbols in this formula, two other important formulas are derived: one for area, one for force. Respectively, they are:

$$A = \frac{F}{P} \qquad F = A \times P$$

However, when using any of these formulas, two of the factors must be known to be able to determine the third unknown factor.

The triangle shown in Figure 1-1-2 is a convenient way to remember the force-area-pressure formulas. It helps you recall the three factors involved: F, A and P. Because the F is above the line in the triangle, it also reminds you that in both formulas indicating division, *F* is always divided by one of the other two factors.

Transmission of Force

Two means of transmitting force are through solids and through liquids. Since this text is on hydraulics, the emphasis is on fluids. Force transmission through solids is presented only as a means of comparison.

Transmission of force through solids. Force applied at one point on a solid body follows a straight line, undiminished, to an opposite point on the body.

Transmission of force through confined liquids. Applied forces are transmitted through bodies of confined liquids in the manner

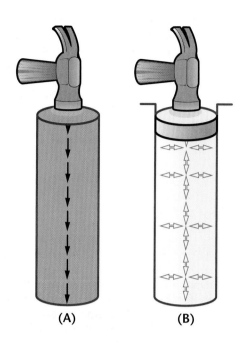

(A)   (B)

Figure 1-1-3. Transmission of force through solids and liquids

described by *Pascal's law*. This law of physics states: "pressure applied to any part of a confined liquid is transmitted without change in intensity to all parts of the liquid." This means that wherever it is applied on the body of liquid, pressure pushes with equal force against every square inch of the interior surfaces of the liquid's container. When pressure is applied to a liquid's container in a downward direction, it will not only act on the bottom surface; but on the sides and top as well. Both principles are illustrated in Figure 1-1-3.

The piston on the top of the tube is driven downward with a force of 100 p.s.i. This applied force produces an identical pressure of 100 p.s.i. on every square inch of the interior surface. Notice the pressure on the interior surface is always applied at right angles to the walls of the container, regardless of its shape. This illustrates that the forces acting within a body of confined liquid are explosive in pattern. If all sides are equal in strength, they will burst simultaneously if sufficient force is applied.

Characteristics of Fluids

The vast difference in the manner in which force is transmitted through confined liquids, as compared with solid bodies, is due to the physical characteristics of fluids — namely shape and compressibility. Liquids have no definite shape; they readily and instantly conform to their container. Because of this characteristic, the entire body of confined fluid tends to move away from the point of the initial force in

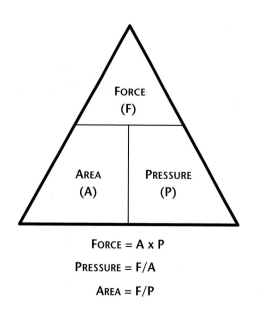

FORCE = A x P

PRESSURE = F/A

AREA = F/P

Figure 1-1-2. Formulas that illustrate the relationship of force, area and pressure

all directions until stopped by something solid such as the walls of the container. Liquids are relatively incompressible. That is, they can only be compressed by approximately 1 percent of their volume. Because liquids lack their own shape and are incompressible, an applied force transmitted through a body of liquid confined in a rigid container results in no more compression than if it were transmitted through solid metal.

The internal resistance of a fluid that tends to prevent it from flowing is called viscosity. The viscosity of a fluid decreases as the temperature increases.

Movement of fluid under pressure. Force applied to a confined liquid can cause the liquid to move only when that force exceeds any other force acting on the liquid in an opposing direction. Fluid flow is always in the direction of the lowest pressure. If the opposing forces are equal, no movement of fluid takes place.

Fluid under pressure can flow into already-filled containers only if an equal or greater quantity simultaneously flows out of them. This is an obvious and simple principle, but one that is easily overlooked.

Effects of temperature on liquids. As in metals, temperature changes produce changes in the size of a body of liquid. With the exception of water, whenever the temperature of a body of liquid falls, a decrease (contraction) in size of the body of fluid takes place. The amount of contraction is slight and takes place in direct proportion to the change in temperature.

When the temperature rises, the body of liquid expands. This is referred to as *thermal expansion*. The amount of expansion is in direct proportion to the temperature. Although the rate of expansion is relatively small, it is important. In a hydraulic system, some provision is usually necessary to accommodate the increase in size of the body of liquid when an increase in temperature occurs.

Mechanical Advantage

By simple definition, mechanical advantage is equal to the ratio of a force or resistance overcome by the application of a lesser force or effort through a simple machine. This represents a method of multiplying forces. In mechanical advantage, the gain in force is obtained at the expense of a loss in distance. Discounting frictional losses, the percentage gain in force equals the percentage loss in distance. Two familiar applications of the principles of mechanical advantage are the lever

and the hydraulic jack. In the case of the jack, a force of just a pound or two applied to the jack handle can raise many hundreds of pounds of load. Note, though, that each time the handle is moved several inches, the load is raised only a fraction of an inch.

Application in hydraulics. The principle used in hydraulics to develop mechanical advantage is simple. Essentially, it is obtained by fitting two movable surfaces of different sizes to a confining vessel, such as pistons within cylinders. The vessel is filled with fluid, and force (input) is applied to the smaller surface. This pressure is then transferred, by means of the fluid, to the larger surface, where a proportional force (output) is produced.

Rate. The rate mechanical advantage is produced by hydraulic means is in direct proportion to the ratio of the size of the smaller (input) area to the size of the larger (output) area. Thus, 10 lbs. of force applied to 1 square inch of surface of a confined liquid produces 100 lbs. of force on a movable surface of 10 square inches. This is illustrated in Figure 1-1-4. The increase in force is not free, but is obtained at

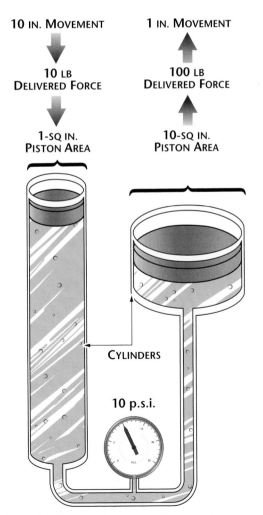

Figure 1-1-4. Hydraulics and mechanical advantage

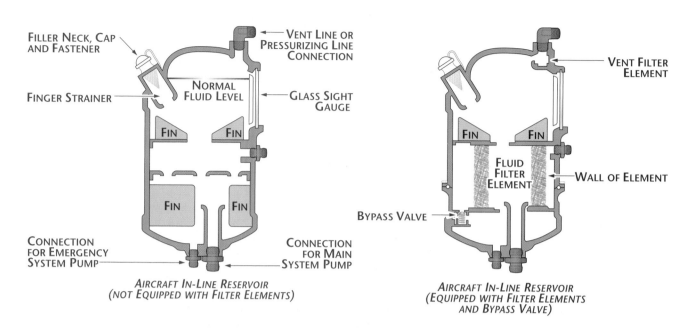

Figure 1-2-1. Typical hydraulic reservoir

the expense of distance. In this case, the ten-fold increase in output force is gained at the expense of a tenfold increase in distance over which the initial force is applied.

The Role of Air in Hydraulics

Some hydraulic components require air as well as hydraulic oil for their operation. Other hydraulic components do not; on the contrary, their performance is seriously impaired if air accidentally leaks into the system.

Familiarization with the basic principles of pneumatics aids in understanding the operation of the hydraulic components requiring air, and those that do not. It aids, also, in understanding how air can upset the normal operation of a hydraulic system if it is present in the system where it must not be.

Air. When used in reference to hydraulics, air is understood to mean atmospheric air. Briefly, air is defined as a complex, indefinite mixture of many gases. Of the individual gases that make up atmospheric air, 90 percent or more is oxygen and nitrogen.

Some knowledge of the physical characteristics of air is quite important to this instruction. Because the physical properties of all gases, including air, are the same, a study of these properties is made with reference to gases in general. It is important to realize, however, though similar in physical characteristics, gases differ greatly in their individual chemical composition. This difference makes some gases extremely dangerous when under pres-

sure or when they come in contact with certain substances.

Air and nitrogen. Air and pure nitrogen are inert gases and are safe and suitable for use in hydraulic systems. Most frequently, the air used in hydraulic systems is drawn out of the atmosphere and forced into the hydraulic system by means of an air compressor. Pure nitrogen, however, is available only as a compressed bottle gas.

Application in hydraulics. The ability of a gas to act in the manner of a spring is important in hydraulics. This characteristic is used in some hydraulic systems to enable these systems to absorb, store and release fluid energy as required. These abilities within a system are often provided by means of a single component designed to produce a spring-like action. In most cases, such components use air, even though a spring might be equally suitable from a performance standpoint. Air is superior to a spring because of its low weight and because it is not subject to failure from metal fatigue, as is a spring. The most common use of air in hydraulic systems is found in accumulators and shock struts.

Malfunctions caused by air. In general, all components and systems that do not require gases in their operation are to some extent impaired by the presence of air. Examples are excessive feedback of loud noises from flight controls during operation and the failure of wheel and rotor brakes to hold. These malfunctions can be readily corrected by *bleeding the system*. Bleeding the system is a controlled way of allowing the air to escape.

Section 2

Aircraft Hydraulic Systems

Hydraulic systems are not new. The basic hydraulic ram principle has been used for centuries. It is still supplying most of our cities with basic water service. Commercially, hydraulic systems are used for everything from brakes to backhoes. It is in the airplane, however, that hydraulic systems have reached a higher degree of sophistication.

Although some aircraft manufacturers make greater use of hydraulic systems than others, the hydraulic system of the average modern aircraft performs many functions. Among the units commonly operated by hydraulic systems are landing gear, wing flaps, speed and wheel brakes, and flight control surfaces.

Hydraulic systems combine the advantages of light weight, ease of installation, simplification of inspection and minimum maintenance requirements. Hydraulic operations are also almost 100 percent efficient, with only a negligible loss due to fluid friction.

All hydraulic systems are essentially the same, regardless of their function. Regardless of application, each hydraulic system has a minimum number of components and some type of hydraulic fluid.

A means of storing hydraulic fluid and minimizing contamination is necessary to any aircraft hydraulic system. Reservoirs and filters perform these functions. The component, which causes fluid flow in a hydraulic system—the heart of any hydraulic system—can be a hand pump, power-driven pump, accumulator or any combination of the three. Finally, a means of converting hydraulic pressure to mechanical motion, linear or rotary, is necessary.

Hydraulic Reservoirs

The hydraulic reservoir is a container for holding the fluid required to supply the system, including a reserve to cover any losses from minor leakage and evaporation. The reservoir can be designed to provide space for fluid expansion, permit air entrained in the fluid to escape, and to help cool the fluid. Figure 1-2-1 shows two typical reservoirs. Compare the two reservoirs item by item and, except for the filters and bypass valve, notice the similarities.

Filling reservoirs to the top during servicing leaves no space for expansion. Most reservoirs are designed with the rim at the filler neck below the top of the reservoir to prevent overfilling. Some means of checking the fluid level is usually provided on a reservoir. This may be a glass or plastic sight gauge, a tube or a dipstick. Hydraulic reservoirs are either vented to the atmosphere or closed to the atmosphere and pressurized. A description of each type follows.

Vented reservoir. A vented reservoir is one that is open to atmospheric pressure through a vent line. Because atmospheric pressure and

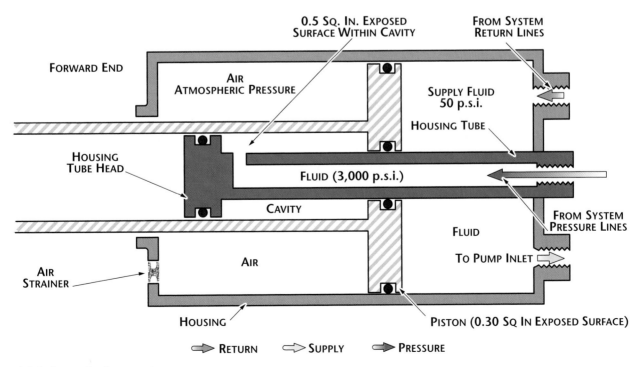

Figure 1-2-2. Pressurized reservoir

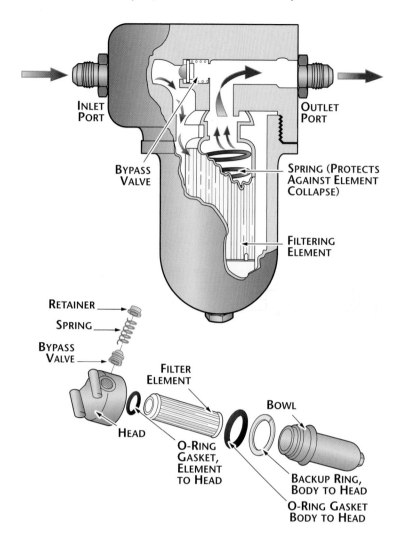

Figure 1-2-3. Typical line filter assembly (exploded view)

Figure 1-2-4. Filter assembly

gravity are the forces that cause the fluid to flow to the pump, a vented reservoir is mounted at the highest point in the hydraulic system. Air is drawn into and exhausted from the reservoir through a vent line. A filter is usually installed in the vent line to prevent foreign material from being taken into the system.

Pressurized reservoir. A pressurized reservoir is sealed from the atmosphere. This reservoir is pressurized either by engine bleed air or by hydraulic pressure produced within the hydraulic system itself. An air pressure regulator is used to reduce the compressor bleed air pressure. The regulator is located between the engine and the reservoir. Pressurized reservoirs are used on aircraft intended for high altitude flight, where atmospheric pressure is not enough to cause fluid flow to the pump. Figure 1-2-2 illustrates a typical pressurized reservoir.

Additional reservoir components. Many reservoirs are constructed with baffles or fins to keep the fluid from swirling and foaming. Foaming can cause air to become entrained in

the system. Filters are incorporated in some reservoirs to filter the fluid before it leaves the reservoir. A bypass valve is used to ensure that the pump does not starve if the filter becomes clogged. A standpipe is used in a reservoir that supplies a normal and an emergency system. The main system draws its fluid from the standpipe, which is located at a higher elevation. This ensures an adequate fluid supply to the secondary system if the main system fails.

Hydraulic Filter

Contamination of hydraulic fluid is one of the common causes of hydraulic system troubles. Installing filter units in the pressure and return lines of a hydraulic system allows contamination to be removed from the fluid before it reaches the various operating components. Filters of this type are referred to as line filters.

Line filter construction. A typical line filter is shown in Figures 1-2-3 and 1-2-4. It has two major parts — the filter case, or bowl, and the filter head. The bowl holds the head that screws into it. The head has an inlet port, outlet port and relief valve. Normal fluid flow is through the inlet port, around the outside of the element, through the element to the inner chamber and out through the outlet port. The bypass valve lets the fluid bypass the filter element if it becomes clogged.

Types of filter elements. The most common filtering element used is the micronic type. It is a disposable unit made of treated cellulose and is formed into accordion pleats. Most filter elements are capable of removing all contaminants larger than 10-25 microns (one micron equals 0.00004 inch). See Figure 1-2-5.

Another type, not commonly used anymore, is the Cuno filter element. It has a stack of closely spaced disks shaped like spoked wheels. The fluid is filtered as it passes between the disks.

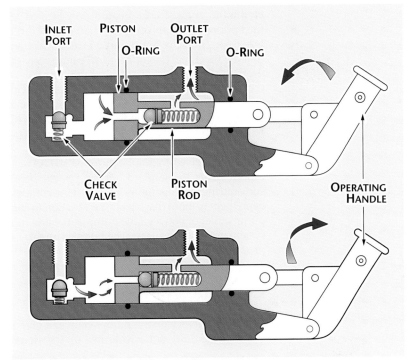

Figure 1-2-6. Double-action hand pump

Figure 1-2-5. Micronic filter element

Hand-Operated Hydraulic Pumps

The heart of any hydraulic system is the pump, which converts mechanical energy into hydraulic energy. The source of mechanical energy may be an electric motor, the engine or the operator's muscle.

Pumps powered by muscle are called hand pumps. They are used in emergencies, as backups for power pumps and for ground checks of the hydraulic system. The double-action hand pump produces fluid flow with every stroke of the handle.

Handle to the right. The double-action hand pump, shown in Figure 1-2-6, consists of a cylinder piston with built-in check valve, piston rod, operating handle and a check valve built into the inlet port. As the handle is moved to the right, the piston and rod also move to the right. On this stroke, the inlet check valve opens as a result of the partial vacuum caused by the movement of the piston, allowing fluid to be drawn into the left chamber. At the same time, the inner check valve closes. As the piston moves to the right, the fluid in the right chamber is forced out into the system.

Handle to the left. When the handle is moved to the left, the piston and rod assembly also move to the left. The inlet check valve now closes, preventing the fluid in the left chamber from returning to the reservoir. At the same time, the piston-head check valve opens, allowing the fluid to enter the right chamber.

In order to have hydraulic fluid available for operation of a hand pump, the hydraulic reservoir must have a standpipe. The standpipe will allow adequate fluid to remain in the tank for emergency operations.

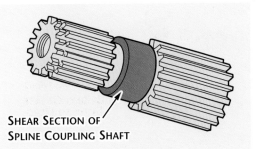

Figure 1-2-7. Pump drive coupling

Power-Driven Hydraulic Pumps

Power-driven pumps receive their driving force from an external power source, such as electric motors, air motors and the aircraft engine. This force is converted into energy in the form of fluid pressure. The four basic types of power-driven hydraulic pumps are *gear, vane, diaphragm and piston*. Piston pumps are further categorized as either *constant delivery* or *variable delivery*. Two types of constant-delivery piston pumps are the angular and cam. All constant delivery or positive displacement pumps must have a pressure-regulating valve.

Pumps are coupled to their driving units by a splined coupling shaft, commonly called a *drive coupling*. As shown in Figure 1-2-7, the shaft is designed with a weakened center section called a *shear section,* with just enough strength to run the pump under normal conditions.. Should some trouble develop causing the pump to turn unusually hard, the shear section will break. This prevents damage to the pump or driving unit.

Hydraulic pumps have a drain port located at the bottom of the pump at the shaft end. If fluid is leaking from the drain port it is an indication that the shaft seal is leaking. Most pumps have a drain port on each side of the mounting base so that the pump may be mounted in any direction for clearance. The drain port has an overboard drain line connected to it, so that any fluid will be drained out of the engine compartment.

If a power-driven hydraulic pump of the correct capacity fails to maintain normal system pressure during the operation of a system component, there is more than likely a restriction in the pump outlet or between the pump outlet and the system pressure regulator. The pump is unable to provide the volume of fluid the system requires.

Constant-Delivery Pumps

Constant-delivery piston pumps deliver a given quantity of fluid per revolution of the drive

Figure 1-2-8. Typical angular piston pump

Figure 1-2-9. Typical rotating-cam piston pump

coupling, regardless of pressure demands. The quantity of fluid delivered per minute depends on pump revolutions per minute (r.p.m.). In a system requiring constant pressure, this type of pump must be used with a pressure regulator.

Gear Pumps

The most common constant-delivery pump is the gear pump. Gear pumps have been used for many years in almost any type of hydraulic system you can imagine. They are still used in many light general aviation aircraft. Classic examples are the hydraulic powerpacs used in the Piper aircraft line. The powerpacs have the pump, reservoir, check and pressure regulator valves, and all other necessary items all in one unit. A powerpac is not a regular maintenance item and should therefore be sent to a Certified Repair Station for repair and overhaul.

While capable of producing extreme pressures, gear pumps have a major drawback. They do not work well in systems that must move a large volume of fluid. To do so would require pumps of ever-increasing size to produce the volume.

Angular Piston Pumps

The basic components of an angular piston pump are shown in Figure 1-2-8. They are:

- A rotating group, consisting of a coupling shaft, universal link, connecting rods, pistons and cylinder block

- A stationary group, consisting of the valve plate and the pump case or housing

The cylinder bores lie parallel to, and are evenly spaced around, the pump axis. For this reason, a piston pump is often referred to as an *axial piston pump*.

Packings on seals are not required to control piston-to-bore leakage. This is controlled entirely by close machining and accurate fit between piston and bore. The clearance is only enough to allow for lubrication by the hydraulic fluid and slight expansion when the parts become heated. Pistons are individually fitted to their bores during manufacture and must not be changed from pump to pump or bore to bore.

Pump operation. As the coupling shaft is turned by the pump power source, the pistons and cylinder block turn along with it because they are interconnected. The angle that exists between the cylinder block and coupling shaft causes the pistons to move back and forth in their respective cylinder bores as the coupling is turned.

During the first half of a revolution of the pump, a cylinder is aligned with the inlet port in the valve plate. At this time, the piston is moving away from the valve plate and drawing hydraulic fluid into the cylinder. During the second half of the revolution, the cylinder is lining up with the outlet port in the valve plate. At this time, the piston is moving toward the valve plate, thus causing fluid previously drawn into the cylinder to be forced out through the outlet port.

Fluid is constantly being drawn into and expelled out of the pump as it turns. This provides a multiple overlap of the individual spurts of fluid forced from the cylinders and results in delivery of a smooth, non-pulsating flow of fluid from the pump.

Cam-Piston Pumps

A cam is used to cause the stroking of the pistons in a cam-piston pump. Two variations are used, the cam rotates and the cylinder block is stationary; in the other, the cam is stationary and the cylinder block rotates. Both cam-piston pumps are described in the following paragraphs.

Rotating-cam pump. The *rotating-cam pump* is the one most commonly used. As the cam turns in a rotating-cam pump (Figure 1-2-9), its high and low points pass alternately and in turn under each piston. It pushes the piston further into its bore, causing fluid to be expelled from the bore. When the falling face of the cam comes under a piston, the piston's return spring pulls the piston down in its bore. This causes fluid to be drawn into the bore.

Each bore has a check valve that opens to allow fluid to be expelled from the bore by the piston's movement. These valves are closed by spring pressure during inlet strokes of the pistons. This fluid is drawn into the bores only through the central inlet passages. The movement of the pistons in drawing in and expelling fluid is overlapping, resulting in a non-pulsating fluid flow.

Stationary-cam pump. The operation and construction of a *stationary-cam pump* are identical to that of the rotating cam except that the cylinder block turns, not the cam.

Variable-Delivery Piston Pumps

A variable-delivery piston pump automatically and instantly varies the amount of fluid delivered to the pressure circuit of a hydraulic system to meet varying system demands. This is accomplished by using a *compensator,* an integral part of the pump, which is sensitive to the amount of pressure present in the pump and in the hydraulic system pressure circuit. When the circuit pressure rises, the compensator causes the pump output to decrease.

Conversely, when circuit pressure drops, the compensator causes pump output to increase. There are two ways of varying output: *demand principle* (cam) and *stroke-reduction principle* (angular).

Demand principle. The demand principle (Figure 1-2-10A) is based on varying pump output to fill the system's changing demands

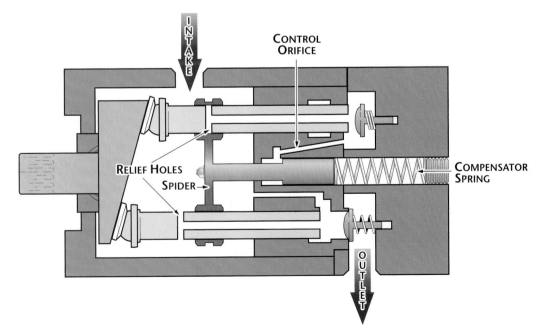

Figure 1-2-10A. Variable-delivery demand-principle cam pump

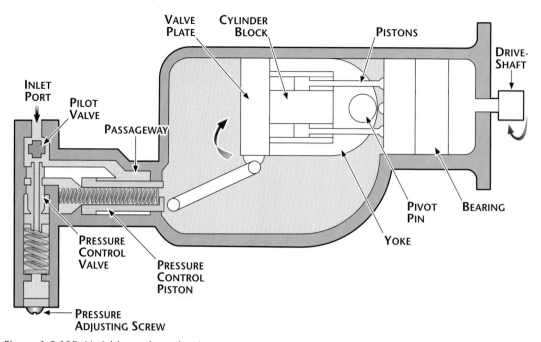

Figure 1-2-10B. Variable stroke-reduction pump

by making the piston stroke effective in varying degrees.

The pistons are designed with large hollow centers. The centers are intersected by cross-drilled relief holes that open into the pump case. Each piston is equipped with a movable sleeve, which can block the relief holes. When these holes are not blocked, fluid displaced by the pistons is discharged through the relief holes into the pump case, instead of past the pump check valves and out the outlet port.

When full fluid flow is required, the sleeves are positioned to block the relief holes for the entire length of piston stroke. When zero flow is required, the sleeves are positioned not to block the flow during any portion of the piston stroke. For requirements between zero and full flow, the relief holes are uncovered or blocked accordingly.

The sleeves are moved into their required positions by a device called a pump compensator piston. The sleeves and compensator piston are interconnected by means of a spider. Fluid pressure for the compensator piston is obtained from the discharge port (system pressure) through a control orifice.

Stroke-reduction principle. The stroke-reduction principle shown in Figure 1-2-10B is based on varying the angle of the cylinder block in an angular pump. This controls the length of the piston's stroke and thus the volume per stroke.

The cylinder block angle change is achieved by using a yoke that swivels around a pivot pin called a *pintle*. Using a compensator assembly consisting of a *pressure-control valve*, *pressure-*control piston* and *mechanical linkage* that are connected to the yoke automatically controls the angle.

As system pressure increases, the pilot valve opens a passageway allowing fluid to act on the control piston. The piston moves, compressing its spring, and through mechanical linkage, moves the yoke toward the zero flow (zero angle) position. As system pressure decreases, the pressure is relieved on the piston, and its spring moves the pump into the full flow position. This results in a control that is very precise, with no drops in system pressure.

Hydraulic Accumulators

The purpose of a hydraulic accumulator is to store hydraulic fluid under pressure. It may be used to:

- Dampen hydraulic shocks that may develop when pressure surges occur in hydraulic systems

- Add to the output of a pump during peak load operation of the system, making it possible to use a pump of much smaller capacity than would otherwise be required

- Absorb the increases in fluid volume caused by increases in temperature

- Act as a source of fluid pressure for starting aircraft auxiliary power units (APUs)

- Assist in emergency operations

Accumulators are divided into types according to the means used to separate the air fluid

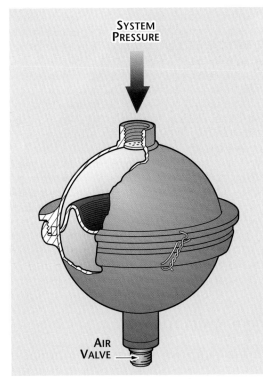

Figure 1-2-11. Diaphragm accumulator

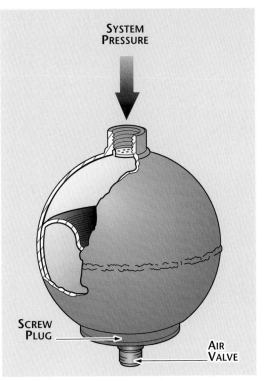

Figure 1-2-12. Bladder accumulator

chambers; these are the *diaphragm, bladder* and *piston accumulators.*

If the air pressure in an accumulator is low it will cause the system pressure to drop rapidly when hydraulic pressure is used for any device. It will also cause rapid cycling of the pressure regulator as the system builds pressure, shuts off, loses pressure and restarts, in an endless cycle.

The air pressure, or preload, of an accumulator is checked, on the ground, two ways. A pressure gauge on the accumulator will indicate the preload when the system fluid pressure is zero. At any other time, it will show the system pressure. The second way requires bringing the system up to operating pressure with an auxiliary pump. After shutting off the pump, actuate any part of the hydraulic system, other than the landing gear, and observe the hydraulic system pressure gauge. System pressure will decrease gradually. A sudden drop to zero will occur when the accumulator preload is reached. That point will be the approximate accumulator preload.

Diaphragm accumulator. The diaphragm accumulator consists of two hollow, hemispherical metal sections bolted together at the center. Notice in Figure 1-2-11 that one of the halves has a fitting to attach the unit to the hydraulic system; the other half is equipped with an air valve for charging the unit with *compressed air* or *nitrogen.* Mounted between the two halves is a synthetic rubber diaphragm that divides the accumulator into two sections. The accumulator is initially charged with air through the air valve to a pres-

Figure 1-2-13. Accumulator bladder

sure of approximately 50 percent of the hydraulic system pressure. This initial air charge forces the diaphragm upward against the inner surface of the upper section of the accumulator.

When fluid pressure increases above the initial air charge, fluid is forced into the upper chamber through the system pressure port, pushing the diaphragm down and further compressing the air in the bottom chamber. Under peak load, the air pressure in the lower chamber forces fluid back into the hydraulic system to maintain operating pressure. Also, if the power pump fails, the compressed air forces a limited amount of pressurized fluid into the system.

Bladder accumulator. The bladder accumulator operates on the same principle and for the same purpose as the diaphragm accumulator, but varies in construction, as shown in Figure 1-2-12, with the bladder itself shown in Figure 1-2-13. The unit

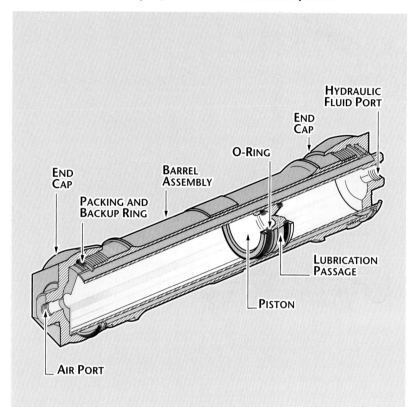

Figure 1-2-14. Piston accumulator (cutaway)

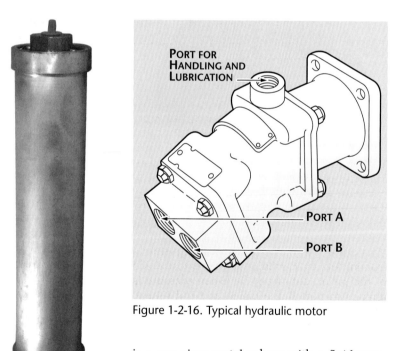

Figure 1-2-15. Piston accumulator

Figure 1-2-16. Typical hydraulic motor

Piston accumulator. The piston accumulator serves the same purpose and operates by the same principles as the diaphragm and bladder accumulators. As shown in Figure 1-2-14, the unit consists of a cylinder and piston assembly with ports on each end. Fluid pressure from the system enters the right port, forcing the piston down against the initial air charge in the left chamber of the cylinder. A high-pressure air valve is located at the left port for charging the unit. A drilled passage from the fluid side of the piston to the outside of the piston provides lubrication between the cylinder walls and the piston. A complete piston accumulator is shown in Figure 1-2-15.

CAUTION: *Release all gas and fluid pressure from an accumulator before removal. Failure to do so can result in serious injury.*

Hydraulic Motors

Hydraulic motors are installed in hydraulic systems to use hydraulic pressure to obtain powered rotation. A hydraulic motor does just the opposite of what a power-driven pump does. A pump receives rotative force from an engine or other driving unit and converts it into hydraulic pressure. A hydraulic motor receives hydraulic fluid pressure and converts it into rotative force.

Figure 1-2-16 shows a typical hydraulic motor. The two main ports through which fluid pressure is received and return fluid is discharged are marked A and B, respectively. The motor has a cylinder block-and-piston assembly in which the bores and pistons are in axial arrangement, the same as in a hydraulic pump. Hydraulic motors can be instantly started, stopped, or reversed under any degree of load; they can be stalled by overload without damage. Reversing the flow of fluid into the ports of the motor changes the direction of rotation of a hydraulic motor.

Hydraulic Actuators

So that fluid pressure produced by a pump can be used to move some object, the pressure must be converted to usable forces by means of an actuating unit. A device called an actuating cylinder is used to impart powered straight-line motion to a mechanism. Hydraulic motors convert the pressure into rotary motion.

Hydraulic systems must also have devices to control or direct the fluid pressure to the various components. Such devices include selector valves, check valves, ratchet valves, irreversible valves, sequence valves and priority valves. Each is described in the paragraphs that follow.

is a one-piece metal sphere with a fluid pressure inlet at the top and an opening at the bottom for inserting the bladder. A large screw-type plug at the bottom of the accumulator is a retainer for the bladder that also seals the unit. A high-pressure air valve is also incorporated in the retainer plug. Fluid enters through the system pressure port. As fluid pressure increases above the initial air charge of the accumulator, it forces the bladder downward against the air charge, filling the upper chamber with fluid pressure.

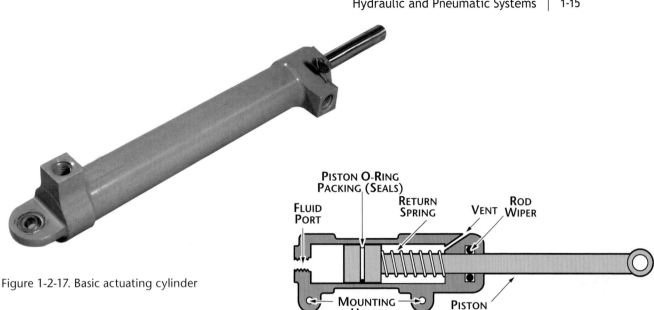

Figure 1-2-17. Basic actuating cylinder

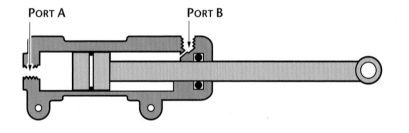

Figure 1-2-18. Single-action actuating cylinder

Figure 1-2-19. Double-action actuating cylinder

Actuating cylinders. A basic actuating cylinder (Figure 1-2-17) consists of a cylinder housing, one or more pistons and piston rods, and one or more seals. The cylinder housing contains a polished bore in which the piston operates and one or more ports through which fluid enters and leaves the bore. The piston and rod form an assembly that moves forward and backward within the cylinder bore. The piston rod moves into and out of the cylinder housing through an opening in one or both ends. The seals are used to prevent leakage between the piston and cylinder bore, and between the piston rod and housing. The two major types of actuating cylinders are *single-action* and *double-action*.

Single-action actuating cylinder. The single-action actuating cylinder, shown in Figure 1-2-18, consists of a cylinder housing with one fluid port, a piston and rod assembly, a piston return spring and seals.

When no pressure is applied to the piston, the return spring holds it and the rod assembly in the retracted position. When hydraulic pressure is applied to the inlet port, the piston, sealed to the cylinder wall by an O-ring, does not allow the fluid to pass. This causes the piston to extend.

As the piston and rod extend, the return spring compresses. A vent on the spring side of the piston allows air to escape. When pressure is relieved, the return spring forces the piston to retract, pushing the fluid out of the cylinder. A *wiper* in the housing keeps the piston rod clean.

The cylinder can be pressure-operated in one direction only. A *three-way control valve* is normally used to control cylinder operation.

Double-action actuating cylinder. The double-action actuating cylinder consists of a cylinder with a port at either end and a piston and rod assembly extending through one end of the cylinder (Figure 1-2-19).

Pressure applied at port A causes the piston to extend, forcing the fluid on the opposite side of the piston out of port B. When pressure is applied to port B, the piston and rod retract, forcing the fluid in the opposite chamber out through port A.

This type of cylinder is powered in both directions by hydraulic pressure. A selector valve is normally used to control a double-action actuating cylinder. A common cause of slow actuation of hydraulic components is an internal leak in the actuating unit.

Selector Valves

Used in hydraulic systems to control the direction of operation of a mechanism, selector

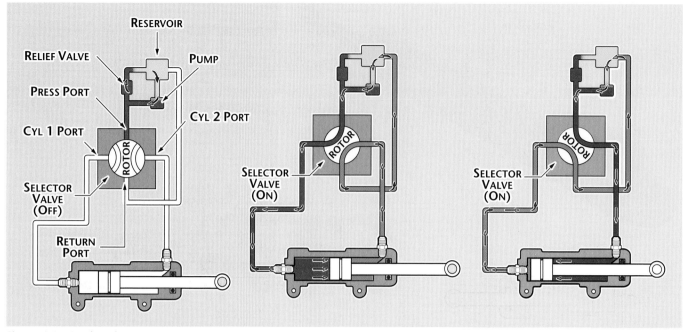

Figure 1-2-20. Closed-center rotor selector valve

valves are also referred to as *directional control valves* or *control valves*. They provide pathways for the simultaneous flow of two streams of fluid, one under pressure into the actuating unit and the other, a *return stream*, out of the actuating unit. The selector valves have various

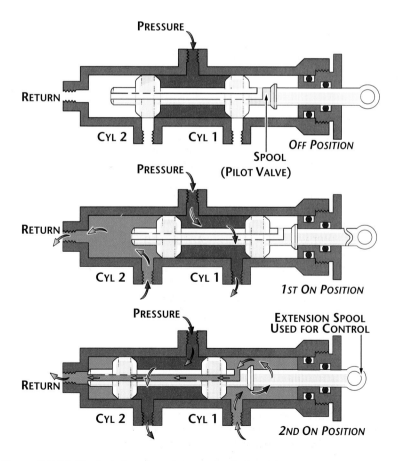

Figure 1-2-21. Typical closed-center spool selector valve

numbers of ports determined by the requirements of the system in which the valve is used. Selector valves with four ports are the most commonly used; they are referred to as *four-way valves*. Selector valves are further classified as *closed-center* or *open-center types*.

Closed-center selector valve. When a closed-center selector valve is placed in the OFF position, its pressure passage is blocked to the flow of fluid. Therefore, no fluid can flow through its pressure port, and the hydraulic system stays at operating pressure at all times. The four-way, closed-center selector valve is the most commonly used selector valve in aircraft hydraulics. There are two types; they are rotor-type and spool-type, closed-center selector valves.

The *rotor-type, closed-center selector valve* is shown in Figure 1-2-20. It has a rotor as its valving device. The rotor is a thick circular disk with drilled fluid passages. It is placed in its various operating positions by relative movement of the valve control handle. In the OFF position, the rotor is positioned to close all ports. In the first ON position, the rotor interconnects the pressure port with the number 1 cylinder port. The number 2 cylinder port is open to return. In the second ON position, the reverse takes place.

The *spool-type, closed-center selector valve* is shown in Figure 1-2-21. This valve has a housing containing four ports and a *spool*, or *pilot*, valve. The spool is made from a round shaft having machined sections forming spaces to allow hydraulic fluid to pass. A drilled passage in the spool interconnects the two end cham-

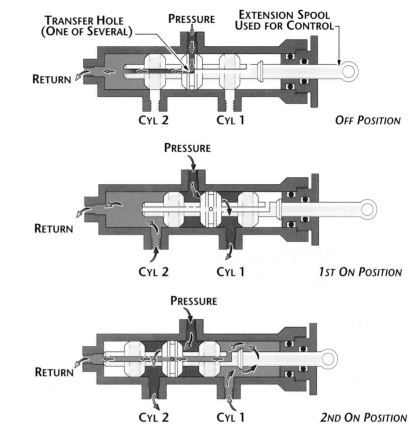

Figure 1-2-23. Typical open-center spool selector valve

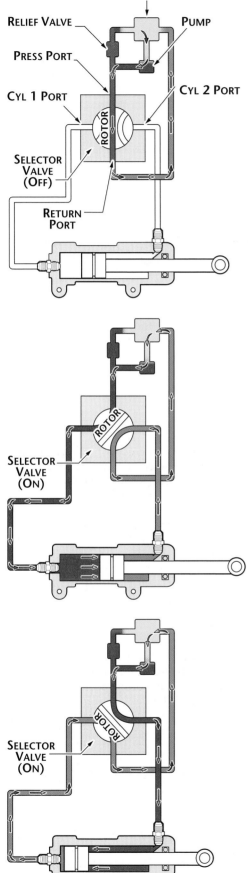

Figure 1-2-22. Open-center rotor selector valve

bers of the selector valve. The large diameters of the spool are the bearing and sealing surfaces and are called *lands*. In operation, the spool valve is identical to the rotor type.

Open-center selector valve. In external appearance, the *open-center selector valve* looks like the closed-center one. Like closed-center valves, open-center selector valves have four ports and operate in one OFF and two ON positions. The difference between the closed-center and open-center valves is in the OFF position. In the closed-center valve, none of the ports are open to each other in the OFF position. In the open-center valve, the pressure and return ports are open to each other when the valve is OFF. In this position, the output of the system pump is returned through the selector valve to the reservoir with little resistance. Hence, in an open-center system, operating pressure is present only when the actuating unit is being operated.

An open-center, rotor-type selector valve is shown in Figure 1-2-22. As you can see, when the valve is in the OFF position, fluid from the pump enters the pressure port, passes through the open center passage in the rotor and back to the reservoir. When the valve is in either of the two ON positions, it functions the same as a closed-center valve.

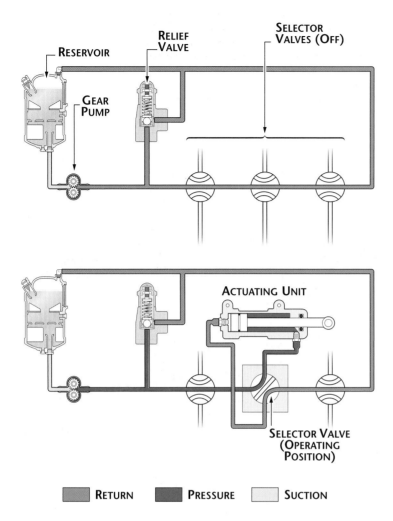

RETURN   PRESSURE   SUCTION

Figure 1-2-24. Basic open-center hydraulic system

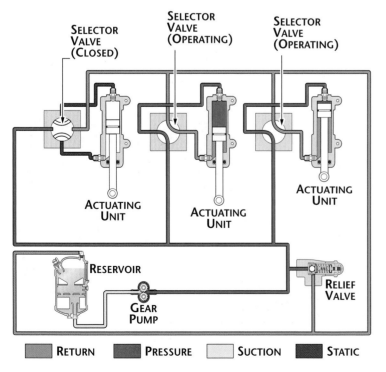

RETURN   PRESSURE   SUCTION   STATIC

Figure 1-2-25. Basic closed-center system

An open-center, spool-type selector valve is shown in Figure 1-2-23. Notice that this valve differs from the closed-center type in that a third land is machined on the spool. This land is used to cover the pressure port when the valve is in the OFF position. It provides an inter-passage in the spool, which allows fluid from the pump to return to the reservoir. Operation in both of the ON positions is the same as the closed-center selector valve.

Hydraulic systems are classified as open-center or closed-center depending upon the type of selector valves used. In an open-center system that has more than one selector valve, the valves are arranged one behind the other (in series).

In a closed-center system, the valves are arranged parallel to each other. An open-center system has fluid flow but no pressure in the system when the selector valve is off.

In a closed-center system, fluid is under pressure throughout the system when the hydraulic pump is operating.

Open-Center System

Figure 1-2-24 shows a basic open-center hydraulic system, which uses a *relief valve* to limit system pressure. As was mentioned earlier, this type of system has fluid flow but no pressure until some hydraulic device is operated. When the selector valves are OFF, fluid flows from the reservoir to the pump through the open-center passage of each valve, and then back to the reservoir. No restrictions exist in the system; therefore, no pressure is present. When one valve is placed in the operating position, the device creates a restriction that the valve controls. Fluid then flows under pressure to that hydraulic device.

Closed-Center System

Figure 1-2-25 shows a basic closed-center system. Fluid is under pressure throughout a closed-center system when the pump is operating. When the selector valves are in the OFF position, fluid cannot flow through the closed centers. This causes pressure to build in the system; it is available at any time a selector valve is turned on. A relief valve is used to keep system pressure from going above a predetermined amount when all valves are off.

Hydraulic Servo

A *servo* is a combination of a selector valve and an actuating cylinder in a single unit. When the operator opens the pilot valve of a servo, it is automatically closed by movement of the servo (or actuating) unit. Hydraulic servos are used in aircraft when precise control is necessary over the distance a component moves.

Typical hydraulic servo. Figure 1-2-26 shows a typical hydraulic servo. In operation, when the pilot valve is displaced from center, pressure is directed to one chamber of the power piston. The other chamber is open to return flow. As the power piston travels, the pilot valve housing travels because the two are attached. The operator is holding the pilot valve itself stationary, and the ports again become blocked by the lands of the pilot valve stopping the piston when it has moved the required distance.

Servo sloppy link. Notice the servo sloppy link in Figure 1-2-26. It is the connection point between the control linkage, pilot valve and servo piston rod. Its purpose is to permit the servo piston to be moved either by fluid pressure or manually. The sloppy link provides a limited amount of slack between connecting linkage and pilot valve. Because of the slack between the piston rod and the connecting linkage, the pilot valve can be moved to an ON position by the connecting linkage without moving the piston rod.

Bypass valve. A bypass valve is provided to minimize the resistance of the servo piston to movement when it must be moved manually. The valve opens automatically when there is no operating pressure on the servo. This allows fluid to flow freely between the chambers on each side of the piston.

Irreversible valve. During normal aircraft operation, external forces from an aircraft's control surfaces, such as rotor blades and ailerons, tend to move servo cylinders. This movement creates a pump-like action in the servo called *feedback*. The *irreversible valve* prevents feedback through the servo to the control stick.

Figure 1-2-27 is a simplified schematic version of an irreversible valve. The broken-line block represents the housing of the valve.

The check valve allows fluid from the pump to flow in the normal direction, as shown by the arrow. Feedback forces tend to move the servo piston opposite to the direction of pump-produced pressure. This tends to force fluid backward through the irreversible valve. The check valve keeps the servo piston from yielding to feedback by locking the rearward flow of fluid. The relief valve is a safety device to limit the pressure produced by feedback-induced movement of the servo piston. It opens to allow fluid to bypass to the return line should the feedback pressure exceed a predetermined safe limit.

Ratchet valve. A *ratchet valve* is used with a double-action actuating cylinder to aid in holding a load in the position where it has been moved. The ratchet valve ensures that there is trapped fluid on each side of the actuating cylinder piston. This is necessary for the cylinder to lock a load against movement in either direction.

A typical ratchet valve is shown in Figure 1-2-28a (next page). It consists of a housing with four ports, a polished bore, two *ball check valves* and a piston. The piston has extensions on either end to unseat the two ball check valves.

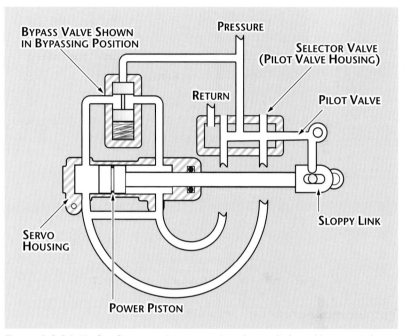

Figure 1-2-26. Hydraulic servo, incorporating sloppy link and bypass valve

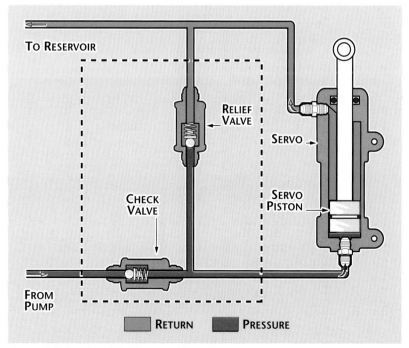

Figure 1-2-27. Simplified irreversible valve

Springs keep these valves on their seats when no pressure is applied to the system.

Valve operation with no pressure. In Figure 1-2-28A, the ratchet valve is shown with no pressure applied. The piston is centered in its bore and both ball check valves are closed. This locks the actuating cylinder in position by trapping all fluid in the cylinder.

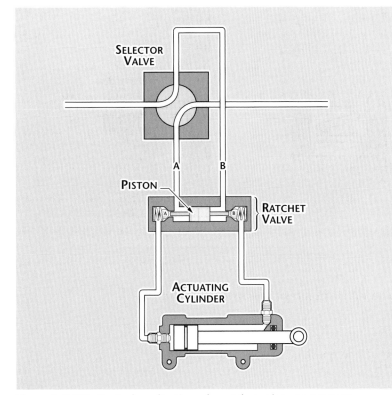

Figure 1-2-28A. Typical application of a ratchet valve, no pressure applied

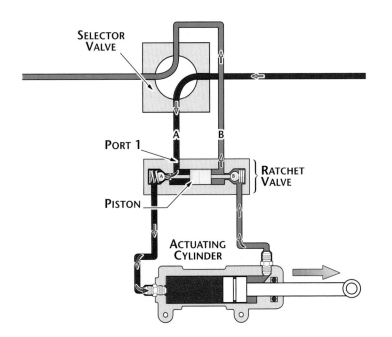

Figure 1-2-28B. Ratchet valve, with pressure applied to port 1

Valve operation with pressure applied. In Figure 1-2-28B, the ratchet valve is shown with pressure applied to port 1. This forces the piston to the right, where it unseats ball check valve B. Pressure entering port 1 also unseats ball check valve A on the left side. Fluid then flows through the ratchet valve, and the piston moves to the right.

Check Valves

A *check valve* is installed in a hydraulic system to control the direction flow of hydraulic fluid. The check valve allows free flow of fluid in one direction, but no flow, or a restricted one, in the other direction.

There are two general designs in check valves. One has its own housing and is connected to other components with tubing or hose. Check valves of this design are called *in-line check valves*. In the other design, the check valve is part of another component and is called an *integral check valve*. Its operation is identical to the in-line check valve. There are two types of in-line check valves: simple and orifice.

Simple in-line check valve. As illustrated in Figure 1-2-29, the simple inline check valve consists of a casing, inlet and outlet ports, and a ball-and-spring assembly. The ball and spring permit full fluid flow in one direction and block flow completely in the opposite direction. Fluid pressure forces the ball off its seat against the spring pressure, permitting fluid flow. When flow stops, the spring forces the ball against its seat, blocking reverse flow.

Orifice in-line check valve. The orifice check valve shown in Figure 1-2-29 is used to allow free flow in one direction and limited flow in the opposite direction. This is accomplished by drilling a passage in the valve seat connecting the inlet side of the valve to the outlet side.

Hydraulic fuse. A hydraulic fuse is an inline device that is designed to remain open to allow a normal fluid flow in the line, but close if the fluid flow increases above an established rate. A quantity-measuring fuse is seen in Figure 1-2-30.

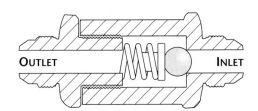

Figure 1-2-29. Simple in-line check valve

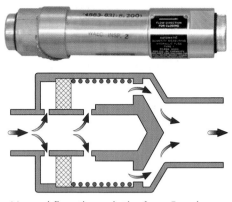

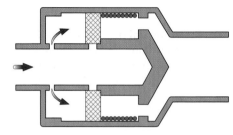

Normal flow through the fuse. Equal pressure across the piston.

Flow is stopped. The pressure drop across the fuse has moved the piston over so that it covers the holes through which the fluid must flow.

Figure 1-2-30. A hydraulic fuse closes when the rate of flow or pressure drop becomes excessive.

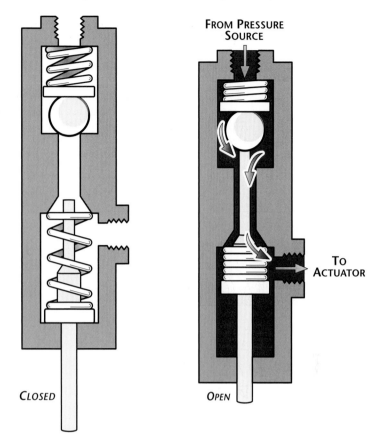

FROM PRESSURE SOURCE

TO ACTUATOR

CLOSED   **OPEN**

Figure 1-2-31. Mechanically actuated sequence valve

A fuse can be found in brake lines that are connected into the main hydraulic system. If one of the lines is broken then most of the main system fluid could be pumped overboard. A fuse in the system would shut down when the flow became excessive, saving the main system functionality.

Sequence Valve

A *sequence valve*, shown in Figure 1-2-31, is placed in a hydraulic system to delay the operation of one portion of that system until another portion of the same system has functioned. For example, it would be undesirable for the *landing gear* to retract before the *gear compartment doors* are completely open. A sequence valve actuated by the fully open door would allow pressure to enter the landing gear retract cylinder.

The sequence valve consists of a valve body with two ports, a ball and seal spring-loaded to the closed position and a spring-loaded plunger. Compressing the plunger spring off-seats the ball and allows the passage of fluid to the desired actuator. The typical sequence valve is mechanically operated, or it can be *solenoid-operated* by means of *microswitches*. In either case, the valve operates at the comple-

tion of one phase of a multi-phase hydraulic cycle.

Priority Valve

A *priority valve* is installed in some hydraulic systems to provide adequate fluid flow to essential units. The valve is installed in the line between a nonessential actuating unit and its source of pressure. It permits free, unrestrained flow of fluid to nonessential units as long as system pressure is normal. When system pressure drops below normal, the priority valve automatically reduces the flow of fluid to the nonessential units.

The priority valve (Figure 1-2-32) resembles a check valve in both external appearance and internal operation. A spring acts against a hollow piston to maintain contact with a valve seat. With no system pressure, the priority valve is in the spring-loaded position, and closed. The piston is against the valve seat. As pressure is applied to the system, fluid passes through the valve seat and also through drilled passages to act against the face of the piston. With normal flow and pressure, the piston moves against the spring tension and allows passage of fluid. If pressure decreases, the spring forces the pis-

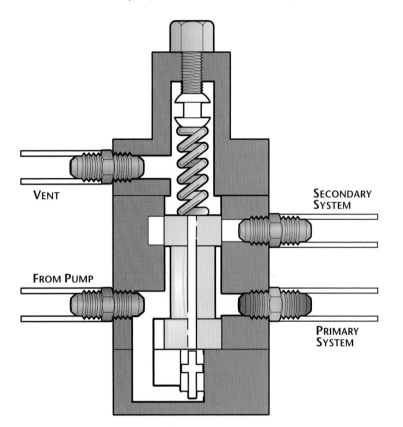

Figure 1-2-32. Typical priority valve

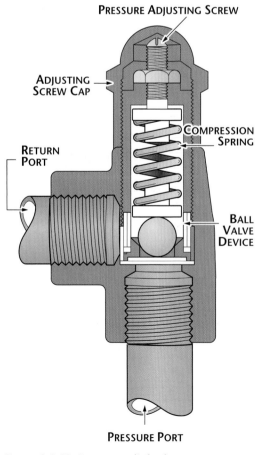

Figure 1-2-33. Pressure-relief valve

ton to seat, assuring a supply of fluid for the essential portion of the system.

Pressure Relief Valves

A *relief valve* is installed in any system containing a confined liquid subject to pressure. The use of relief valves falls into one or more of these three categories:

- In the first category, a relief valve is used to protect a hydraulic system if the pump compensator fails. The relief valve is adjusted to open at a pressure slightly higher than normal system operating pressure.

- In the second category, a relief valve is used to protect a system subject to pressure increases caused by thermal expansion. Thermal relief valves installed between system return manifold and those portions of the system where pressure can be trapped. Thermal relief valves relieve pressure above the setting of the system pressure relief valve before it builds up high enough to do any damage.

- In the third category, a relief valve is used as the sole means of pressure control in a hydraulic system.

Relief valves. The configurations for relief valves are either *two-port* or *four-port*. Both types operate in the same way. The main reason for additional ports is convenience in connecting the plumbing. Only the two-port pressure relief valve is described in this text. If it is necessary to adjust several pressure regulating valves in the same system, it is important to start with the unit requiring the highest setting first. Then adjust all remaining valves in descending order of their relief pressure.

Two-port relief valve. A typical two-port relief valve is shown in Figure 1-2-33. It consists of a housing with an inlet and an outlet port, a valving device, a compression spring and an adjustment screw. When the hydraulic system is pressurized, the pressure acts against the valving device, in this case, a ball. The ball is held against its seat by a coil spring. When the fluid pressure is great enough against the ball to overcome the force of the spring, the ball is unseated and allows fluid to pass.

The exact pressure at which this takes place is called the *cracking pressure*. This pressure can be adjusted to any desired pressure by means of the *pressure adjustment screw*. Fluid passing the valving ball flows into return lines and back to the reservoir.

Before any adjusting is done on a relief valve, the system unloader valve must be disabled temporarily. Otherwise it will not allow the system to reach the maximum relief valve setting.

Blow down valve (flap overload valve). A type of relief valve installed in the normal flap down line to provide a blowback feature. This relief valve prevents flaps from being lowered at an airspeed which could cause structural damage. If the pilot tries to extend the flaps when the airspeed is too high, the opposition caused by the airflow will open the overload valve and return the fluid to the reservoir.

Pressure Reducer

A pressure reducer provides more than one level of pressure in a system that has a single hydraulic pump. The reducer (Figure 1-2-34) consists of a three-port housing, piston, poppet and spring, adjusting spring and adjusting screw. A *poppet* is a valving device with a flat face. The three ports of the housing are input pressure port, reduced-pressure port and return port.

Withholding pressure. The pressure reducer operates on the principle of withholding pressure rather than relieving it. With no pressure in the system, the adjusting spring tension holds the poppet open. As system pressure builds up, fluid passes through the poppet to the reduced-pressure port. When the pressure

acting against the piston exceeds the force of the adjusting spring in the pressure reducer, the poppet moves to close the inlet port. Further buildup of system pressure does not affect the reduced pressure until it decreases enough to allow the inlet to be opened by spring tension.

Relieving pressure. Pressure reducers also relieve increased pressure resulting from thermal expansion. As the pressure at the reduced pressure port increases, the piston moves against the adjusting spring, opening the return port and relieving the excess pressure.

Pressure Regulator

A constant displacement hydraulic pump must have a pressure regulator to unload the pump when the system has reached the working pressure, and flow demands are low.

Unloading valve. The unloading valve in a closed hydraulic system provides a low-pressure path for the fluid from the pump to the reservoir, when there are no flow requirements. An unloading valve is shown in Figure 1-2-36.

Flow from the pump is directed through the regulator to the system, until all the cylinders and actuators are moved to their desired positions and no further fluid flow or pressure is required. The flow is then directed to the accu-

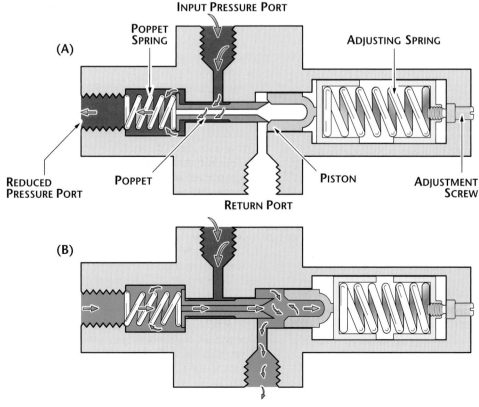

Figure 1-2-34. Pressure reducer

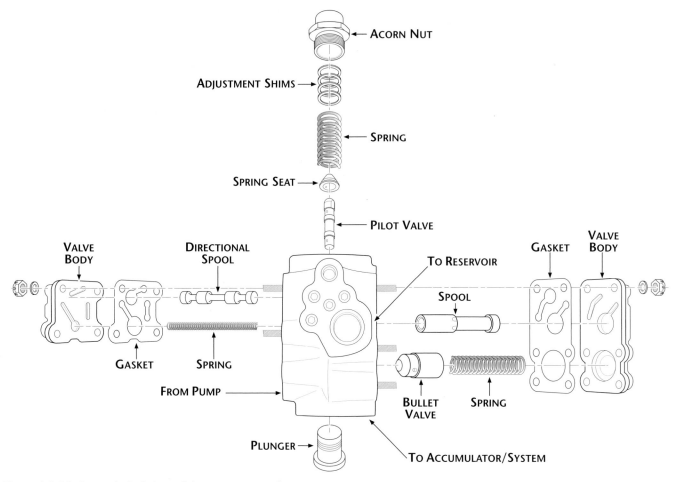

Figure 1-2-35. An exploded view of the pressure regulator

mulator, and pressure builds to a specific value as the fluid side of the accumulator is filled by compressing the air in the accumulator. An exploded view of the valve is shown in Figure 1-2-35.

Figure 1-2-36. The pressure regulator unloads the pump during periods of low demand.

Fluid pressure then shifts the unloading valve to the open bypass, or "kicked out", position. In this position, the flow from the pump is directed back to the reservoir and the pump is said to be unloaded, and will turn without resistance.

The unloading valve works in conjunction with the system accumulator. An inline check valve keeps the accumulator, and the system, pressurized for use by the system components.

The inline check valve keeps the accumulator from discharging into the reservoir. A simple hydraulic system with an external pressure regulator is shown in Figure 1-2-37.

Fluid flow from the accumulator to the system will cause the system to bleed off pressure until the bottom range of the unloading valve setting is reached. This will cause the bullet valve to shift, blocking the path to the return and directing the flow to the system and the accumulator. The valve is now "kicked in" and the pump will provide fluid flow to the system. This cycle will continue as the demands on the system change.

Modern, compact, hydraulic systems use a pump that has a built-in regulator to control the pump output.

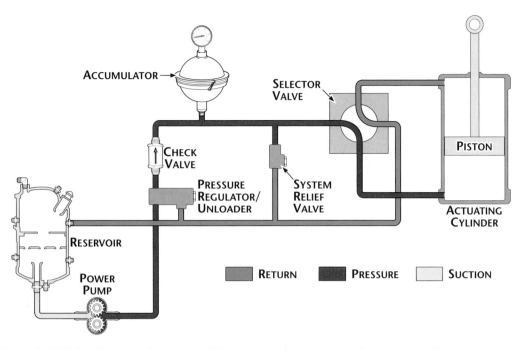

Figure 1-2-37. Simple hydraulic system with an external pressure regulator or unloader

Pressure Switches

A *pressure switch* is designed to open or close an electrical circuit in response to a predetermined hydraulic pressure; the switch activates a warning or protective device. At a set minimum pressure, the switch can turn on a light to warn the pilot, turn a pump off, or activate a solenoid-controlled valve. Types of pressure switches, piston and diaphragm, are described in the following paragraphs.

Piston pressure switch. The piston pressure switch (Figure 1-2-38) consists of a housing, a cylinder bore and piston, an adjustable spring for loading the piston, a microswitch and linkage for transmitting movements of the piston to the microswitch. The housing has a pressure port for connection to system pressure and an electrical receptacle for connecting the switch to an electrical circuit.

Diaphragm pressure switch. The *diaphragm pressure switch* consists of a housing, a diaphragm, an adjustable spring to load the diaphragm, a microswitch and linkage for transmitting movements of the diaphragm to the microswitch shown in Figure 1-2-39. The housing has ports for the same functions as those in the piston switch.

Pressure switch operation. The two types of pressure switches have the same operating principles; only the piston one is covered here.

Fluid pressure enters the pressure port and moves the face of the piston against the adjustment spring. When the pressure becomes great

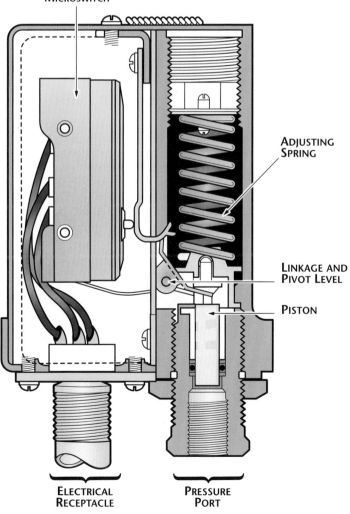

Figure 1-2-38. Piston pressure switch

Figure 1-2-39. Diaphragm pressure switch

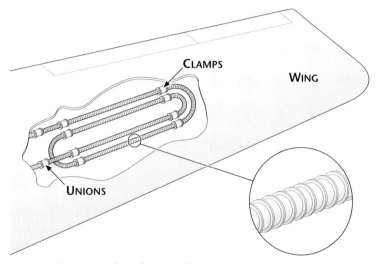

Figure 1-2-40. Fin tubing heat exchanger

enough to overcome the force of the spring, the piston moves and causes the pivot lever to rotate. The movement of the lever is transmitted through the linkage to the microswitch button. This closes the electrical circuit.

Heat Exchangers

Heat exchangers are used in many different parts of the aircraft. Some are used to cool the engine oil and at the same time they heat the fuel to prevent ice crystals from building up in the fuel filters. Hydraulic systems also use heat exchangers to cool the fluid.

Hydraulic oil coolers.. When hydraulic fluid runs through high-pressure pumps at high rates or flow, it becomes hot. That heat can cause some of the components whose internal tolerances are very tight to stick or become unpredictable in operation.

Many of the systems use a cooler that is mounted in one of the wing fuel tanks. They are simply tubes that radiate heat into the fuel by way of cooling fins. This can be seen in Figure 1-2-40.

Section 3
Seals and Gaskets

Seals and gaskets are used throughout aircraft plumbing systems to prevent leaks when two components are joined together. The material from which the seals are manufactured varies depending upon the fluid or gas being conducted and the operating pressure range of the system. Using the proper type of seal and exercising care during installation are two of the most important phases of plumbing maintenance. Lack of care during this phase of maintenance is one of the most frequent causes of system failure or leaks.

The seals or packings used in hydraulic systems are manufactured from rubber, leather, Teflon, metal, or a combination of any of these. Two types of rubber, natural and synthetic, are used for making hydraulic seals; however, only synthetic rubber seals can be used with mineral-base hydraulic fluid. Examples of some of the different kinds of seals used in plumbing systems are shown in Figure 1-3-1 and discussed in the following paragraphs.

Seal Identification

O-rings. O-rings are circular rubber seals with a circular cross-section. They are almost universal in usage and are discussed in their own section.

V-rings (chevron). The use of V-rings is rather limited in hydraulic systems; however, they are used in some shock struts. A V-ring can seal in only one direction and can be used to seal surfaces regardless of whether there is movement between the parts.

U-rings. Similar to V-rings in design and function, U-rings are used to seal pistons and shafts on some master brake cylinders.

Cup seals. Another type of seal, used frequently on master brake cylinders and drum type wheel cylinders, is a cup seal. They are effective in controlling leaks in only one direction, and when installed the lip of the cup must be facing the fluid to be contained.

Oil seals. Composite seals made from both rubber and metal are called oil seals, and they are used to seal hydraulic pump and motor drive shafts. Their outer body, or case, is made from pressed steel and is force-fitted into the component housing. Inside the metal case is a lipped rubber seal and a spring. The rubber seal is securely anchored against movement to the metal case, and the spring encircles the lip, holding it firmly to the surface it seals and is commonly referred to as a Garloc seal. During installation, the housing must be free from foreign matter or burrs, and the seals must be seated squarely with proper special installation tools.

Wiper seals. Scrapers or wiper seals are made of metal, leather, or felt and are used to clean and lubricate the exposed portion of piston shafts. When installed and operating properly, wiper seals prevent dirt from entering the system and aid in preventing piston shafts from binding.

Gaskets

A gasket is a piece of material placed between two parts where there is no movement. The gasket is used as a filler to compensate for irregularities on the surfaces of the two mating parts. Many different materials are used for making gaskets. For use in hydraulic systems the gaskets may be made from treated paper, synthetic rubber, copper, or aluminum.

O-ring gaskets. The most common type of gasket used in aircraft hydraulic systems is the O-ring. When used as a gasket the O-ring has the same advantage as when used as a seal, as explained in a previous paragraph.

Crush washers. The second most commonly used gasket is the crush washer, used in hydraulic systems and made from aluminum or copper. Fittings using these washers have concentric grooves and ridges that bear against or crush the washer. These grooves and ridges seal the washer and fitting as the connecting parts are tightened together.

Fabricating gaskets. Some types of gaskets can be field-fabricated as long as the bulk material conforms to the required specifications. When you cut replacement gaskets from bulk material, the most important consideration is the exact duplication of the thickness of the original gasket.

Installing gaskets. Like seals, gaskets must be examined before installation to ensure their serviceability. The component surfaces to be connected must be thoroughly cleaned. During assembly, care must be taken not to crimp or twist the gaskets. When tightening the components, the gaskets must not be compressed into the threads where they can be cut, damaged, or block mating surfaces from being flush. When removing old gaskets, be careful not to scratch or gouge the sealing surface.

Hydraulic Seals

Seals reduce or prevent internal and external leakage between two objects. Hydraulic seals are used throughout aircraft hydraulic systems to cut down on internal and external leakage of hydraulic fluid, thereby preventing loss of system pressure. Two general types of seals used in hydraulic and pneumatic systems are the *dynamic type* and the *static type*. A dynamic seal is used between two moving parts of a unit. A static seal is used between two stationary parts.

Most of the dynamic and static seals used on today's aircraft are manufactured in the form of O-rings and are called *packings*. However, conditions will arise when special nonstandard seals will have to be made for specific uses on the aircraft. See Figure 1-3-2.

A hydraulic seal may consist of more than one component, such as an O-ring and a backup ring or possibly an O-ring and two backup rings. Hydraulic seals between nonmoving fittings are called gaskets; hydraulic seals inside a sliding or moving part are called packings.

Composition. Seals are composed of several different types of materials. The material depends on the use of the seal and the type of fluid it will come in contact with. The seals used in petroleum-base hydraulic fluid (MIL-H-5606) are made from synthetic rubber (nitrile). Seals in Skydrol® fluid systems are made from ethylene-propylene.

Advances in aircraft design make it necessary to develop O-ring seals out of compounds that could meet the changing conditions. Hydraulic O-rings were originally established under AN (Air Force-Navy) specification numbers (6227, 6230 and 6290) for use in MIL-H-5606 fluid at operating temperatures ranging from -65° to 160°F.

When aircraft designs raised the operating temperatures, newer compounds were developed under MS specifications with an operating range of -65° to 275°F. These newer compounds are used to manufacture the MS28775 and MS28778 O-ring seals (packings). The MS28775 O-ring replaces AN6227 and AN6230 and MS28778 replaces AN6290; however, O-ring replacement should only be done by part number as specified in the technical manual.

Storage. Packings and gaskets should be stored in a dark, cool, dry place; they should be kept away from excessive heat, strong air currents, dampness and dirt. Do not expose them to electric motors or other equipment that gives off heat and ozone.

"O" RING

"T" RING

"D" RING

"V" RING

Figure 1-3-1. Seals used in plumbing systems

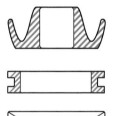
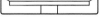
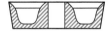

Figure 1-3-2. Examples of nonstandard seals

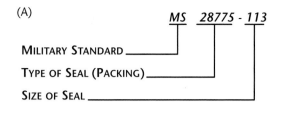

(A)

MS 28775 - 113

MILITARY STANDARD _____
TYPE OF SEAL (PACKING) _____
SIZE OF SEAL _____

KZ5330-833-7491
PACKING PREFORMED SYN. RUBBER
I EACH (MS28778-5)
DISC-38329
A-5/07
SSR 810-B-90
MFD. DATE 4-07 CURE DATE 507
STILLMAN RUBBER CO. (MFGR/CONTR)
MIL-G-5510A

(B)

Figure 1-3-3. (A) Part number of standard O-ring seal and (B) manufacturer's cure date

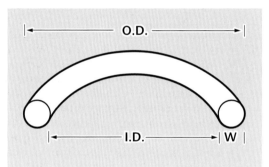

Figure 1-3-4. When an O-ring is measured, three dimensions are needed; outside diameter, cross-section diameter and inside diameter.

O-Rings

Identification. O-rings are manufactured according to military specifications. They are identified from the technical information printed on the O-ring package. Because the size of O-rings cannot be positively identified visually without the use of special equipment, O-rings are packaged in individual sealed envelopes labeled with all the necessary data. Colored dots, dashes and stripes, or combinations of dots and dashes on the surface of the O-ring are no longer used to identify O-rings. The part number of a standard O-ring seal is illustrated at Figure 1-3-3.

If the seal's part number cannot be found in the technical order, its size may help determine the part number. The size of an O-ring seal may be determined by measuring the seal's width (W), the inside diameter (ID), or the outside diameter (OD). Figure 1-3-4 illustrates the various dimensions that can be measured. If two are known, the third can easily be determined. For example, the outside diameter can be found if the width and inside diameter are known. Twice the width plus the inside diameter will equal the outside diameter. Once the measurements of the seal are known, cross-reference it in the supply catalogs.

O-ring shelf life. Most O-ring age limitation is determined by its cure date, anticipated service life and replacement schedule. Age limitation of synthetic rubber O-rings is based on the fact that the material deteriorates with age. O-ring age is computed from the cure date. Shelf life of O-rings is further discussed in the "Hardware and Materials" chapter in *Introduction to Aircraft Maintenance*.

Removal. Some seals may be removed by squeezing the seal between the thumb and forefinger; this will force the O-ring ring out of the groove. Then the entire seal is removed. Seals may also be removed with a tool; however, you must carefully choose the correct tool. A variety of tools may be used on any given job, but they should be made from soft metal, such as brass and aluminum. Also, tools made from phenolic rod, plastics and wood may be used. Avoid using pointed or sharp-edged tools that might scratch or mar surfaces or damage the O-rings.

Storage. Proper storage practices must be observed to protect O-rings. Most synthetic rubbers are not damaged by several years of storage under ideal conditions; however, their enemies are heat, light, moisture, oil, grease, fuels, solvents, thinners, strong drafts, or ozone (type of oxygen formed from an electrical charge). Damage by exposure is magnified when rubber is under tension, compression, or stress. When storing O-rings, avoid the following conditions:

- Stacking parts improperly and storing them in improper containers, which can cause defects in shape.

- Applying force to the O-ring corners and edges and squeezing rings between boxes and storage containers, which can cause creases. Storing rings under heavy parts, which can cause compression and flattening.

- Using staples to attach identification, which can cause punctures. Hanging O-rings from nails or pegs which can cause the rings to become dirty and develop defective shapes (O-rings should be kept in their original envelopes).

- Allowing rings to become oily or dirty because of fluids leaking from parts stored above and adjacent to O-ring surfaces.

- Applying adhesive tape directly to O-ring surfaces (a torn O-ring package should be secured with pressure-sensitive, moisture-proof tape, but the tape must not contact the O-ring surface)

- Keeping overage parts because of improper storage arrangements or unreadable identification (O-rings should be arranged so that older seals are used first)

Backup Rings

Types. Backup rings support O-rings and prevent them from wearing and causing leakage. The two types of backup rings are the Teflon type and the leather type. Teflon backup rings are generally used with packings and gaskets; however, leather backup rings may be used with gasket-type seals in systems operating up to 1,500 p.s.i. Teflon rings are made from a fluorocarbon-resin material which is tough, friction-resistant and more durable than leather. Teflon backup rings do not deteriorate with age, can tolerate temperatures greater than those encountered in high-pressure hydraulic systems, and are unaffected by any other system fluid or vapor.

Identification. Backup rings are not color-coded or marked and must be identified from package labels. The dash number which follows the specification number on the package shows the size and, in some cases, relates directly to the dash number of the O-ring for which the backup ring is dimensionally suited. For example, the single spiral Teflon ring MS28774-6 is used with the MS28775-006 O-ring, and the double spiral Teflon ring MS28782-1 is used with the AN6227B-1 O-ring.

Installation. Installation of backup rings is covered in the "Hardware and Materials" chapter of *Introduction to Aircraft Maintenance.*

Storage. Precautions similar to those for O-rings must be taken to prevent contamination of backup rings and damage to hydraulic components. Teflon backup rings may be stocked in individual sealed packages like the O-rings or several may be stored on a cardboard mandrel. If unpackaged rings are stored for a long time without using mandrels, an overlap may develop. To prevent this condition, stack Teflon rings on a mandrel of a diameter comparable to the desired diameter of the spiral ring. Stack and clamp the rings with their coils flat and parallel.

Do not store leather backup rings on mandrels as this can possibly stretch and distort the original shape of the leather ring. Leather rings should be stored in sealed individual packets.

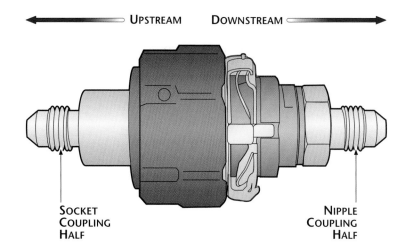

Figure 1-3-5. Typical quick-disconnect coupling

Shaft Seals

Most shaft seals on pumps have two flat surfaces that rotate against each other. One flat surface is made from a soft material such as bronze or carbon and is called the sealing ring. The other flat surface is made from stainless steel and is called the mating ring. The mating ring rotates with the drive shaft. Although all makes of shaft seals differ in design, they perform the same function; they seal a moving part — the shaft. The shaft seal must limit the leakage of fluid trying to escape, but not stop it altogether. This is because the shaft seal acts like a bearing and must be lubricated like one. The lubricant comes from the slight leakage of fluid past the two mating surfaces. Never scratch or damage the two mating surfaces because this will cause them to leak.

Quick-Disconnect Couplings

Quick-disconnect couplings of the self-sealing type are used at various points in many hydraulic systems. These couplings are installed at locations where frequent uncoupling of lines is required for inspection and maintenance. Figure 1-3-5 shows a typical quick-disconnect coupling. Each coupling assembly consists of two halves held together by a union nut. Each half contains a valve, which is held open when the coupling is connected; this allows fluid to flow in either direction through the coupling. When the coupling is disconnected, a spring in each half closes the valve. This prevents loss of fluid and entrance of air. The union nut has a quick-lead thread, which allows the coupling to be connected or disconnected by turning the nut.

Various types of union nuts are used in hydraulic systems. For one type, a quarter turn

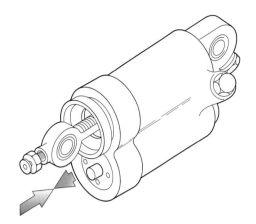

Figure 1-4-1. Piston-type displacement damper

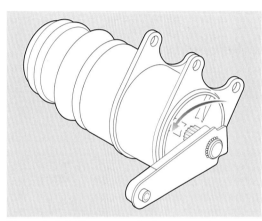

Figure 1-4-2. Vane-type displacement damper

of the union nut locks or unlocks the coupling. For another type, a full turn is required. Some couplings require wrench tightening; others are connected and disconnected by hand. Some installations require that the coupling be secured with safety wire; others do not require any form of safetying. Because of these differences, all quick-disconnect couplings must be installed according to the instructions in the applicable maintenance manual.

Section 4

Dampening and Absorbing Units

Hydraulic Dampers

A damper is a device that controls the speed of relative movement between two connected objects. Usually one end of the damper is connected to a fixed member; the other end, to a movable part. The reacting parts of the damper

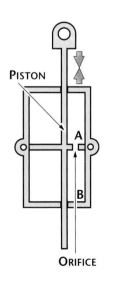

Figure 1-4-3. Piston-type displacement damper

move against considerable resistance, which slows the speed of relative movement between the objects.

Hydraulic dampers operate either by displacing fluid (displacement dampers) or by shearing fluid (shear dampers).

Displacement dampers

The two types of displacement dampers are the piston type (Figure 1-4-1) and the vane type (Figure 1-4-2.) Though different in construction, both types have the same basic design characteristics — a sturdy metal container with a sizable inner space divided into two or more chambers. The chambers vary in size according to the position of the parts within the damper. The chambers must be completely filled with fluid to operate properly.

Piston-type displacement dampers. In this type damper, the piston and rod assembly divides the space within the damper housing into two chambers. See Figure 1-4-3. Seal rings on the piston prevent fluid leakage between the chambers. An orifice permits fluid to pass with restricted flow from one chamber to the other. A filler port (not shown) services the damper with fluid.

As the piston is forced up, chamber A decreases in proportion to the distance the piston is moved. Simultaneously, chamber B increases by a comparable size. The hydraulic fluid displaced from chamber A flows through the restricting orifice into chamber B. When the piston is moved toward the left, reverse changes occur in the chamber sizes and in the direction of fluid flow. The restriction of the fluid flow by the orifice slows the rate of speed at which a given amount of force can move the damper piston. The rate at which a damper moves in response to a force is called damping rate or timing rate.

In some dampers, the opening is a fixed size, and the timing rate is not adjustable. In other dampers, the orifice size is adjustable to allow for timing adjustments. The three types of piston dampers are the nose landing gear damper, the tail rotor pedal damper and the rotor blade damper.

Nose landing gear damper. The nose landing gear of an aircraft has a tendency to shimmy when the aircraft is taxiing at any appreciable speed. This type damper is used to eliminate wheel shimmy without interfering with the normal steering movements on the nose wheel. See Figure 1-4-4.

Rotor blade damper. Piston-type dampers are used on helicopter rotor head assemblies

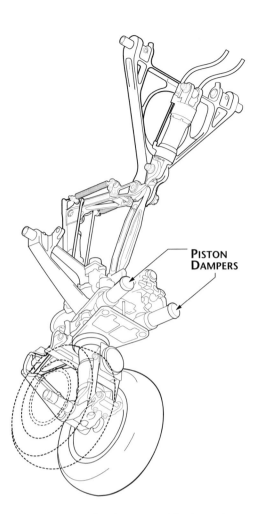

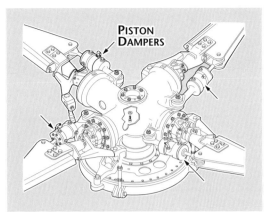

Figure 1-4-5. Main rotor assembly with piston-type rotor blade dampers

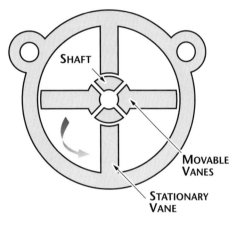

Figure 1-4-4. Nose landing gear with piston damper

Figure 1-4-6. Vane-type displacement damper

and tail rotor hub assemblies to control lead-lag movements of rotor blades. Note how the dampers are connected in the illustration at Figure 1-4-5. Lead-lag movements occur when three or more blades are in a set, and they are hinged to the rotor head.

Vane-type displacement dampers. A typical vane-type damper consists basically of a cylindrical housing having a polished bore with two stationary vanes (called abutments) and a shaft supporting two movable vanes. (See Figure 1-4-6) Together, the four vanes split the cylinder bore lengthwise into four chambers. The two stationary vanes are attached to the damper housing. The two movable vanes, along with the shaft, make up a unit called a wing shaft, which rotates between the abutments. One end of the wing shaft is splined and protrudes through the damper housing. A lever arm attached to the splined end rotates the wing shaft.

The damper chambers are completely filled with fluid. At any instant of damper motion, the fluid is subjected to forced flow. As the wing shaft rotates, fluid between the chambers flows through an opening within the

wing shaft, which interconnects the four chambers. Then, a restraining force is developed in the damper, dependent on the velocity of fluid flow through the orifice. Slow relative movement between the wing shaft and damper housing causes a low-velocity flow through the opening and little resistance to damper arm rotation. A more rapid motion of the wing shaft increases the speed of fluid flow and thus increases resistance to damper arm rotation.

The timing rate of vane-type dampers can be adjusted by a timing adjustment, centrally located in the exposed end of the wing shaft. This adjustment sets the effective size of the opening through which fluid flows between chambers; it determines the speed of movement with which the damper will respond to an applied force. The vane-type mechanism is sensitive to changes in fluid viscosity caused by changes in fluid temperature. Most vane-type dampers have a thermostatically operated compensating valve to provide consistent timing rate performance over a wide range of temperatures. The two types of vane-type dampers are the nose landing gear damper and the stabilizer bar damper.

Figure 1-4-7. Nose landing gear with vane damper

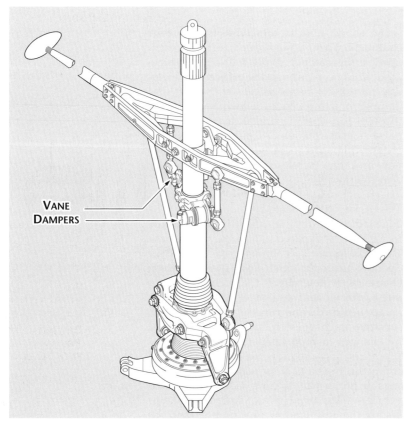

Figure 1-4-8. Stabilizer bar with vane dampers

Nose landing gear damper. A vane-type damper is used on the nose landing gear of some aircraft to eliminate the shimmy tendency. See Figure 1-4-7.

Stabilizer bar damper. Some helicopters have vane-type dampers that control the degree of sensitivity with which a helicopter responds to movements made by the pilot on the flight controls. See Figure 1-4-8. These dampers are mounted on a bracket attached to the helicopter mast (main shaft). The damper arms are interconnected with other parts of the flight control system.

Shear Dampers

In dampers operating on the shear principle, fluid is not forced out of one space and into another space within the damper as it is in displacement dampers. Instead, action on the fluid involves tearing (shearing a thick film of highly viscous fluid into two thinner films that move with resistance in opposite directions. Highly viscous fluid is thick-bodied, syrupy and sticky.

In a shear damper, two reacting parts are free to slide or rotate past each other as the damper operates. The surfaces facing each other are relatively smooth; between them is a preset gap of a few thousandths of an inch. This gap is filled with highly viscous fluid. As the parts of the damper move relative to each other, the film of fluid in the gap between them shears into two thinner films. Each film sticks to and moves along with one of the parts. It is the friction within the fluid itself that causes resistance to movement of the parts to which the films stick.

To better understand this principle, imagine a puddle of syrup spilled on a relatively smooth tabletop. A sheet of paper placed on top of the puddle would move with considerable drag. This is very much like what happens between the parts of a shear damper as the damper operates. The two types of aircraft shear dampers are the rotary type and the linear type.

Rotary-type shear damper. A typical rotary-type shear damper consists of two members that are free to rotate together. See Figure 1-4-9. Each of the members is attached to one of the two objects whose relative movement the damper will restrain.

One of the damper members has a flange-like section that fits between these two objects. See Figure 1-4-10. Bearing points ensure that the flange is centered between the two surfaces. The spaces between the flange surfaces and the other two surfaces are filled with highly viscous fluid. A spring-loaded piston applies pressure to a supply of replenishment fluid to ensure that the spaces are always completely filled with fluid.

Linear-type shear damper. A typical linear-type shear damper, shown in Figure 1-4-11, consists essentially of two telescoping tubular members. The members have a means of connecting the unit to the two objects between which movement is to be restrained.

The telescoping tubes have a predetermined space between them, and the space is held to uniform thickness by means of bearing points (Figure 1-4-11). The springs at the ends of the inner tubular member provide it with a centering tendency that makes the damper double acting. The spring-loaded piston keeps the space between the tubular members filled with fluid.

Servicing shear dampers. Servicing is generally confined to keeping the damper clean, keeping attachment points lubricated, and refilling the chamber, in which the spring-loaded piston operates, with fluid. Only high-viscosity silicone-type fluid is used for servicing dampers operating on the shear principle. Dampers will not operate satisfactorily with any other commonly available fluid.

Shock Struts

A shock strut can be thought of as a combination suspension unit and shock absorber. It performs functions in an aircraft similar to those performed in an automobile by the chassis spring and the shock absorber.

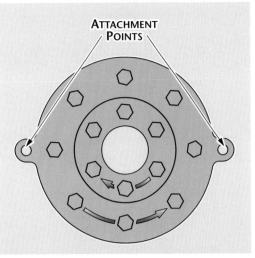

Figure 1-4-9. Rotary-type shear damper

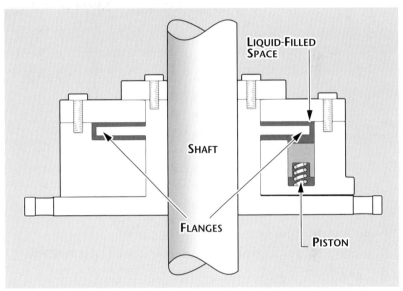

Figure 1-4-10. Schematic of a rotary damper

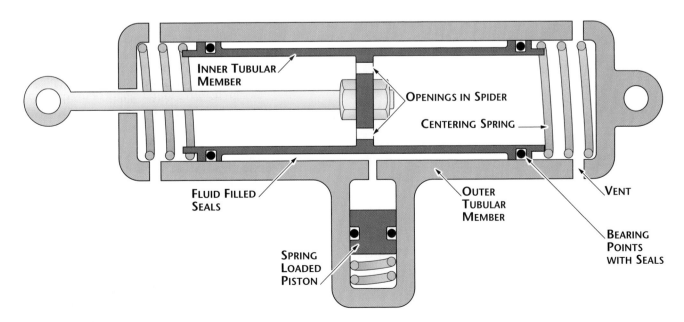

Figure 1-4-11. Schematic of a linear damper

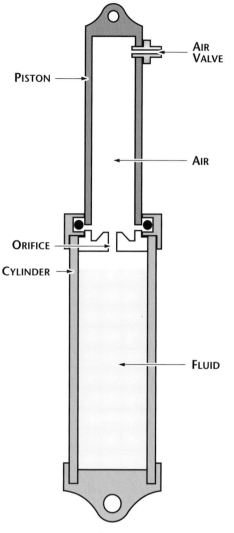

Figure 1-4-12. Simple shock strut (extended)

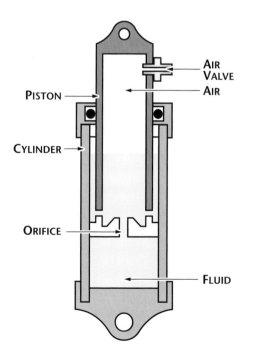

Figure 1-4-13. Simple shock strut (compressed)

The two major types of shock struts are the *mechanical type* and the *pneudraulic type*. In the mechanical type, a rubber or spring mechanism performs the cushioning operation. Mechanical types will be discussed further in the section on landing gears.

In the pneudraulic type, air and hydraulic fluid accomplish this. The two types of pneudraulic struts are the simple type and the complex type.

Simple Shock Struts

The basic parts of a simple shock strut are two telescoping tubes: a piston and a cylinder. A simple shock strut is installed with the piston uppermost and the cylinder filled with fluid. See Figure 1-4-12. An orifice in the piston head permits fluid to pass from one chamber to the other. When a shock strut has sufficient fluid the space above the fluid is then filled with air. When the aircraft is landing and the shock strut is compressing, fluid is forced through the orifice into the piston. The movement of fluid through the orifice, together with the compression of the air, absorbs the energy of the descending aircraft's motion.

When the load on the shock strut is lightened, the shock strut extends. See Figure 1-4-13. The force exerted by the compressed air in the shock strut and, during takeoff, by the weight of the lower tube and attached landing gear causes this extension. When the shock strut is extending, fluid in the piston passes through the orifice into the cylinder.

Complex Shock Struts

A complex shock strut (Figure 1-4-14) works in essentially the same manner as a simple one; however, it contains, besides two telescoping tubes, a number of parts that provide a more effective damping action than a simple strut. Design features found singly or in combination in complex-type shock struts are the metering pin, plunger and floating piston.

Metering pin. The metering pin changes the effective size of the orifice to vary the rate of fluid flow from one chamber of the shock strut to the other. The diameter of the metering pin varies along its length; it is almost equal at the ends and smaller in the middle. The unanchored end of the metering pin is located in the orifice when the shock strut is fully extended. The large diameter of the pin at this end provides a high resistance to fluid flow, a condition that is require during landing. The small diameter portion of the metering pin is located within the orifice when the shock strut is in the taxi position (partially compressed). This provides the low resistance to fluid flow that is required for taxiing. The portion of the metering pin nearest its anchored end lies within the orifice when the shock strut is completely compressed. The large diameter of the metering pin at this end provides increased resistance to fluid flow. The design of the pin at this end ensures against bottoming of the shock strut during unusually hard landings. The gradual increase in the diameter of the pin toward the anchored end prevents a sudden change in resistance to fluid flow.

Plunger. Some complex shock struts are mounted on the aircraft with their cylinders uppermost (See Figure 1-4-15.) In such a unit, a plunger anchored in the cylinder extends downward into the piston. The plunger forces fluid out of the piston and into the cylinder during the shock strut compression. The plunger is hollow; fluid enters and leaves its interior through an orifice and holes in its walls.

Functions of Shock Struts

Shock struts perform three major functions. They support the static load (deadweight) of the aircraft, cushion the jolts during taxiing or towing of the aircraft and reduce shock during landing.

Supporting static loads. The normal load of a parked aircraft is static; meaning the force present is fixed. The pressure of the air and fluid within a shock strut tends to keep the shock strut fully extended. However, air pressure in a shock strut is not enough to keep the strut fully extended while supporting the static load of an aircraft. Therefore, a shock strut gives under load and compresses until the air pressure builds enough to support the aircraft.

Cushioning jolts. As an aircraft taxies, the uneven surface of the runway causes the aircraft to bob up and down as it moves forward (sometimes air currents contribute to this effect). The inertia of the aircraft fuselage in opposition to such up-and-down movement causes the force of the taxi load to fluctuate. This bouncing motion is held within limits by the damper-like action of the shock strut. This dampening results from resistance created by the back-and-forth flow of fluid through the orifice as the shock strut extends and compresses.

Reducing shock. The aircraft will continue to descend at a high rate when landing, even after the wheels touch the ground. In the few remaining inches that the fuselage can move

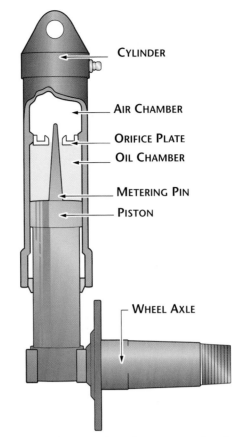

Figure 1-4-14. Metering pin-type complex shock strut

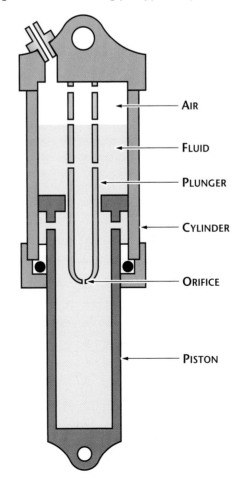

Figure 1-4-15. Plunger-type complex shock strut

SERVICE INSTRUCTIONS

PIPER STRUT SER. NO.

EXTEND STRUT. RELEASE AIR. REMOVE OIL LEVEL PLUG. FILL UPRIGHT STRUT WITH MIL-H-5606 HYDRAULIC OIL, COMPRESS STRUT FULLY, REPLACE PLUG, INFLATE AND MAINTAIN 2.00 INCHES VISIBLE STRUT EXTENSION UNDER STATIC LOAD.

Figure 1-4-16. This service tag includes all information necessary to service this particular strut with fluid and air pressure.

toward the ground after the wheels touch; the descent of the aircraft must be stopped. To do this, the shock strut must remove a great amount of energy from the downward movement of the aircraft. The impact force is very great compared to the force exerted by the mere weight of the aircraft. The shock strut removes some of the energy of motion and impact force by converting energy into heat and dissipating the heat into the atmosphere. The resistance to fluid flow offered by the orifice is the principal means of developing the heat. Also, the temperature of the air inside the strut rises as the air is compressed.

The speed of a descending aircraft while landing causes over compression of the air in the shock strut. As a result, the air pressure is greater than that needed to support the static load of the aircraft. The excess pressure tends to extend the shock strut and bounce the aircraft back into the air. For comfort and control of the aircraft, this rebound has to be held to the lowest level possible. The most common means of counteracting rebound involves the use of a shock strut annular space or snubber.

The annular space is a chamber that surrounds the polished piston surface that lies within the cylinder. The space has no definite volume; the volume depends on the amount the shock strut is extended or compressed. The annular space is at minimum size when the shock strut is completely extended and at maximum size when the strut is completely compressed. As the shock strut extends, fluid passes from the piston into the annular space. Compression of the shock strut forces fluid from the annular space back into the piston. Transfer of fluid into or out of the annular space takes place through transfer passages in the wall of the piston. The fluid moves with some resistance, which varies with the size of the transfer passages.

In simple shock struts, the transfer passages are merely holes. In many complex shock struts,

the passages are provided with a snubber valve or rebound control valve. Such a valve allows fluid to flow more freely into the annular space during shock strut compression than it flows out during extension.

Maintenance

Shock struts should be frequently checked for leakage, proper air pressure, secure attachment and cleanliness. The exposed portion of the shock strut piston should be cleaned frequently with a clean, lint-free cloth moistened with hydraulic fluid. Specific instructions for servicing with hydraulic fluid and air pressure are stamped on the nameplate of the shock strut and are given in the applicable aircraft manual. With a few exceptions, a single port in the shock strut serves as a filler hole for both hydraulic fluid and air. An air valve assembly screws into the port.

Servicing instructions. Most shock struts have a tag affixed that bears the service instructions for that strut. These instructions should include the type of fluid, as well a how to fill the strut and how to inflate it. See Figure 1-4-16.

CAUTION: *Always be sure to release the air pressure before attempting to remove the air valve core or assembly. Rock the aircraft and depress the valve core several times with a suitable metal tool to ensure that all pressure is released. Air pressure could blow out the air valve core or assembly when either of them is loose.*

Section 5

Hydraulic Contamination

All modern aircraft contain hydraulic systems that operate various mechanisms. Aircraft hydraulic systems are designed to produce and maintain a given pressure over the entire range of required fluid flow rates. The pressure used in most commercial aircraft is 3,000 p.s.i. The primary use of hydraulic fluids in aircraft hydraulic systems is to transmit power, but hydraulic systems perform other functions. Hydraulic fluid acts as a lubricant to reduce friction and wear. Hydraulic fluid serves as a coolant to maintain operating temperatures within limits of critical sealant materials, and it serves as a corrosion and rust inhibitor. Critical functions of hydraulic systems may be impaired if the hydraulic system fluid is allowed to become contaminated beyond acceptable limits.

Hydraulic fluid contamination is defined as any foreign material or substance whose presence in the fluid is capable of adversely affecting the system performance or reliability. Contamination is always present to some degree, even in new, unused fluid. Contamination must be below the level that adversely affects system operation. Hydraulic contamination control consists of requirements, techniques and practices that minimize and control fluid contamination.

Hydraulic Contamination Control Program

Hydraulic contamination in aircraft and related support equipment is a major cause of hydraulic system and component failure. Every technician who performs hydraulic maintenance should be aware of the causes and effects of hydraulic contamination. You should follow correct practices and procedures to prevent contamination.

Fluid sampling. The contamination level of a particular system is determined by analysis of a fluid sample drawn from the system. Hydraulic system fluid sampling is accomplished on a periodic basis according to the applicable inspection program.

You should perform analysis of hydraulic systems if extensive maintenance or crash damage occurs. You should perform the analysis when a metal-generating component fails, an erratic flight control function or a hydraulic pressure drop is noted, or there are repeated and/or extensive system malfunctions. Analysis should be performed when there is a loss of system fluid, or when the system is subjected to excessive temperature. You should perform analysis of the hydraulic system anytime hydraulic contamination is suspected.

Maintenance procedures. Hydraulic fluid contamination controls ensure the cleanliness and purity of fluid in the hydraulic system. Fluid sampling and analysis is performed periodically. The condition of the fluid depends, to a large degree, on the condition of the components in the system. If a system requires frequent component replacement and servicing, the condition of the fluid deteriorates proportionately.

Replacement of aircraft hydraulic system filter elements takes place on a scheduled or conditional basis, depending upon the requirements of the specific system. Many filter elements look identical, but all of them are not compatible with flow requirements of the system.

If the hydraulic system fluid is lost to the point that the hydraulic pumps run dry or cavitate, you should change the defective pumps, check filter elements and decontaminate the system as specified in the manufacturers maintenance manual.

All portable hydraulic mules should receive a periodic maintenance check. When the portable hydraulic test stand is not in use, it should be protected against contaminants such as dust and water. You should ensure that correct hoses are used and that they are approved for the type of fluid being used. Properly cap hoses when they are not being used. At the very minimum ground hydraulic power systems need to be as clean as the airplanes on which it is used.

Use only approved lubricants for O-ring seals. Incorrect lubricants will contaminate a system. Many lubricants look alike, but few are compatible with hydraulic fluids. Most maintenance or service manuals should list approved lubricants. If they do not, then use the system fluid.

Maintenance practices. Good housekeeping and maintenance practices help eliminate problems caused by contamination. Often, you cannot see harmful grit. Use only authorized hydraulic fluid, O-rings, lubricants, or filter elements. When dispensing hydraulic fluid, make sure you use an authorized fluid service unit. Check to make sure that the hydraulic fluid can is clean before it is installed. After use, dispose of all empty hydraulic fluid cans and used hydraulic fluid in accordance with federal hazardous material (HAZMAT) instructions. Keep hydraulic fluid in a closed container at all times.

Remove exterior contaminants by using approved wiping cloths. Lint-free wiping cloths should be used on surfaces along the fluid path. If possible, have the replacement component on hand for immediate installation upon removal of defective components. This leaves the system open for the shortest time possible. Replace filters immediately after removal. If possible, fill the filter bowl with proper hydraulic fluid before you install it to minimize the induction of air into the system. Do not reset differential pressure indicators if the associated filter element is loaded and in need of replacement.

Store O-rings, tubing hoses, fittings and components in clean packaging. Do not open or puncture individual packages of O-rings or backup rings until just before you use them. Do not use used or unidentifiable O-rings. Replace seals or backup rings with new items when they have been disturbed. Use the correct O-ring installation tool when you install O-rings over threaded fittings to prevent threads from damaging the O-ring.

If packages of tubing, hoses, fittings, or components are opened when received or found opened, decontaminate their contents. Decontaminate the system if you suspect it is contaminated (including water). Keep the working area where hydraulic components are repaired, serviced, or stored clean and free from moisture, metal chips and other contaminants. Perform required periodic checks on equipment you use to service hydraulic systems.

Keep portable hydraulic test stand reservoirs above three-quarters full. Seal all hydraulic lines, tubing, hoses, fittings and components with approved metal closures. You should not use plastic plugs or caps because they are possible contamination sources. Install quick-disconnect dust covers. Store unused caps and plugs in a clean container.

Before using a portable hydraulic test stand clean all connections, interconnect the pressure and return lines of the stand and circulate the hydraulic fluid through the test stand filters before connecting portable hydraulic test stands to aircraft. This will give you the cleanest possible connection.

Types of Contamination

There are many different forms of contamination, including liquids, gasses and solid matter of various composition, size and shape. Normally, contamination in an operating hydraulic system originates at several different sources. The rate of its introduction depends upon many factors directly related to wear and chemical reaction. Contamination removal can reverse this trend. Production of contaminants in the hydraulic system increases with the number of system components. The rate of contamination from external sources is not readily predictable. A hydraulic system can be seriously contaminated by poor maintenance practices that lead to introducing large amounts of external contaminants. Poorly maintained service equipment is another source of contamination.

Contaminants in hydraulic fluids are classified as *particulate* and *fluid contamination*. They may be further classified according to their type, such as organic, metallic solids, non-metallic solids, foreign fluids, air and water.

Particulate contamination. The type of contamination most often found in aircraft hydraulic systems consists of solid matter. This type of contamination is known as *particulate contamination*.

The size of particulate matter in hydraulic fluid is measured in *microns* (millionths of a meter).

The largest dimensions (points on the outside of the particle) of the particle are measured when determining its size. The relative size of particles, measured in microns, is shown in Figure 1-5-1.

Contamination of hydraulic fluid with particulate matter is a principal cause of wear in hydraulic pumps, actuators, valves and servo valves. Contamination increases the rate of erosion of the sharp spool edges and general deterioration of the spool surfaces. Because of the extremely close fit of spools in servo valve housings, the valves are particularly susceptible to damage or erratic operation when operated with contaminated hydraulic fluid.

Organic contamination. *Organic solids,* or *semisolids,* are one of the particulate contaminants found in hydraulic systems. They are produced by wear, oxidation, or *polymerization* (a chemical reaction). Organic solid contaminants found in the systems include minute particles of O-rings, seals, gaskets and hoses. These contaminants are produced by wear or chemical reaction.

Oxidation of hydraulic fluids increases with pressure and temperature. *Antioxidants* are blended into hydraulic fluids to minimize such oxidation. Oxidation products appear as organic acids, asphaltics, gums and varnishes. These products combine with particles in the hydraulic fluid to form *sludge*. Some oxidation products are oil soluble and cause an increase in hydraulic fluid viscosity, while other oxidation products are not oil soluble and form sediment. Oil oxidation products are not abrasive.

These products cause system degradation because the sludge or varnish-like materials collect at close-fitting, moving parts, such as the spool and sleeve on servo valves. Collection of oxidation products at these points causes sluggish valve response.

Metallic solid contamination. Metallic solid contaminants are usually found in hydraulics systems. The size of the contaminants will range from microscopic particles to those you can see with the naked eye. These particles are the result of the wearing and scoring of bare metal parts and plating materials, such as silver and chromium. Wear products and other foreign metal particles, such as steel, aluminum and copper act as metallic catalysts in the formation of oxidation products. Fine metallic particles enter hydraulic fluid from within the system. Although most of the metals used for parts fabrication and plating are found in hydraulic fluid, the major metallic materials found are ferrous, aluminum and chromium particles.

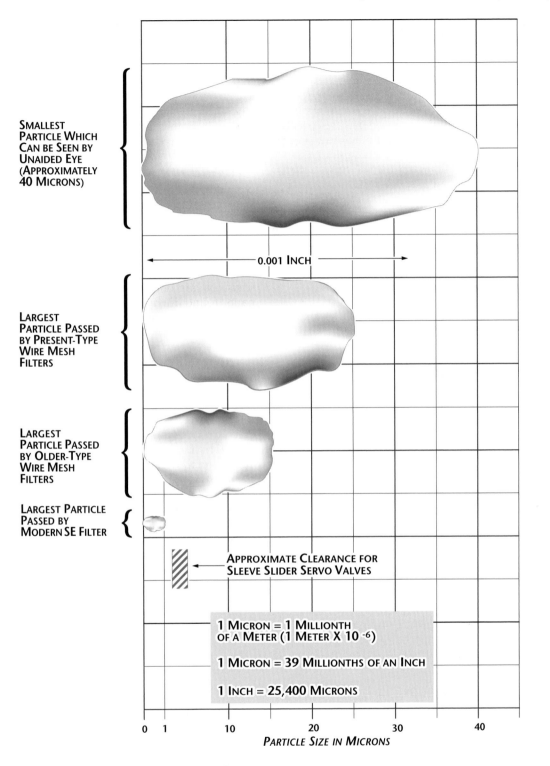

Figure 1-5-1. Graphic comparison of particle sizes

Hydraulic pumps usually contribute the most contamination to the system because of their high-speed internal movement. Other hydraulic systems produce hydraulic fluid contamination due to body wear and chipping.

Hydraulic actuators and valves are affected by contamination. Large metallic or hard nonmetallic particles collect at the seal areas. These particles may groove the inside wall of the actuator body due to a scraping action. Smaller particles act as abrasives between the seals and the actuator body, causing wear and scoring. Eventually, the fluid leaks and the seals fail because the seal extrudes into the enlarged gap between the piston head and the bore of the actuator body. Once wear begins, it increases at a faster rate because wear particles add to the abrasive material. In a similar manner, metallic or nonmetallic parts may lodge in the poppets and poppet-seat portions of valves and cause system malfunction by holding valves open.

Inorganic solid contamination. The inorganic solid contaminant group includes dust, paint particles, dirt and silicates. These and other materials are often drawn into hydraulic systems from external sources. The wet piston shaft of a hydraulic actuator may draw some of these foreign materials into the cylinder past the wiper and dynamic seals. The contaminant materials are then dispersed in the hydraulic fluid. Also, contaminants may enter the hydraulic fluid during maintenance when tubing, hoses, fittings and components are disconnected or replaced. To avoid these problems, all exposed fluid ports should be sealed with approved protective closures.

Glass particles from glass bead peening and blasting are another contaminant. Glass particles are particularly undesirable because glass abrades synthetic rubber seals and the very fine surfaces of critical moving parts.

Fluid Contamination

Air, water, solvents and foreign fluids can contaminate hydraulic fluid.

Air. Hydraulic fluids are adversely affected by *dissolved, entrained,* or *free air.* Air may be introduced through improper maintenance or as a result of system design. Air is sometimes introduced when changing filters. You can minimize this kind of contamination by putting hydraulic fluid into the filter holder before reassembling the filter. By doing this, you have introduced less air into the hydraulic system. The presence of air in a hydraulic system causes a *spongy* response during system operation. Air causes cavitation and erodes hydraulic compo-

nents. Air also contributes to the corrosion of hydraulic components.

Water. Water is a serious contaminant of hydraulic systems. Corrective maintenance actions must be taken to remove all free or emulsified water from hydraulic systems. Hydraulic fluids and hydraulic system components are adversely affected by dissolved, emulsified or free water. Water may be induced through the failure of a component, seal, line or fitting, poor or improper maintenance practices and servicing. Water may also be condensed from air entering vented systems.

The presence of water in hydraulic systems can result in the formation of undesired oxidation products, and corrosion of metallic surfaces will occur. This oxidation will also cause hydraulic seals to deteriorate and fail, resulting in leaks. If the water in the system results in the formation of ice, it will reduce fluid flow and impede the operation of valves, actuators, or other moving parts within the system. This is particularly true of water located in static circuits or system extremities and subject to high-altitude, low-temperature conditions. *Microorganisms* will grow and spread in hydraulic fluid contaminated with water. These microorganisms will clog filters and reduce system performance.

Solvent. Solvent contamination is a special form of foreign fluid contamination. The original contaminating substance is a *halogenated* (chlorinated) solvent, introduced by improper maintenance practices. It is extremely difficult to stop this kind of contamination once it occurs, but using the right cleaning agents when performing hydraulic system maintenance can prevent it. Chlorinated solvents, when allowed to combine with minute amounts of water, hydrolyze to form *hydrochloric acids*. These acids attack internal metallic surfaces in the system, particularly those that are ferrous, and produce a severe rust-like corrosion that is virtually impossible to arrest. Extensive component overhaul and system decontamination are generally required to restore the system to an operational status (See Table 1-5-1).

Foreign fluids contamination. Contamination of hydraulic fluid occurs when the wrong fluids get into the system, such as oil, engine fuel, or incorrect hydraulic fluids. If you think that contamination has occurred, the system must be checked by chemically analyzing fluid samples. This analysis is generally conducted by an outside contractor, or vendor, which verifies and identifies the contaminant and directs decontamination procedures. The manufacturer's maintenance manuals have information on fluid compatibilities.

NONHALOGENATED CLEANING SOLVENTS	
PRODUCT	MANUFACTURER
PD-680	Texaco Co.
Safety Kleen Solvent	Safety Kleen Corp.
Stoddard Solvent	Chevron Oil Co.
Odorless Mineral Spirits	Ashland Chemical
White Mineral Spirits	Chevron Oil Co.
Safety Solvent	Various
HALOGENATED CLEANING SOLVENTS	
Freon TF	
Freon 113	
Trichloretheylene	

Table 1-5-1. These are some of the cleaning fluids a technician may encounter. The maintenance manual is the final reference in all cases.

The effects of foreign fluid contamination depend upon the nature of the contaminant. The compatibility of the construction materials and the system hydraulic fluid with the foreign fluid must be considered when dealing with contamination. Other effects of this type of contamination are hydraulic fluid reaction with water and changes in flammability and viscosity characteristics. The effects of contamination may be mild or severe, depending upon the contaminant, how much is in the system and how long it has been in the system.

Sampling points. A fluid sampling point is a physical point in a hydraulic system from which small amounts of hydraulic fluid are drawn to analyze for contamination. Sampling points include air bleed valves, reservoir drain valves, quick-disconnect fittings, removable line connections and special valves installed for this specific purpose.

To determine the contamination level, a single fluid sample is required. This sample must be representative of the working fluid in the system, and most maintenance manuals and inspection handbooks will list the places they want a sample to be taken from.

Sampling

Most hydraulic system fluid analysis vendors provide a "kit" to take the samples. This not only expedites the analysis process by providing the correct shipping and packing materials, it also provides the correct amount of fluid for the sample. There may be additional directions on the sample bottle in the kit. Most vendors can provide an analysis in one working day.

Filter bowl contents analysis. Hydraulic fluid samples obtained from filter bowls and/or elements cannot be used to determine system contamination levels.

Evaluate filter bowl residues by following the procedures in applicable manuals. As you gain experience about normal contaminates for specific aircraft systems and hours of operation, you will be able to evaluate filter bowl residue. Through experience, analysis of main pressure line and case drain filter bowl residues is useful in verifying failure of the upstream hydraulic pump, as large amounts of metal usually show up in these particular assemblies. Residue in other filter assemblies is affected by so many other components and factors that analysis is difficult. Filter bowl residues should be analyzed only as a means of identifying or verifying suspected component failure. Examine residue from those filter assemblies directly downstream from the component.

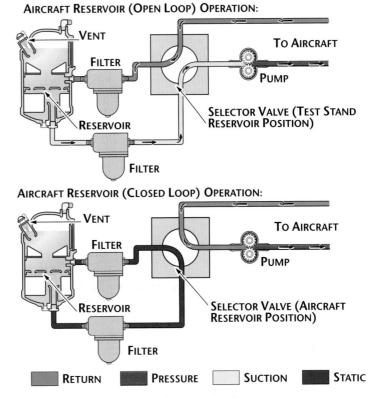

Figure 1-5-2. Fluid flow during decontamination

Decontamination

System decontamination is a maintenance operation performed when a system contains fluid that is unacceptable because of contamination. The fluid may be contaminated with foreign matter or it is not considered acceptable for service for some other reason. The purpose of decontamination is to remove foreign matter from the operating fluid or to remove the contaminated fluid itself. Before you can decontaminate an affected system, replace any failed or known contamination-generating components. Other components of the system are not to be disturbed, unless required.

Methods

There are four basic methods used to decontaminate aircraft hydraulic systems. The methods are *recirculation cleaning, flushing, purging* and *purifying.*

Recirculation cleaning. Recirculation cleaning is a decontamination process in which the system to be cleaned is powered from a clean external power source. The system is cycled so it produces a maximum interchange of fluid between the powered system and the service equipment used to power it, as shown in Figure 1-5-2. When decontaminating a system, the contaminated fluid is circulated through the hydraulic filters in the aircraft system and in the portable hydraulic test stands.

Decontamination that uses the recirculation cleaning method is a filtration process. It can remove only that foreign matter that is retained by the filter elements normally found in the equipment. A key factor in recirculation cleaning is the use of high-efficiency, three-micron (absolute) filter elements. Absolute filter elements have no fluid bypass when the filter clogs. The filters have a large dirt-holding capacity in the portable test stands used for this purpose. In a single fluid pass, these filters remove all particulate matter larger than three microns and a high percentage of the other particles down to submicron size. Recirculation cleaning is effective in removing hard particulate matter from hydraulic fluid that is otherwise serviceable.

The filters are NOT capable of removing water, other foreign fluids, or dissolved solids. Therefore, recirculation cleaning is limited to decontamination of systems found to have a particulate level in excess of system specifications, whose fluid is considered otherwise acceptable. For specific procedures on recirculation cleaning, you should refer to the applicable maintenance manual.

Use recirculation cleaning to remove excessive particulate matter that results from normal component wear, limited component failure, or external sources. Clean the system by powering it with an external portable hydraulic test stand. Operate the aircraft systems so maximum interchange of fluid is produced between the aircraft and the test stand.

Test stands used for recirculation cleaning must be equipped with three-micron (absolute) filtration. If contamination is severe, or if aircraft filters are suspected of being loaded or damaged, or if differential pressure indicators have been activated, install new (or cleaned and tested) filter elements in the aircraft before you begin cleaning. Set up and operate the test stand in a manner compatible with the requirements of the specific aircraft and system being powered. Adjust the test stand output pressure and flow volume for normal operation of the aircraft system being recirculation cleaned.

Sample and analyze the system after the cycling of components. If the contaminant level shows improvement but is still unacceptable, repeat the recirculation cleaning process. If no improvement is observed, attempt to determine the source of contamination. System flushing may be required.

Flushing. Flushing is a decontamination method in which contaminated system fluid

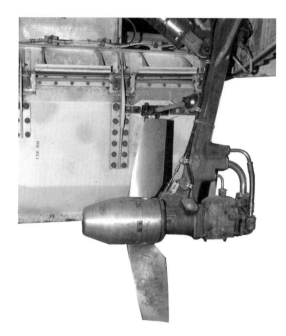

Figure 1-5-3. Flushing all subsystems in most cases means also flushing the emergency hydraulic pump, like this Ram Air Turbine (RAT). Used on large aircraft, they are operated only in an extreme flight emergency hydraulic failure.

is removed to the maximum extent practicable and then discarded. It is a draining process that is generally accomplished by powering the aircraft system with a portable hydraulic test stand. The contaminated return-line fluid from the aircraft is then allowed to flow overboard into a suitable receptacle for disposal. In effect, filtered fluid from the portable hydraulic test stand is used to displace contaminated fluid in the system and to replenish it with clean serviceable fluid. The amount of fluid removed and replaced during system flushing varies. It depends upon such factors as the nature of the contaminant, layout of the system and the ability to produce flow in all affected circuits. Portions of operating systems are often dead ended. Fluid found in these portions is static and not affected by the normal system fluid flow. Remove contaminated fluid in these circuits and associated components by partially disassembling the unit. Drain and totally flush the unit.

Generally, system flushing continues until analysis of the return line fluid from the system being decontaminated indicates that the fluid is acceptable. If there is severe contamination, considerable quantities of hydraulic fluid may be expended, making it important to closely monitor the portable hydraulic test stand reservoir level and replenish it as required.

Flushing effectively decontaminates systems containing water, large amounts of gelatinous-type materials, or fluid that is chemically unacceptable (containing chlorinated or other solvents). This type of fluid contamination or degradation cannot be remedied by conventional filtration. In severe cases of particulate contamination, such as those that result from major component failure, flushing techniques may more easily correct the problem than will recirculation cleaning.

Detailed procedures for flushing hydraulic systems are found in the aircraft maintenance manuals. The basic procedures are discussed in the following text and will give you some idea of the procedures used when flushing aircraft hydraulic systems. Normally, flushing requires you to remove fluids that are found to be chemically or physically unacceptable, or fluids contaminated with water, other foreign fluids, or particulate matter not readily filterable because of its nature or the quantity involved. Use an external portable hydraulic test stand to power the contaminated system and accomplish flushing. Allow return fluid from the aircraft to flow overboard into a waste container for disposal. Aircraft subsystems should be operated to produce maximum displacement of aircraft fluids by cleaned, filtered fluid from the portable test stand. See Figure 1-5-3.

Drain, flush and service the reservoirs or other fluid storage devices in the contaminated system before system flushing. If you know that the contamination originated at an aircraft pump, drain and flush the hoses and lines directly associated with the pump output. Case drains should be drained and flushed separately.

If the aircraft filters are suspected of being loaded, install new or cleaned and tested filter elements in the aircraft hydraulic filters before flushing. Adjust the test output pressure and the flow volume for normal operation of the aircraft system being flushed. Monitor the reservoir level in the portable test stand continuously during the flushing operation. Use approved fluid-dispensing equipment to replenish the reservoir before the level decreases to the half-full point. Depletion of the service equipment reservoir fluid may result in cavitation or failure of the test stand pump. Upon successful completion of system decontamination, service the system to establish proper reservoir fluid level and to bleed entrapped air.

Purifying. Purification of hydraulic fluid requires special equipment and as a process is not used much outside of the military. It is the process of removing air, water, solid particles and chlorinated solvents from hydraulic fluids.

Purging. Purging is a decontamination process in which the aircraft hydraulic system is drained to the maximum extent practicable and the removed fluid discarded. Then, a suitable cleaning agent is introduced into the hydraulic system and circulated as effectively as possible to dislodge or dissolve contaminating substances. The cleaning operation is followed by complete removal of the cleaning agent, then replacing it with new hydraulic fluid. After purging the system, flushing and recirculation cleaning is performed to ensure adequate decontamination.

Selection of Method

The type of contamination present in a system determines the method by which a system is decontaminated. Normally, recirculation cleaning is the most effective decontamination method, considering maintenance man-hours and material requirements. This method should be used whenever possible. However, if a system is contaminated by some substance other than readily filterable particles, it may be necessary to flush the system, or in certain very extreme cases, to purge it.

Contamination control sequence. System decontamination is one operation of a contamination control sequence that includes hydraulic fluid sampling and analysis. Decontamination is performed when the results of sampling and analysis indicate an unacceptable contamination level. Then, additional testing determines when an acceptable level is reached.

The control sequence will either be listed in the maintenance or service manual, or will be part of an inspection program.

Hydraulic Fluids

The first MIL SPEC hydraulic fluid was MIL-H-7644. It was a vegetable based fluid principally composed of caster oil and alcohol. It was colored blue for identification. MIL-H-7644 was used back when most hydraulic systems consisted of brakes only. It was compatible with natural rubber seals. About the only use today for MIL-H-7644 is in the restoration of antique aircraft; and only then when petroleum based seals are not available.

The next major development in hydraulic fluid was the advent of MIL-H-5606. MIL-H-5606 consists of petroleum products with additive materials to improve viscosity (temperature characteristics), inhibit oxidation and act as an anti-wear agent. The oxidation inhibitor was included to reduce the amount of oxidation that occurs in petroleum-based fluids when they are subjected to high pressure and high temperature, and to minimize corrosion of metal parts due to oxidation and resulting acids. The temperature range of MIL-H-5606 is between -65°F to +275°F. It is dyed red so it can be distinguished from incompatible fluids. The fluid is compatible with synthetic rubber seals and is flammable, as most petroleum products are. MIL-H-5606 is still in use in some smaller general aviation aircraft.

MIL-H-83282 replaced MIL-H-5606 as the fluid of choice in larger aircraft. It is dyed red so it can be distinguished from incompatible fluids. MIL-H-83282 has a synthetic hydrocarbon base and contains additives to provide the required viscosity and anti-wear characteristics, which inhibit oxidation and corrosion. It is used in hydraulic systems having a temperature range of -40°F to +275°F. Flash point, fire point and spontaneous ignition temperature of MIL-H-83282, which is fire resistant, exceeds that of MIL-H-5606 by more than 200°F. The fluid extinguishes itself when the external source of flame or heat is removed. Hydraulic fluid MIL-H-83282 is compatible with all materials used in systems using MIL-H-5606. It maybe combined with MIL-H-5606 with no adverse effect other than a reduction of its fire-resistant properties.

MIL-H-81019 is an ultra-low temperature hydraulic fluid. It is used in aircraft when extremely low surrounding temperatures are expected. MIL-H-81019 consists of petroleum products with additive materials to improve its viscosity (temperature characteristics), increase its resistance to oxidation, inhibit corrosion and act as an anti-wear agent. It is dyed red so it can be distinguished from other incompatible hydraulic fluids. It is not normally mixed with MIL-H-5606 or MIL-H-83282. MIL-H-81019 is designed to operate in hydraulic systems having a temperature range between -90°F to +120°F.

The search for a non-flammable hydraulic fluid led to the development of a phosphate ester based fluid called Skydrol 7000®. The fluid was not compatible with any petroleum-based product and used butyl rubber seals. It was colored light green for identification. Skydrol 7000® was discontinued in 1973.

As jet aircraft became a way of life, Skydrol 7000® was replaced with Skydrol 500®. Colored purple to distinguish it from the earlier product. It was discontinued in 1958. Skydrol 500® has under gone a series of modifications during its lifetime. These modifications were specified as Skydrol 500® type I, type II, type III and type IV. All current phosphate ester-based hydraulic fluids are colored purple.

The current Skydrol fluids are LD-4 and 500B-4. Fluid compatibility is generally not a prob-

lem, because the prior versions have not been manufactured since 1981. In addition to the two type-IV fluids in current use, Skydrol 5 was introduced in 1995. It is less toxic, easier on paint and more stable at high temperatures. It has better high-temperature stability (275°F) and lower density than type-IV fluids, and is compatible with them. The lower density is a plus for modern passenger aircraft. As an example, a Boeing 777 carries 180 gallons of hydraulic fluid and the lower density saves 124 lbs. in weight. Higher-temperature stability contributes to a doubling of the useful life of the fluid.

The color of used phosphate ester fluid is not an indication of its condition. Condition can only be established by analysis.

Fluid analysis. The fluid analysis procedure includes several tests, the most important of which is acidity, measured by the neutralization number titration. The phosphate ester decomposes under heat stress, especially with moisture present, and forms organic acids. When the neutral number reaches a value of 1.5, the fluid is spent and must be replaced. The manufacturer puts an additive in the fluid to react with the organic acid to neutralize it. As long as this additive is present, the neutral number will not increase. There is enough additive present to give current fluids a life of at least 3,000 hours at their rated temperature. The fluid manufacturer recommends that aircraft operators sample their systems at "C" check intervals to look at acid levels.

Other tested items are for percent moisture, particulate contamination and percent Cl (chlorine).

Checking Aircraft Hydraulic Fluid Levels

There are specific procedures for checking hydraulic fluid levels in each model of aircraft. These procedures must be followed to make sure the system operates at the required fluid level. An indicating device at the system reservoir generally determines fluid level. The type of indicator used varies with the aircraft model. Sight-glass, gauge and piston-style indicators are commonly used.

Remember, you need to follow the procedures contained in the applicable technical manuals when you actually service hydraulic systems and components.

Use the correct fluids for each piece of fluid-dispensing equipment, and mark the equipment to indicate the type of fluid. Use the specified hydraulic fluid to service hydraulic systems.

Take precautions to avoid accidental use of any other fluid. Do not leave hydraulic fluid in an open container any longer than necessary, particularly in dusty environments. Exposed fluid will readily collect contaminants, which could jeopardize system performance.

Do not reuse hydraulic fluid drained from hydraulic equipment or components. Dispose of drained fluid immediately so it will not be accidentally reused.

Applying Hydraulic Power

Before you connect a test stand to an aircraft system, make sure that all personnel, work stands and other ground-handling equipment are clear of flight control surfaces, movable doors and other units. Stay clear of these areas when either electric power or hydraulic pressure is applied to the aircraft. Sudden movement can cause injury or damage.

> NOTE: *Refer to the applicable maintenance manual for specific procedures to follow when applying external electric and hydraulic power.*

Before connecting the hydraulic test stand to the aircraft, set the test stand controls to the required positions and values to complete the aircraft tests. Operate the test stand to confirm the settings. Reduce the volume adjustment to minimum flow and shut down the stand. Connect the test service hoses to the aircraft ground power quick-disconnects, making sure that all connectors are clean before connection. Mate all the attached dust caps and plugs to protect against their contamination during test stand operation.

Do not kink or damage test stand hoses when connecting them to aircraft systems.

> NOTE: *Use the procedures found in the applicable manufacturers manual to actually power the aircraft hydraulic system.*

Operational checks. When operating the test stand, you need to periodically check the condition of system fluid through the sight glass. If you see evidence of air, bleed the system at both the test stand and air bleed points in the aircraft until the fluid appears clear. Also, you need to monitor the filter differential pressure indicators, particularly those associated with the three-micron filter assemblies. In some cases, loaded filter indicators may extend due to cold starting conditions. Reset the indicator and continue to monitor it until the equipment reaches the normal operating temperature. In case of an emergency (for example, a ruptured hydraulic hose in aircraft), you should open the

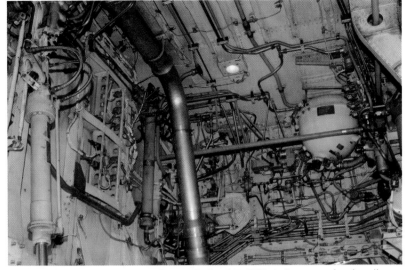

Figure 1-5-4. As you can see from this Boeing 757 main gear wheel well, trying to bleed air from hydraulic lines could be a time-consuming task.

bypass valve to relieve pressure and stop the flow of hydraulic fluid to the aircraft.

Shutdown procedure. In aircraft equipped with pressurized reservoirs, hydraulic accumulators, or surge dampers, a reverse flow of fluid through the aircraft filters could damage the system. You need to use the correct shutdown procedures.

Remove the external hoses from the aircraft hose ports. Connect one end to the hose storage manifold disconnects on the test stand. Do not drag the hose ends on the tarmac or expose them to contamination. Install all dust caps and plugs, including those at the aircraft quick disconnects. Close all the access doors to protect instruments and controls.

Air Bleeding

Air bleeding is a service operation in which entrapped air is allowed to escape from a closed hydraulic system. For specific air bleed procedures for each model aircraft, you should refer to the applicable maintenance manual. Excessive amounts of free or entrained air in an operating hydraulic system result in degraded performance, chemical deterioration of fluid and premature failure of components. Therefore, when a component is replaced or a hydraulic system is opened for repairs, that system must be bled of air to the maximum extent possible upon repair completion.

Hydraulic fluid can hold large amounts of air in solution. Fluid, as received, may contain dissolved air or gasses equivalent to 6.5 percent by volume, which may rise to as high as 10 percent after pumping. Dissolved air generates no problem in hydraulic systems so long as it stays dissolved, but when it comes out of solution (as extremely minute bubbles), it becomes *entrained* or free air. Free air could enter a system during component installation, filter element installation, or opening the system during repairs.

Free air is harmful to hydraulic system performance. The compressibility of air acts as a soft spring in series with the stiff spring of the oil column in actuators or tubing, resulting in degraded response. Also, because free air can enter fluid at a very high rate, the rapid collapse of bubbles may generate extremely high local fluid velocities that can be converted into impact pressures. This is the phenomenon known as *cavitation*. Cavitation causes pump pistons and slide valve metering lands to wear rapidly, commonly causing component failure.

Any maintenance operation that involves breaking into the hydraulic system introduces air into the system, as shown in Figure 1-5-4. The amount of such air can be minimized by pre-filling replacement components with new, filtered hydraulic fluid. Because some residual air may still be introduced, all maintenance of this type is followed by a thorough air bleed of the system. Most hydraulic systems are designed to self-scavenge free air back to the system reservoir. Pressurized reservoirs have air bleed valves to remove this air.

Air bleed valves are sometimes found at high points in the aircraft circulatory system, filter assemblies and remote system components such as actuators. These valves make the removal of free air easier. Refer to the applicable maintenance manual for the location and use of additional bleed points. In systems not equipped with additional bleed points, you may have to loosen line connections temporarily at strategic points in the system, which permits removal of entrapped air from remote or dead-end points. When you bleed a system in this manner, be careful to avoid excessive loss of hydraulic fluid and prevent the induction of air or contaminants into the system.

An effective means for measuring the air in your system is known as the reservoir sink check. In this method, the fluid level in the aircraft reservoir is checked with the system, both pressurized and non-pressurized. The presence of air or any compressible gas in the system causes the pressurized reading to be lower (reservoir sink), indicating the need for possible maintenance action. This check is particularly effective when performed after a long aircraft down period, in which case dissolved air has had lots of time to come out of solution.

Many hydraulic systems are designed in such a way that they tend to automatically bleed air out of the system when actuators or motors are operated. If you see a system component, such as a flap move quicker with each actuation after the installation of a new actuator or pump, it indicates that the air is being worked out of the system. Continue to operate the system at allowable intervals until the component is moving at desired speed.

All air bleed operations must be followed by a check of the system hydraulic fluid level. Fluid replenishment may be required, depending upon the amount of air and fluid purged from the system.

Contamination Control

The direct connection between hydraulic service equipment and the systems or components being checked or serviced is necessary to minimize the introduction of external contaminants. Test units that are not properly configured, maintained, or used may severely contaminate an aircraft's hydraulic system.

Cleanliness. Always keep external fluid connections, fittings and openings clean and free of contamination. When not in use, protect fittings or hose ends using metal dust caps or other approved closures. You can use clean, polyethylene bags if you do not have the approved metal closures, providing the bags are adequately secured and are protected from physical damage and the entrance of water. When equipment is not being used, store it in clean, dry areas.

Section 6

Typical Corporate Aircraft Hydraulic Systems

The example hydraulic system presented here is for a Gulfstream business jet. It is a modern multi-faceted system that uses most of the principles that will be encountered in virtually any complex hydraulic system. If you will follow the illustrations as you read the text, you will not only be able to apply most of the preceding information in this chapter, but will also very quickly learn how to follow a system diagram.

Hydraulic Systems

The aircraft has five hydraulic subsystems, as shown in Figure 1-6-1. The hydraulic system is designed for use with Hy-Jet IV Skydrol

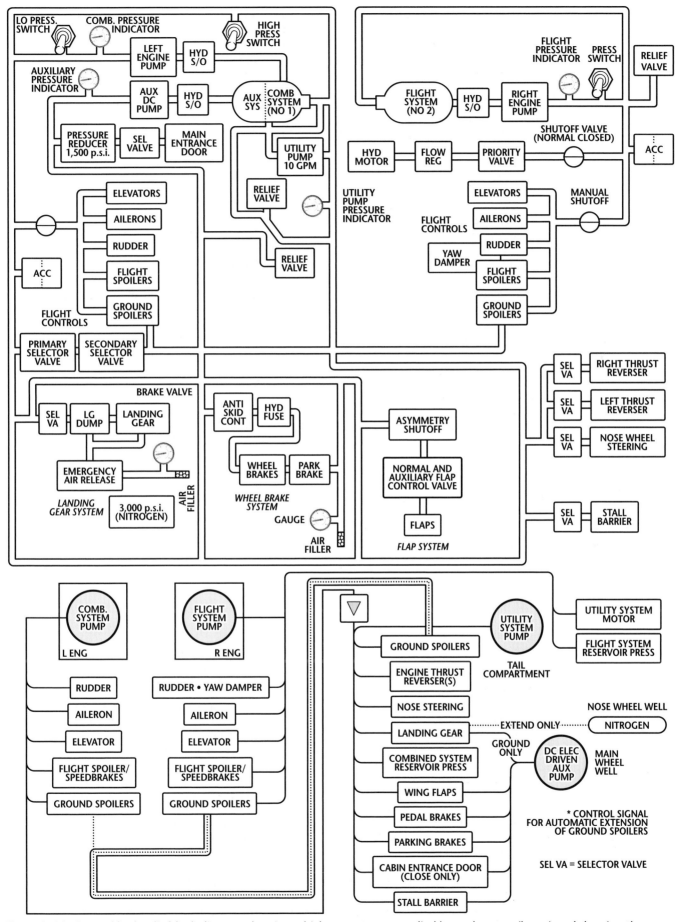

Figure 1-6-1. General hydraulic block diagram showing which systems are supplied by each pump (lower) and showing the complete hydraulic system (upper)

HYDRAULIC SUBSYSTEMS
Combined - 1,500 ± 200/3,000 ± 200 p.s.i.
Flight - 1,500 ± 200/3,000 ± 200 p.s.i.
Utility - 3,000 ± 200 p.s.i.
Auxiliary - 3,000 ± 200 p.s.i.
Emergency Landing Gear - 3,000 + 0 - 50 p.s.i. (Nitrogen @ 70°F).

Table 1-6-1. Different pressures for a Gulfstream jet.

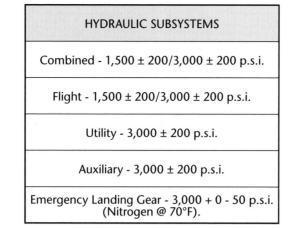

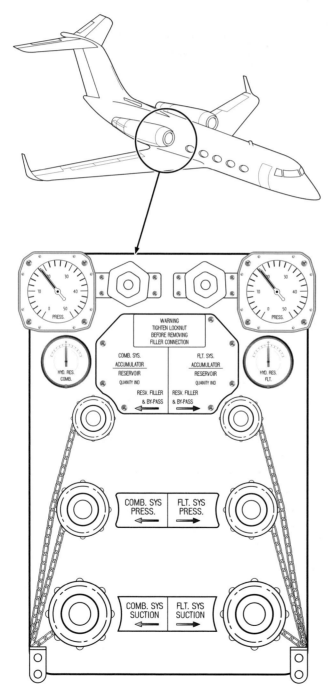

Figure 1-6-2. Hydraulic service panel

hydraulic fluid, functional in temperatures ranging from -54° to 107°C (-65° to 225°F).

The combined and flight hydraulic systems are pressurized by engine-driven pumps. A hydraulic motor driven by the *flight system pump* pressurizes the utility hydraulic system and an electrically driven pump pressurizes the *auxiliary system*. In addition the *emergency pneumatic system* provides for emergency extension of the landing gear. The different pressures are listed in Table 1-6-1.

Reservoirs and fluid quantity. The combined and the flight system reservoirs are located in the left and right sides of the tail compartment, respectively. The system fluids do not intermix, and each reservoir is pressurized by its own system. The total fluid capacity of the combined reservoir is 5 1/2 gallons, which includes 3 3/4 gallons in the combined compartment of the reservoir and the remaining 1 3/4 gallons in the auxiliary compartment. The total fluid capacity of the flight system reservoir is 1 1/4 gallons.

The combined and flight system reservoirs are pressure-filled by means of individual filler lines located at the hydraulic service panel (Figure 1-6-2) on the underside of the aircraft just forward of the tail compartment door. They can also be serviced with a replenishing pump mounted in the tail compartment (Figure 1-6-3). A manually operated valve selects the reservoir to be filled.

Both reservoirs are pressurized by system hydraulic pressure to ensure adequate fluid flow to the pump. Reservoir fluid level is indicated on the *hydraulic reservoir quantity indicator* located on the copilot's outboard knee panel and at the hydraulic service panel. Power required for operation is 28 VDC from the Essential DC Bus through two circuit breakers labeled COMB HYD QTY and FLT HYD QTY on the copilot's circuit breaker panel. A visual sight gauge is installed on the aft end of each reservoir to clearly show the FULL and REFILL levels. No electrical power is needed, but the system must be pressurized to obtain an accurate reading of the sight gauge. The proper fluid level is shown when the piston within the sight gauge indicates FULL.

Hydraulic pump compensation. The combined and flight system pumps incorporate compensating solenoids that control the pressure output. When a solenoid is energized, the respective pump produces 1,500 p.s.i.; when de-energized, the pump produces 3,000 p.s.i.

A pressure switch controls the combined pump in the flight system (Figure 1-6-4). In the cruise condition (landing gear and flaps up) and with the flight system pressure above 800 p.s.i., the combined system pump solenoid is energized and pump output is 1,500 p.s.i. With the landing gear and/or flaps extended, or if flight system pressure is below 800 p.s.i., the combined system pump solenoid is de-energized and pump pressure is 3,000 p.s.i. The combined system low-pressure and high-pressure switches control the flight system pump (Figure 1-6-4 and Table 1-6-2).

Combined Hydraulic System

The combined hydraulic system (Figure 1-6-5) supplies 3,000 p.s.i. during takeoff and landing to operate the flight controls and the landing gear, stall barrier, wing flaps, wheel brakes, nose wheel steering, ground spoilers and thrust reversers. During flight, the combined system supplies 1,500 p.s.i. to operate the flight controls; namely elevators, stall barrier, ailerons, rudder and speed brake/flight spoilers. Power to operate the combined system is

28 VDC from the Essential DC Bus through the COMB HYD CONT circuit breaker- on the pilot's circuit breaker panel.

The variable-volume pump driven by the left engine generates combined system pressure. The pump draws fluid from the combined reservoir through a shutoff valve controlled by the L FIRE PULL T-handle. This handle is pulled only for an engine fire. As pressure builds to 800 p.s.i. after the left engine starts, a pressure switch extinguishes the yellow COMB HYD caution light on the *master warning lights panel.* Normal operating pressures of 1,500 p.s.i. or 3,000 p.s.i. are indicated on the left side of the *hydraulic pressure indicator* on the copilot's inboard knee panel. The indicator is powered by 26 VAC, 400 Hz from the left engine instrument Inverter Bus through a circuit breaker labeled COMB HYD PRES on the pilot's auxiliary circuit breaker panel.

During takeoff and landing, with the flaps extended, the combined pump solenoid is de-

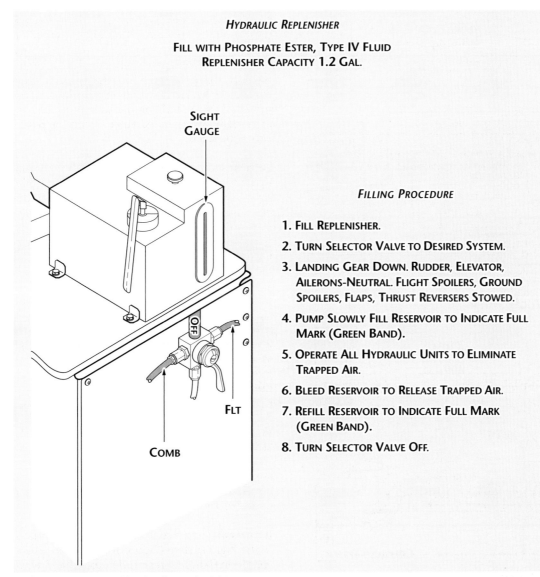

Hydraulic Replenisher

Fill with Phosphate Ester, Type IV Fluid
Replenisher Capacity 1.2 Gal.

Sight Gauge

Filling Procedure

1. Fill Replenisher.
2. Turn Selector Valve to Desired System.
3. Landing Gear Down. Rudder, Elevator, Ailerons-Neutral. Flight Spoilers, Ground Spoilers, Flaps, Thrust Reversers Stowed.
4. Pump Slowly Fill Reservoir to Indicate Full Mark (Green Band).
5. Operate All Hydraulic Units to Eliminate Trapped Air.
6. Bleed Reservoir to Release Trapped Air.
7. Refill Reservoir to Indicate Full Mark (Green Band).
8. Turn Selector Valve Off.

OFF

FLT

COMB

Figure 1-6-3. Manual hydraulic replenishing pump

energized causing the system to be pressurized to 3,000 p.s.i. When the flaps are retracted and the aircraft is in a clean configuration, the pump pressure is compensated to 1,500 p.s.i. for actuation of the primary flight controls. If the flight system pump fails, the combined system will up compensate to 3,000 p.s.i. An accumulator in the system dampens pressure surges. A relief valve, set at 3,850 p.s.i., will protect the system from excessive pressures. A thermal switch senses reservoir fluid temperature and illuminates the yellow COMB HYD HT caution light if fluid exceeds 220°F. combined system pressure failure (less than 800 p.s.i.) will cause the amber COMB HYD caution light on the master warning lights panel to illuminate.

Flight Hydraulic System

The variable-volume pump driven by the right engine generates the flight hydraulic system pressure (Figure 1-6-6). The system supplies 1,500 p.s.i. to operate the flight controls, yaw damper and ground spoilers. The pump draws fluid from the flight system reservoir through a shutoff valve controlled by the R FIRE PULL T-handle. As pressure builds to 800 p.s.i. after the right engine start, a pressure switch extinguishes the yellow FLT HYD caution light on the master warning lights panel. Power for operation of the flight system is 28 VDC from the Essential DC Bus through a circuit breaker labeled FLT HYD CONT on the pilot's circuit breaker panel.

GEAR/FLAP POSITION	PRESSURE IN COMBINED SYSTEM	FLIGHT SOLENOID CONDITION	FLIGHT PUMP PRESSURE
Up	Above 800 p.s.i. Below 800 p.s.i.	Engerized De-energized	1,500 p.s.i. 3,000 p.s.i.
Down	Above 2,600 p.s.i. Below 2,200 p.s.i.	Engerized De-energized	1,500 p.s.i. 3,000 p.s.i.

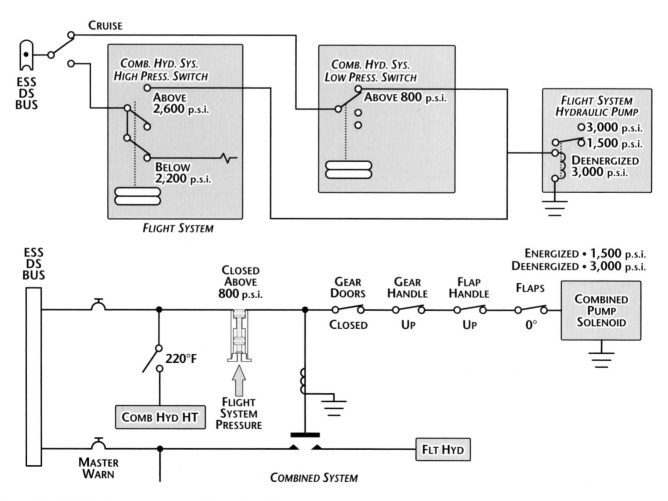

Figure 1-6-4. Hydraulic pump compensation schematic

The pressure output of the pump is either 1,500 or 3,000 p.s.i., as indicated on the right side of the hydraulic pressure indicator on the copilot's inboard knee panel. The indicator is powered by 26 VAC, 400 Hz from the left engine instrument inverter bus through a circuit breaker labeled FLT HYD PRESS on the pilot's auxiliary circuit breaker panel. The pump normally provides 1,500 p.s.i. If a combined hydraulic system failure occurs, the flight system pump is up compensated to 3,000 p.s.i. An accumulator in the system dampens pressure surges, and a relief valve, set at 3,850 p.s.i., protects the flight system from excessive pressure. A thermal switch senses reservoir fluid temperature and illuminates the amber FLT HYD HT caution light on the master warning lights panel and prevents operation of the utility system if

fluid temperature reaches 220°F. A failure of the flight system is indicated by illumination of the amber FLT HYD caution light.

Combined and flight system shutoff valves.
An electrically operated gate-type shutoff valve, which is normally open, is located just forward of and beneath each system reservoir. If a hydraulic leak or fire is discovered, pulling the left or right FIRE PULL T-handle can actuate the shutoff valves. These T-handles are located at the center of the glareshield panel. Each of the valves shuts off the hydraulic fluid through the suction line to the engine-driven pump.

When one of the T-handles is pulled, a limit switch is actuated which energizes the close windings of the split field DC motor in the

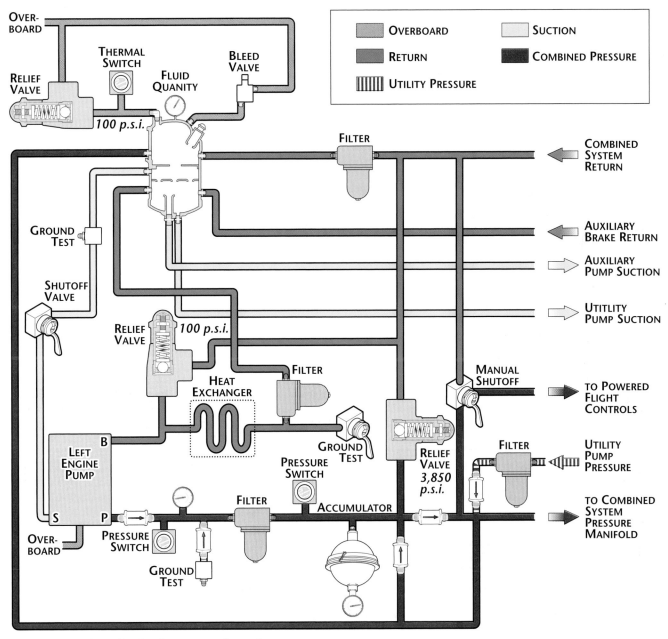

Figure 1-6-5. Combined hydraulic system schematic

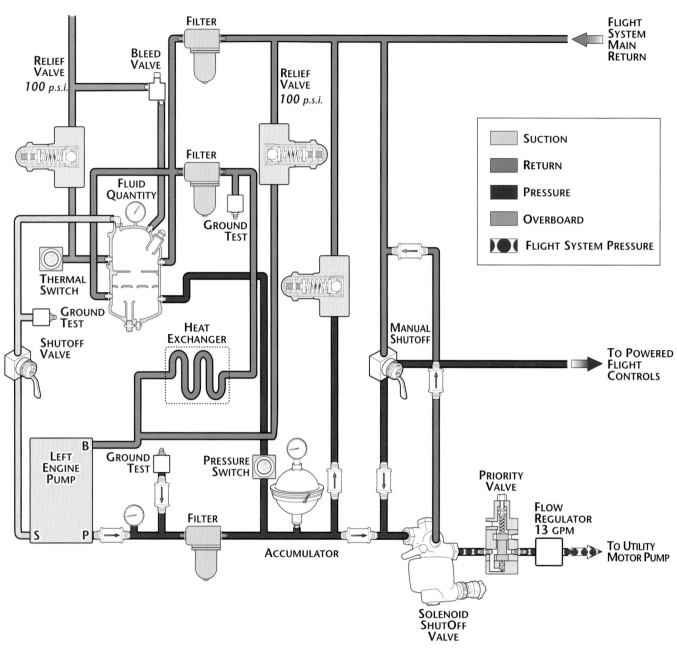

Figure 1-6-6. Flight hydraulic system schematic

shutoff valve. Electrical power is supplied to the valves by the essential DC bus through two circuit breakers, labeled as L HYD S/O and R HYD S/O. The circuit breakers are located on the pilot's circuit breaker panel.

Utility Hydraulic System

The utility hydraulic system (Figure 1-6-7) is a backup for the combined system. Combined system failure during takeoff or landing opens a solenoid shutoff valve to direct flight system pressure to operate the hydraulic motor-driven utility pump. A priority valve permits pressure flow to the hydraulic motor that drives the utility pump if flight system pressure is above 2,000 p.s.i., thereby ensuring operating pres-

sure for the primary flight controls. If flight system pressure drops below 1,730 p.s.i. while the utility pump is operating, the utility pump will cease operation. Utility system pressure is indicated on the UTIL (left side of the UTIL/AUX HYD PRES indicator on the copilot's inboard knee panel. The indicator is powered by 26 VAC, 400 Hz from the right engine instrument inverter bus through a circuit breaker labeled UTIL HYD PRESS on the pilot's auxiliary circuit breaker panel.

When the utility system is operating, the pump draws fluid from the combined reservoir to supply 3,000 p.s.i. to all subsystems normally actuated by the combined system, except the primary flight controls. Operation of the utility system is automatic when the combined system

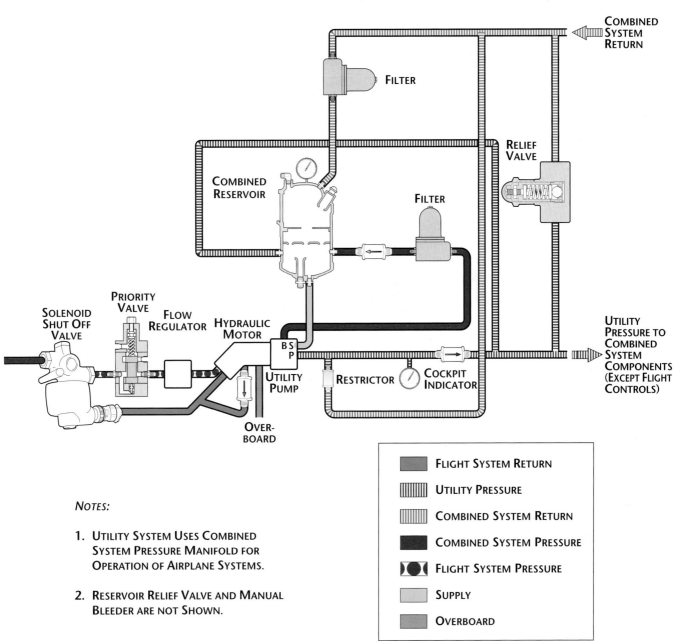

Figure 1-6-7. Utility hydraulic system schematic

NOTES:

1. UTILITY SYSTEM USES COMBINED SYSTEM PRESSURE MANIFOLD FOR OPERATION OF AIRPLANE SYSTEMS.

2. RESERVOIR RELIEF VALVE AND MANUAL BLEEDER ARE NOT SHOWN.

Legend:

- FLIGHT SYSTEM RETURN
- UTILITY PRESSURE
- COMBINED SYSTEM RETURN
- COMBINED SYSTEM PRESSURE
- FLIGHT SYSTEM PRESSURE
- SUPPLY
- OVERBOARD

fails due to failure of the pump or another system component, provided:

- Combined reservoir fluid is available

- Landing gear and/or flap levers are not in the UP position

- Flight system pressure is greater than 2,000 p.s.i. and its fluid temperature does not exceed 220°F

- The UTILITY HYD PUMP Switch is in either the NORM or the OVERRIDE position

Utility hydraulic pump switch. The UTILITY HYD PUMP switch is used to control the operational mode of the utility hydraulic pump. The switch is a three-position switch with the positions NORM, OVERRIDE and OFF. Selecting the OFF position will prevent any operation of the utility hydraulic pump and will illuminate the amber UTIL HYD OFF indicator on the master warning light panel. Power for the "OFF" indication is 28 VDC from the Essential DC Bus through a circuit breaker labeled UTIL HYD PUMP OFF on the pilot's circuit breaker panel. The switch should be selected to OFF if combined fluid is lost to prevent dry operation of the utility pump. In the NORM position, the switch allows the utility hydraulic pump to operate automatically. The OVERRIDE position disengages the utility system solenoid shut-off valve to the open position to allow the utility hydraulic system to operate anytime the flight hydraulic system pressure is above 2,000 p.s.i.

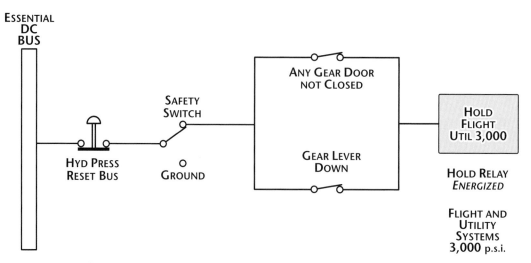

Figure 1-6-8. Hydraulic pressure reset circuit diagram

Combined, Flight and Utility Pump Control Circuit Operation

With both engines operating, the normal ground configuration for hydraulic pressure is; combined system - 3,000 p.s.i., flight system - 1,500 p.s.i., and utility system - 0 p.s.i.

After the airplane is airborne, with the landing gear handle and FLAP lever in the up position, the flaps up and all three gear doors closed, the electrical circuits configure to the clean flight (cruise) configuration. In this configuration:

1. The combined system pump solenoid is energized through the flight system pressure switch and the pump is compensated to 1,500 p.s.i.

2. The utility system pump solenoid shutoff valve is energized closed.

3. The combined system low-pressure switch maintains the flight system pump solenoid energized. The pump continues to be compensated to 1,500 p.s.i.

With both engines operating during the normal flight configuration, hydraulic pressure is; combined system - 1,500 p.s.i., flight system - 1,500 p.s.i., and utility system - 0 p.s.i.

With an engine failure during takeoff, immediate gear retraction is essential. Assuming a left engine failure (engine windmilling) the combined hydraulic system is incapable of performing this function as engine windmilling speeds are such as to keep system pressure above the combined hydraulic system low-pressure switch setting.

To overcome this condition, the combined system high-pressure switch is used to energize

the Hold relay as system pressure drops below 2,200 p.s.i. With the gear lever down or any gear door not closed, the relay remains energized even if pressure fluctuation causes the high-pressure switch to open and close. This results in de-energizing the flight hydraulic system pump and the utility system solenoid shutoff valve. The flight system pump compensates to 3,000 p.s.i. and powers the utility system to supply 3,000 p.s.i. to operate the landing gear.

Once the gear is up with the doors closed, as the flaps are retracted to the up position (cruise condition), the following occurs:

1. Power through the flight system pressure switch energizes the utility solenoid shut-off valve. Utility system pressure drops to 0 p.s.i.

2. Power through the combined system low-pressure switch (pressure above 800 p.s.i.) energizes the flight system pump solenoid, and flight system pressure returns to 1,500 p.s.i.

With the airplane airborne and the left engine failed, the flight configuration for hydraulic pressure is combined system greater than 1,000 p.s.i. (engine windmilling), flight system 1,500 p.s.i., and utility system 0 p.s.i. When the landing gear or FLAP lever is moved from the UP position, the flight system pump again compensates to 3,000 p.s.i. and the utility shutoff valve opens. The utility system pump supplies 3,000 p.s.i. for landing gear operation.

If the combined system fails during flight, the following occurs:

1. The combined system high-pressure switch opens as hydraulic pressure drops to 2,200 p.s.i. This is of no consequence to the electrical circuit in the cruise condition.

2. The combined system low-pressure switch opens as hydraulic pressure drops below 800 p.s.i. The flight system pump compensates to 3,000 p.s.i.

3. The COMB HYD warning light illuminates when the combined system pressure drops below 800 p.s.i.

At this time, the utility pump shutoff valve solenoid remains energized, receiving power through the flight system pressure switch and the UTILITY HYD PUMP switch. As soon as either the landing gear lever or the FLAP lever is moved from the UP position, the utility pump shutoff valve opens (solenoid de-energized), and the utility system, delivering 3,000 p.s.i., acts as a backup system for the failed combined hydraulic system.

If the flight hydraulic system fails during flight, the following occurs:

1. The flight system pressure switch opens when pressure drops below 800 p.s.i. This de-energizes the combined system pump solenoid and the combined system is compensated to 3,000 p.s.i.

2. The FLT HYD warning light illuminates.

3. The utility system shut-off valve remains closed, its solenoid energized by power through the combined system low-pressure switch. The utility pump does not operate. The combined hydraulic system, now operating at 3,000 p.s.i., supplies sufficient power to operate all the hydraulic systems.

Hydraulic pressure reset button. During approach or clean flight, should the landing gear lever be selected down prior to extending the wing flaps, the electrical system senses a momentary drop below 1,500 p.s.i. in the combined system. This is below the combined system high-pressure switch setting of 2,200 p.s.i. Therefore, the Hold Relay is initially energized by power from the high-pressure switch.

With the Hold relay energized, the seemingly abnormal configuration for hydraulic pressure is now: combined system - 3,000 p.s.i., flight system - 3,000 p.s.i., and utility system - 3,000 p.s.i. Depressing the HYD PRESS RESET button (Figure 1-6-8) will de-energize the Hold relay and permit the flight pump to return to 1,500 p.s.i., to energize the utility system solenoid shutoff valve to close and to achieve the correct system pressures of: combined - 3,000 p.s.i., flight - 1,500 p.s.i., and utility - 0 p.s.i.

Auxiliary Hydraulic System

The auxiliary hydraulic system (Figure 1-6-9) receives pressure from an electrically driven pump (located on the aft bulkhead in the left

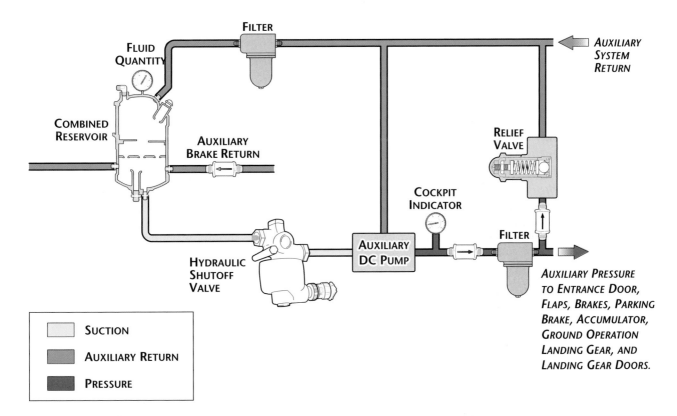

Figure 1-6-9. Auxiliary hydraulic system schematic

main wheel well), which draws fluid from the auxiliary section of the combined reservoir. The pump provides 3,000 p.s.i. for actuation of wing flaps, brakes and parking/emergency brakes. On the ground, the main entrance door, landing gear and landing gear doors are also actuated by the auxiliary hydraulic system. Pressure in the system is indicated on the AUX (right) side of the UTIL/AUX HYD PRES indicator on the copilot's knee panel. The indicator is powered by 26 VAC, 400 Hz from the Right Engine Instrument Inverter Bus through a circuit breaker labeled AUX HYD PRESS on the pilot's auxiliary circuit breaker panel. A relief valve, set at 3,850 p.s.i., protects the system from excessive pressure.

If the auxiliary pump electric motor overheats, the yellow AUX HYD HT indicator on the master warning light panel will illuminate.

NOTE: *Current draw of the auxiliary pump electric motor is very high. Without an engine operating or DC external power connected, limit its use to very short intervals.*

Either of the AUX HYD PUMP switches, one on the pilot's side panel and one on the copilot's inboard knee panel normally operate the auxiliary pump. The pump can also be started with the DOOR CONTROL switch on the pilot's circuit breaker panel, the OUTSIDE DOOR switch in the outside accessory panel or the auxiliary system pump control switch on the ground service valve. Power required for operation is 28 VDC from the Battery Tie Bus

through a circuit breaker labeled AUX HYD PUMP on the pilot's circuit breaker panel.

NOTE: *When operating AUX hydraulic pump for extended time periods with external power connected, select BAT 1 and BAT 2 switches OFF. With the battery switches OFF, the external DC through the Essential DC Bus will power the AUX hydraulic pump.*

Brake accumulator. A brake accumulator, located in the nose wheel well, provides hydraulic pressure for PARK/EMER brakes. Two gauges are connected to the air side of the accumulator; one is located in the wheel well, below the accumulator, and the other is located on the copilot's inboard knee panel. When properly pre-charged with dry nitrogen, 1,200 p.s.i. ± 25 p.s.i. at 70°F, the accumulator will hold approximately 22.5 cubic inches of fluid. This is sufficient for five or six applications of the PARK/EMER brake handle. The accumulator is installed in the auxiliary hydraulic power system; therefore, anytime the auxiliary pump operates, it charges the accumulator with approximately 3,000 p.s.i. hydraulic pressure. A check valve is installed in the pressure line to maintain the pressure for future use after the auxiliary pump is shut off.

Section 7

Aircraft Pneumatic Systems

The use of pneumatics is so integrated into the overall operation of other aircraft systems that it is difficult to look at it as a stand alone system.

Pneumatic systems are sometimes used for:

- Brakes
- Opening and closing doors
- Driving hydraulic pumps, alternators, starters, water injection pumps, etc.
- Operating emergency devices
- Heating/air conditioning systems
- De-ice systems/rain-removal systems
- Pressurization systems
- Engine starting
- Gyro instrument systems

The type of unit used to provide pressurized air for pneumatic systems is determined by the system's air pressure requirements.

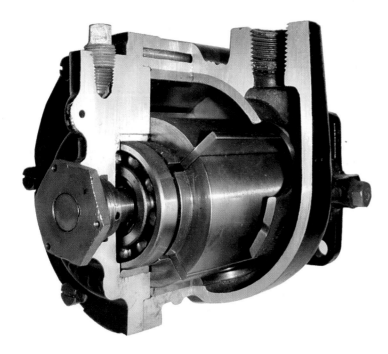

Figure 1-7-1. This cutaway of a wet pump clearly shows the inner workings of a sliding vane pump.

Pressure Systems

High pressure system. For high pressure systems, air is usually stored in metal bottles at pressures ranging from 1,000 to 3,000 p.s.i., depending on the particular system. This type of air bottle has two valves, one of which is a charging valve. The other valve is a control valve. It acts as a shutoff valve, keeping air trapped inside the bottle until the system is operated.

The high pressure storage cylinder has a definite disadvantage. Since the system cannot be re-charged during flight, operation is limited.

On some aircraft, permanently installed air compressors have been added to re-charge air bottles whenever pressure is used for operating a unit.

High pressure systems are normally considered to be emergency pressure systems. How the air is used to activate brakes and landing gear is covered in the respective chapters.

Medium pressure system. A medium pressure pneumatic system (100-150 p.s.i.) draws *bleed air* from a turbine engine compressor section. In this case, air leaves the engine through a takeoff and flows into tubing, carrying air first to the pressure-controlling units and then to the operating units.

Early turbine engines could produce large amounts of bleed air, and airframe manufacturers put it to good use. The various systems operated by bleed air on today's turbine aircraft are very efficient. This efficiency has allowed engine designers to reduce the amount of bleed air available, thus reserving it for producing power from the engine. The reduction of bleed air volume increases fuel efficiency and power available from the engines.

Systems that operate by bleed air are so varied that they will be discussed in their respective chapters.

Low pressure system. Many aircraft equipped with reciprocating engines obtain a supply of low-pressure air from *vane-type pumps*. These pumps are typically mounted on the accessory case and driven by the aircraft engine. Low pressure systems are generally used to operate gyro instruments and pneumatic deicers. They can produce up to 8" Hg suction and/or up to 20 p.s.i. pressure, depending on the pump and the system requirements.

System supply pumps. Vane pumps are manufactured in two types; so called *wet pumps* and *dry pumps*.

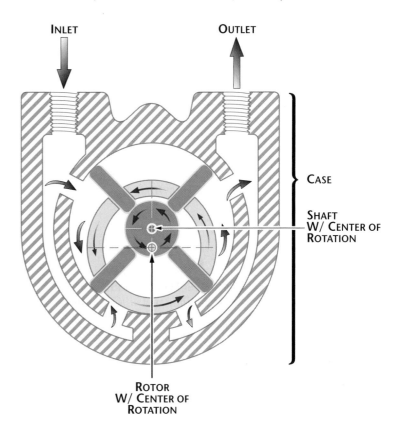

Figure 1-7-2. Operational diagram of a sliding vane pump

Wet pumps. Figure 1-7-1 is a cutaway of a typical wet pump. Wet pumps typically have steel sliding vanes and are lubricated, cooled, and sealed against air leakage by engine oil. Because of this, each wet pump must have a *vapor separator* on the pressure line. This allows the oil vapors, or mist, to *coalesce* onto the inside of the separator and drain back into the engine crankcase.

Wet pumps have all but been replaced in favor of dry pumps in today's airplanes.

Dry pumps. The basic design of a sliding vane dry pump is the same as a wet pump, without the oil. Instead of having steel sliding vanes, dry pumps use carbon vanes. This makes them self-lubricating and allows the vanes to *wear in* and *self seal*.

Figure 1-7-2 shows a schematic view of one of these pumps, which consists of a housing with two ports, a drive shaft, and two vanes. The drive shaft and the vanes contain slots so the vanes can slide back and forth through the drive shaft. The shaft is eccentrically mounted in the housing, causing the vanes to form four different sizes of chambers. In the position shown, the inlet is the largest chamber and is connected to the supply port. As depicted in the illustration, outside air can enter the inlet of the pump.

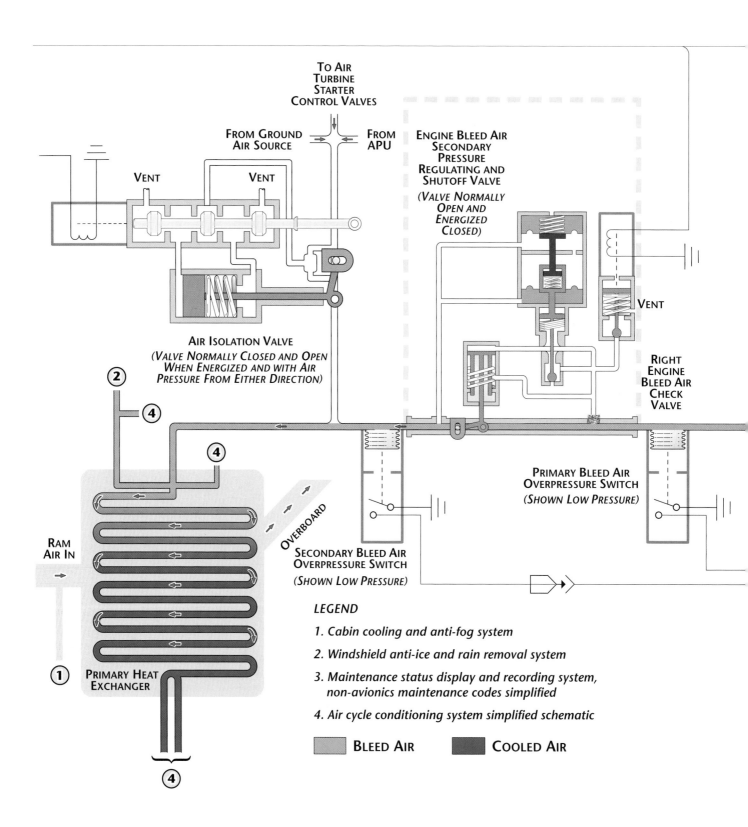

Figure 1-7-3. All controls necessary in a complete bleed air system are illustrated here. The schematic is for bleed air only and stops where operating systems start.

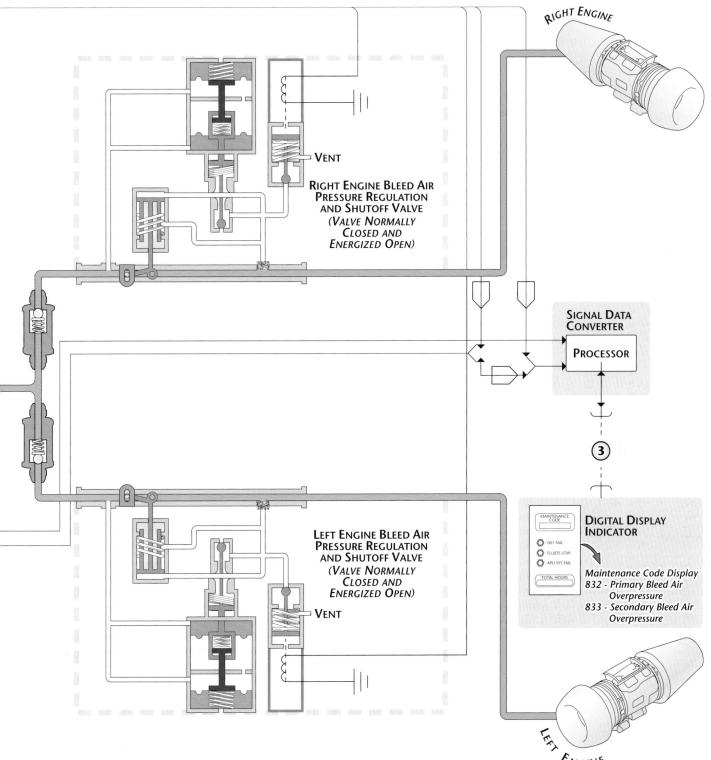

RIGHT ENGINE

VENT

RIGHT ENGINE BLEED AIR
PRESSURE REGULATION
AND SHUTOFF VALVE
(VALVE NORMALLY
CLOSED AND
ENERGIZED OPEN)

SIGNAL DATA
CONVERTER

PROCESSOR

③

DIGITAL DISPLAY
INDICATOR

MAINTENANCE
CODE

001 FAIL
FLUIDS LOW
APU SYS FAIL

TOTAL HOURS

Maintenance Code Display
832 - Primary Bleed Air
Overpressure
833 - Secondary Bleed Air
Overpressure

LEFT ENGINE BLEED AIR
PRESSURE REGULATION
AND SHUTOFF VALVE
(VALVE NORMALLY
CLOSED AND
ENERGIZED OPEN)

VENT

LEFT ENGINE

When the pump begins to operate, the drive shaft rotates and changes positions of the vanes and sizes of the chambers. There are four such chambers in this pump, and as each goes through the cycle of intake, compression, and exhaust, continuous pressure is produced. Thus, the pump delivers to the pneumatic system a continuous supply of compressed air at from 1 to 20 p.s.i.

Vacuum and air pressure pumps. In theory, vane type pressure and vacuum pumps are the same. Suction is produced at the intake port, while pressure is produced at the exhaust port. Most installations on general aviation airplanes manufactured in the last several years use pneumatic systems that consist of vacuum only, and use vane pumps that are designed to produce vacuum only. Other sliding vane air pumps are designed for producing air pressure, while yet others will provide both. Pumps designed for one system are not interchangeable with another. Pressure systems are normally found on light twin and corporate type airplanes that are certified for instrument flight and have deice equipment installed. Combination pressure and vacuum systems are generally used on larger airplanes that require more air volume for the deice, leaving less pressure for the instruments.

Cabin Pressurization Pumps

Cabin pressurization in another place where the lines between pneumatic systems and the systems they supply become a bit blurred.

There are primarily three ways that an airplane can get air pressure for raising the cabin altitude; bleed air from a turbine engine, a centrifugal supercharger mounted on a reciprocating engine, and a positive displacement air pump (Roots type supercharger) driven as an accessory.

Bleed air systems. Figure 1-7-3 shows the layout of a typical bleed air system. Most systems will have all, or most, of the controls shown in the illustration. If you follow the schematic of the engine bleed air pressure regulators, you can readily figure out how they work. The manual adjustment allows bleed air pressure to move the piston connected to the butterfly type main valve. It also allows for a balance between *upstream* and *downstream* pressures, with a vent to allow for changes in demand and shutoff in the event of a failure of the regulator.

As is normal in most aircraft systems, there is a backup regulator for the complete system, as well as a left and right regulator. The overpressure and isolation valves are self-explanatory.

Figure 1-7-4. A Roots-type supercharger is a positive-displacement air pump

Roots-type supercharger. Roots superchargers are similar in operation to a gear pump. The rotors take air and move it around the outside of the case and exhaust it 180° later. Just like a gear pump, it is a positive displacement pump, as each revolution of the rotors moves a specific amount of air. Pressure is built up by restricting the exhaust rate, or leakage, while the supercharger continues to add more air. Typically, the superchargers have a gear drive system that is driven by an accessory drive pad on the engine. Some have been hydraulically driven by using a variable speed transmission system. Roots cabin superchargers were the system of choice just prior to the advent of the turbine engine bleed air systems. Many piston engine corporate type airplanes still use this type of system. A Roots supercharger is shown in Figure 1-7-4.

Centrifugal superchargers. Centrifugal superchargers, used for cabin pressurization, are basically the same as external superchargers used to boost the engine induction system. They have used an accessory drive system that incorporated either a two-speed drive, or system for driving the compressor. Typically, the drive system is controlled by the cabin pressurization controller and varies by the air demand placed on the system.

Pneumatic System Components

Pneumatic systems are often compared to hydraulic systems, but such comparisons can only hold true in general terms. Pneumatic systems do not utilize reservoirs, hand pumps, accumulators, regulators, or engine-driven or electrically driven power pumps for building normal pressure. But similarities do exist in some components.

High pressure system relief valves. Relief valves are a necessary part of a high pressure system. They act as pressure-limiting units and prevent excessive pressures from bursting lines and blowing out seals. They operate similar to a hydraulic relief valve with one major exception; the surplus air is simply vented overboard instead of being returned.

Medium pressure system relief valves. Bleed air systems seldom use relief valves, relying instead on redundant overpressure shut off valves that work in conjunction with the pressure regulation and shutoff valves.

Low pressure system relief valves. On low pressure systems no relief valve is really necessary; the pressure regulating valve performs that function.

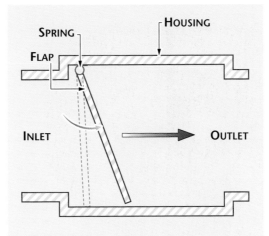

Figure 1-7-5. Flap type pneumatic check valve

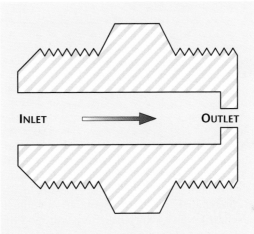

Figure 1-7-6. Orifice type restrictor with a large inlet port and a small outlet port

Control valves. Control valves are also a necessary part of any pneumatic system. In high pressure systems the control valves are typically a poppet type valve, whereas in medium pressure systems generally rely on a butterfly-type valve.

Check valves. Check valves are used in both hydraulic and pneumatic systems. Air enters the inlet port of the check valve, compresses a light spring, forcing the check valve open and allowing air to flow out the outlet port. But if air enters from the outlet, air pressure closes the valve, preventing a flow of air out the inlet port. Thus, a pneumatic check valve is a one-direction flow control valve. See Figure 1-7-5.

Restrictors. Restrictors are a type of control valve used in pneumatic systems. An orifice restrictor can be used to change the operating characteristics on an item by restricting its inflow (Figure 1-7-6). The small outlet port reduces the rate of airflow and the speed of operation of an actuating unit.

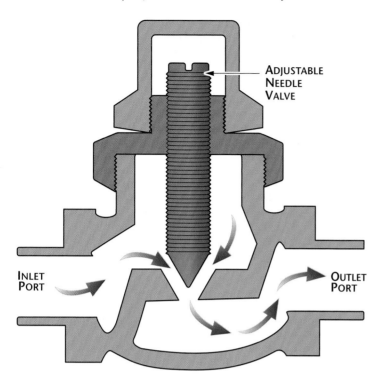

Figure 1-7-7. Variable restrictor

Variable restrictor. Another type of speed-regulating unit is the variable restrictor (Figure 1-7-7). Variable pneumatic restrictors are used where maintenance adjustments must be made to correct airflow volume. It contains an adjustable needle valve, which has threads around the top and a point on the lower end. Depending on the direction turned, the needle valve moves the sharp point either into or out of a small opening to decrease or increase the size of the opening. Since air entering the inlet port must pass through this opening before reaching the outlet port, this adjustment also determines the rate of airflow through the restrictor.

Filters. Pneumatic systems need to be protected from dirt and airborne particles. While bleed air systems benefit from the turbine engine air cleaner, compressor and vacuum pumps on reciprocating engines do not. These systems must rely on separate air filtration devices. These can be simple felt/foam type wrappers for a minimum instrument vacuum system, or complete system filters as shown in Figure 1-7-8. These filters are part of the regular inspection system and must be changed as specified in the inspection program.

Moisture separators. Moisture builds up in pneumatic systems that use mechanical compressors. A moisture separator is required to reduce the chance of any ice or water building up in the other components.

When the pump is shut down the moisture sump is vented overboard. The vent is con-trolled by a solenoid dump valve and a pressure switch in the compressor discharge line. The pressure switch detects the loss of pressure and energizes the dump valve solenoid to the open position, venting the sump.

Air bottles. High pressure air bottles are normally used to provide landing gear extension and brakes in an emergency situation. The air bottle usually stores enough compressed air for several applications of the brakes. A high-pressure air line connects the bottle to an air valve, which controls operation of the emergency brakes. See Figure 1-7-9.

If the normal brake system fails, the air bottle control handle is placed in the *on* position. The valve then directs high-pressure air into lines leading to the brake assemblies. A shuttle valve isolates the original hydraulic lines.

Lines and tubing. Lines for pneumatic systems consist of rigid metal tubing and flexible rubber hose. Fluid lines and fittings are covered in detail in another chapter.

Pneumatic Power System Maintenance

Maintenance of the pneumatic power system consists of servicing, troubleshooting, removal and installation of components and operational testing.

An air compressor's lubricating oil level should be checked daily in accordance with the applicable manufacturer's instructions. The oil level is indicated by means of a sight gauge or dipstick. When re-filling the compressor oil tank,

Figure 1-7-8. A typical filter for a dry air/vacuum pump system

the oil (type specified in the applicable instructions manual) is added up to the specified level. After the oil is added, ensure that the filler plug is torqued and safety wire is properly installed.

The pneumatic system should be purged periodically to remove the contamination, moisture, or oil from the components and lines. Pressurizing it and removing the plumbing from various components throughout the system purges the system. Removal of the pressurized lines will cause a high rate of airflow through the system causing foreign matter to be exhausted from the system. If an excessive amount of foreign matter, particularly oil, is exhausted from any one system, the lines and components should be removed and cleaned or replaced.

Bleed air compressed air systems are different only in the source of air pressure. Servicing the balance of the system is the same for both sources of air pressure.

Upon completion of pneumatic system purging and after re-connecting all the system components, the system air bottles should be drained to exhaust any moisture or impurities, which may have accumulated there.

After draining the air bottles, service the system with nitrogen or clean, dry compressed air. The system should then be given a thorough operational check and an inspection for leaks and security.

Figure 1-7-9. High pressure air bottle

2

Cabin Environmental Systems

Section 1

Cabin Environmental Systems

Control of the cabin environment is an extremely important part of modern aircraft operations. Naturally, we all want to breathe comfortably, be warm and stay cool. Gone are the days of bundling up in leather flying suits just to ride in an airplane. Today's sophisticated passengers demand the utmost in traveling comfort. Indeed, the failure of an operator to consider the comfort of his passengers can spell the demise of his business.

Cabin environmental systems start with the basic necessity of life: oxygen. Enough oxygen must be available at the correct pressure altitude, not just to sustain life, but to do it comfortably.

Oxygen is essential to life. A reduction of essential oxygen supplies can produce important changes in body functions, thought processes and degree of consciousness. The sluggish condition of mind and body caused by a deficiency in, or lack of, oxygen is called *hypoxia*. There are several causes of hypoxia, but the one that concerns aircraft operations is the decrease in partial pressure of oxygen in the lungs.

The rate at which the lungs absorb oxygen depends upon the oxygen pressure. The pressure that oxygen exerts is about 1/5 of the total air pressure at any one given level. At sea level, this pressure value (3 p.s.i.) is sufficient to saturate the blood. However, if the oxygen pressure is reduced, either from the reduced atmospheric pressure at altitude or because the percentage of oxygen in the air breathed decreases, then

Learning Objectives

REVIEW
- types, operation, inspection and overhaul of cabin heaters

DESCRIBE
- how pressurization systems work and their sources, control and care

EXPLAIN
- how air- and vapor-cycle air conditioning systems and their components operate
- features of gaseous and solid oxygen systems and their components, inspection and maintenance

Left. Passengers in early aircraft, like this Ford Trimotor from the 1930s (inset), had little more than heat to provide cabin comfort. Modern jet airliners experience a wide range of temperatures and pressures during each flight and need much more extensive cabin systems.

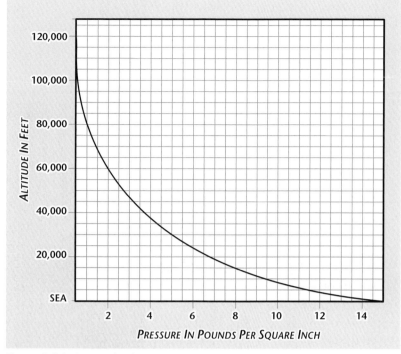

Figure 2-1-1. Atmospheric pressure

COMPOSITION OF THE ATMOSPHERE IN % BY VOLUME	
Nitrogen	78.08%
Oxygen	20.95%
Argon	0.93%
Carbon Dioxide	
Neon	
Helium	
Methane	0.04%
Krypton	
Hydrogen	
Xenon	
Ozone	

Table 2-1-1. Composition of the atmosphere

the quantity of oxygen in the blood leaving the lungs drops, and hypoxia follows.

The Need for Oxygen

From sea level to 7,000 ft. altitude, the oxygen content and pressure in the atmosphere remain sufficiently high to maintain almost full saturation of the blood with oxygen and ensure normal body and mental functions.

At higher altitudes, there is a decrease in barometric pressure, resulting in decreased oxygen content of the inhaled air. Consequently, the oxygen content of the blood is reduced.

At 10,000 ft. above sea level, oxygen saturation of the blood is about 90 percent. Long exposure at this altitude will result in headache and fatigue (so-called *altitude sickness*). Oxygen saturation drops to 81 percent at 15,000 ft. above sea level. This decrease results in sleepiness, headache, blue lips and fingernails, impaired vision and judgment, increased pulse and respiration, and certain personality changes.

At 22,000 ft. above sea level, the blood saturation is 68 percent, and convulsions are likely to occur. Remaining without an oxygen supply at 25,000 ft. for 5 minutes where the blood saturation is down to 50-55 percent will cause unconsciousness.

Composition of the Atmosphere

The atmosphere (air) is composed mainly of nitrogen and oxygen. It also contains smaller quantities of other gases, such as carbon dioxide, water vapor and ozone (See Table 2-1-1).

Oxygen and its importance cannot be overestimated. Without oxygen, life as we know it cannot exist. Oxygen occupies 21 percent of the total mixture of atmospheric gases.

Carbon dioxide (CO_2) is also an essential gas, but for different reasons. A small quantity in the atmosphere is utilized by the plants to manufacture the complex substance which many animals use as food. Carbon dioxide also helps in the control of breathing in man and other animals. The lungs require a small amount of carbon dioxide be present. If too much carbon dioxide is exhaled, the body goes into a state of hyperventilation. A person will seem to be unable to get enough air and will continue to breath heavily, exhaling even more carbon dioxide. Once the CO_2 level in the lungs reaches a critical level, the person's system will cause them to faint, or pass out. Once in an unconscious state, normal breathing will resume and the CO_2 level in the system will reestablish itself. Hyperventilation can be prevented by using a paper bag or other method to cause re-inhalation of the exhaled air. This temporarily increases the intake of CO_2 and rebalances the system quickly.

The amount of water vapor in the air is variable, but it rarely contains over 5 percent, even in very moist conditions. Vapor is not the only form of moisture found in the air — water droplets and ice particles are also present. The moisture in the air absorbs energy from the sun. Therefore, the amount of water in the air can be directly related to weather conditions.

Ozone is a variety of oxygen that contains three atoms of oxygen per molecule, rather than the usual two. The major portion of the ozone in the atmosphere is formed by the interaction of oxygen and the sun's rays near the top of the ozone layer.

Ozone is of great consequence, both to living creatures on Earth and to the circulation of the upper atmosphere. Ozone is important to living organisms because it filters out most of the sun's ultraviolet radiation, which research continues to indicate has myriad consequences for all life on Earth.

Because of the importance of the ozone layer, the use of chemicals that deplete the ozone layer must be restricted. Typical among the ozone-depleting chemicals is Freon, particularly R-12, the refrigerant gas used in most air-conditioning systems in the past.

The gases of the atmosphere have weight, just like solid matter. The weight of a column of air stretching from the surface of the earth out into space is called the *atmospheric pressure*. If this column is 1 sq. inch in area, the weight of air at sea level is approximately 14.7 lbs. The atmospheric pressure, therefore, can be stated as 14.7 p.s.i. at sea level. Another common way of stating the atmospheric pressure is to give the height of a column of mercury, which weighs the same as a column of the atmosphere of the same cross-sectional area. When measured this way, the atmospheric pressure at sea level is normally *29.92 inches Hg (inches of mercury)*. The weather is measured in millibars, with *1,013.2 millibars* being standard sea level pressure.

The atmospheric pressure decreases as altitude increases. The reason for this is quite simple: the column of air that is weighed is shorter. How the pressure changes for a given altitude is shown in Figure 2-1-1. The decrease in pressure is a rapid one. At 50,000 feet, the atmospheric pressure has dropped to almost $^1/_{10}$ of the sea level value. At a few hundred miles above Earth, the air has become so rarefied (thin) that the atmosphere can be considered nonexistent.

Temperature and Altitude

Earth's atmosphere is a mixture of gases that reaches well over 350 miles (560 kilometers) high, getting ever thinner until it blends into space. The layers of the atmosphere are divided based on how temperature changes with height (See Figure 2-1-2). Those divisions are as follows:

Troposphere. The *troposphere* extends from the surface of the Earth to 5-9 miles (26,400-47,520 ft., or 8-14.5 km) high. It contains half the molecules of the Earth's atmosphere. This is the layer where most clouds form and weather occurs. It

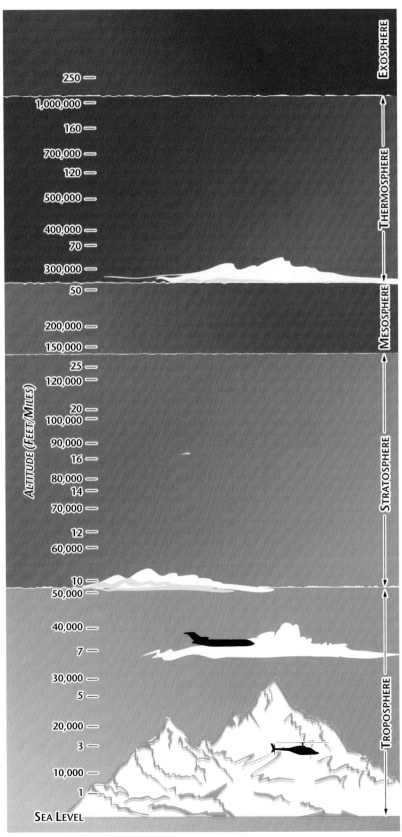

Figure 2-1-2. Layers of the atmosphere

is also the region that most commercial aviation operates in. Temperature decreases with altitude (to about -52°C), because Earth's surface absorbs the energy from the Sun, heats faster than the atmosphere, then gives off heat into the atmosphere.

Stratosphere. The *stratosphere* starts above the troposphere and extends to 31 miles (163,680 ft, or 50 km) high. Temperature increases (to about -3°C) with altitude, because the ozone layer, located at about 15.5 miles (25 km) from the surface of the Earth, absorbs the ultraviolet radiation from the Sun, which warms the atmosphere. About 99 percent of the molecules in Earth's atmosphere are within the troposphere and the stratosphere.

Mesosphere. The *mesosphere* starts above the stratosphere and extends to about 53-56 miles (279,840-295,680 ft., or 85-90 km) high. Meteors burn up in this layer of the atmosphere. Temperature decreases with altitude to about -90°C.

Thermosphere. The *thermosphere* starts above the mesosphere and extends to about 372 miles (1,964,160 ft. or 600 km) high. Temperature rises with altitude and, depending on the activity of the Sun, could reach over 1,500°C. The atmosphere is so thin here that even a small increase in energy from the Sun can cause a large increase in temperature.

Exosphere. The *exosphere* starts above the thermosphere and continues until it merges with space.

Hypoxia

It is obvious that a means of preventing hypoxia and its ill effects must be provided for all living creatures who undergo high-altitude travel. When the atmospheric pressure falls below 3 p.s.i. (approximately 40,000 ft.), even breathing pure oxygen is not sufficient.

The low partial pressure of oxygen, low ambient air pressure and temperature at high altitude make it necessary to create the proper environment for passenger and crew comfort. The most difficult problem is maintaining the correct partial pressure of oxygen in the inhaled air. The military achieves this by using pressure oxygen masks in high-performance airplanes. This can be achieved by using oxygen, pressurized cabins or pressure suits. Obviously, pressure suits are best left to the military and to the astronauts.

In general aviation, pressurization of the aircraft cabin is the accepted method of protecting persons against the effects of hypoxia. Within a pressurized cabin, people can be transported comfortably and safely for long periods of time, particularly if the cabin altitude is maintained at 8,000 ft. or below, where the use of oxygen

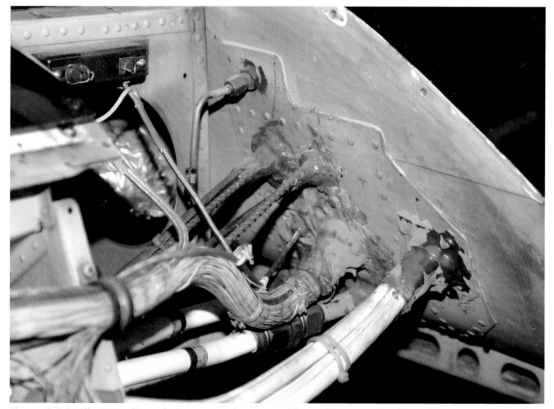

Figure 2-2-1. All wiring, lines, fittings, etc., that extend from the pressurized area into unpressurized areas must be sealed against leakage.

equipment is not required. However, the flight crew must be aware of the danger of decompression and must be trained to meet an emergency should it occur.

Section 2

Pressurization Systems

Aircraft are flown at high altitude to provide better fuel economy and to avoid severe turbulence and weather. Smaller aircraft that do not have pressurization systems are limited to lower altitudes.

Several functions must be accomplished by a cabin pressurization system to allow for passenger comfort and safety. A cabin pressurization system must be capable of maintaining a *cabin pressure altitude* of approximately 8,000 ft. at the maximum operating altitude of the aircraft. The system should be designed to exchange the cabin air at a reasonable rate to eliminate odors and to remove stale air. This air exchange should take place without any noticeable change to cabin pressure.

In the typical pressurization system, the cabin, flight compartment and baggage compartments are incorporated into a sealed unit that is capable of containing air under a pressure higher than outside atmospheric pressure. Any work performed in the sealed areas must be kept sealed. Checking pressure sealing is part of a normal inspection (see Figure 2-2-1).

Pressurized air is pumped into the sealed fuselage by cabin superchargers in piston-powered airplanes, and by bleed air systems in jet aircraft. Air is released from the fuselage by a device called an *outflow valve* (Figure 2-2-2). The supercharger systems provide a constant inflow of air to the pressurized area, while the pressurization provided by bleed air systems may be varied as required. The outflow valve, by regulating the air exit, is the major controlling element in the pressurization system.

The flow of air through an outflow valve is determined by the degree of valve opening. This valve is controlled by an automatic system that can be set by the flight crew. In the event of a malfunction of the automatic controls, manual controls are also provided.

The operating maximum altitude of the aircraft is determined by the maximum amount of pressurization allowed by the pressurization system. The amount of pressurization provided by the system is limited by several factors. The

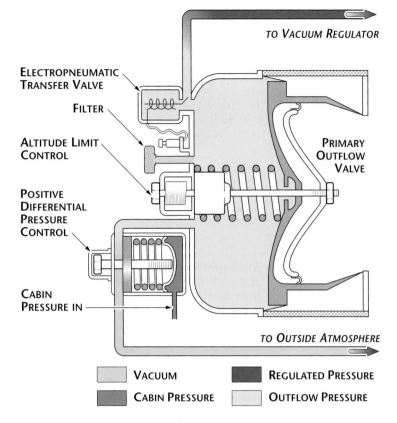

Figure 2-2-2. An outflow valve maintains the cabin pressure in a pressurized fuselage.

primary limiting factor is the *maximum differential pressure* that the fuselage is designed for. The differential pressure of the cabin is the *differential pressure ratio* between the inside and outside pressure. If the differential pressure becomes too great, the cabin structure may fail.

A second factor that limits the degree of pressurization is the amount of pressure that is available. As the aircraft altitude increases, the pressure supplied to the turbocharger, supercharger or compressor is decreased. Because the inlet pressure is decreased, the outlet pressure will also be decreased. While the same thing will happen with a bleed air system, it will happen at a much higher altitude.

Pressurization Problems

There are many complex technical problems associated with pressurized aircraft. Perhaps the most difficult problems are in the design, manufacturing and selection of structural materials that will withstand the great differential in pressure that exists between the inside and outside of a pressurized aircraft when flying at high altitudes. If the weight of the aircraft structure were of no concern, it would be a relatively simple matter to construct a fuselage that could withstand tremendous pressures.

Figure 2-2-3. A pressurization safety valve from a Cessna 421. Larger aircraft use the same principal, only on a larger scale and designed to handle a larger amount of airflow.

Pressurized aircraft are built to provide a cabin pressure altitude of not more than 8,000 ft. at maximum operating altitude.

The atmospheric pressure at 8,000 ft. is approximately 10.92 p.s.i., at 40,000 ft. it is nearly 2.72 p.s.i. If a cabin altitude of 8,000 ft. is maintained in an aircraft flying at 40,000 ft., the differential pressure that the structure will have to withstand is 8.20 p.s.i. (10.92 p.s.i. minus 2.72 p.s.i.). If the pressurized area of this aircraft contains 10,000 sq. inches, the structure will be subjected to a bursting force of 82,000 lbs., or approximately 41 tons. In addition to designing the fuselage to withstand this force, a safety factor of 1.33 must be added. The pressurized portion of the fuselage will have to be constructed to have an ultimate strength of 109,060 lbs. (82,000 x 1.33), or 54.50 tons. On a large airplane, the outside diameter of the fuselage will increase, some startlingly so. Some will grow by several inches.

Preventing over-pressurization and the subsequent destruction of the airplane is the job of the *pressurization safety valve* (Figure 2-2-3). The safety valve works similarly to an outflow valve, except that it is preset to a maximum cabin pressure differential. Should the outflow valve malfunction, the safety valve will not allow the cabin pressure to exceed the safe limit. Some safety valves are also used as *cabin pressure dump valves* to relieve any residual pressure after landing. When used as a dump valve, they are normally controlled by the landing gear squat switch.

Some systems use a delivery air duct check valve to prevent the loss of pressurization if the cabin air compressor or the engine is shut down.

Another design problem is the pressurization cycle: a pressurization and depressurization of the system equal one cycle. Many aircraft are cycle-life limited. Once the aircraft reaches its established number of cycles, it is mandatory to retire it from service. Newer aluminum alloys and more dependable, stronger fasteners (rivets) have, in many cases, done away with mandatory airframe retirement.

Pressure vessel inspections are part of the Aging Fleet Inspection programs that have been developed since the Aloha Airlines incident in 1988.

Pressurization Systems

Terms and definitions. To understand the operating principles of pressurization and air conditioning systems, it is necessary to learn some terms and definitions. These are:

- Absolute pressure — Pressure measured along a scale that has zero value at a complete vacuum

- Absolute temperature — Temperature measured along a scale that has zero value at that point where there is no molecular motion (–273.1°C or –459.6°F)

- Adiabatic — A word meaning no transfer of heat. The adiabatic process is one in which no heat is transferred between the working substance and any outside source.

- Aircraft altitude — The actual height above sea level at which an aircraft is flying

- Ambient temperature — The temperature in the area immediately surrounding the object under discussion

- Ambient pressure — The pressure in the area immediately surrounding the object under discussion

- Standard barometric pressure — The weight of gases in the atmosphere sufficient to hold up a column of mercury 760 mm high (approximately 30 inches) at sea level (14.7 p.s.i.)

- Cabin altitude — Used to express cabin pressure in terms of equivalent altitude above sea level

- Differential pressure — In aircraft air conditioning and pressurizing systems, the difference between cabin pressure and atmospheric pressure

- Gauge pressure — A measure of the pressure in a vessel, container or line, as compared to ambient pressure or the pressure that reads on the gauge

- Ram-air temperature rise — The increase in temperature created by the ram compression on the surface of an aircraft traveling at a high rate of speed through the atmosphere. The rate of increase is proportional to the square of the speed of the object.

Basic requirements. Basic requirements for the proper functioning of cabin pressurization are:

- A source of compressed air for pressurization and ventilation. Cabin pressurization sources can be engine-driven compressors, independent cabin superchargers or air bled directly from the engine.

- A means of controlling cabin pressure by regulating the outflow of air from the cabin (cabin-pressure regulator and an outflow valve)

- A method of limiting the maximum pressure differential to which the cabin pressurized area will be subjected. Pressure relief valves, negative (vacuum) relief valves and dump valves are used to accomplish this.

- A means of regulating (in most cases, cooling) the temperature of the air being distributed to the pressurized section of the airplane. This is accomplished by the refrigeration system, heat exchangers, control valves, electrical heating elements and cabin temperature-control system.

- The sections of the aircraft that are to be pressurized must be sealed to reduce air leakage to a minimum. It must also be capable of safely withstanding the maximum pressure differential.

In addition to the components just discussed, various valves, controls and allied units are necessary to complete a cabin pressurizing system. When auxiliary systems, such as windshield rain-clearing devices, pressurized fuel tanks and pressurized hydraulic tanks are required, additional shut-off valves and control units are necessary.

Temperature Scales

Celsius. Celsius is a scale on which 0°C represents the freezing point of water, and 100°C is equivalent to the boiling point of water at sea level.

Fahrenheit. Fahrenheit is a scale on which 32°F represents the freezing point of water, and 212°F is equivalent to the boiling point of water at sea level.

Sources of Cabin Pressure

Cabin pressurization on aircraft equipped with reciprocating engines is normally provided by superchargers or turbochargers.

The use of engine turbochargers and superchargers for cabin pressurization is quite simple, but it does have several disadvantages. First, that the cabin air may be contaminated with fumes from the engine exhaust, lubricating oils and exhaust gases. Second, the pressure developed at high altitudes is very low and may be insufficient to pressurize the cabin. The third disadvantage is that the engine may be deprived of the supercharging that it needs to operate efficiently at high altitude.

With gas turbine engines, the cabin can be pressurized by bleeding air from the engine compressor. Usually, the air bled from an engine compressor is sufficiently free from contamination and can be used safely for cabin pressurization. It is generally obtained from two places on each engine: at an intermediate stage and the last stage of the compressor. Under normal conditions, all pressure would come from the intermediate stage bleed; at low-power settings, the last stage bleed would be used. Bleed air is usually called *customer air* by the airframe manufacturers.

Positive-Displacement Cabin Compressors (Superchargers)

Included in this group are reciprocating compressors, vane-type compressors and Roots blowers. The first two are not very suitable for aircraft cabin pressurization because of the large quantity of oil present in the air delivered to the cabin.

The action of a *Roots type blower* (Figure 2-2-4) is basically like that of a gear pump. It takes a predetermined volume of air, compresses it, then delivers it to the cabin duct. It is called a *positive-displacement supercharger*. The most common use of a Roots blower is to supercharge diesel engines. Roots blowers are also the supercharger of choice of hot-rod dragsters.

Figure 2-2-4. The Roots supercharger shown is a hydraulically driven unit.

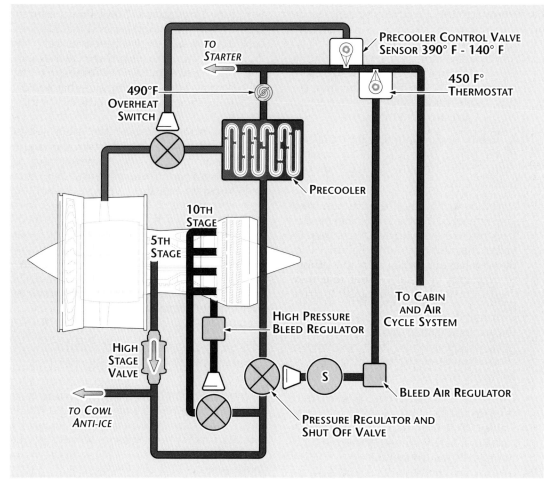

Figure 2-2-5. A diagram of a bleed-air supply system for a transport-category aircraft

The rotors are mounted in an airtight casing on two parallel shafts. The lobes do not touch each other or the casing, and both rotors turn at the same speed. A common misconception is that the air being compressed travels through the center of the lobes and is exhausted as compressed air. That is not the way it works. Being a constant displacement pump, like a gear-type oil pump, the air travels around the outside of the case and is dumped into a plenum at the exit. Pressure is built up because the blower can deliver more air than the system can use; each rotation adds more air, and the pressure increases. Pressure is controlled in a cabin system by the outflow valve setting. As an engine supercharger, it is controlled by a pressure-actuated pop-off valve.

If you remember your basic physics, any time air is compressed by a pump, there is a rise in temperature. The supercharger housing is usually finned on the external surface to provide some cooling area. The cooling effect is sometimes further increased by shrouding the supercharger housing and passing a stream of outside air through it. Air-cooling is also used to reduce the temperature of internal parts. The cooling air is ducted through drilled passageways into the rotor

cavities and is expelled at the inlet side of the supercharger cover.

To achieve an oil-free delivery of air, the supercharger bearings use labyrinth seals that permit a small amount of air leakage. Any drops of oil that may have gotten past the rubber seals are thus blown back into the bearing area.

Because of the air pulsations caused by the rotors, all positive-displacement compressors emit a shrill noise during operation. Silencers are used to reduce the noise level.

Centrifugal cabin compressors. The cabin supercharger is essentially an air pump. It has a centrifugal impeller similar to a turbo-supercharger in the induction system of a reciprocating engine. Outside air at atmospheric pressure is admitted to the supercharger and is then compressed by the high-speed impeller and delivered to a distribution system.

Engine-driven superchargers used on most older reciprocating-engine transport aircraft required a variable-ratio drive mechanism. The gear ratio of these superchargers was automatically adjusted to compensate for changes of engine r.p.m. or outside atmospheric pressure. Normally, the gear ratio is eight to 10 times engine speed when operating at normal cruising conditions. The drive ratio is at a maximum when operating at high altitude with low engine r.p.m. The system was similar to the variable-speed drives that are used on large aircraft generators today.

Turbocompressors. For a variety of reasons, bleed air was not used directly from early jet and turbine engines. Instead, the bleed air was used to drive a turbine wheel, which in turn drove a centrifugal compressor. They were called turbocompressors and were located in the engine nacelles or in the fuselage. There could be as many as four turbocompressors in an aircraft. These units operated similarly to the turbo-superchargers used on the piston-engined airplanes of today, except that they used bleed air instead of exhaust gases to drive the pump.

These early installations used a complicated supercharger control system that was designed to maintain a constant speed, hence a constant pressure.

Bleed air systems. With modern pressurization systems in airplanes that use turbine engines (jet or turboprop), the pressurized air is taken from the engine compressor. It is bled off the compressor in one or two places: usually from an intermediate stage and the last compressor stage. Air from the intermediate stage is considered low-pressure air and is supplied by what is called a *low-stage bleed port*. Air from a last stage is naturally taken from the *high-stage bleed port* (see Figure 2-2-5).

Low-pressure air is used during takeoff, climb and cruise conditions. High-pressure bleed air is used during slow flight, descent and any time the low-pressure air supply is inadequate. Both bleed ports are never open at the same time, and the changeover is automatic.

Bleed air from a turbine engine is not free power, nor is it especially cheap. Modern engine design takes into consideration the air pressures and volumes needed to run the engines at rated power and design them to do just that. A certain amount of customer air is allowed for, but the supply is not extravagant. Any time air is bled from the compressor, there is an equal amount of reduction in power output. So much so that some airplane designs will use bleed air supplied by the APU during takeoff and climb, allowing full power to be available for the main engines. Most aircraft operators have a prescribed set of parameters that the environmental system will operate under. Anytime that pressurization can be reduced or temperatures allowed to moderate, there is a corresponding increase in fuel efficiency. Bleed air costs money.

Auxiliary air pressure. There are three other sources of air pressure that may be considered as part of this same system; they are APU air supply, ground air supply from an external air supply cart and ram air. All three are controlled from the same controller system that controls bleed air; just the source is different.

Bleed air control. A bleed air system has its own control system. While they vary from one design to another, most are a combination of electrical and air pressure activation. Unlike the cabin temperature controls, the bleed air control panel is generally located in a cockpit overhead panel.

Cabin Pressure Control System

The cabin pressure control system provides cabin pressure regulation, excess and negative pressure relief. The system also provides the means for selecting the desired cabin altitude in the isobaric and differential range. An outflow valve, cabin pressure controller, a safety valve and a negative pressure relief valve are the typical components of a pressurization system.

Pressurization Valves

Outflow valve. The outflow valve is the principal control of the pressurization system. The valve, located in a pressurized portion of the fuselage, is normally located under the floor panels or on the rear pressure bulkhead. The outflow valve allows the air to vent overboard through openings in the aircraft construction. Small aircraft normally have one outflow valve while larger aircraft may have up to three.

One type of outflow valve, operated by an electric motor, is a butterfly type valve. The motor receives amplified electrical signals from the

Figure 2-2-6. The outflow valve is the main control of a pressurization system.

Figure 2-2-7. The cabin pressure regulator provides fully automatic cabin pressurization control.

Figure 2-2-8. Outflow valve: (A) fully closed, and (B) partially open.

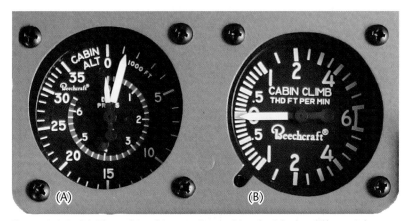

Figure 2-2-9. (A) The cabin altimeter is combined with the differential pressure gauge, (B) The cabin rate of climb is a separate gauge.

pressurization controller to open and close the valve during pressurized flight.

Another type is a pneumatic outflow valve, controlled by pressure and vacuum signals from the pressurization controller. The controller uses cabin air pressure, along with pneumatic or bleed air to operate the valve. This type of outflow valve is shown in Figure 2-2-6.

Before flight, the outflow valve is fully open. The squat switch circuit electrically holds a

solenoid valve in the outflow control line in the open position. At take-off, the squat switch changes position and the outflow valve starts to close.

As the aircraft climbs in altitude, the outflow valve gradually moves toward the closed position, providing a greater restriction to the venting of cabin pressurization air. During cruise flight the position of the outflow valve controls the cabin altitude, set by the flight crew on the cabin pressurization controller. The speed at which the outflow valve opens and closes, sets the cabin rate of climb. That rate of climb is adjusted by the flight crew, or is preset by the airframe manufacturer.

In case the outflow valve should fail, an automatic cabin pressure relief valve is used. The valve will automatically open if the cabin pressure reaches a preset valve. This valve may be part of the outflow or a separate unit.

Cabin pressure controller. The cabin pressure controller controls cabin altitude by regulating the position of the outflow valve. The controller usually provides either fully automatic or manual control of pressure within the aircraft. Normal operation is automatic, requiring only the selection of the desired cabin altitude and the rate of climb of the cabin pressure.

The controller looks very much like an altimeter, which has several added adjustment knobs for altitude and rate of climb, as seen in Figure 2-2-7. The dial is graduated increments up to approximately 10,000 ft., cabin altitude. Usually, one pointer indicates the desired cabin altitude and is set with the cabin altitude knob. Another pointer, the cabin rate of climb, is set with the rate knob. In some instances, there is another pointer or a rotating scale, which indicates the corresponding aircraft pressure altitude.

The cabin pressure controller controls cabin pressure to a selected value in the isobaric range and limits cabin pressure to a preset differential value in the differential range. The isobaric range maintains the cabin at a constant pressure altitude during flight. It is used until the aircraft reaches the altitude at which the difference between the pressure inside and outside the cabin is equal to the highest differential pressure for which the fuselage is structurally designed.

Differential control prevents the maximum differential pressure, based on the structural design of the fuselage, from being exceeded. This pressure is determined by the structural strength of the cabin and by the relationship of the cabin size to the probable areas of rupture, such as window areas and doors.

The controller adjusts an internal, electric or pneumatic, signaling device that compares the existing cabin pressure with an aneroid or evacuated bellows. If the cabin altitude does not correspond to that adjustment, the aneroid causes an appropriate signal to modulate the outflow valve open or closed. When the aneroid senses that the cabin altitude equals the adjustment, the signals to the outflow valve stop. As long as other factors do not change, the outflow valve stays at the position to maintain desired cabin pressure. The controller can sense any change, such as a variance of aircraft altitude, and re-adjust the outflow valve as necessary. Fully closed and partially open outflow valves are seen in Figure 2-2-8.

The rate control determines how fast the outflow valve opens and closes during climb and descent. In some controllers, the rate signal is partially automatic. The barometric setting compensates the controller for the normal errors encountered during flight. The rate setting improves the accuracy of the controller and, for example, protects the cabin from partial pressurization while landing.

Several instruments indicate the status of the pressurization process to the flight crew. The first, the cabin differential pressure gauge, indicates the difference between the inside and outside air pressures, and the maximum allowable differential pressure. Secondly, the cabin altimeter, which indicates the pressure altitude of the cabin, serves as a check on the performance of the system. In some installations, a single combination gauge is used. Depending on the type of aircraft the cabin altitude may be between 6000 ft and 8000 ft. The third instrument indicates the cabin rate of climb or descent. A combined cabin altimeter and differential pressure gauge, as well as a cabin rate of climb is shown in Figure 2-2-9.

Cabin air pressure safety valve. The cabin air pressure safety valve, seen in Figure 2-2-3, is a pressure relief valve. The valve prevents cabin pressure from exceeding a predetermined differential pressure. Upon landing the squat switch circuitry will drive the valve to the full open position, to prevent any residual cabin pressure from being retained in the aircraft.

Negative pressure relief valve. Many pressurized aircraft have a negative pressure relief valve. This valve may be part of the outflow valve or be an individual component. The valve can be as simple as a hinged flap on the rear pressure bulkhead that opens inward when outside air pressure is greater than cabin pressure. During pressurized flight, the internal cabin pressure holds the flap closed. The negative pressure relief valve prevents the cabin altitude from being higher than the aircraft altitude.

In some installations, a manually operated relief valve serves as a backup when all other controls fail. During emergency depressurization, the manual control will equalize the cabin pressure to the ambient atmospheric pressure by opening the outflow valve.

Testing Pressurization Systems

Checking the cabin pressurization system consists of the following:

- A check of pressure regulator operation
- A check of pressure relief and dump valve operation
- A cabin static pressure test
- A cabin dynamic pressure test

Cabin pressurization is achieved by connecting a test system to the fuselage test ports and introducing the correct amount of pressure into the fuselage (Figure 2-2-10).

Always follow the aircraft manufacturer's instructions, as well as the instructions for the test system.

To check the pressure regulator, connect an air test stand and a manometer (a gauge for measuring pressure, usually in inches of Hg) to the appropriate test adapter fittings. With an external source of electrical power connected, position the system controls as required. Then pressurize the cabin to 7.13 inches Hg, which is equivalent to 3.5 p.s.i.

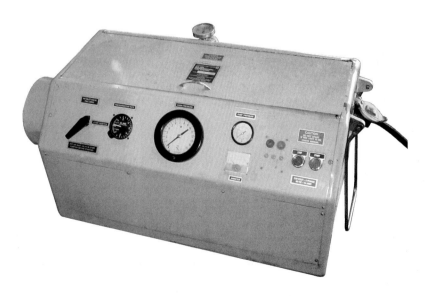

Figure 2-2-10. A ground cabin pressurization test system

NOTE: *The pressurization settings and tolerances presented here are for illustrative purposes only. Consult the applicable maintenance manual for the settings for a particular make and model aircraft.*

Continue to pressurize the cabin, checking to see that the cabin pressure regulator maintains this pressure.

The complete check of the pressure-relief and dump valves consists of three individual checks:

1. With the air test stand connected to pressurize the cabin, position the cabin pressure selector switch to dump the cabin air. If cabin pressure decreases to less than 0.3 in. Hg (0.15 p.s.i.) through both the pressure-relief and dump valves, the valves are dumping pressure properly.

2. Using the air test stand, re-pressurize the cabin. Then position the manual dump valve to dump. A lowering of the cabin pressure to 0.3 in. Hg (0.15 p.s.i.) and an airflow through the pressure-relief and dump valves indicate that the manual dumping function of this valve is satisfactory.

3. Position the master pressure-regulator shut-off valve to "all off" (a position used for ground testing only). Then, using the air test stand, pressurize the cabin to 7.64 inches Hg (3.75 p.s.i.). Operation of the pressure-relief and dump valves to maintain this pressure indicates that the relief function of the cabin pressure-relief and dump valves is satisfactory.

The cabin static pressure test checks the fuselage for structural integrity. To perform this test, connect the air test stand and pressurize the fuselage to 10.20 inches Hg (5.0 p.s.i.). Check the aircraft skin exterior for cracks, distortion, bulging and rivet condition.

Pressure-checking the fuselage for air leakage is called a *cabin pressure dynamic pressure test.* This check consists of pressurizing the cabin to a specific pressure using an air test stand. Then, using a manometer, determine the rate of air pressure leakage within a certain time limit specified in the aircraft's maintenance manual. If leakage is excessive, large leaks can be located by sound or by feel. Small leaks can be detected using a bubble solution or an ultrasonic leakage tester.

A careful observation of the fuselage exterior, prior to its being washed, may also reveal small leaks around rivets, seams or minute skin cracks. A telltale stain will be visible at the leak area.

Cabin Pressurization Troubleshooting

Troubleshooting consists of three steps:

1. Establishing the existence of trouble

2. Determining all possible causes of the trouble

3. Identifying or isolating the specific cause of the trouble

Troubleshooting charts are frequently provided in aircraft maintenance manuals for use in determining the cause, isolation procedure and remedy for the more common malfunctions which cause the cabin air conditioning and pressurization systems to become inoperative or uncontrollable.

These charts usually list the most common system failures. Troubleshooting charts are organized in a definite sequence under each trouble, according to the probability of failure and ease of investigation. To obtain maximum value, the following procedures are recommended when applying a troubleshooting chart to system failures:

1. Determine which trouble listed in the table most closely resembles the actual failure being experienced in the system.

2. Eliminate the possible causes listed under the trouble selected, in the order in which they are listed, by performing the isolation procedure for each until the malfunction is discovered.

3. Correct the malfunction by following the instructions listed in the correction column of the troubleshooting chart.

Section 3

Air-Cycle Air Conditioning Systems

We are all familiar with commercial and residential air conditioning systems that supply us with cooled and dehumidified air. When we want heat, we turn off the air conditioning (cooler) and turn on the heater (a different system). An air-cycle air conditioning system is not quite like that. It is a system where pressurization, heating, cooling and dehumidifying all take place seamlessly within the same unit.

If you remember your physics, then you remember that compressing air heats it up, and

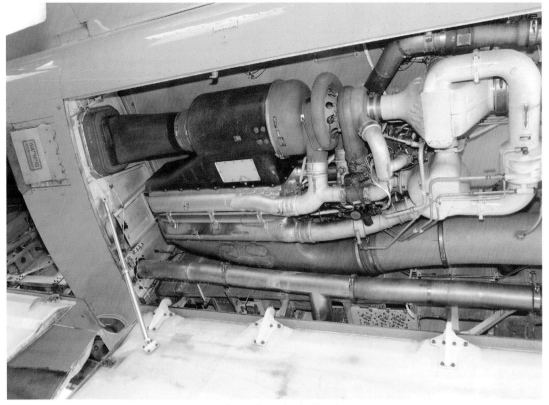

Figure 2-3-1. A complete bootstrap air-cycle pack in the belly of a Boeing 757

that expanding compressed air cools it down. This is the basic concept behind air-cycle air conditioning system operation. Though the idea is simple, it has taken the technology of turbine engines and bleed air systems to make air-cycle systems work effectively (Figure 2-3-1).

Air Conditioning

The temperature of the air supply from an engine-driven compressor exceeds that of the outside air by several hundred degrees. This air requires at least some cooling before it enters the cabin pressure area. After a portion of the heated air has been routed through air-cycle refrigeration components and cooled, it is blended with the remaining hot air to attain the desired air output temperature. Then the blended air is distributed to the cabin areas.

Air-Cycle Operation

To understand the operation of an air-cycle air conditioning system, the terms air-cycle and bootstrap must be clearly defined.

Air-Cycle. A simple *air-cycle system* cools air by transferring heat from the compressed air to ambient air, and by extracting work from the compressed air as it is being expanded. This extraction of work removes energy from the air and results in a reduction of the air's temperature.

The simplified example in Figure 2-3-2 supplies cool air by utilizing a source of compressed (or bleed) air, a heat exchanger, a cooling (air expansion) turbine and a fan that performs work by inducing ambient air to flow across the heat exchanger.

First, ambient air is compressed, then passed through the heat exchanger. A heat exchanger is physically and functionally similar to an automobile's water radiator, or to an oil cooler

As the hot, compressed air traverses the heat exchanger's passages, heat energy is transferred from the compressed air, via the cooling fins, to the ambient (fresh) air that is ventilating the exchanger. This heat transfer lowers the energy level of the compressed air considerably and is accompanied by a temperature drop to a point just slightly above that of the outside ambient air.

Since the air is still compressed, even more energy (heat) can be extracted by putting the air to work driving an *expansion turbine*. In Figure 2-3-2, a load is applied to the turbine by coupling the suction fan to it, ensuring that the air will expend a substantial amount of energy driving the turbine. This causes a further drop of air temperature. The fan, in return, ventilates the heat exchanger with ambient air. Finally, cool, compressed air exits the expansion turbine, ready for distribution.

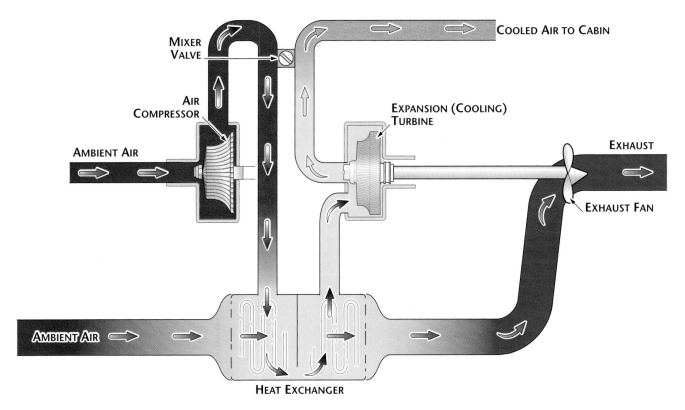

MIXER VALVE

AIR COMPRESSOR

AMBIENT AIR

EXPANSION (COOLING) TURBINE

COOLED AIR TO CABIN

EXHAUST

EXHAUST FAN

AMBIENT AIR

HEAT EXCHANGER

Figure 2-3-2. A simple air-cycle refrigeration system

A contrast to the air-cycle system is the more familiar vapor-cycle system commonly employed by domestic refrigerators, auto air conditioners and several modern private and corporate aircraft. The vapor-cycle system utilizes a closed refrigeration system in which an entrapped gas (such as Freon) is pressurized and cooled so that it condenses into a liquid. The liquefied gas expands and evaporates to cool a heat exchanger in the refrigerated area, and the gas is then recompressed and recycled. Thus, the vapor-cycle system's refrigerant is continuously changing from the gaseous phase to the liquid phase, then back to the gaseous phase, etc. The air-cycle system's refrigerant is air, and it always remains in the gaseous phase.

Air-cycle definitions. There are three terms to remember:

- *Air-cycle machine* refers to the expansion and compression turbines.

- *Air-cycle pack* refers to the entire unit, including heat exchangers and necessary piping.

- *Air-cycle system* is everything all the way out to the gasper.

Bootstrapping. *Bootstrap* implies self-help, as in the expression "pulling yourself up by your bootstraps." The turbine/heat exchanger fan arrangement of Figure 2-3-2 has some qualities of a bootstrap unit, for work expended to remove heat from the compressed air is recovered and used to power the suction fan that ventilates the system's heat exchanger. However, the true bootstrap unit is a compressor/expansion turbine assembly similar to the one used in the system shown in Figure 2-3-3.

After ambient air has been compressed and passed through a primary heat exchanger, it is routed to the bootstrap unit compressor. The bootstrap unit compresses the air a second time, elevating the air pressure appreciably above that of the primary compressor air output and elevating the air temperature to a level that may approximate or exceed that of the primary compressor air output.

Next, the highly compressed air is passed through a secondary heat exchanger (where it is cooled again by the heat-transfer process); then it is routed back to the expansion turbine segment of the bootstrap unit. The air decompresses as it passes through and drives the expansion turbine, and the turbine (attached to the same shaft as the compressor) recovers sufficient energy from the expanding air to drive the bootstrap compressor. This not only supplies more highly compressed air to the expansion turbine, but it extracts more heat from the air.

The air that exits from the expansion turbine is cool and ready for distribution. A system incorporating a compressor/expansion turbine unit

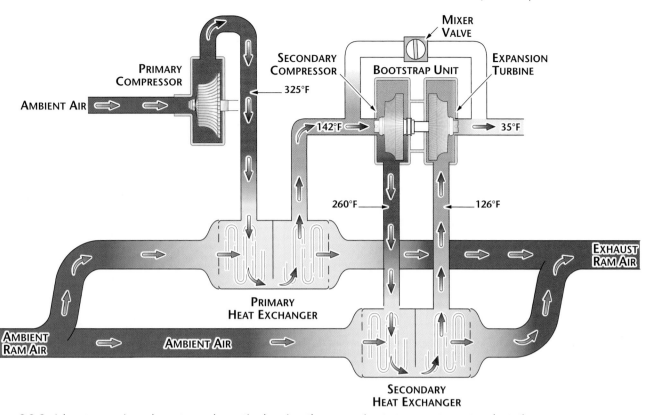

Figure 2-3-3. A bootstrap air-cycle system schematic showing the approximate temperature at each station

is called a bootstrap system. Figure 2-3-1 is a bootstrap air conditioning pack on a Boeing 757.

In Figures 2-3-2 and 2-3-3, it can be assumed that the conditioned air was used, then exhausted from the conditioned area. An air-cycle system that provides a continuous supply of fresh conditioned air and exhausts the used air back to ambient atmosphere is called an *open air-cycle system*. However, if the air-cycle system reclaims the used air for re-conditioning and recirculation, it is called a *closed air-cycle system*.

A bootstrap air-cycle system differs from a *simple air-cycle system* because it has a secondary air compressor/expansion turbine unit and a secondary heat exchanger.

Bootstrap systems are used in many systems where the bootstrapping helps reduce the power demand on the compressor. Modern air-cycle systems use this method because bleed air comes at the price of performance. Bootstrapping helps conserve bleed air and saves both power and fuel. See Figure 2-3-1.

The function of an air-cycle air conditioning system is to maintain a comfortable air temperature within the cabin and crew stations. The system will increase or decrease the temperature of the air, as needed, to obtain the desired value. Most systems are capable of producing an air temperature of 70-80°F with normally anticipated outside air temperatures.

This temperature-conditioned air is then distributed so that there is a minimum of *stratification* (hot and cold layers). The system, in addition, must provide for the control of humidity, it must prevent the fogging of windows, and it must maintain the temperature of wall panels and floors at a comfortable level. At altitude, outside air temperatures (OAT) of -50°F are not unusual.

In a typical system, the air temperature is measured and compared to the desired setting of the temperature controls. Then, if the temperature is not correct, hot or cold air is mixed together in a *mixer valve* to create a uniform temperature in the cabin. In summary, an air conditioning system is designed to perform any or all of the following functions:

- Supply ventilation air
- Supply heated air
- Supply cooling air
- Supply air for pressurization

Ventilation Air

Ventilation air is obtained through ram air ducts installed in the aircraft structure. Air entering these openings usually passes through the same duct system that is used for heating and cooling.

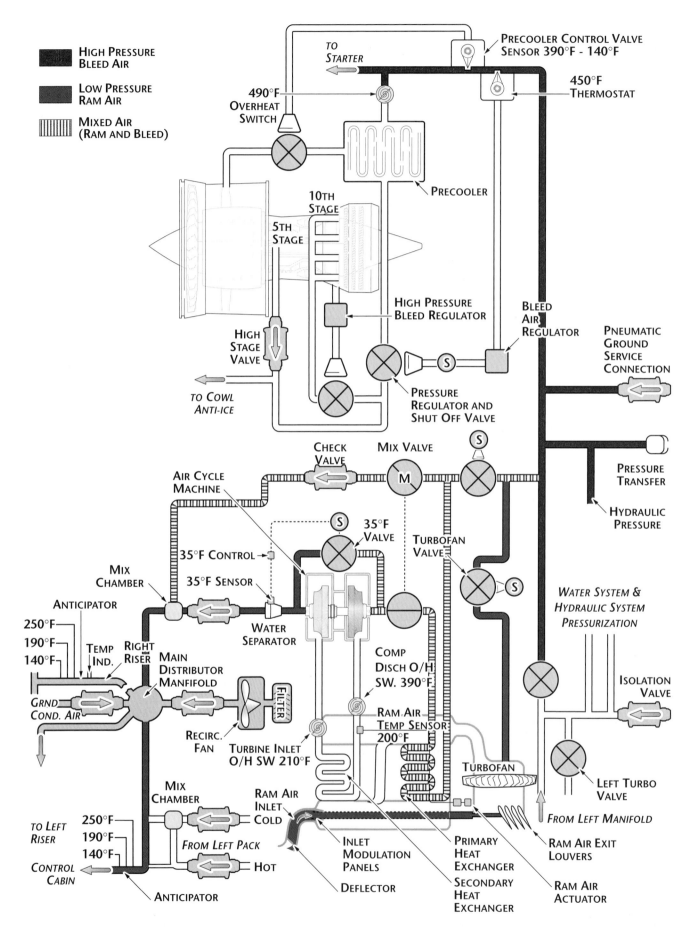

Figure 2-3-4. A working schematic of an air-cycle system

As a design safety feature, ram air ducts must be able to clear the cabin of smoke and noxious fumes in a matter of seconds, should the need arise.

Recirculating air. While FAA regulations set standards for the number of air changes per hour inside the cabin, it also allows for a certain amount of air to be recirculated. Recirculated air naturally doesn't need to be heated or cooled as much as new air. This reduces bleed air requirements and makes more power available from the engines. More power means that the airplane's speed can be maintained at a lower power setting; in essence, backing off the throttle. That saves fuel, thereby saving money.

When ventilating air from an APU or ground servicing equipment is used, the same fans that help with the recirculating air will be used.

The air-cycle cooling system is the system of choice for turbine-powered airplanes and helicopters.

Air-cycle Operation

A valve controls the compressed airflow through the expansion turbine. To increase cooling, the valve is opened to direct a greater amount of the compressed air to the turbine. When no cooling is required, the turbine air is shut off.

Other valves, operated in conjunction with the turbine air valve, control the flow of ambient air through the heat exchanger. The overall control effect of these valves is to increase the heat exchanger's cooling airflow at the same time increased cooling is obtained at the turbine.

Use of the air-cycle system imposes an increased load on the turbine engine compressor. As more cooling is demanded from the turbine, a greater amount of high-pressure air is bled from the engine.

Bleed air is not unlimited, nor is it free. Turbine engine bleed air reduces the engine power available by removing compressed air that cannot be used for combustion. Engine manufacturers build into an engine design the amount of bleed air that can be used. It is known as customer air, and is the air used by the airframe designer for airplane systems.

This description of the operation of an air conditioning system is intended to provide an understanding of the manner in which the system is controlled, the functions of the various components and subassemblies, and their effect on total system operation. Figure 2-3-4 is a schematic of a typical system. Frequent reference to the schematic should be made during the following operational description.

Figure 2-3-5. A jet pump for increasing airflow in a gasper

The system is composed of a *primary heat exchanger, primary heat-exchanger bypass valve, flow limiters, refrigeration unit, main shut-off valve, secondary heat exchanger, refrigeration-unit bypass valve, ram-air shutoff valve and an air temperature control system.* A cabin-pressure regulator and a dump valve are included in the pressurization system.

Air for the cabin air conditioning and pressurization system is bled from the compressors of both engines. The engine bleed lines are cross-connected and equipped with check valves to ensure a supply of air from either engine.

Jet Pump. The *jet pump* principal is very straightforward. A venturi is installed into the center of a duct with airflow space around it. When air pressure (bleed air) is induced to flow through the venturi, a low pressure area is created by the venturi outflow, naturally creating a low pressure area behind it. The low pressure then sucks a large volume of air around the venturi to fill the space. This action increases the total airflow dramatically.

The principal uses of a jet pump are twofold:

1. To increase high pressure air available and use less bleed air in the process.

2. To move large volumes of air with less duct pressure and power loss.

The jet pump process is also incorporated in several different valve designs that may go by any of the following names:

- Bleed air regulator
- Flow control valve
- Air ejector
- Pack valve
- Jet pump

By looking at the photograph of a jet pump in Figure 2-3-5 it is easy to get an idea of how it

Figure 2-3-6. On this system trainer you can see the complete air expansion system.

works: expanding air creates a low pressure area that sucks more air through the screened inlet.

Sonic venturi. While dealing with bleed air distribution, sometimes the speed and pressure needs to be controlled without creating a complete sub-system to handle it. A *sonic venturi* is used in some systems as an additional duct pressure limiting device when dealing with bleed air. When the speed of duct flow gets too great, a sonic venturi will allow a small amount of the airflow to exceed the speed of sound, thereby creating a pressure wave inside the duct. If you remember your aerodynamics that means that all flow behind the pressure wave will be reduced and will be subsonic.

Flow limiting. A flow-limiting nozzle is incorporated in each supply line to prevent the complete loss of pressure in the remaining system if a line ruptures, and to prevent excessive hot air bleed through the rupture.

In reading the schematic in Figure 2-3-5 , the initial input of hot air is indicated from one engine only.

Primary heat exchanger. Air from the engine manifolds is ducted through a flow limiter to the primary heat exchanger and its bypass valve simultaneously. Cooling air for the heat exchanger is obtained from an inlet duct and is exhausted overboard.

The air supply from the primary heat exchanger is controlled to a constant temperature of approximately 300°F by the heat-exchanger mix valve. The mix valve is automatically controlled by upstream air pressure and a downstream temperature-sensing element. These provide temperature data to cause the valve to maintain the constant temperature by mixing hot engine bleed air with the cooled air from the heat exchanger.

Secondary heat exchanger. The air is next routed to the refrigeration-unit mixer valve, to the compressor section of the refrigeration unit, then to the secondary heat exchanger. The mixer valve automatically maintains compartment air at any preselected temperature between 60°F and 125°F by controlling the amount of hot air that bypasses the refrigeration unit and mixes with the refrigeration unit output.

Cooling air for the secondary heat exchanger core is obtained from an inlet duct. After cooling the cabin air, the cooling air is exhausted overboard.

Expansion turbine. As the cabin air leaves the secondary heat exchanger, it is routed to the expansion turbine, which is rotated by the air pressure exerted on it. In performing this function, the air is further cooled before entering the *water separator*, where the moisture content of the air is reduced. Figure 2-3-6 shows a complete expansion turbine from an air-cycle system trainer.

Water separator. If the temperature of the air is below the dew point, fog will discharge from the cabin vents. It happens because water vapor is not removed from the cool, low-pressure air before it is discharged into the cabin. Under continuous operation, the air passes through a Dacron® polyester *coalescer*. The purpose of a coalescer bag is to give the water vapor a place to collect. This is accomplished by causing the air to swirl as it enters the bag. By doing so, the heavier water content slings to the outside, where it collects on the coalescer bag. The condensed water drains to the bottom of the separator and is collected for draining overboard. (See Figure 2-3-7). In most systems, the drain water is sprayed into the heat-exchanger cooling system to help lower the air temperature.

Some newer air-cycle systems do not have a coalescer system. Instead, they have a high-pressure system that consists of a condenser, a water extractor and a reheater. The cool outlet air is allowed to condense, and the water forms droplets, which pass to a water extractor. A helix, similar to a centrifugal air cleaner, spins the air and removes the droplets by centrifugal force. The water is collected and sprayed into the ram air feeding the heat exchangers. This increases the efficiency of the heat exchangers. By eliminating the coalescer bag, scheduled maintenance is eliminated, and that part of the system goes *on-condition maintenance.*

The 35° valve. If you reflect for a moment, it would seem obvious that if the coalescer bag were to freeze, the complete system would shut down. A simple method has been devised to prevent this. It is commonly called the 35°

Figure 2-3-7. The coalescer bag is housed inside the stainless steel housing.

valve. It functions just as its name implies: it maintains 35°F at the entrance to the water separator to prevent freezing.

Gasper. From the water separator, the air is routed through the temperature sensor to the cabin. Air enters the cabin spaces through a network of ducts and diffusers and is distributed evenly throughout the spaces. This duct system is called a gasper. Most systems incorporate directional, overhead vents that can be rotated by the cabin occupants to provide additional comfort. The King Air distribution system in Figure 2-3-8 is, in principal, similar to most systems, regardless of the origin of the warm or cooled air.

Alternate ram air. An *alternate ram-air system* is provided to supply the cabin with ventilating air if the normal system is inoperative or the cabin areas must be rid of foul odors or fumes which might threaten comfort, visibility or safety.

Temperature Control

Temperature is controlled by either a pneumatic or electronic controller, depending on the age of the aircraft system. In the older pneumatic systems, the various mixing valves were controlled by an air pressure signal routed through lines to the various controllers. From there, electrical signals were sent to the drive motors in the valves. Next came electrical controllers, with some electronic components integrated into them.

The controllers in today's cabin systems are all solid state electronic. Most have a microprocessor built in, the function of which is to integrate the myriad actions that manage the cabin atmosphere. The solid state controllers also have built-in BITE functions. BITE allows maintenance information and automated troubleshooting to be available from the cockpit.

In newer equipment, the process of getting out the charts and trying to figure out what went wrong with the system are gone.

The controller system discussed here is about half pneumatic and half electronic. We are not going to get into the interior workings of the controller, but will approach it from the standpoint of what it does, instead of how it work.

Before going further, there are a couple of basic electricity components that need reviewing:

- **Thermistor.** A thermistor is a resister that changes with its exposure to heat. It

can act like an on/off switch at a specified temperature. It can also provide a resistance that varies inversely with heat.

- **Wheatstone bridge.** A Wheatstone bridge is an electrical component that can take varying electrical inputs and develop electrical outputs based on the inputs.

Temperature controller. Most passenger airplane cabin systems have more than one controller: one for the crew cabin and from one to three units for the passenger cabin. Most controllers work the same. An example is shown in Figure 2-3-9.

A typical controller is a fully automatic feedback unit that receives input from the temperature selector and from temperature sensors. The control processes the information

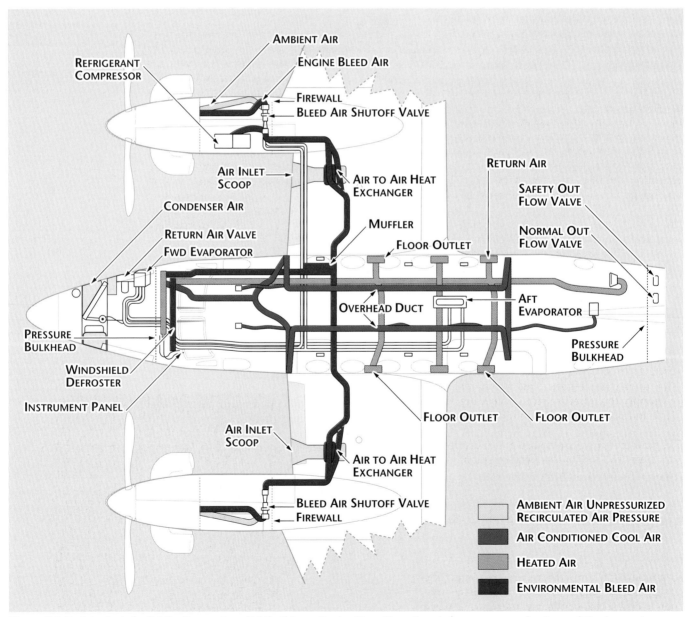

Figure 2-3-8. A typical air distribution system. While this particular King Air system is for a vapor cycle air conditioning system, the distribution system is similar to most airplanes.

and then sends an output signal that positions the temperature control mix-valve to maintain the selected cabin temperature. The crew can bypass the automatic controls by directly adjusting the power to the actuators.

The temperature controller utilizes a Wheatstone bridge network, two arms of which are the main cabin temperature sensor and the cabin temperature selector. A difference in the selected and sensed temperatures is detected as an unbalance of the bridge. This error is amplified and used to control the power output to the mixer-valve actuator. The valve is repositioned so as to restore the balance of the bridge, thus restoring the sensed cabin temperature to the selected cabin temperature.

Temperature limiting. The basic control operation is modified by the inputs from two other temperature sensors. The duct *anticipator sensor* is located in the common manifold in the mix-bay. It produces an output signal that is a function of changes of cabin supply air temperature.

Any change of supply air temperature results in an anticipator signal which is in opposition to heat change, but which decays with time. This results in a stabilizing effect during response to selection changes and adequate inlet temperature regulation during transient bleed-air conditions. In essence, it slows the rate of bleed air change during temperature transition.

The duct *temperature limiting sensor* is also located in the common manifold. This sensor, together with a *preset reference resistor,* forms a part of an auxiliary Wheatstone bridge in the controller. The output of this bridge is blocked by a diode, as long as the duct temperature does not exceed a threshold value of about 140°F.

Once this threshold is exceeded, the bridge produces a cooling signal, which increases with duct temperature. This signal prevents the supply temperature from exceeding about 160°F and protects the distribution ducting.

The temperature sensors used are of the thermistor type: The resistance of each element varies inversely with the temperature.

The duct limit sensor has a single element, while the cabin temperature sensors have two elements connected in series.

The anticipators are used where signals related to changes of temperature are required. The anticipator consists of four thermistor elements; one series-connected pair is thermally impeded, while the other series-connected pair is un-lagged.

The un-lagged open sensor elements respond very rapidly to changes in temperature when directly exposed to the duct air, while the lagged elements are enclosed by thermal insulation to delay their response to temperature changes. The unbalance of the anticipator bridge circuit is thus a function of immediate changes of duct air temperature.

In addition to the temperature limiting and anticipator sensors, there are other thermal switches in the common manifold to detect overheat conditions. These switches are independent of the automatic temperature-control systems.

If supply air exceeds 190°F, power is applied to the appropriate mix-valve to drive it to the full cold position. As a backup to the 190°F switch, the second switch, set at 250°F, will cause the fast-acting pack shut-off valve to close.

In either case, the duct overheat warning light (on the crew person's panel) is energized. The warning light will stay on until the system is reset by the flight crew. This can only be done after the thermal switch has cooled to below its trip temperature by approximately 20-30°F.

Temperature control system. The control system is a fully automatic feedback control that receives input information from the temperature selector and from various temperature sensors. The control processes this information and produces an output signal which positions the temperature-control mix-valve to maintain the selected cabin temperature.

The automatic control can be bypassed by the flight crew, who can position the mix-valve by directly controlling the power to the valve actuators. Figure 2-3-10 is an operational schematic of a cabin temperature controller.

Figure 2-3-9. Temperature controller

The temperature controller also utilizes a Wheatstone bridge network, two arms of which are the main cabin temperature sensor and the cabin temperature selector. A difference in the selected and sensed temperatures is detected as an unbalance of the bridge. This error is amplified and used to control the power output to the mix-valve actuator. The mix-valve is repositioned to restore the balance of the bridge (i.e. restore the sensed cabin temperature to the selected cabin temperature).

Temperature control panel. Each primary temperature selector consists of a variable resistor and switch assembly, enclosed within an aluminum alloy housing, a dial assembly, a control knob and an electrical connector to facilitate connection to the aircraft wiring harness.

The variable resistor and switch assembly consists of a variable wire-wound resistor, two normally open switches with actuators, one normally closed switch with actuator and momentary and detent mechanism, all operated by a common shaft and enclosed within a housing.

Two air mix-valve indicators are supplied to show the position of the appropriate temperature control valve. The signal is sent from the position transmitter situated on the valve itself.

Included on the panel are two *duct overheat* lights that illuminate when duct temperatures

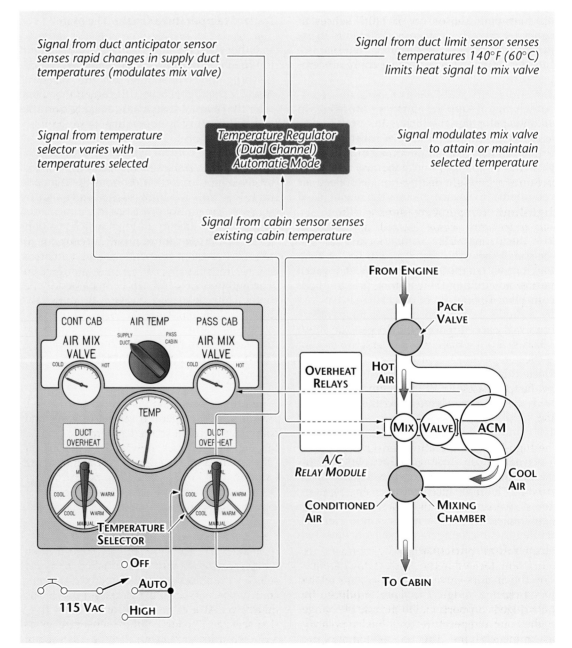

Figure 2-3-10. Temperature control panel

reach 190°F. After correcting the overheat condition, the system may be returned to normal by pushing the reset switch.

The air temperature for the supply duct and the passenger cabin can be read on the temperature gauge located in the center of the panel. The temperature bulbs are located in the passenger cabin distribution duct (supply duct). The passenger cabin temperature bulb is located with the cabin temperature sensor in the forward right-hand side of the cabin (pass cabin).

Temperature sensors. Temperature sensors monitor the temperatures in the cabin and flight deck. They also provide reference signals to the temperature regulators when in the automatic mode of operation to control the heating and cooling of the air conditioning package.

The high-limit sensor set at 140°F senses a higher than normal temperature in the ducts and sends a reference signal to the controller to restrict the hot air from entering the manifold.

The temperature anticipator senses any rapid changes in temperature in the supply duct. It will also anticipate the restoration of the temperature selected to minimize overshooting of the pre-selected temperature. This occurs when the temperature selected and the temperature in the flight deck/cabin are not equal.

High-limit temperature sensor. The *high-limit temperature sensor*, located downstream from the air mix-valve, forms one arm of the *high-limit bridge*. The signal produced by this bridge is referenced to a fixed-bias voltage and the algebraic sum of these potentials is applied to the pre-amplifier.

At normal duct temperatures, the pre-amplifier input is approximately zero, and operation of the temperature control subsystem is not affected. At approximately 140°F in the supply duct, the sensor will limit auto control and drive the air mix-valve toward the cold position.

The high-limit temperature sensor consists of a thermally open sensing element, with a single thermistor bead mounted in a probe-type housing. The resistance of the sensing element varies inversely with any change in ambient temperature.

Temperature anticipator. The temperature anticipator, located in the air duct downstream from the air mix-valve, forms two arms of the rate-of-change bridge. The signal produced by this bridge is proportional to the rate of change in duct air temperature and has a polarity determined by the direction of temperature change.

This signal is applied to the amplifier to oppose the effect of the error signal produced by the control bridge. Temperature overshoot is minimized by damping the anticipator response to the setting of the primary temperature selector.

Each temperature anticipator consists of a thermally open sensing element and a thermally impeded sensing element mounted in a common probe-type housing. The resistance of each sensing element varies inversely with any change in ambient temperature.

Since the response of the thermally open sensing element to changing temperature lags behind the response to the thermally impeded sensing element, the relationship of the two resistances is a function of transient duct air temperature changes.

Control temperature sensor. The *control temperature sensor* forms one arm of the temperature-control bridge, and it senses the existing cabin or flight deck temperature. When the cabin air temperature and the temperature selected by the controller reach the same value, the control bridge is in balance and the temperature-control subsystem is in standby mode (quiescence).

At any time that the sensor senses a temperature change in the cabin/flight deck, the control bridge will go out of balance and produce an error signal to the regulator.

The control temperature sensor consists of a thermally open sensing element with two thermistor beads wired in series, mounted in a probe-type housing. The resistance of the sensing element varies inversely with any change in ambient temperature.

Automatic Temperature Control

When the selector knob is moved to AUTO, a switch in the temperature selector closes and completes a circuit to the temperature regulator. Setting the knob pointer for a particular cabin temperature adjusts a potentiometer fixed to the knob shaft. This potentiometer serves as a reference resistance in the regulator temperature-control bridge.

The cabin temperature sensor provides the resistance in the other leg of the bridge. If cabin temperature is the same as that asked for by the selector, the silicon-controlled rectifier-actuator control prevents any current passing on to the mix-valve. At a cabin temperature other than that selected, the temperature sensor provides a higher or lower resistance in the other leg of the control bridge.

Figure 2-4-1. A heat intensifier on an exhaust system

As a result, the regulator-actuator control completes a circuit to valve, so cabin temperature agrees with selected temperature.

The anticipator bridge and the duct temperature-limit bridge sense conditioned air temperature to slow down changes requested by the control bridge and prevent duct overheat. The actuator control moves the mix-valve so that cabin temperature changes slowly and without raising duct temperature above limits.

Section 4

Cabin Heaters

While there are still a few brave souls who fly open-cockpit airplanes in winter, the rest of us like to have some heat in the cabin. The idea of dressing in layers of wool, topped-off by sheepskin-lined leather flying suits isn't for everyone.

Cabin heaters range from simple heat-transfer systems that use engine heat through a heat exchanger, to combustion heaters that use gasoline from the airplane's fuel system to produce cabin heat. The common thread is that they all use the combustion of fuel to produce heat; therefore, they must be inspected regularly and maintained correctly.

Exhaust Gas Heaters

A relatively simple heating system used on a small general aviation aircraft utilizes the engine exhaust gases as a heat source. A *hot-air muff,* or jacket, is installed around the muffler. Air routed through the hot-air muff picks up heat by convection through the muffler material. The heat muff also supplies hot air for carburetor de-icing. The heated air is routed to the cabin through Aeroduct® tubing and controlled by the heater valve opening. The danger of exhaust-gas contamination of the heated air is greater in this type of heater than in most all other types of heating systems (see Figure 2-4-1).

A heating system consists of a heating unit and the necessary ducting and controls. The units, ducts and controls will vary considerably from system to system. Most controls have one thing in common, though: pull for heat and push for no heat.

Flexible ducting. *Flexible ducting* is used in almost all heater, carburetor heater, vent and engine installations to carry air, both hot and cold, from one place to another. Properly named Aeroduct and manufactured by the Aeroquip Corporation, flex duct is an impregnated wire-wound material that is rated by the temperature it is designed to withstand, and by its internal wire winding. It comes in four ranges and is described as follows:

- CAT – temperature-rated from -65°F to +300°F, it is made of a single ply of black neoprene impregnated fiberglass. It has a copper-coated steel spiral inside, with a fiberglass cord wrapped around the outside.

- SCAT – temperature-rated from -80°F to +450°F, it is made of a single ply of red silicone rubber-impregnated fiberglass. It also has a copper-coated steel spiral inside and a fiberglass cord wrapped around the outside.

- CEET – temperature-rated from -65°F to +350°F, it is made of two plies of silicone rubber-impregnated fiberglass with the same wire coil between the plies, along with a single fiberglass cord wrapped around the outside. CEET is similar to CAT in color.

- SCEET – temperature-rated from -80°F to +500°F, it is made of two plies of red silicone rubber with the wire spiral between the plies and the same fiberglass cord around the outside.

The fiberglass cord wound around the outside is to hold the inner wire coil in place. The principal differences are whether the wire windings are molded into the fabric or not, and the temperature range. Do not use CAT or SCAT in areas subject to wind load, because the flapping will cause the wire coils to shift, and the tubing will collapse. Figure 2-4-2 shows a typical use of SCEET hose.

Inspection. Exhaust gas heaters demand a very complete inspection to eliminate the danger of carbon monoxide poisoning. Some heat intensifiers have airworthiness directive (AD) notes against them that require even more complete tests during inspection. A good example is the pressure decay tests that must be performed on Bonanza heaters. The AD requires that the air intake and exhaust ducts on the heater core be plugged and the core pressurized. The rate of air escaping from the core can then be measured to check for leaks.

In order to inspect many heat-muff installations, it will be necessary to remove the cover that contains the hose flanges. This can generally be accomplished by removing screws or by removing a long, channel-shaped clip.

Any leakage, cracking, heat damage or serious corrosion must be repaired. It will frequently be necessary to send the part to a certified repair station for replacing and re-welding parts.

Combustion Heaters

One of the most popular cabin combustion heaters is the product line manufactured by Janitrol. Janitrol has produced a bewildering variety of heater types for almost any kind of airplane you can imagine. Not all are interchangeable; in fact, most are not even close.

Because of the wide variety of models, we can only cover the basics here. To cover all of the actual heaters would require an extensive collection of maintenance, parts and overhaul manuals. That, however, does not mean that you can repair or overhaul a heater without the correct manual. Always have one on hand when doing maintenance on a heater.

Description and Operation

The heater assembly is cylindrical in shape and is fabricated of heat-resistant alloy steel. A combustion chamber and radiator assembly, welded gas-tight, form the principal part of the heater. At one end of the combustion chamber and radiator assembly is the fuel inlet, combustion-air inlet and exhaust outlet.

A stainless steel wrap-around jacket with a seam-sealing joint encloses the combustion chamber and radiator assembly.

A removable, spray-type combustion head covers the inlet end of the combustion chamber. A

Figure 2-4-2. Aeroduct tubing connecting ram air to the heat muff

spark plug and ground electrode are mounted in the combustion head to provide a spark gap. The fuel-spray nozzle is mounted in a nozzle holder and feed assembly that is attached to the combustion head.

A combustion-air relief valve is mounted on the combustion-air inlet. The relief valve is spring-loaded and adjustable. It connects to the exhaust outlet, thus allowing excess combustion air to bypass the heater.

Operation. Heat is produced by burning a fuel-air mixture in the combustion chamber. The fuel usually passes through a fuel filter, pressure regulator and solenoid valve before entering the heater. Aviation gasoline, or kerosene, is injected into the combustion chamber

Figure 2-4-3. A heater ignition unit

through the spray nozzle. The resulting cone-shaped fuel spray mixes with combustion air and is ignited by a spark.

Electric current for ignition is supplied by an ignition unit that converts direct current (DC) to high-voltage oscillating current (see Figure 2-4-3). This provides a continuous spark across the gap between the spark plug and ground electrode. A shielded lead connects the ignition unit to the spark plug. Most combustion heaters are controlled by a thermostat cycling switch that cycles fuel on and off to the heater as demand requires.

Combustion air enters the combustion chamber tangent to its surface. This arrangement imparts a whirling, or spinning, action to the air, thus producing a whirling flame. The flame is stable and sustains combustion under the most adverse conditions, because it is whirled around itself many times. Therefore, ignition is

continuous and the combustion process is self-piloting. The burning gases travel the length of the combustion chamber and pass into the radiator, then travel the length of the radiator and out the exhaust.

Excess combustion airflow is prevented by either an air relief valve, or a differential pressure regulator.

Ventilating air passes through the heater between the jacket and radiator and between the radiator and combustion chamber. Consequently, the ventilating air comes into contact with three heated, cylindrical surfaces. Figure 2-4-4 shows a complete heater assembly.

Inspection and Service

Because heater inspections and overhauls are based on heater hours of operation and not airplane hours, most installations have an hour meter installed. Figure 2-4-5 shows a combustion heater installed in a Piper aircraft. Notice that the hour meter (commonly called a Hobbs meter) is installed in a prominent position.

100-hour inspection (heater time). Check the shielded lead connections at the spark plug and ignition unit for security and possible damage.

Examine the fuel and air connections for any evidence of leakage.

500-hour inspection (heater time). Overhaul the heater. At this overhaul, replace all parts that show evidence of damage or excessive wear.

AD requirements. Most heaters have AD notes that include additional inspections and/or inspection requirements. Always research the AD notes before starting any work on combustion heaters.

Troubleshooting information for heaters is listed in Table 2-4-1..

Overhaul

Heater overhaul is not a difficult process. It consists mainly of disassembly, inspection, parts repair or replacement, reassembly and test.

Cleaning combustion chamber. Either of the following methods may be used to clean the combustion chamber:

- Soak the combustion chamber and radiator assembly overnight in an Oakite® M-3 stripper solution. This solution can be made by mixing one pound of Oakite®

Figure 2-4-4. Aircraft heater assembly

with each gallon of water used. The solution should be kept at a temperature of 190-210°F during the soaking period. Rinse the heater thoroughly with water after it is removed from the solution.

- Use a stainless steel brush (not ordinary steel) or a sandblast cleaner to remove any accumulation of carbon or other foreign material from the inside of the combustion chamber. If sandblast is used, clean out sand after this operation is complete.

Cleaning spray nozzle. Because clean burning is critical to the safe operation of a heater, follow directions when cleaning the fuel nozzle. Never use a wire or welding-tip cleaner to try and clean a fuel nozzle. It will damage the hole size, thus skewing the calibration. Figure 2-4-6 is an exploded view of a fuel-nozzle assembly.

1. Disassemble the spray nozzle by unscrewing and removing the fuel strainer and two-piece core from the nozzle body. The strainer should be only finger-tight. Remove the core with a screwdriver.

2. Clean these parts by immersing them in Stoddard® Solvent.

3. If immersion in solvent does not thoroughly clean the parts, use a small, soft, non-metallic brush to aid in cleaning.

4. If brushing fails to remove dirt particles from the grooves in the core and the orifice in the body, clean the grooves and orifice with a stick made by sharpening a soft piece of wood (a matchstick is satisfactory).

5. After removal of foreign material, rinse the parts carefully in clean solvent and dry them with filtered, compressed air.

6. When reassembling the spray nozzle, tighten the core with a screwdriver, and tighten the strainer with the fingers.

7. Place the spray nozzle in a protective envelope (cellophane or waxed paper) until ready for installation in the heater.

Cleaning and inspection of spark plug. Before cleaning, examine the spark plug for evidence of cracked or broken porcelain and for arcing or carbon tracks inside the well of the spark plug. If cracked or broken porcelain is found, no further examination is necessary, and the spark plug should be discarded. Arcing or carbon tracks may be caused by shorting of the spark plug or by dirt on the spring connector seated in the well of the spark plug. In either case, the fault should be corrected before reinstalling the old spark plug in the heater or before using a new spark plug.

Figure 2-4-5. Installed view of heater assembly with an hour meter

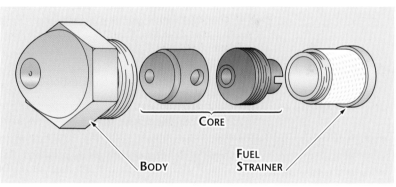

Figure 2-4-6. Exploded view of spray nozzle

1. Wipe out the inside of the well of the spark plug with a clean cloth, dampened with carbon tetrachloride. This will remove grease and carbon deposits.

2. Clean the spark plug by grit-blasting. Close the well of the spark plug with a stopper to keep out dirt during cleaning.

3. Do not use a metal tool for scraping these parts, as this will upset the flow characteristics of the nozzle.

4. All parts of the nozzle assembly must be kept together, as nozzle parts are not individually interchangeable. Extreme caution should be used to keep the nozzle free of dirt and other foreign material.

TROUBLE	PROBABLE CAUSE	REMEDY
Heater does not light	IGNITION SYSTEM FAILURE:	
	1. No power to ignition unit	Close switches, replace burned out fuses, repair open circuits
	2. Inoperative vibrator in ignition unit	Replace vibrator
	3. Faulty spark plug	Replace spark plug
	4. Faulty shielded lead	Replace shielded lead
	5. Worn ground electrode	Replace electrode
	6. Ignition unit inoperative	Overhaul or replace ignition unit
	INSUFFICIENT FUEL:	
	1. Fuel solenoid not energized	Close switches, replace burned out fuses, or repair open circuits
	2. Fuel supply pressure may be low	Increase fuel pressure to value required
	3. Fuel filter clogged	Replace filter element or clean if new one is not available
	4. Spray nozzle clogged	Clean spray nozzle
	5. Fuel pressure regulator faulty	Overhaul the regulator
	6. Fuel solenoid inoperative	Replace solenoid valve
	INSUFFICIENT COMBUSTION AIR:	
	1. Leaks or obstruction in combustion air supply line.	Repair leaks or remove obstructions
	2. Faulty combustion air relief valve	Overhaul or replace relief valve
Heater is cycled off and on by limit switch	1. Limit switch out of calibration or faulty	Calibrate or replace switch
	2. Cycling switch (if used) out of calibration or faulty	Calibrate or replace switch
	3. Ventilating air stream may be obstructed	Remove obstructions
Backfiring, pulsating combustion, or smoky exhaust	1. Fouled spark plug	Clean or replace spark plug
	EXCESSIVE FUEL FLOW INTO HEATER:	
	1. Spray nozzle dirty or loose	Overhaul nozzle
	2. Spray nozzle is oversize	Check markings on nozzle. If nozzle is oversize, replace with proper size nozzle
	3. Fuel pressure regulator may be faulty	Overhaul regulator
	4. Restriction in exhaust line	Remove restriction
	5. Insufficient combustion air	Correct as instructed above

Table 2-4-1 Troubleshooting chart for combustion heaters

5. Do not brush off the outside of the radiator assembly, as this may make satisfactory visual inspection more difficult.

Inspection of Miscellaneous Parts

- Inspect the ground electrode for excessive erosion. If this condition is evident, replace the ground electrode.

- Examine the head gasket and nozzle-holder gasket. Replace if damaged.

- Check all screws and nuts for visual damage. Replace whenever threads are stripped or other damage is evident.

Inspection of combustion chamber and radiator assembly. Slight scaling and discoloration of the combustion chamber and radiator assembly is a normal condition found on heaters that have been in use for any length of time. The scale will not be mottled distinctly but will be dull, dark gray, and a dark gray powder can be rubbed off scaly areas. This condition does not constitute grounds for rejecting the heater, since considerable life can still be expected if there are no soft spots in the metal where it has been subjected to severe overheating.

Types of Damage to Expect

Damage to a heater may be classified as any of the following:

- Soft, spongy metal as a result of overheating
- Deformation as a result of backfiring
- Fatigue cracks
- Pin holes

Soft and spongy metal. Severe overheating will result in a general weakening of the metal that results in soft and spongy spots. These are usually found directly opposite the crossover passages. They can be detected by tapping lightly with a ball peen hammer. This will give a slightly soft or spongy response in contrast to a solid feel when tapping on live metal. These soft spots will usually have a dull, dark gray appearance, indicating considerable oxidation on the surface, and they will be surrounded by an area where slight oxidation has occurred. The presence of such soft spots is reason for rejecting the combustion chamber and radiator assembly.

Deformation as a result of overheating. Extreme overheating will result in distortion of the outer wall of the radiator near the crossovers. This type of deformation is abrupt and does not extend far from the crossover area. Soft spots and extreme oxidation will also be present. The presence of this type of deformation is reason for rejecting the combustion chamber and radiator assembly.

Deformation as a result of backfiring. Deformation caused by backfiring usually pushes the inner wall of the radiator in toward the combustion chamber of the heater. This condition may have occurred to a comparatively new heater and does not mean that the heater must be discarded. This type of deformation will usually be evidenced by a gradual change in contour extending the full length of the pass. However, if this deformation causes an increase of more than 10 percent in ventilating air pressure drop across the heater, the heater should be discarded.

Fatigue cracks. Fatigue cracks are usually found near the crossover passages. Some cracks may be large enough to be seen during visual inspection, while others can be found only by a leakage test of the combustion chamber and radiator assembly. Should a visual examination reveal cracks, repair these before making the leakage test.

Pin holes. Pin holes usually cannot be seen by visual inspection, but evidence of pin holes will be revealed during a leakage test of the combustion chamber and radiator assembly. Should a visual inspection reveal pin holes, repair these before making the leakage test.

Leakage Test of Combustion Chamber and Radiator Assembly

If a visual inspection does not reveal pin holes or cracks, make a leakage test of the combustion chamber and radiator assembly. With all openings closed and an air pressure of 6 p.s.i. applied to the interior, submerge the combustion chamber and radiator assembly in a tank of water. Air bubbles will reveal any leaks.

Repair. Pin holes and small cracks may be welded if they are in areas accessible to the welder. Successful repair work hinges on the ability of the inspector to intelligently analyze the condition of the metal where the weld is to be made, as well as on the skill of the welder.

Welding is not successful when the old weld or parent metal is in, or nearing, a state of complete oxidation (soft or spongy metal). When this condition exists, it will be found that a new crack usually develops around the outer edge of the new weld.

CAUTION: *Any heater found to have been damaged by overheating has been operating in a system where there is some control func-*

tioning improperly. The spray nozzle should be checked for a loose core, and the fuel-pressure regulator, limit switch and fuel solenoid valve should be checked for proper operation and calibration. Since overheating is most likely caused by lack of ventilating air, the system should be checked for possible restriction in the ventilating airstream. A heater should never be reinstalled or replaced in a system where evidence of faulty operation exists, until the trouble has been corrected.

An inspector must take into consideration the fact that the presence of cracks or pin holes may be the forerunner of a general breakdown of the metal. Where the general condition of the metal is bad, it would be unwise to spend the time repairing a combustion chamber and radiator assembly that might develop other leaks after a few hours of operation.

Cracks and pin holes sometimes develop in an original weld. This often puts the damaged area in an inaccessible place as far as repair is concerned. A new weld should always be made on the same side as the original weld. Trying to weld a crack or hole in a seam or where two pieces of metal are joined together will not be successful unless it is made on the same side as the original weld.

Clean the area to be welded by brushing with a stainless steel brush (not ordinary steel) or by sandblasting. All sand must be removed after sandblasting. The area should then be wiped with a 30-percent solution of nitric acid. The weld can be made either with an acetylene torch or by the Heliarc® method. If you do not have the requisite welding skills, send the part to be repaired to a CRS certified for the process.

After final repair, repeat the leakage test.

Operational Test Equipment Requirements

- A blower capable of delivering 600 lbs. per hour at a static pressure of 10.0 inches of water
- A filtered, controlled fuel supply
- A direct-current (DC) power source — either a battery or a generator may be used
- An ignition unit
- A shielded lead assembly
- A ventilating-air orifice plate with a 2.5-inch diameter orifice
- Two 10-inch water manometers
- A heater test set (Figure 2-4-7)

Operational Test Procedure

1. Close both blast gates and start the blower.

2. Adjust the blast gates to provide a combustion-air pressure of 4.0 inches of water and a ventilating-air pressure differential of 4.7 inches of water.

3. Turn on the ignition to the heater.

4. Turn on the fuel supply and check for leaks.

5. Adjust the fuel pressure.

6. The heater should ignite within 10 seconds. Should the heater fail to ignite, refer to the service manual.

7. After the heater begins operating, readjust the fuel pressure, combustion-air pressure and ventilating-air pressure differential to the values specified in Step 2.

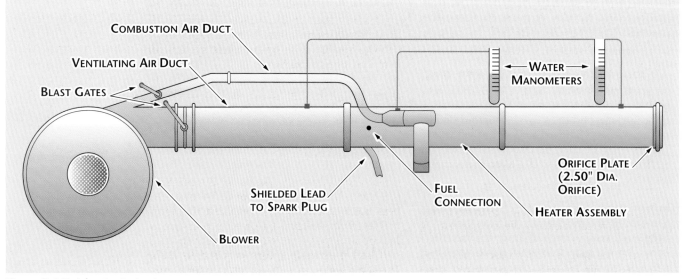

Figure 2-4-7. A heater test set

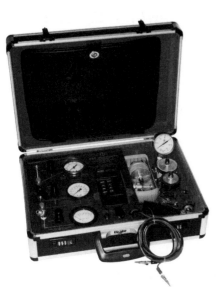

Figure 2-4-8. Leak tester for complying with the AD note test procedures.

8. Operate the heater for several minutes, then turn it on and off several times to check ignition dependability. Wait approximately one minute between ignition trials.

9. Reduce the combustion-air pressure to 2.5 inches of water; then repeat the above test procedure at the reduced combustion-air pressure.

Leakage Test of Heater Assembly

This test is to be conducted only after the heater has passed the operational test. It is also a test mandated by an AD note.

Equipment required. The following things must be present to complete a heater assembly leakage test. See Figure 2-4-8:

- A source of compressed air
- A 15-inch mercury manometer
- Two manual shut-off valves
- Rubber tubing and clamps for connecting the compressed-air supply line to the combustion-air inlet tube of the heater
- Two rubber expansion plugs to close the exhaust tube
- A plug for the fuel-inlet connection
- A plug for the drain connection

Test connections. It is essential to perform these functions before commencing the leakage test:

1. Install the two manual shut-off valves in series in the compressed-air supply line, to provide positive shut-off control.

2. Connect the mercury manometer downstream from both valves.

3. Remove the combustion-air relief valve and tube from the heater.

4. Connect the compressed-air supply line to the combustion-air inlet tube of the heater with rubber tubing and clamps.

5. Close the two openings in the exhaust tube with the rubber expansion plugs.

6. Plug the fuel-inlet connection.

7. Plug the drain connection.

Leakage Test Procedure

1. Apply air pressure until a reading of 8.0 inches of mercury is obtained on the manometer.

2. Close both manual shut-off valves to "lock" the air pressure in the combustion chamber and radiator assembly of the heater.

3. Time the pressure drop. The maximum allowable pressure drop is 5.0 inches of mercury in four minutes' time.

4. If the pressure drop is greater than that specified above, make certain that there are no leaks at the test connections by checking each connection with soap suds.

Section 5

Vapor-Cycle Air Conditioning System (Freon)

Vapor-cycle cooling systems are used on most small- and medium-sized general aviation aircraft. Some aircraft have a vapor-cycle cooling system with an air-cycle system for heat and pressurization.

An aircraft Freon system is similar, in principle, to the automotive or home air conditioner. It uses similar components and operating principles and, in most cases, depends upon the electrical system for power.

Vapor-cycle systems make use of the scientific fact that a liquid can be vaporized at any temperature by changing the pressure acting upon it. Water, at sea-level barometric pressure of 14.7 pounds per square inch ambient (p.s.i.a.), will boil if its temperature is raised to 212°F. The same water, in a closed tank under a pressure of 90 p.s.i.a., will not boil at less than 320°F.

If the pressure is reduced to 0.95 p.s.i.a. with a vacuum pump, the water will boil at 100°F. If the pressure is reduced even more, the water will boil at a still lower temperature (i.e. At 0.12 p.s.i.a., water will boil at 40°F). Water can be made to boil at any temperature if the pressure corresponding to the desired boiling temperature can be maintained.

Refrigeration Cycle

The basic laws of thermodynamics state that heat will flow from a point of higher temperature to a point of lower temperature. If heat is to be made to flow in the opposite direction, some energy must be supplied. The method used to accomplish this in an air conditioner is based on the fact that, when a gas is compressed, its temperature is raised and, similarly, when a compressed gas is allowed to expand, its temperature is lowered.

To achieve the required reverse flow of heat, a gas is compressed to a pressure high enough so that its temperature is raised above that of the outside air. Heat will now flow from the higher-temperature gas to the lower-temperature surrounding air (heat sink), thus lowering the heat content of the gas. The gas is now allowed to expand to a lower pressure. This causes a drop in temperature that makes it cooler than the air in the space to be cooled (heat source).

Heat will now flow from the heat source to the gas, which is then compressed again, beginning a new cycle. The mechanical energy required to cause this apparent reverse flow of heat is supplied by a compressor. A typical refrigeration cycle is illustrated in Figure 2-5-1.

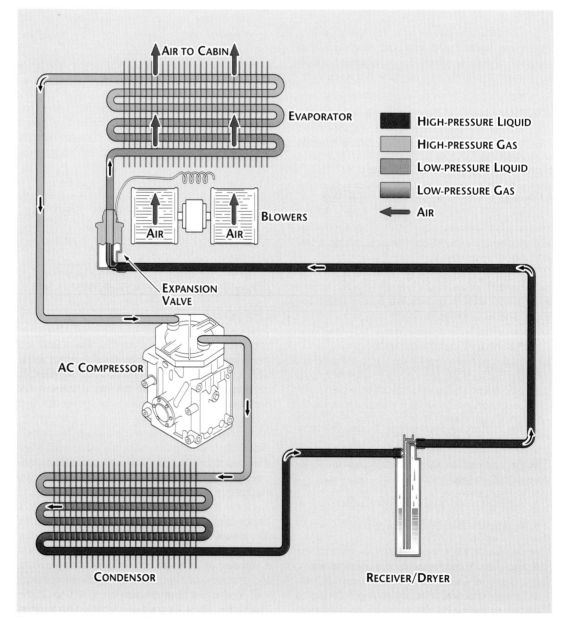

Figure 2-5-1. A typical refrigeration system in schematic layout

This refrigeration cycle is based on the principle that the boiling point of a liquid is raised when the pressure of the vapor around the liquid is raised. The cycle operates as follows: a liquid refrigerant confined in the receiver at a high pressure is allowed to flow through the *expansion valve* into the *evaporator*. The pressure in the evaporator is low enough so that the boiling point of the liquid refrigerant is below the temperature of the air to be cooled. Heat flows from the space to be cooled to the liquid refrigerant, causing it to boil (to be converted from liquid to a vapor). Cold vapor from the evaporator then enters the compressor, where its pressure is raised, thereby raising the boiling point. The refrigerant, at a high pressure and high temperature, flows into the condenser. Here, heat flows from the refrigerant to the outside air, condensing the vapor back into a liquid. The cycle is repeated to maintain the cooled space at the selected temperature.

Liquids that will boil at low temperatures are the most desirable for use as refrigerants. Comparatively large quantities of heat are absorbed when liquids are changed to a vapor. For this reason, liquid Freon is used in most vapor-cycle refrigeration units, whether in aircraft, automotive or home air conditioners.

Freon is a fluid that boils at a temperature of approximately 39°F under atmospheric pressure. Similar to other fluids, the boiling point may be raised to approximately 150°F under a pressure of 96 p.s.i.a. These pressures and temperatures are representative of one type of Freon. Actual values will vary slightly with different types of Freon. The type of Freon selected for a particular aircraft will depend upon the design of the Freon-system components installed.

Freon, similarly to other fluids, has the characteristic of absorbing heat when it changes from a liquid to a vapor. Conversely, the fluid releases heat when it changes from a vapor to a liquid. In the Freon cooling system, the change from liquid to vapor (evaporation, or boiling) takes place at a location where heat can be absorbed from the cabin air, and the change from vapor to liquid (condensation) takes place at a point where the released heat can be ejected to the outside of the aircraft. The pressure of the vapor is raised prior to the condensation process, so that the condensation temperature is relatively high. Therefore, the Freon — condensing at approximately 150°F — will lose heat to the outside air, which may be as hot as 100°F.

The quantity of heat that each pound of refrigerant liquid absorbs while flowing through the evaporator is known as the *refrigeration effect.* Each pound flowing through the evaporator is able to absorb only the heat needed to vaporize

it, if no superheating takes place. Superheating is defined as raising the temperature of a gas above that of the boiling point of its liquid state. If the liquid approaching the expansion valve were at exactly the temperature at which it was vaporizing in the evaporator, the quantity of heat that the refrigerant could absorb would be equal to its latent heat. That is, the amount of heat required to change the state of a liquid, at the boiling point, to a gas at the same temperature.

When liquid refrigerant is admitted to the evaporator, it is completely vaporized before it reaches the outlet. Since the liquid is vaporized at a low temperature, the vapor is still cold after the liquid has completely evaporated. As the cold vapor flows through the balance of the evaporator, it continues to absorb heat and becomes superheated.

The vapor absorbs *sensible heat* (heat which causes a temperature change when added to, or removed from, matter) in the evaporator as it becomes superheated. This, in effect, increases the refrigerating effect of each pound of refrigerant. This means that each pound of refrigerant absorbs not only the heat required to vaporize it, but also an additional amount of sensible heat, thereby superheating it.

Freon System Components

The major components of a typical Freon system are the *compressor, condenser, expansion valve* and *evaporator* (see Figure 2-5-1). Other minor items may include the *condenser fan, receiver* (Freon storage), *dryer, surge valve and temperature controls.* These items are interconnected by appropriate tubing to form a closed loop in which the Freon is circulated during operation.

Compressor. The principle of operation of the system can be explained by starting with the compressor. The compressor increases the pressure of the Freon when it is in vapor form. This elevated pressure raises the condensation temperature of the Freon and produces the force necessary to circulate the Freon through the system (Figure 2-5-2).

The compressor is normally driven by an electric motor drive mechanism. Most compressors are the piston-type. The compressor is designed to act on Freon in a gaseous state and, in conjunction with the expansion valve, maintains a difference in pressure between the evaporator and the condenser. If the liquid refrigerant were to enter the compressor, compressor damage would occur, as liquid is non-compressible. This type of malfunction is called *slugging.* Automatic controls and proper operating procedures must be used to prevent slugging.

Condenser. The Freon gas is pumped to the condenser for the next step in the cycle. At the condenser, the gas passes through a heat exchanger, where outside (ambient) air removes heat from the Freon (Figure 2-5-3). When heat is removed from the high-pressure Freon gas, a change of state takes place, and the Freon condenses to a liquid. It is this condensation process which releases the heat the Freon picks up from the cabin air. The flow of ambient air through the condenser unit is ordinarily modulated by controllable inlet or outlet doors, according to cooling requirements. A condenser-cooling air fan or air ejector is often used to help force the ambient air through the condenser; this item is important for operation of the system on the ground.

Receiver. From the condenser, the liquid Freon flows to the receiver, which acts as a reservoir

Figure 2-5-2. A Freon-system compressor

Figure 2-5-3. The condenser discharges the heat to ambient outside air.

for the liquid refrigerant. The fluid level in the receiver varies with system demands. During peak cooling periods, there will be less liquid than when the load is light. The prime function of the receiver is to ensure that the thermostatic expansion valve is not starved for refrigerant under heavy cooling load conditions.

The receiver is frequently a receiver/filter/dryer unit with an included *sight glass* (Figure 2-5-4). The filter/dryer is essentially a sheet-metal housing with inlet and outlet connections, containing alumina desiccant, a filter screen and a filter pad. The alumina desiccant acts as a moisture-absorber so that dry Freon flows to the expansion valve. A conical screen and fiberglass pad act as a filtering device, removing contaminants.

Scrupulously clean refrigerant at the expansion valve is a must because of the critical clearances involved. Moisture may freeze at the expansion valve, causing it to hang up, with a resulting starvation or flooding of the evaporator.

In some systems a sub-cooler is used to reduce the temperature of the liquid refrigerant after it leaves the receiver, the purpose is to prevent premature vaporization of the liquid.

Sight glass. To aid in determining whether servicing of the refrigerating unit is required, a liquid-line sight glass or liquid level gauge is installed in the receiver/dryer, or in a line between the filter/dryer and the thermostatic expansion valve. The sight glass consists of a fitting with a window, permitting a view of fluid passage through the line. In most systems, the sight glass is constructed as a part of the filter/dryer.

During refrigeration unit operation, a steady flow of Freon refrigerant observed through the sight glass indicates that sufficient charge is present. If the unit requires additional refrigerant, bubbles will be present in the sight glass.

Expansion valve. The liquid Freon flows to the expansion valve for the next step in the operation. The Freon coming out of the condenser is high-pressure liquid refrigerant. The expansion valve lowers the Freon pressure, thus lowering the temperature of the liquid Freon. The cooler liquid Freon makes it possible to cool cabin air passing through the evaporator.

The expansion valve, mounted near the evaporator, meters the flow of refrigerant into the evaporator. Efficient evaporator operation depends upon the precise metering of liquid refrigerant into the heat exchanger for evaporation. If heat loads on the evaporator were constant, an orifice size could be calculated and used to regulate the refrigerant supply. A practical system, however, encounters varying heat loads, therefore requiring a refrigerant throttling device to prevent starvation or flooding of the evaporator, which would affect the evaporator and system efficiency. This variable-orifice effect is accomplished by the thermostatic expansion valve, which senses evaporator conditions and meters refrigerant to satisfy them. By sensing the temperature and the pressure of the gas leaving the evaporator, the expansion valve precludes the possibility of flooding the evaporator and returning liquid refrigerant to the compressor.

The expansion valve (Figure 2-5-5) consists of a housing containing inlet and outlet ports. Refrigerant flow to the outlet ports is controlled by the positioning of a metering valve pin. Valve-pin positioning is controlled by the pressure created by the remote sensing bulb, the superheat spring setting and the evaporator-discharge pressure supplied through the external equalizer port.

The remote sensing bulb is a closed system filled with refrigerant, and the bulb is attached to the evaporator. Pressure within the bulb corresponds to the refrigerant pressure leaving the evaporator. This force is felt on top of the diaphragm in the power-head section of the valve, and any increase in pressure will cause the valve to move toward an *open* position. The bottom side of the diaphragm has the forces of the superheat spring and evaporator-discharge pressure acting in a direction to close the valve pin. The valve position, at any instant, is the result of these three forces.

If the temperature of the gas leaving the evaporator increases above the desired superheat valve, it will be sensed by the remote bulb. The pressure generated in the bulb is transmitted to the diaphragm in the power section of the valve, causing the valve pin to open. A decrease in the temperature of the gas leaving the evaporator will cause the pressure in the remote bulb to decrease, and the valve pin will move toward the closed position.

The superheat spring is designed to control the amount of superheat in the gas leaving the evaporator. A vapor is said to be superheated when its temperature is higher than that necessary to change it from a liquid to a gas at a certain pressure. This ensures that the Freon returning to the compressor is in the gaseous state.

The equalizer port is provided to compensate for the effect the inherent evaporator pressure drop has on the superheat setting. The equalizer senses evaporator-discharge pressure and reflects it back to the power-head diaphragm, adjusting the expansion-valve pin position to hold the desired superheat value.

Figure 2-5-4. Receiver/dryer/filter/sight glass, all in one unit

Figure 2-5-5. Thermostatic expansion valve

Evaporator. The next unit in the line of cooling flow after the expansion valve is the evaporator, which is a heat exchanger, forming passages for cooling air flow and for Freon refrigerant — in essence, a small radiator. Air to be cooled flows through the evaporator (Figure 2-5-6).

The Freon changes from a liquid to a vapor at the evaporator by, in effect, boiling, and the pressure of the Freon is controlled to the point where the boiling (evaporation) takes place at a temperature that is lower than the cabin air temperature. The pressure (saturated pressure) necessary to produce the correct boiling temperature must not be too low; otherwise, freezing of the moisture in the cabin air will block the air passages of the evaporator (normally called *freezing-up the evaporator*). As the Freon passes through the evaporator, it is entirely converted to the gaseous state.

Essential to obtaining the maximum cooling and preventing liquid Freon from reaching the compressor, the evaporator is designed so that heat is taken from the cabin air; therefore, the cabin air is cooled. All the other components in the Freon system are designed to support the evaporator, where the actual cooling is done.

After leaving the evaporator, the vaporized refrigerant flows to the compressor and is compressed. Heat is being withdrawn through the walls of the condenser and carried away by air circulating around the outside of the condenser. As the vapor condenses to a liquid, it gives up the heat which was absorbed when the liquid changed to a vapor in the evaporator. From the condenser, the liquid refrigerant flows back to the receiver, and the cycle is repeated.

Freon, CFCs and Pollution

Since Chlorofluorocarbons (CFCs) were developed in the 1930s, they have been widely used as air conditioner (A/C) and refrigerator coolants, aerosol can propellants, electronic parts cleaners and foam-blowing agents. CFC-12 (also called Freon or R-12) has been the most common type of coolant used in automobile A/C systems, and it accounted for 20 percent of all U.S. CFC consumption in 1992. CFCs have been useful in so many applications because they are non-flammable, non-toxic and extremely stable in the environment.

Unfortunately, as stable CFC molecules rise into the stratosphere, they are split by the Sun's ultraviolet radiation. Chlorine molecules react with and split ozone molecules, depleting the ozone layer. The ozone layer, 10-30 miles above the Earth's surface, protects biological life from the sun's harmful UV rays. CFCs have also been identified as greenhouse gases, i.e. contributors to global warming.

Most automobile and aircraft systems manufactured prior to 1994 use Freon in their A/C systems. Since that time, several non-CFC refrigerants, such as hydrofluorocarbons (HFCs), have been developed. Since 1995, the most common substitute for CFCs in these A/C systems is HFC-134a (or R-134a). Unlike CFCs, HFCs contain no chlorine and do not harm the ozone layer, though they are considered greenhouse gases.

Hydrochlorofluorocarbons (HCFCs) are also used as CFC-12 substitutes in A/C systems. HCFCs do contain chlorine, but their ozone depletion capability is significantly less than CFCs. HCFC production is scheduled to be banned by the year 2030.

Regulations Governing Use of CFC-12, HCFCs and HFCs

Section 609 of the 1990 Clean Air Act Amendments implemented regulatory requirements for personnel and facilities servicing automobile or aviation A/C units.

CFC-12 and HCFC regulations. The following are some examples of rules limiting the use and disposal of CFC-12 and HCFC.

- It is illegal to vent CFCs or HCFCs to the atmosphere. Penalties of up to $25,000 per day, per violation can be levied, and prison terms can be given to anyone who knowingly vents CFC-12 or HCFCs into the atmosphere. The regulations require that CFC-12 and HCFCs be recycled. However, it is not illegal to use in-stock, recycled or remanufactured stocks of these chemicals.

- All facilities servicing motor vehicle or aircraft A/C systems must certify to EPA that they have acquired and are properly using approved CFC- or HCFC-recycling equipment.

- Technicians who service motor vehicle A/C systems must be certified.

- Sales of refrigerant are restricted to certified technicians.

- The EPA requires that facilities with refrigerant-recycling equipment keep records of the name and address of the facility to which any refrigerant is sent for reclamation. These records must be kept for three years. The facility must also have records showing that all persons authorized to operate any recycling equipment are currently certified.

Figure 2-5-6. Either a fan or ram air blows through the evaporator to cool the cabin

- It is illegal to vent HCFCs into the atmosphere. These chemicals must be collected during servicing.

Technician certification. To become certified, all technicians servicing automobile or aircraft A/C systems must complete an EPA-approved refrigerant-recycling course.

The EPA's list of approved technician-certification programs can be obtained from the Stratospheric Ozone Protection Hotline (SOPH). The certification usually involves studying a booklet received from a certification center, then taking a test by mail or at a central location.

Approved Collection and Recycling Equipment

At the present, the EPA's collection and recycling requirements can be met only with equipment carrying Underwriters Laboratories, Inc., (UL) certification or with equipment at least as stringent as the Society for Automotive Engineers (SAE) J-series standards. Before purchasing new equipment, make sure it is approved. Figure 2-5-7 is an example of a combined service and reclamation unit for both R-12 and R-134a.

Pollution Prevention Tips

All Montreal Protocol participant countries ceased Freon production by January 1996. As supplies of stockpiled and recycled Freon decrease, the cost is increasing markedly. Proper collection and efficient recycling are incentives to prevent losses of Freon to the atmosphere.

The following general practices are for certified technicians. They are also things an A&P technician should know.

Leaks. Detect refrigerant leaks with a simple visual inspection of the hoses, connections and condenser. Visible oil leaks usually indicate a leak in the system. A leaking A/C system can also be detected by draining the refrigerant from the system and then monitoring its ability to hold a vacuum: pressure loss indicates a leak. Electronic sniffers can also be used for leak detection. Fluorescent leak-detection systems are available, but many compressor manufacturers advise against their use, due to the abrasive nature of the dye particles. Use of flame-detection systems is strongly discouraged, as lethal phosgene gas (mustard gas), can be produced when CFC-12 comes in contact with an open flame. Also, avoid using leak-detection products containing Freon.

Figure 2-5-7. Service and reclamation of both refrigerants in one unit

Maintenance and repair. Evacuate all refrigerant prior to A/C maintenance or repair. Make it a policy to encourage customers to have leaking A/C systems repaired, rather than topped off with refrigerant. However, leak repair is not required under federal law.

Manifold hoses. To prevent leakage, manifold hoses must have shut-off valves within 12 inches of the ends of each line.

Cross-contamination. Avoid cross-contamination of refrigerants. Cross-contamination often occurs when the A/C system has been partially charged with a refrigerant other than the type designated on the system information label. Cross-contaminated A/C systems will suffer reduced performance, damage from chemical breakdown and lubrication problems. Additionally, if a cross-contaminated system is connected to recovery/recycling equipment, it can foul components such as filters and dryers, which will then have to be replaced. Furthermore, the cross-contaminated refrigerant can be passed from contaminated recovery/recycling equipment to other systems. Be wary of complex A/C service histories and makeshift or damaged fittings that may indicate cross-contamination. Another consideration is that all cross-contaminated refrigerant must be sent to a refrigerant-recycling facility for separation and purification, which will increase operating costs.

Retrofitting. When retrofitting an older A/C system, it is critical to have a certified A/C technician evacuate the old refrigerant. As an A&P, you may not remove or replace an item that is in the charged system. To do so would release the gas into the atmosphere. Once vacuumed, you may remove and replace any item in the system. Once installed, you will have to call the certified A/C technician back to recharge the system and leak-check it for you.

If any large amount of work must be done on a system, the customer may be a candidate for conversion to a non-CFC system. For conversion to an R-134a system, there are STC'd kits that are available in the aftermarket.

Air Conditioning System Maintenance

Inspections. At each regularly scheduled inspection, or any time the cowling is opened up to the point you can see the equipment, visually inspect the system. The principal items to look for are:

- Any signs of oil leakage from the compressor or hoses

- Any chafing of hoses

- Condition of the drive belt or condition of the electrical wiring or hydraulic hoses, if motor-driven

- Security of attachment

- Anything unusual

Servicing. In contrast to a hydraulic system, where the lines contain fluid at all times, a Freon loop contains both liquid and vapor at the same time. This, in addition to the fact that it is difficult to deduce exactly where in the system the liquid will be at any one instant, makes it difficult to check the quantity of Freon in the system.

Regardless of the amount of Freon in the complete system, the liquid level can vary significantly, depending on the operating conditions.

To check the Freon level, it is necessary to operate the refrigeration unit for approximately five minutes to reach a stable condition. If the system uses a sight glass, observe the flow of Freon through the sight glass. A steady flow indicates that a sufficient charge is present. If the Freon charge is low, bubbles will appear in the sight glass.

When a service technician adds Freon to a system, as much oil should be added as was lost with the Freon being replaced. It is impossible to determine accurately the amount of oil left in a Freon system after partial or complete loss of the Freon charge. However, based on experience, most manufacturers have established procedures for adding oil. The amount of oil to be added is governed by the following determinants:

Figure 2-5-8. Hoses used for R-12/R-22 have different ends than hoses used for R-134a.

1. The amount of Freon to be added

2. Whether the system has lost all of its charge and has been purged/evacuated

3. Whether a topping charge is to be added

4. Whether major components of the system have been changed

Usually, $^1/_4$ ounce of oil is added for each pound of Freon added to the system. When changing a component, an additional amount of oil is added to replace that which is trapped in the replaced component. The oil is a special grade of refrigeration oil.

Oil for lubrication of the compressor, expansion valve and associated seals must be sealed in the system.

Manifold set. Whenever a Freon system is opened for maintenance, the Freon and oil must be removed for recycling. Replenishment of the Freon and oil is a must, and this requires the use of a special set of gauges and inter-connected hoses. A separate manifold set should be maintained for R-12/R-22 and R-134a. As a matter of fact, the systems are set up so that hoses that work on one won't fit the other (Figure 2-5-8). The R-12/R-22 hoses have smaller fittings than the R-134a hoses. The oil vacuumed from, or installed in, each system is not compatible with the other. Do not allow them to become mixed. Should they become mixed, the Freon vacuumed from the system must be recycled as contaminated and not mixed with other recycled material.

The low-pressure gauge is a compound gauge, meaning it will read pressures on either side of atmospheric. It will indicate from about 30 inches of mercury gauge pressure (below atmospheric) to about 60 p.s.i. gauge pressure above atmospheric.

The high-pressure gauges usually have a range from zero up to about 600 p.s.i. gauge pressure.

The low-pressure gauge is connected on the manifold directly to the low-side fitting. The high-pressure gauge, likewise, connects directly to the high-side fitting. The center fitting of the manifold can be isolated from either of the gauges or the high- and low-service fittings by the hand valves. When these valves are turned fully clockwise, the center fitting is isolated. If the low-pressure valve is opened (turned counter-clockwise), the center fitting is opened to the low-pressure gauge and the low-side service line. The same is true for the high-side service line when the high-pressure valve is opened.

Special hoses are attached to the fittings of the manifold valve for servicing the system. The high-pressure charging hose attaches to the service valve in the high side, either at the compressor discharge, the receiver dryer or on the inlet side of the expansion valve.

The low-pressure hose attaches to the service valve at the compressor inlet, or at the discharge side of the expansion valve. The center hose attaches to the vacuum pump for evacuating the system, or to the refrigerant supply for charging the system. Charging hoses used with Schrader valves must have a pin to depress the valve.

When not using the manifold set, be sure the hoses are capped to prevent moisture contaminating the valves.

Purging the system. After the system has been opened for maintenance or parts replacement, it must be purged. Purging is a process for an A/C-certified technician. We can no longer simply bleed out a bit of Freon through the recharging manifold.

Evacuating the system. If even a small amount of water is present in the system may completely block the lines. If water is present in the system, it will freeze when the system starts to operate. This frozen moisture will clog the expansion valve, causing the system to malfunction. Anytime the system is opened, it must be evacuated to remove the moisture from the system.

Evacuation of the system is accomplished by connecting the center hose of the manifold set of a vacuum pump. The vacuum applied to the system will cause the moisture to vaporize and be drawn out of the system.

A typical pump used for evacuating air-conditioning systems will pump 0.8 cubic ft. of air per minute and will evacuate the system to about 29.62 inches of mercury (gauge pressure). At this pressure, water will boil at 45°F. Pumping down, or evacuating, a system usually requires about 60 minutes' pumping time.

Recharging

Purging hoses. Opening the high-pressure valve will permit Freon to flow into the system. The low-pressure gauge should begin to indicate that the system is coming out of the vacuum.

1. Close both valves.

2. Start the engine, setting the r.p.m. at about 1,250.

3. Set the controls for full cooling.

4. With the Freon container upright to allow vapor to escape, open the low-pressure valve to allow vapors to enter the system.

CAUTION: *If the can is turned upside down in order to speed up the process, there is a danger of having liquid refrigerant enter the compressor. Liquid refrigerant will damage the compressor to the point it will need to be replaced.*

5. As many pounds of refrigerant should be put into the system as called for by the specifications.

6. Close all valves, remove manifold set and perform an operational check.

Section 6

Oxygen Systems

Types of oxygen. Aviators' breathing oxygen is supplied in two types: Type I and Type II. Type I is gaseous oxygen, and Type II is liquid oxygen (LOX). Aviators' breathing oxygen is 99.5 percent pure. The water vapor content must not be more than 0.02 milligrams per liter when tested at 21.1°C (70°F) and at sea-level pressure. Gaseous oxygen is the choice for civilian aircraft. LOX is used for military aircraft and will not be covered here.

Oxygen used for welding and other commercial purposes is called *technical oxygen*. The moisture content of technical oxygen is not as rigidly controlled as is breathing oxygen; therefore, the technical grade should never be used in aircraft oxygen systems.

The extremely low moisture content required of breathing oxygen is set not to avoid physical injury to the body, but to ensure proper operation of the oxygen system. Air containing a high percentage of moisture can be breathed indefinitely without any serious ill effects. The moisture affects the aircraft oxygen system in the small orifices and passages in the regulator. Freezing temperatures can clog the system with ice and prevent oxygen from reaching the user. Extreme precautions must be taken to safeguard against the hazards of water vapor in oxygen systems.

Characteristics of oxygen. Oxygen combines with most of the other elements. The combining of an element with oxygen is called *oxidation*. Combustion is simply rapid oxidation. In almost all oxidations, heat is given off. In combustion, the heat is given off so rapidly it does not have time to be carried away. The temperature rises extremely high, and a flame appears.

Oxygen does not burn, but it does support combustion. Therefore, combustible materials burn more readily and more vigorously in an oxygen-rich environment than in air.

In addition to existing as a gas and a liquid, oxygen can exist as a solid. Liquid oxygen is pale blue in color. It flows like water, and weighs 9.52 lbs. per gallon.

Gaseous oxygen systems. Gaseous oxygen systems are used primarily in multi-place aircraft, where the systems are used only periodically.

Handling oxygen. The pressure in gaseous oxygen-supply cylinders should not be allowed to fall below 50 p.s.i. If the pressure falls much below this value, moisture is likely to accumulate in the cylinder. See Figure 2-6-1.

All oxygen under pressure is potentially very dangerous if handled carelessly. In particular, oxygen and petroleum do not mix. The mixture can cause the petroleum to ignite. Personnel servicing or maintaining oxygen systems and components must be meticulously careful about preventing grease, oil, hydraulic fluid or similar hydrocarbons, as well as other contamination, from coming in contact with lines, hoses, fittings and equipment. This contact presents a fire and explosion hazard.

If, because of hydraulic leaks or some other malfunction, components of the oxygen system do become externally contaminated, they should be cleaned using only approved oxygen-system cleaning compounds. Some service instructions specify the use of a variety of cleaning compounds. The preferred compound is oxygen-system cleaning compound conforming to Military Specification MIL-C-8638.

The following safety precautions should be adhered to:

- Under no circumstances should a non-approved cleaning compound be used on any oxygen lines, fittings or components.

- When handling oxygen cylinders, the valve protection cap should always be in place. Before removing the cap and opening the valve, ensure that the cylinder is firmly supported. A broken valve may cause a pressurized cylinder to be propelled like a rocket.

- Do NOT use oxygen in systems intended for other gases, or as a substitute for compressed air.

- Cylinders being stored for use on gaseous oxygen servicing trailers or any other use must always be properly secured. Do not handle cylinders or any other oxygen equipment with greasy hands, gloves or other greasy materials. The storage area should be located so that oil or grease from other equipment cannot be accidentally splashed or spilled on the cylinders.

All gaseous oxygen systems include the following:

- Containers (cylinders) for storing the oxygen supply

- Tubing to route the oxygen from the main supply to the user(s)

- Various valves for directing the oxygen through the proper tubing

- A metering device (regulator) to control the flow of oxygen to the user

- A gauge(s) for indicating oxygen pressure

- A mask to direct the oxygen to each user's respiratory system

Figure 2-6-1. Most walk-around O_2 regulators have a mark to indicate the desired amount of gas to leave in the bottle.

Cylinders

Gaseous oxygen cylinders are high-pressure cylinders. They may be steel, steel wrapped in composite material (lightweight) or aluminum. They are designed to operate at pressures of 1,800-2,400 p.s.i. The most common types of cylinders in general aviation are DOT-3AA-1800, DOT-3HT-1850 or DOT-E-8162 (composite) cylinders.

The main advantage of the high-pressure cylinder is that it minimizes space used for storing gaseous oxygen. All high-pressure oxygen cylinders are painted green, in accordance with the established federal color codes.

Cylinders come equipped with a manually operated handwheel valve. Opening the handwheel-operated valve assembly releases the contents of the cylinder. Most high-pressure valves are called top-seated. When opening a high-pressure valve, open it slowly at first so the pressure has a lesser impact on the equipment. Then, open the valve completely, seating it at the top so gas will not leak around the shaft. When closing a valve, close it completely and it will bottom-seat to prevent leaks.

The valve is equipped with a fusible metal safety plug and a safety disc to release the contents of the cylinder if the pressure becomes excessive because of high temperatures. The safety plug is filled with a fusible metal, designed to melt at temperatures ranging from 208°-220°F (97.8°-104.5°C).

Because oxygen is a gas, it will expand when the temperature rises, increasing a cylinder's internal pressure. To make sure a cylinder is filled to the proper capacity, it is necessary to use a temperature chart to calculate the proper pressure. Naturally, the recharge pressure will be less in a very hot environment, lower in a cold one. Figure 2-6-2 shows a temperature chart for servicing a system.

The cylinder and valve assembly is connected to the oxygen tubing by silver soldering the tubing to a coupling and securing the valve outlet with a coupling nut.

Leak-testing Gaseous Oxygen Systems

This test is performed at different times, depending on the inspection requirements for the particular type of aircraft. The system is allowed to cool, usually one hour after filling, before the pressures and temperatures are recorded. After several hours have elapsed, they are recorded again. Some manufacturers recommend a six-hour wait, others a 24 hour wait. The recorded pressures are then corrected for any change in temperature since filling.

When oxygen is being lost from a system through leakage, the gauge reading will be less than that shown on the pressure/temperature-correction chart. Leakage can often be detected by listening for the distinct hissing sound of escaping gas. If the leak cannot be located audibly, it will be necessary to use a specially compounded leak-test liquid. To make this check, apply the test solution to areas suspected of leakage. Watch for bubbles. Make the solution thick enough to adhere to the contours of the fittings.

Any leak, no matter how small, must be found and repaired. With a small leak that continues over a period of time, the surroundings and atmosphere become saturated. Such conditions are especially dangerous, because personnel may not be aware that oxygen enrichment exists. Oxygen-enriched conditions are almost always present in poorly ventilated areas.

No attempt should be made to tighten a leaking fitting while the system is charged.

Draining an Oxygen System

Oxygen bottles should not be emptied below 50 p.s.i. or left open to the atmosphere. Moisture may enter the bottle and create corrosion. When it is necessary to drain the system, it can be done by inserting a filler adapter into the filler valve and opening the shut-off valves. Do not drain the system too rapidly, as this will cause condensation within the system. An alternate method of draining the system is to open the emergency valve on the demand-oxygen regulator. Perform this job in a well-ventilated area and observe all fire precautions.

Cleaning an Oxygen System

Always keep the external surfaces of the components of the oxygen system clean and free of corrosion and contamination with oil or grease. As a cleaning agent, use anhydrous (waterless) ethyl alcohol, isopropyl alcohol (anti-icing fluid) or any other approved cleaner. If mask-to-regulator hoses are contaminated with oil or grease, the hoses should be replaced.

Cylinder Testing

All pressure cylinders are subject to periodic testing by a Department of Transportation-approved testing station. Naturally, this includes oxygen cylinders. The actual DOT requirements are found in CFR 49, part 173.34.

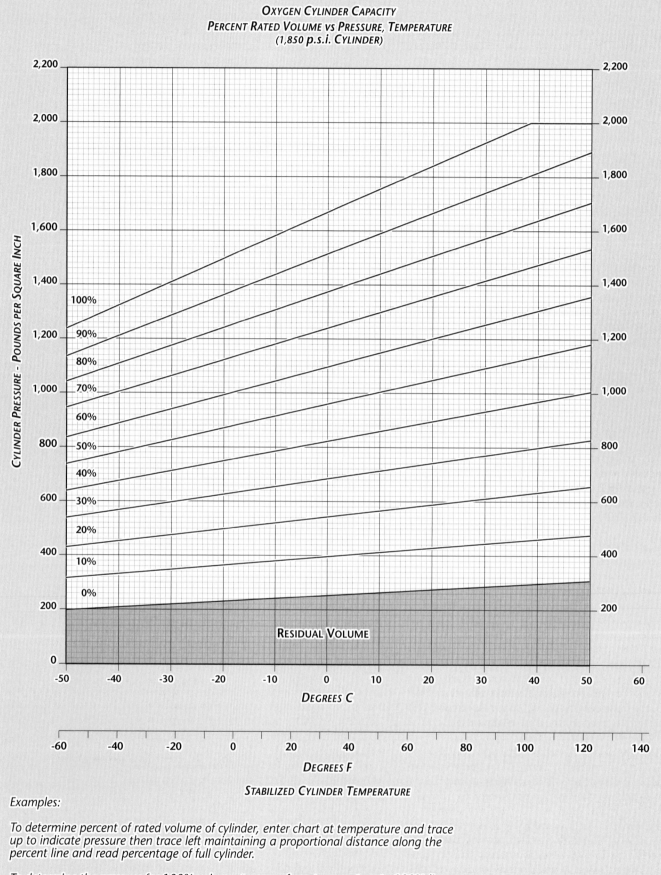

OXYGEN CYLINDER CAPACITY
PERCENT RATED VOLUME VS PRESSURE, TEMPERATURE
(1,850 p.s.i. CYLINDER)

Examples:

To determine percent of rated volume of cylinder, enter chart at temperature and trace up to indicate pressure then trace left maintaining a proportional distance along the percent line and read percentage of full cylinder.

To determine the pressure for 100% volume, trace up from temperature to 100% line and trace across to cylinder pressure.

Figure 2-6-2. Oxygen-servicing temperature chart

The basic testing requirements and mandatory retirement times are as follows:

- Cylinders must be hydrostatically tested to five-thirds of their working pressure.

- DOT-3HT cylinders must be tested every three years and retired after 4,380 fillings, or 24 years.

- DOT-3AA cylinders must be tested to 5/3 of their working pressure every five years

- DOT-E-8162 cylinders must be tested every three years and retired after 10,000 fillings, or 15 years.

After manufacture, all oxygen cylinders are marked by stamping on the cylinder (see Figure 2-6-3). All cylinders have at least the following markings stamped into them:

1. Department of Transportation specification

2. Cylinder serial number

3. Registered owner

4. Date of manufacture

5. Name of the current owner

6. Retest markings

The retest markings should show the month, facility, year and Plus-rating star stamp (see Figure 2-6-3). The + symbol (Plus-rating) indicates that the cylinder qualifies for a 10 percent overfill rating, while the star symbol (star stamp) indicates that the cylinder meets the requirements for the 10-year retest.

Regulators

The success or failure of high-altitude flight depends primarily on the proper functioning of the *oxygen breathing regulator*. Acting as a metering device, the regulator is the heart of the oxygen system. To perform successfully in an aircraft system, a regulator must deliver the life-supporting oxygen in the quantities demanded throughout its entire range of operation.

Continuous-flow regulator. A continuous-flow regulator is just that: Oxygen flows continuously into the mask. Most walk-around bottles and oxygen supplied by an activated oxygen candle are continuous-flow types. Because of the continuous flow, inhaling is not a problem, but exhaling means the oxygen in the mask gets exhaled also. While not very efficient in the use of oxygen, they are uncomplicated and work well for short-duration requirements.

Most smaller constant-flow bottles don't use a cylinder gauge. The regulators are preset, and

the flow rate is non-adjustable. Constant-flow systems for smaller general aviation airplanes generally have both a flow-rate and a cylinder-pressure gauge, and the flow-rate is usually adjustable. These types are more efficient than the basic bottle-and-mask system, because of the adjustability. Some will have a low-pressure regulator that adjusts itself as the altitude changes, making them somewhat more efficient.

Diluter-demand system. The diluter-demand system is more complicated, but it is much more efficient in its use of oxygen. A diluter-demand regulator works just as its name implies: out-flowing oxygen stops at the regulator until the wearer takes a breath. When you inhale, the regulator opens, and oxygen, mixed with ambient air, enters the mask. Upon exhalation, the regulator closes and stops the flow of oxygen. This makes for a much more efficient use of oxygen, because you are not exhaling oxygen from the supply system. By virtue of aneroid-controlled supply valves, diluter-demand regulators also are self-adjusting for altitude. Figure 2-6-4 shows a diluter-demand regulator.

Because oxygen is odorless, colorless and tasteless, these regulators need a method of indicating when they are working properly. The indicator, called an eyeball or a blinker, is part of the regulator and is visible to the crew. It consists of a small indicator that blinks each time the person wearing the mask inhales and the regulator causes oxygen to flow. The procedure is almost foolproof: no blink means no flow.

Naturally, a diluter-demand system is much more complicated, and component maintenance is beyond the realm of an A&P techni-

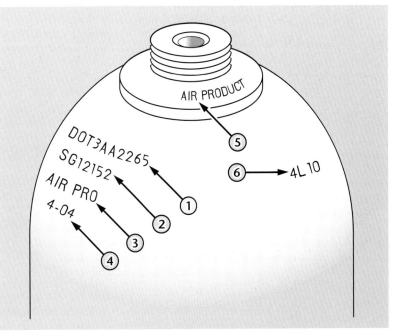

Figure 2-6-3. Cylinder stampings, showing the cylinder's history

Figure 2-6-4. A diluter-demand regulator. The blinker (left) winks when a breath is drawn.

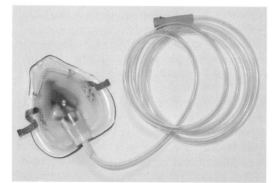

Figure 2-6-5. A simple constant-flow oxygen mask

cian. However, A&Ps are responsible for system maintenance. Each MMM or SRM will have a section devoted to oxygen-system maintenance.

Although A&P technicians are not primarily responsible for maintenance of regulators, they are responsible for performing operational checks in the aircraft, as well as for removal and installation.

Pressure-demand regulator. The third type of regulator is a pressure-demand type. It works similarly to the diluter-demand, except that it is based on pressure rather than inhalation. Special crew training is required for use of these types of systems. Pressure-demand systems are principally for military and research aircraft and are not likely to be found in general aviation or commercial aircraft.

Oxygen Masks

Masks vary from the extremely simple constant-flow units that fit over the mouth and nose (Figure 2-6-5) to complex, full-face masks that are certified for smoke and fume cabin emergencies. Within this mix are the plastic

drop-down masks used with oxygen-candle emergency systems and the quick-donning diluter-demand masks that must be available for cockpit crew (Figure 2-6-6).

Tubing

Two types of tubing are used in aircraft oxygen systems: low-pressure aluminum alloy tubing is used in lines carrying pressures up to 450 p.s.i., and high-pressure stainless steel tubing is used in lines carrying pressure above 450 p.s.i. Some systems are still around that use copper tubing, which was the original material of choice.

All stainless and copper tubing joints are silver-soldered.

> **NOTE:** *Some of the newer aircraft are equipped with high-pressure oxygen lines made of high-strength aluminum alloy.*

Lines running from the filler valve to each of the cylinders are called *filler lines*. Those running from the cylinders to the regulators are called *distribution lines* or *supply lines*.

Oxygen lines, like all other lines in the aircraft, are identified by strips of colored tape. The color code for oxygen lines is green and white with the words Breathing Oxygen printed in the green portion, while black outlines of rectangles appear in the white portion.

Resistance to fatigue failure is an important factor in oxygen line design, because the line pressure in a high-pressure system will, at times, exceed 1,800 p.s.i. and at other times be as low as 300 p.s.i. Because of these varying pressures and temperatures, expansion and contraction occur all the time. These fluctuations cause *metal fatigue,* which must be guarded against in both the design and the construction specifications for tubing. Steps are taken during installation to prevent fatigue failure of the tubing. Tubing is bent in smooth coils wherever it is connected to an inflexible object, like a cylinder or a regulator. Every precaution is taken to prevent the accidental discharge of compressed oxygen because of faulty tubing or installation. Although simple in construction and purpose, tubing is the primary means by which oxygen is routed from the cylinders to the regulator stations.

High-pressure tubing is usually stainless steel or seamless copper tubing and is manufactured in accordance with strict specifications. It has an outside diameter of 3/16 inch and a wall thickness of 0.035 inch. For application in high-pressure oxygen installations, copper tubing is type-N (soft-annealed), and is pressure-tested at not less than 3,000 p.s.i.

High-pressure tubing is used between the oxygen cylinder valve and the filler connection in all systems, between the cylinder valve and the regulator inlet in high-pressure systems and between the cylinder valve and pressure reducer in reduced high-pressure systems.

To connect high-pressure copper tubing, adapters and fittings are silver-soldered to the tubing ends. Due to the high pressures involved, the security (leak-tightness) of all high-pressure lines relies primarily on a metal-to-metal contact of all its fittings and connections. A fitting properly silver-soldered to the end of a length of copper tubing will not come loose or leak. Proper cleaning of all lines that are either repaired or contaminated is a must. Any contamination must not be allowed in the system, because it could cause serious problems for the people using it. Manufacturers have mandatory guidelines for cleaning and purging the system, and they must be followed.

Some of the later models of airplanes use aluminum alloy or stainless steel tubing in high-pressure oxygen system installations. Replacement tubing should be manufactured of the same type material as the original tubing or a suitable substitute as specified in the MMM.

Thread sealer. Teflon tape is normally used as a thread sealer. Never use Permate or Zinc Chromate compounds as they contain a petroleum product.

Valves

Various types of valves are installed in gaseous oxygen systems. Among the most commonly used are check valves, pressure reducing valves and filler valves.

Check valves. Check valves are installed at various points in the oxygen system. Their purpose is to permit the flow of oxygen in one direction only. Check valves are located in the system to prevent the loss of the entire oxygen supply in the event a cylinder or line is ruptured.

Oxygen system check valves. Various styles of single-, dual- and triple-check valves are available. The arrow (or arrows) embossed on the valve casting indicates the direction of flow through the valve.

Pressure-reducing valves (or pressure reducers) are used in certain oxygen systems for the purpose of reducing high cylinder pressure to a working low pressure. In most installations, the pressure reducers are designed to reduce the pressure from 1,800 p.s.i to a working pres-

sure of 60-70 p.s.i. They are always located in the oxygen-distribution lines between the cylinders and the flight station outlet.

Filler valves. All oxygen systems are designed so the entire system can be serviced (refilled) through a common filler valve. The filler valve is generally located so that it may be reached by a man standing on the ground or wing (Figure 2-6-7). The filler valve contains a check-valve, which opens during the filling operation and closes when filling is completed. A dust cap keeps out dust, dirt, grease and moisture.

Gauges. Gauges are used in gaseous oxygen systems to indicate the oxygen pressure in pounds per square inch (p.s.i.). All systems are equipped with at least one gauge that indicates the amount of oxygen in the cylinder(s). The gauge also indicates, indirectly, how much longer the oxygen will last.

The volume of any gas compressed in a cylinder is directly proportional to the pressure. If the pressure is half, the volume is half, etc. Therefore, if 900 p.s.i of oxygen remains in a 1,800 p.s.i system, half the oxygen is left.

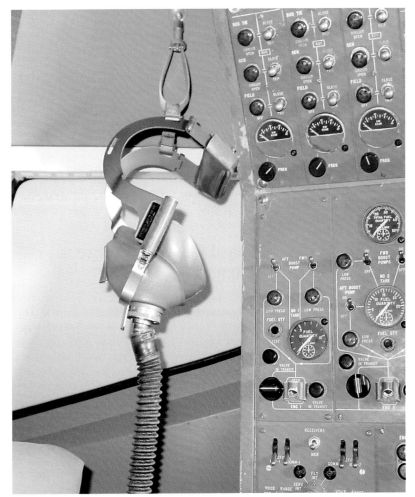

Figure 2-6-6. A quick-donning oxygen mask used on the flight deck

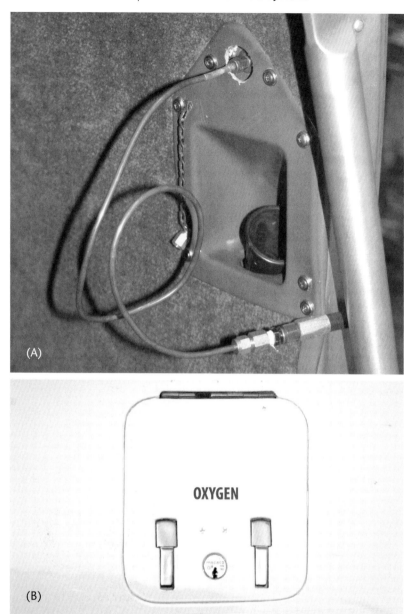

Figure 2-6-7. The filler valve for some King Air oxygen systems is located (A) just inside the door opening, while other systems have a special access panel on (B) the outside of the fuselage.

Typical Gaseous Oxygen Systems

System operation. Figure 2-6-8 is an operational schematic of a Beech King Air. The pressure manifold, which is equipped with internal check-valves, receives oxygen flow from the cylinders, directs the flow into a common line and routes it to the pressure reducer. The manifold assembly also connects to a filler line, allowing the three cylinders to be recharged simultaneously from an external supply. The pressure reducer decreases the pressure to 65 p.s.i.

Incorporated on the low-pressure side of the pressure reducer is a relief valve, which connects through tubing to an overboard dis-

charge indicator. In the event of excessive pressure developing within the low-pressure section of the pressure reducer, the excess pressure will flow through the relief valve and out the overboard discharge line. This flow will rupture the green disc in the discharge indicator, giving a visual indication of a malfunctioning pressure reducer.

A line from the high-pressure side of the pressure reducer connects to a gauge in the cockpit. This gauge can give the pilot an indication of pressure in the three storage cylinders.

The oxygen flows from the low-pressure side of the reducer to the three regulators. A flexible hose attached to each regulator is for attachment of the oxygen mask.

Portable Oxygen Systems

Portable oxygen systems include walk-around cylinders, emergency oxygen and survival kits. These systems are used primarily to maintain crew functions in the event of failure of the fixed oxygen systems. All are small, lightweight, high-pressure, self-contained gaseous systems, which are readily removed from the aircraft.

Walk-around cylinders. Walk-around cylinders are standard equipment on many transport aircraft and are used separately or in addition to a permanently installed oxygen system. Each system consists of a reducer and regulator assembly mounted directly on a small oxygen cylinder.

Figure 2-6-9 illustrates a high-pressure walk-around oxygen system. A short flexible breathing tube, clamped to the outlet of the regulator at one end and fitted with a connector at the other end, provides the necessary assembly for the attachment of the demand mask tube.

System Maintenance

The maintenance procedures discussed in this section are general in nature. Consult the applicable service manual prior to performing any maintenance. Routine maintenance includes servicing of cylinders, checking the system and regulators for leaks, operationally checking the system and troubleshooting malfunctions.

Malfunctions may become apparent during inspections, testing or actual use of the oxygen system. The remedies for some malfunctions will be quite obvious, while others may require extensive time and effort to pinpoint the actual cause. The effectiveness of corrective action

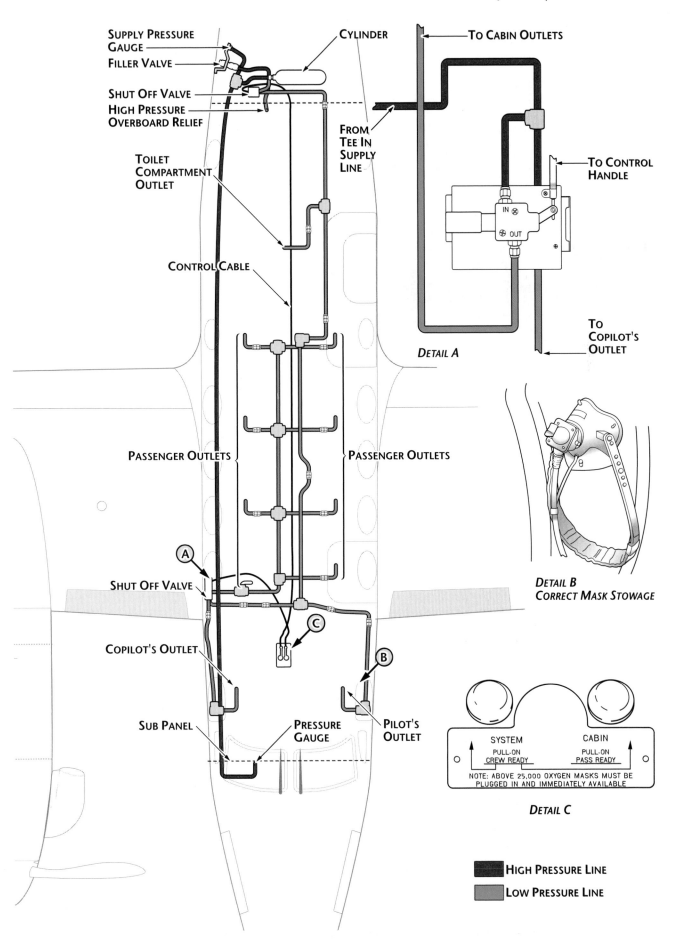

SUPPLY PRESSURE GAUGE
FILLER VALVE
SHUT OFF VALVE
HIGH PRESSURE OVERBOARD RELIEF
TOILET COMPARTMENT OUTLET
CONTROL CABLE
CYLINDER
FROM TEE IN SUPPLY LINE
TO CABIN OUTLETS
TO CONTROL HANDLE
IN
OUT
TO COPILOT'S OUTLET
DETAIL A
PASSENGER OUTLETS
PASSENGER OUTLETS
DETAIL B
CORRECT MASK STOWAGE
A
SHUT OFF VALVE
C
COPILOT'S OUTLET
B
SUB PANEL
PRESSURE GAUGE
PILOT'S OUTLET

SYSTEM
PULL-ON
CREW READY
CABIN
PULL-ON
PASS READY
NOTE: ABOVE 25,000 OXYGEN MASKS MUST BE PLUGGED IN AND IMMEDIATELY AVAILABLE

DETAIL C

HIGH PRESSURE LINE
LOW PRESSURE LINE

Figure 2-6-8. Reduced high-pressure oxygen system schematic

will be dependent on an accurate diagnosis of the malfunction.

Troubleshooting. Troubleshooting of the gaseous oxygen system, as with other systems, is the process of locating a malfunctioning component or unit in a system. To troubleshoot intelligently, you must be familiar with the system and know the function of each component within the system. You can study the schematic diagrams of the system provided in the MMM to gain a mental picture of the location of each component in relation to other components. By learning to interpret these diagrams, you can save time in isolating malfunctioning components. The schematic diagram does not indicate the location of components in the aircraft; however, it will provide the means to trace the oxygen flow from the cylinder through each component to the mask.

Installation diagrams provided in either the MMM or the Illustrated Parts Breakdown (IPB) will assist you in locating the particular component in the aircraft.

The MMMs provide a variety of troubleshooting charts, which are intended to aid you in discovering the cause of malfunction and its remedy.

Oxygen system purging. When an oxygen system has been left open for servicing or repair, the system must be purged of any moisture or contaminates. The simplest method is to connect the system to a full oxygen bottle, connect several facemasks, then open the bottle and allow the oxygen to fill the system and drain it. This process should be repeated at least three times before returning the system to service. Dry nitrogen can also be used to purge the system in the same manner.

Replacing lines, fittings or components. When oxygen system lines or components are replaced, it is very important that all traces of oils or petroleum products be removed. Stabilized trichloroethylene or acetone can be used to clean tubing or fittings. After they are cleaned, they should be dried with heat or by blowing them with dry air or dry nitrogen.

> **CAUTION:** *Only thread lubricants and sealants that are specially formulated for oxygen systems should be used when assembling the parts of the system. Only oxygen compatible lubricants that meet specification MIL-G-27617 should be used, any other type of material can result in a hazardous condition.*

When the parts are dry, clean, new protective plugs and covers should be used to close all openings. Purge the components before returning the system to service.

Daily Inspections

These inspections are visual inspections followed by a functional test. Inspections and tests are performed in conjunction with the inspection requirements for the aircraft.

> **CAUTION:** *Once again, make certain that when working with oxygen, all clothing, tube fittings and equipment are free of oil, grease, fuel, hydraulic fluid or any combustible liquid. Fire and/or explosion may result when even slight traces of combustible material come in contact with oxygen under pressure.*

Testing diluter-demand regulators. To perform the functional test, proceed as follows:

1. Place supply valve control lever in the ON position.

2. Place the diluter control lever in the NORMAL OXYGEN position.

3. Connect the oxygen hose to the quick-disconnect, place the mask to the face and inhale. Proper regulator operation will be indicated by the flow-indicator assembly showing white during inhalation and black during exhalation.

   While at ground level, the regulator will not normally supply oxygen from the supply system to the mask. Therefore, the emergency pressure-control lever must be used to check the oxygen supply function of the regulator at low altitudes. The emergency lever is spring-loaded at the NORMAL position, and will return to NORMAL when released.

4. Hold the emergency pressure-control lever in the TEST MASK position and observe the flow indicator. Flow indica-

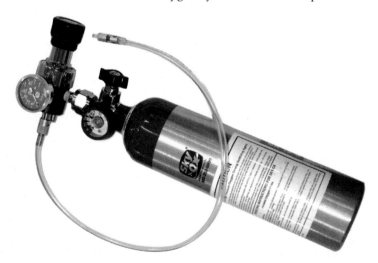

Figure 2-6-9. A portable walk-around oxygen bottle designed for cabin crew use. Walk-around systems use a simple mask and a constant-flow regulator.

tor should be white, indicating a flow through the regulator.

Upon completion of the functional test, secure the regulator as follows:

1. Disconnect the mask from the supply hose.

2. Ensure that the emergency pressure control lever returns to the NORMAL position.

3. Place the diluter-control lever in the 100-percent position.

4. Place the supply valve control lever in the OFF position.

Solid Oxygen Systems

Solid oxygen systems are designed to provide emergency oxygen to passengers and cabin crew on larger passenger aircraft. The flight crew uses gaseous oxygen and quick-donning full-face masks in an emergency. As a safety precaution on commercial flights, each time a flight officer leaves the flight deck for any reason, the remaining flight officer must don his mask.

Chemical oxygen. Chemical *oxygen generators,* commonly called *oxygen candles,* are actually composed of sodium chlorate ($NaClO_3$) and additional materials to support the combustion process. To produce oxygen, the sodium chlorate candle is ignited and starts to decompose rapidly (burn). The gas given off by the sodium chlorate decomposition, after filtering, is medically pure oxygen.

The amount of oxygen generated by the process is a design feature of the candle and cannot be regulated. Once started, it burns to the end of the chlorate supply. Generally, about 15 minutes worth of oxygen supply is produced.

Installation modules. Oxygen generators are installed in overhead passenger service panels on most airliners. Should emergency oxygen be needed because of a cabin pressurization failure, the masks will drop from a door on the service panel. They can be released by the pilot, or dropped automatically from the pressurization loss. Each module will service between three and five oxygen masks. Figure 2-6-10 shows a customer-service panel lowered to reveal the oxygen candle and oxygen masks.

Each mask will hang by its supply hose, which is *shanked* (shortened by folding and tied). Firmly yanking down on the hose straightens the supply hose and fires the squib on the oxygen generator. The squib (a pyrotechnic primer)

will ignite the chlorate candle, and oxygen will start flowing. Once the chlorate candle is burning, oxygen flow will continue until the chlorate is consumed. The outflow cannot be regulated.

A burning oxygen candle gets hot. Even though the unit has interior insulation, the outside of the canister can be 250°F or more. The installation is designed to accommodate the temperature without danger of fire. Oxygen candles are inert below 400°F and cannot be accidentally activated by that temperature. If they were accidentally activated during a cabin fire, their oxygen output is so consistent that it would not affect the fire catastrophically.

Servicing Gaseous Oxygen Systems

Gaseous oxygen systems are serviced using a special procedure that helps conserve oxygen. A fully charged 250 cubic foot (cf) capacity oxygen cylinder may have enough capacity to fill two aircraft systems before the cylinder pressure drops too low to bring the third refill up to pressure. You would need another full 250-cf cylinder to refill the third system. In a short time, there would be a large collection of cylinders that are unusable, yet are 75 percent full of oxygen.

Special oxygen service carts using the cascade system solve the problem. Each cart mounts six 250-cf oxygen bottles, all connected to a manifold. Because of prior usage, each bottle has a

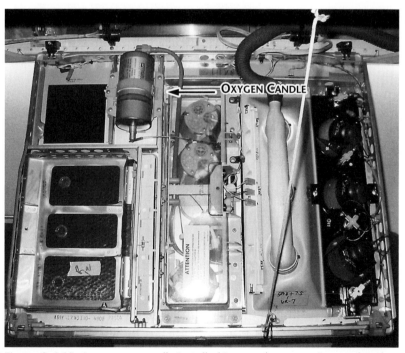

Figure 2-6-10. An oxygen candle installed in a customer-service overhead panel

Figure 2-6-11. A gaseous oxygen service cart that uses the cascade system

Figure 2-6-12. The business end of an oxygen service cart, showing the tags on each cylinder for recording the remaining cylinder pressure

different pressure, and each bottle has its pressure recorded on a logbook system. A gaseous oxygen service cart is shown in Figure 2-6-11.

To start the filling operation, the bottle with the lowest pressure is turned on just enough to purge the filler hose with oxygen and blow out any accumulation of debris. Once purged, the fill hose is connected to the system fill fitting.

Once connected, the cylinder with the lowest pressure is slowly opened and oxygen is allowed to transfer to the aircraft system until all flow stops. Then the cylinder valve is closed and the current bottle pressure is recorded, superseding the previous log entry. The next lowest-pressure bottle is now slowly opened and it, too, is allowed to flow until all flow stops. This stair-stepping, or cascading, system is repeated until the aircraft system reaches its proper pressure. By using this system, the higher-pressure supply cylinders are drawn down less. Their higher pressure is used to boost the system pressure, instead of supplying the majority of the volume.

Because all supply cylinders are connected to a manifold, there is no requirement that higher-pressure cylinders be in a specific location on the service cart. All you have to do as an A&P technician is to be sure and purge the line, fill the system to the correct pressure, and re-record the remaining cylinder pressures in the log for the next refill (Figure 2-6-12).

Oxygen Safety

Cylinders. Although not flammable, oxygen is considered as such in most safety programs. Being an oxidizer, oxygen can greatly accelerate any fire. Even fires that might be controllable by themselves will go out of control when an excess supply of oxygen is present. Do not attempt to fight an out-of-control fire by yourself. Call the fire department and tell them there is a fire that is oxygen-fueled. That will alert them to bring the correct equipment for putting it out. Observe all of the following:

• Never allow any type of oil or petroleum product anywhere near excess oxygen. The atmosphere produced by excess oxygen can self-ignite and cause an intense fire.

• Handle oxygen cylinders carefully, and open the valves slowly.

• If opened too quickly, it is possible for a regulator diaphragm to rupture, allowing high-pressure gas to escape through a damaged regulator.

- When replacing a storage cylinder, keep the cylinder protective cap screwed onto the cylinder until it is firmly secured.

- Before attaching a regulator to a full cylinder, briefly crack the cylinder valve to blow dust and debris from the fitting.

- When storing high-pressure cylinders, always chain or strap them to something stable, like the beam of a building.

- Never use oxygen to inflate anything.

Oxygen generators. Oxygen generators are normally safer to handle than oxygen cylinders. There are still a few basic rules that must be followed:

- Never remove the safety cap when handling oxygen candles until the maintenance instructions tell you to do so.

- Follow the aircraft maintenance instructions for replacing all solid oxygen candles.

- It is safer to store oxygen candles in their shipping container than to just put them on the shelf.

- There are specific regulations for shipping oxygen candles. They are considered a hazardous material.

- NEVER ship an oxygen candle in an aircraft cargo compartment. To do so can, and has, caused at least one violent airplane accident.

- During an inspection, always follow the established procedure. To vary from it could cause you to trip the trigger on the squib, starting the irreversible oxygen generation process to start.

CFRs for Oxygen

The federal regulations dealing with oxygen systems for both reciprocating and turbine aircraft are:

- 23.1441 — Oxygen equipment and supply

- 23.1443 — Minimum mass flow of supplemental oxygen

- 23.1445 — Oxygen distribution system

- 23.1447 — Equipment standards

- 23.1449 — Means of determining use of oxygen

- 23.1450 — Chemical oxygen generators

- 23.1451 — Protection of oxygen equipment

- 125.219 — Oxygen for medical use by passengers

- 135.91 — Oxygen for medical use by passengers

- 135.157 — Oxygen equipment requirements

3
Landing Gear Systems

This G-III is making a typical short runway landing at a general aviation airport. The main wheels have just touched the runway and left a little rubber smoke; the nosewheel is still in the air, and the thrust reversers are just coming unlocked. As soon as the nose-wheel touches down the brakes will come on and the anti-skid will start working. This airplane has every system described in this chapter and are all working at the same time.

Section 1

Aircraft Landing Gear

Landing gear for aircraft is quite varied. Depending upon the aircraft size, its design and the area of its operation, no two systems will look exactly the same.

Types of Landing Gear

For land operation, the landing gear may be placed in two general categories. These are fixed and retractable.

Fixed landing gear. Fixed landing gear is normally associated with light single engine aircraft. The main gear is supported either by the wings or the fuselage, depending upon the design. Where wings are used, their structure is built up to distribute the weight into the wing spars. On a fuselage mounted landing gear, the load normally is transmitted into a box structure that holds both landing gear. See Figure 3-1-1 for an example of fixed landing gear.

Learning Objectives:

REVIEW
- design and operation of landing gear types including multiple wheeled

DESCRIBE
- steps of landing gear inspection and maintenance process
- designs, sizes and construction of tires
- typical wheel assemblies and maintenance tasks
- aircraft anti-skid system purpose and operation
- characteristics of each type of brake, brake system and assembly

EXPLAIN
- how to identify, inspect and classify repairability of tire tubes
- tire maintenance preventative and repair procedures
- maintenance practices for brake systems, assemblies and components

Left. This G-III is making a typical short runway landing at a general aviation airport.

Figure 3-1-1. Fixed landing gear on (A) a high wing design, and (B) a low wing design

The design of the device used to absorb the landing load may further classify fixed landing gear. These may be a bungee cord type, flex type, or an air-oil shock strut.

The flex type may use a steel strut or gear leg, which will flex under loads. Some of these flexible gear legs have been built out of fiberglass. The air oil shock strut utilizes a tube filled with hydraulic oil and air pressure. The chapter on hydraulics explains the workings of an air-oil cylinder.

Most of the fixed landing geared aircraft make use of a nose wheel. Normally these are shock strut type gear. This gear is normally attached to the engine mount. Nose gear of this type is usually steerable with the rudder pedals. In some instances, the nose wheels are not steerable, but caster instead.

Conventional landing gear. The conventional landing gear layout was adopted very early in aviation. Two wheels mounted on struts up front carried most of the weight, while a tail skid in the rear had the advantage of creating drag during landings and takeoffs. The extra drag on the tail had the advantage of helping to keep the airplane pointed in the right direction. It also helped slow the airplane during landing, thus brakes were either not used or a minimal design was all that was necessary for control.

These designs also brought about the fabled *three point landing.* To three point an airplane

Figure 3-1-2. A typical example of conventional landing gear is this Cessna 180. They are also popular as float planes.

Figure 3-1-3. A Boeing 757 with gear and flaps down. Notice the tilt angle of the landing gear trucks.

the pilot had to bring about a stall while only inches above the runway. The airplane would then touchdown with the main gear and tail skid at the same time. As mentioned above, the drag from the tail skid helped maintain directional control.

As the need for all-weather airports became the norm, more runways became paved and tail skids lost their purpose. In their place, manufacturers simply installed a tailwheel. Figure 3-1-2 is a typical Cessna 180 tail wheel airplane. While it was a simple fix, on hard surface runways tailwheels offered no real advantages; indeed some disadvantages appeared. One of the major disadvantages was the *ground loop*. A ground loop is a loss of directional control brought about by a rapid swing of the tail during landing. With the CG being behind the main gear, the loss of control is very rapid. The common result is that the airplane tips over sideways to the point that a wing tip scrapes the ground. If it scraped hard enough a wing could suffer a broken spar.

As airplanes acquired engines that are more powerful and grew in weight, steerable tailwheels became necessary. Today's tailwheel is normally steerable by the rudder pedals through a cable system. With the loss of the tail skid drag, brakes became necessary for steering as well as stopping. With the continuing increase in weight and higher landing speeds, brake systems became more complicated.

Retractable landing gear. Retractable landing gear is normally found on high performance aircraft. This eliminates the drag created by the fixed gear and makes the airplane perform better. Virtually all retractable gears

are of the nose gear configuration and utilize air-oil shock struts.

To raise and lower the gear, hydraulic or electrical power is commonly used. Only very few types of light aircraft use electrical power. Heavier aircraft typically use hydraulic power because of the energy required to raise and lower the gear. On a large transport category airplane the weight of the landing gear and its supporting structure can approach one-third the weight of the entire airplane. Figure 3-1-3 shows a Boeing 757 on a final approach for landing.

Like the fixed gear, structural provisions are built into the fuselage or wings to support the loads imposed by the retracting gear. This often requires beams, box structures, or a buildup of the spars. The added structure necessary to attach the landing gear generally extends from one side of the fuselage to the other. The structure is called a *carry through*.

Figure 3-1-4. The movement of most retractable landing gear is upward and inward. This places the pivot point in the wing structure and provides a wider tread.

Figure 3-1-5. The nose wheel well is generally the main supporting structure for the nose of the aircraft.

In addition to the structure necessary to support the spar, a place must be provided to stow the gear in the retracted position. This area is known as the *wheel well*. On most aircraft, the wheel wells have doors that are sequenced with the gear. Most main gear movement is up and inward as shown in Figure 3-1-4. This main gear is attached to a beam that is fastened to the rear wing spar and the fuselage. When the gear is retracted, the wheel rests in the fuselage wheel well.

The nose gear is generally placed in a structural box that supports the gear and forms the nose gear wheel well. A view of this is shown in Figure 3-1-5. This gear retracts forward. The nose gear, because of its ability to steer, is required to have a centering device on the landing gear so the gear, when retracted, will fit in the space required.

Seaplanes and amphibians. Some aircraft operations require that landings be made on water. Some aircraft are amphibious, which means they have a landing gear and provisions for water landing. This is usually accomplished by the design of the hull and the use of retractable landing gear. The hull is the fuselage built to withstand the forces of landing in addition to the strains of flight. To stabilize the aircraft in the water, pontoons are sometimes placed towards the tip of the wing. Since the hull or fuselage must support the aircraft in the water, the landing gear often retracts into the fuselage above the water line as in Figure 3-1-6.

Normally the hull portion has compartments that are watertight. This prevents a leak from filling the hull; only a compartment gets flooded. On aircraft such as this, corrosion is of

Figure 3-1-6. A Grumman Widgeon amphibian seaplane

primary concern and special precautions must be taken with the aluminum structure.

Other aircraft used on the water are often land planes converted to water use. These aircraft require floats installed in place of the wheels.

The floats may be made of aluminum or fiberglass. The basic construction of the float is made up of stringers and bulkheads covered with skin. To aid in maneuvering while taxiing, water rudders are placed at the back of the float. To keep the pilot from over controlling the airplane, these are retractable during takeoff and landing.

Some floats are additionally equipped with wheels. The wheels are retracted during water landings and are extended when landing on the ground. This is accomplished by a hydraulic system. Float planes are sometimes able to land on grass without wheels. However, this procedure will depend upon the manufacturer and his recommendations. See Figure 3-1-7 for a typical set of floats that would be mounted on a fixed gear aircraft.

Skiplanes. For arctic or snow operations, skies are often required. Two basic types are most common. One kind is the wheel replacement type that requires the wheel and brake assembly to be removed and the ski is installed on the axle. This type of installation will require cables to limit the ski travel on the axle and to hold the ski in proper rig of a few degrees upward at the tip of the ski. Skies are either made of aluminum or fiberglass.

On some installations, it is desirable to have the ability of landing on snow or dry runway. For these installations, a retractable ski is used. These may be operated manually or hydraulically, depending on the design.

Helicopter landing gear. Helicopters use a variety of landing gear types. These include skid-type gear, fixed-wheel gear and retractable-wheel gear.

The most popular landing gear for light helicopters is the *skid-gear*. This type of gear supports the aircraft during landings by being attached to the main structure of the fuselage. Normally the cross tubes of the skid-gear absorb moderate landing loads with the spring action of the tube. In some instances, a hydraulic dampener is used to help cushion the load. Although these gears can absorb rather heavy compression loads, they are not stressed for tension loads. On the bottom of the skids, shoes are attached for the purpose of preventing damage to the skid tube due to abrasion caused by forward speed. Several weld beads are usually placed on the shoe to protect it. See Figure 3-1-8.

Figure 3-1-7. These are amphibian floats mounted on a standard airplane. Note the retracted water rudders. *Photo courtesy of Cessna*

Figure 3-1-8. Skid type helicopter landing gear

Figure 3-1-9. Frequently helicopters will land on a dolly so they can be moved in and out of the hanger.

Figure 3-1-10. Ground handling wheels attached to a skid landing gear

The skid-gear is quite simple and easy to maintain but does present some problems with ground handling since wheels must be placed on the skids to move the helicopter by hand or towing. For this reason a special dolly or wheeled platform is often used. See Figure 3-1-9.

An option is available for temporary wheels on skid-gear on some helicopters. They are called ground handling wheels. Attachment is commonly with an overcenter arrangement that makes the wheel assemblies readily removable (Figure 3-1-10).

When the helicopter is not exposed to landings in unimproved areas, most operators prefer a wheeled undercarriage. Most use a nosewheel configuration with a caster-type nosewheel. When the wheeled gear is an option, it is a normally a fixed type of landing gear. However, in the more sophisticated helicopters, retractable landing gear is not unheard of. Figure 3-1-11 is an illustration of a Sikorsky S-76 with retractable gear.

It has two main gears and a nose gear. The nose gear consists of an air/oil shock strut, a shimmy damper, a trunnion, and two wheel assemblies. This is attached to a structural bulkhead of the aircraft.

The main gear consists of a beam, tire/wheel, and brake assembly on each side. These attach to the main beam assemblies of the helicopter structure. An air/oil shock strut is attached to each gear beam and the helicopter fuselage. The brake system is used for differential braking and as parking brakes.

Multiple wheeled landing gear. As aircraft weight increases, it is common practice to use two or more wheels per gear leg (Figure 3-1-12). Multiple wheels spread the aircraft's weight over a larger area, in addition to providing a safety margin if one tire should fail. It is also common practice to use smaller dual wheels and tires rather than single larger wheels and tires. Although the rotational speed is higher during takeoff and landing, the smaller units are easier to fit inside the wings and nacelles when they are retracted.

Larger transport aircraft need even more footprint to take the landing loads created by the very size of the aircraft. A system of multiple axles called *trucks* makes eight or more main gear wheels practical.

Bogies and trucks. When two wheels are attached to a beam on one strut, the attached mechanism is referred to as a *bogie*. Seldom is one bogie used by itself, therefore it becomes a *truck* (two bogies). Figure 3-1-13 shows two bogies on a single truck. The number of wheels that are included on a truck, as well as the number of trucks, is determined by the gross design weight of the aircraft and the surfaces on which the loaded aircraft may be required to land.

The tricycle arrangement of the landing gear is made up of many assemblies and parts. These

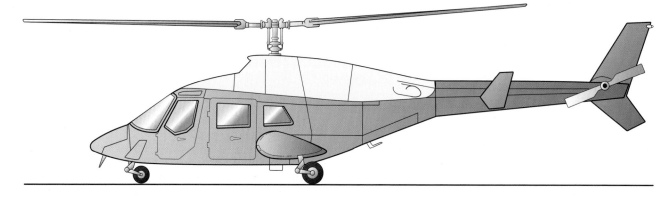

Figure 3-1-11. Bell 430 helicopter with retractable landing gear

Figure 3-1-12. Dual main landing gear wheel arrangement

Figure 3-1-13. Bogie truck main landing gear assembly

consist of air/oil shock struts, main gear alignment units, support units, retraction and safety devices, auxiliary gear protective devices, nose wheel steering system, aircraft wheels, tires, tubes and aircraft brake systems. The airframe technician should know all about each of these assemblies, their inspection procedures and their relationships to the total operation of the landing gear.

Bungee cord landing gear. *Bungee cord*, also called *shock cord*, has been used in landing gear systems for many years. Reserved for small airplanes, the system is simplicity itself. A shock cord is nothing more than many strands of rubber (similar to a rubber band) wrapped around and around a mandrel until the bundle cross section is about 3/4-inch in diameter. Then the bundle is covered with woven fabric. The end result is a large, strong, fabric covered rubber band. The hardware section of Book One, *Introduction to Aircraft Maintenance* explains how to interpret the date code woven into the covering.

The strut for attaching the bungee cords is a welded assembly of two tubes, one sliding inside the other. Each has welded fittings for attaching the cord. See Figure 3-1-14. Fabric or leather covers are typically fastened around the cords to help protect them from the elements. It is unlikely you will ever be called upon to replace any bungee cord shocks, but it could happen. If it should then bear this in mind: the cords are very stiff, strong and difficult to install without special tooling. It should NOT be attempted without the tooling to do the job.

Leaf type gear legs. The so-called Cessna type gear legs have been used on a great many different makes of airplanes. As for the gear leg itself, it is the most simple of all. It is one piece of spring steel. The top end is clamped to the fuselage landing gear carry-through structure, while the lower end has attachment holes for the axle assembly. Tapered shims can be used to correct the toe in/toe out somewhat. No further adjustment is possible, or necessary. The spring steel legs need to be inspected for cracks. A 10-power

Figure 3-1-14. These are bungee cord shock struts on a homebuilt. Notice that the assembly contains two shock cords hooked on the lower tube, wrapped around the upper attach points, then hooked again on the lower tube.

Figure 3-1-15. A classic non-retractable shock strut landing gear system

magnifying glass is standard equipment for a visual inspection. Normally, each airframe manufacturer will specify what NDT procedures are necessary for a thorough inspection.

Shock struts, non-retractable. Non-retractable shock struts are the next most complicated system. This type of landing gear has been the favorite of the Piper Aircraft Corp. for many years. It has been used on several models of trainer aircraft, as well as a series of inexpensive single engine airplanes (Figure 3-1-15).

The inner workings of air-oil shock struts are covered in the hydraulics section of this textbook. These types of installations are simple, strong and actually require little maintenance. The principal maintenance required is to check the struts oil level, and top it off with MIL-H-5606 hydraulic fluid if it is low. The safest, practical way to check the oil level is to jack up

the aircraft, release the air pressure and compress the strut. The air pressure will have to be released as per instructions. The oil level may then be checked visually by removing the plug. The oil level should be even to the bottom of the filler hole. The maintenance manual explains the method of topping off the fluid level. Once topped off, refilling the strut with air is necessary. There is no fixed air pressure. The air pressure is adjusted by measuring the extension of the inner tube with the airplanes weight setting on it. That measurement must be obtained from the Aircraft Maintenance Manual. It is in the range of three to five inches.

Not enough extension will allow the strut to bottom out when landing and taxiing. Too much air pressure will put additional strain on the structure when landing and taxiing by not allowing the strut to absorb the load properly.

Air valves. Two of the various types of high-pressure air valves currently in use on shock struts are illustrated in Figure 3-1-16. One valve (Figure 3-1-16A) contains a valve core and has a 5/8-inch swivel hex nut. It is used for medium pressure installations. The other air valve (Figure 3-1-16B) has no valve core and has a 3/4-inch swivel hex nut. Using a wrench to turn the tapered valve seat extending through the body closes it. It is called a *Schrader valve*.

> **WARNING:** *All pressurized struts are potentially dangerous to handle. Follow the manufacturer's safety instructions to the letter.*

When some airplanes are jacked up and the strut extended, the inner tube is retained in the outer tube only by the torque links (*scissors*). Do not try to disconnect them with the strut extended. Do not extend the strut to its full length with the scissors removed, as the inner tube will be free to slide out of the strut housing.

(A)   (B)

Figure 3-1-16. (A) Medium air valve, and (B) high-pressure (Schrader) air valve

Some shock struts have an internal sleeve, or spacer, to eliminate this problem by providing an extension stop.

Always grease the torque links as specified by the service schedule. Not greasing them properly will accelerate the bushing wear and require their replacement.

Retractable Landing Gear

One of the first major innovations in aircraft performance was the invention of retractable landing gear. By getting the gear out of the slipstream, streamlining was improved dramatically and speed and range improved proportionally.

The retractable landing gear has added greatly to the complexity of modern aircraft design. Some retract inward, some outward and some back. Only a few nose gears fold forward. Some retract into the wings, some into the fuselage and some do both.

As explained in the section on wheels, most gear systems today use more than one wheel per main gear leg. While most systems are operated hydraulically, some general aviation aircraft still retract the gear electrically. Some also use an electrically operated hydraulic system (Power Pack). Both Piper and Cessna manufacture retractable gear airplanes that use electrically powered hydraulic power packs for gear retraction. It simplifies the airplane systems. On larger airplanes that have additional requirements for hydraulic power, gear systems are almost exclusively hydraulic. Most have the ability to operate from more than one hydraulic system, not including an emergency extension system.

Basic retractable systems. Figure 3-1-17 is a drawing of a retractable main gear that would be found on a transport category airplane. It shows the main items that are required for a retractable landing gear to operate. For practical purposes all systems, though they may look different, will have the same basic components arranged in the same manner. Figure 3-1-18 is another drawing showing the same parts on a retractable nose wheel.

Electrically Operated Retraction System

The Beechcraft King Air C90 is an example of an electrically operated landing gear system. All three landing gear struts are conventional air-oil units.

Nose gear. The nose gear is steerable through direct linkage between the rudder pedals and

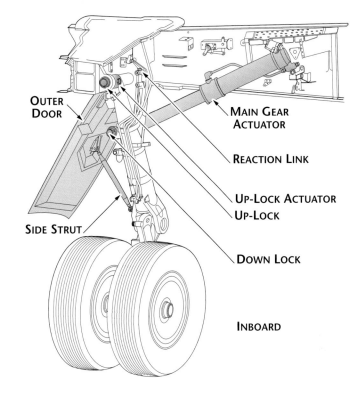

Figure 3-1-17. A drawing of a Boeing 737 main gear showing the basic items necessary for a complete retraction system; all retractable systems will have these components.

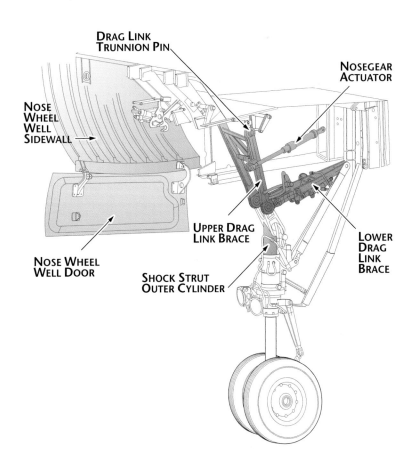

Figure 3-1-18. Similar to Figure 3-1-17, only showing a classic transport category nose gear retractable system

Figure 3-1-19. The squat switch is visible on the upper torque link of this fully extended main gear leg.

Figure 3-1-20. Gear position handle and the three green gear down lights

the nose gear steering linkage. A single cylinder shimmy damper is installed on the nose gear strut to dampen out all vibrations. After takeoff, a straightening roller centers the nose wheel as the nose gear is retracted aft into the wheel well. Nose gear doors mechanically close during retraction and open during extension.

Main gear. The main gear retracts forward into the nacelle wheel wells. The main gear doors operate by mechanical linkage and remain open when the gear is extended. During ground operations, the main gear is held open by mechanical downlock hooks.

A two-position switch on the copilot's sub-panel controls the electric gear. During ground operation the control switch is held in position by a solenoid operated downlock hook that is released when the airplane is in flight and the *squat switch* (Figure 3-1-19) is activated.

Squat switch. This switch is mounted on one of the main gear housings and connected to the torque links (Figure 3-1-19). When the airplane's weight is on the wheels the switch is opened, disarming the retract system. When airborne, the gear portion of the switch is armed, allowing retraction. Squat switches may have several poles and may be used for other functions, i.e. pressurization dump valve operation. All retractable landing gear, electrical or hydraulic, will have a squat switch in the system.

Cockpit controls and position indicators. With retractable landing gear, we need some warning systems not only to tell us that the gear is up, but also when it is (or is not) down and locked. There are also things that can be built into the warning systems that can help keep us from making a mistake: warning horns. Anytime the throttle is closed and the wheels are not extended, the warning horn will sound. Large aircraft systems have considerably more safety checks built into the gear warning systems than light airplanes. We will cover those systems later in the chapter.

The King Air cockpit systems have five gear position lights, three green and two red (Figure 3-1-20). The green lights, one for each gear illuminate when the gear is down and locked. The two red lights are located in the landing gear control handle. They will illuminate when a gear is not down and locked, when one or more landing gear is in transit, not down and locked, or when the *warning horn* sounds. When the gear is up and locked, all lights will be out.

Warning horn. Two microswitches, one for each throttle, are located on the control pedestal.

Their purpose is to sound the gear warning horn any time the throttles are retarded to land and the gear is not down and locked.

Landing gear motor and actuator. Figure 3-1-21 shows a layout of a chain and torque tube, electrically driven retract system. The landing gear motor is a split-field, series wound and 28-volt DC motor with a dynamic braking feature. A split-field device contains one field for running in one direction, while the other field runs in the opposite direction; thus extension and retraction without any gear shifting. The unit is located under the floorboards ahead of the main center section spar.

The gear motor drives a gearbox, which in turn drives two torque tubes that run out to the main gear actuators. A duplex chain whose sprocket is attached to one of the torque shafts from the gearbox motor drives the nose gear actuator. Limit switches are adjacent to the torque tubes and ensure that motor rotation stops when the desired travel is reached. Friction clutches and circuit breakers provide even more protection against overrun.

Emergency extension. All retractable landing gear systems need a backup emergency system. The King Air C90 uses an emergency engage handle and a ratchet lever that is located next to the pilots seat. To use the emergency extension system the circuit breaker first needs to be pulled. Any possibility of the motor starting to run during manual extension would be catastrophic.

Then the handle is engaged and turned clockwise to disconnect the motor and lock the emergency system to the gearbox. Once all of this is engaged correctly, the handle is pumped up and down to activate the continuous ratchet, which turns the torque shafts and lowers the gear.

Once on the ground, the advice is to change nothing until the original malfunction is found and corrected. You do not want the gear to start folding up and the whole airplane to come down on you. Jack stands and safety stands would be in order before doing a gear retraction test.

Light Aircraft Hydraulic Landing Gear

A unique thing about the King Air 90 series is that it was produced with both electrically and hydraulically operated landing gear systems. This gives us an opportunity to examine both systems on the same airframe.

Nose and main gear. Actually, the main and nose gear assemblies are not that much differ-

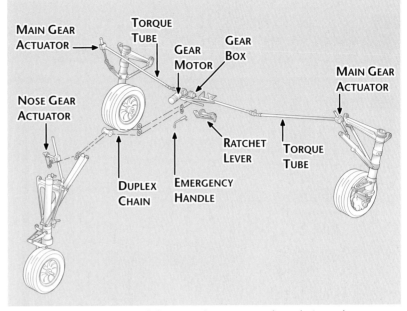

Figure 3-1-21. This pictorial drawing shows a complete chain and torque tube operated, electrically driven retract system.

ent from the ones on the mechanical retraction system. In essence, the difference is in the actuators and the internal power pack system. The major difference between the King Air hydraulic power pack and other light aircraft power pack systems is the use of an 800 p.s.i. accumulator. Most small aircraft systems do not use an accumulator.

System operation. The hydraulic system is a closed center system (Figure 3-1-22). When gear up is selected, a four-way solenoid operated valve is energized. This opens the pressure and return lines and allows the pressure from the accumulator to start activating the up actuators through the gear up line. When the system pressure reaches the low pressure setting, the power pack starts running, building the pressure back towards the system high pressure setting. The nose gear has an internal lock that is locked by cylinder extension and unlocked by initial system pressure. Once unlocked, the cylinder pressure pulls up on the drag link and starts to retract the gear. The main gear is unlocked by retract pressure on the locking link, and then the gear is retracted.

Once retracted, system pressure increases to the maximum setting; then is shut off by a pressure switch. The landing gear is held in place by system pressure alone. The accumulator, fully charged from the retract cycle, maintains pressure even though there may (will) be some internal leakage through the valve. When the system pressure bleeds down to the cut-in point, the power pack will automatically start and run just long enough to rebuild system pressure to the max cut-off point, and then disconnect. The secret to a system like this (if

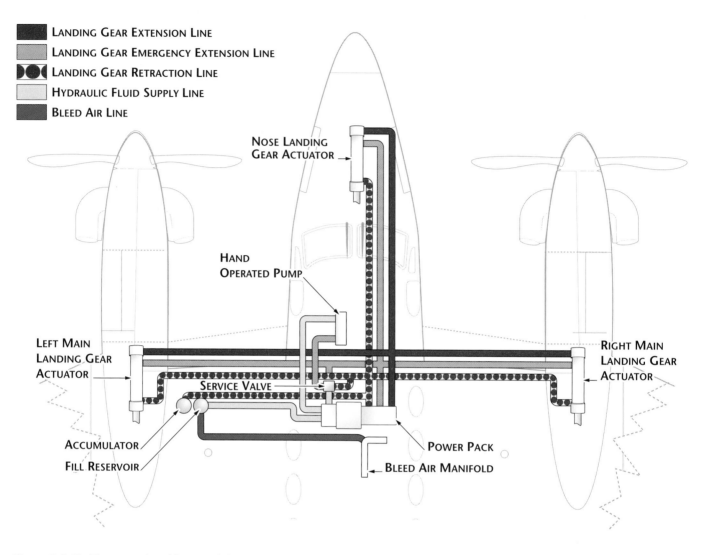

LANDING GEAR EXTENSION LINE
LANDING GEAR EMERGENCY EXTENSION LINE
LANDING GEAR RETRACTION LINE
HYDRAULIC FLUID SUPPLY LINE
BLEED AIR LINE

NOSE LANDING GEAR ACTUATOR

HAND OPERATED PUMP

LEFT MAIN LANDING GEAR ACTUATOR

RIGHT MAIN LANDING GEAR ACTUATOR

SERVICE VALVE

ACCUMULATOR

FILL RESERVOIR

POWER PACK

BLEED AIR MANIFOLD

Figure 3-1-22. The operational layout of the King Air C90 hydraulic gear system

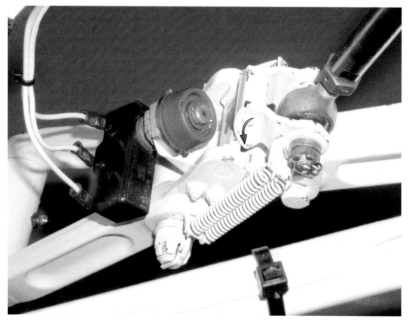

Figure 3-1-23. A main gear drag link getting ready to lock. When the block goes into the hook the gear will be locked down. The lock will also trip the switch and indicate a gear down light.

there is a secret) is that the cut-in point for the hydraulic power pack is high enough that the gear has not yet begun to sag below the wheel well when it cuts in.

Gear extension is almost the reverse of the up operation. Selecting down with the gear control handle will switch the solenoid valve positions, allowing the down lines to become pressurized and the up lines to turn into returns, and the gear will start down. Once the nose gear reaches the end of its travel, the drag link will go over center, locking it extended.

At the same moment, the internal lock in the cylinder will lock, securing the drag link. Meanwhile, the main gear will extend the drag links to their fullest, then activate the positive hook latch with the last bit of cylinder movement (Figure 3-1-23). As the pressure increases to maximum, the gear-operating handle will kick into neutral position and the power pack will shut off.

Air Transport Hydraulic Landing Gear

Large aircraft hydraulic landing gear systems all have one thing in common; they are very heavy.

The heavy landing gear retract fine, but present a design problem when they extend. The heavy landing gear will try to free fall, pulling the actuating cylinder piston faster than the system pressure can fill the down cylinder. The result is an imbalance of pressures and a rather extreme load on the structure as the gear falls half out of the airplane.

There are two methods used to rectify the problem:

- A cross-flow valve that allows fluid from the actuator UP side to flow to the actuator DOWN side until the system fluid flow can catch up with itself. When the fluid flow matches the demand in the down side of the system, the cross-flow valve shuts off and normal fluid flow to the return line continues. Cross-flow valves are used on older technology systems.

- Most current systems use a flow limiting system like the Boeing 737 system described below. Rather than create an entire new valve, the flow limiting devices become part of the selector valve system. They both reduce the shock loads brought about by a landing gear in free fall, and try to make the gear stay together during extension and retraction.

Landing gear hydraulic schematic (UP). With the landing gear control lever placed in the up position, the UP line pressurizes. The main and nose gear are hydraulically actuated to retract simultaneously. The pressure then goes to the modular packages. See Figure 3-1-24.

The flow limiters, in conjunction with the transfer cylinder, allow the lock actuators to unlock and start the retraction cycle before full pressure is built up in the nose and main gear actuators. This is a momentary motion, and then the nose and main gear actuators supply the force necessary to retract the gear. At the main gear uplock actuators, pressure is applied to retract the piston.

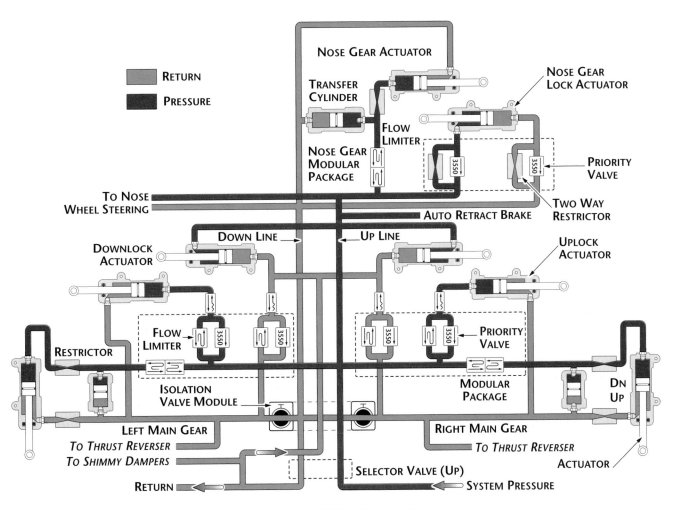

Figure 3-1-24. A schematic of the landing gear hydraulics of a B737 while retracting

When the gear gets to the retract position, a roller on the strut strikes the uplock hook and allows the actuator piston to pull the hook into the locked position. The pressure applied to the nose gear lock actuator to start the retraction cycle also locks the nose gear in the up and locked position.

Hydraulic pressure from the up line is applied to both brake metering valves to stop wheel rotation before the wheels enter the wheel well area.

Landing gear hydraulic schematic (DN). With the landing gear control lever in the down position, hydraulic pressure is ported to the down line. Hydraulic pressure is applied to the DN port of the Gear Actuators, unlock port of the uplock actuator, lock port of the down lock actuator, and through the Nose Gear Modular Package to the Nose Gear Lock Actuator. The pressure at the DN port of the Gear Actuators also is felt at the transfer cylinder. See Figure 3-1-25.

The return fluid leaving the upside of the Gear Actuators is being restricted by the flow limiters in the modular package. The limiting action and the movement of the piston inside the transfer cylinder momentarily causes a transfer of pressure from the DN port to the up port of Gear Actuators. The piston differential area causes the gear to momentarily retract. This momentary action allows the Lock Actuators to unlock the gear.

The transfer cylinder piston reaches the end of its travel and the pressure on the up side of the actuators drop. The gear then extends in the normal fashion. As the rear reaches its full extend position, the Lock Actuators force the lock strut into an over-center position. Hydraulic pressure holds the gear in the down and locked position; in the absence of hydraulic pressure, spring bungees will hold the gear in the locked position.

Hydraulic pressure from the down line also supplies the thrust reversers and the nose wheel steering system.

Landing Gear Rigging and Adjustment

Occasionally it becomes necessary to adjust the landing gear switches, doors, linkages,

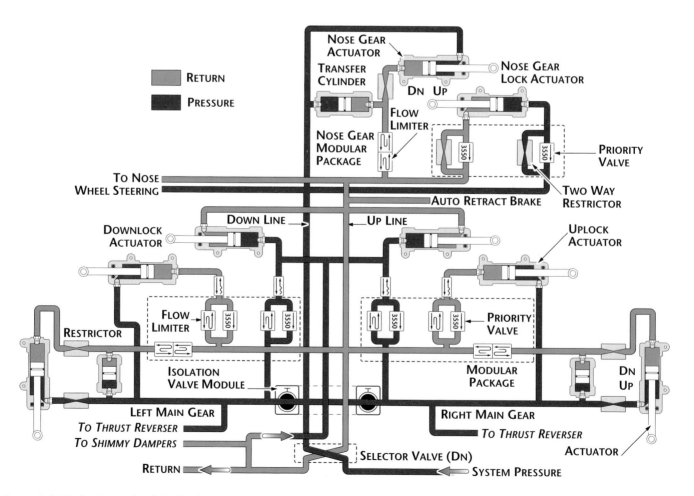

Figure 3-1-25. A schematic of the landing gear hydraulics of a B737 while extending

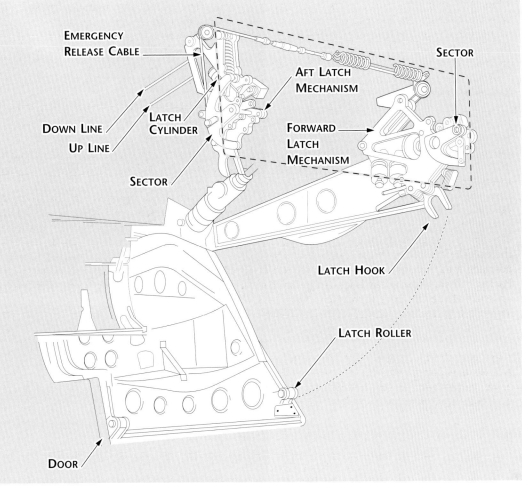

Figure 3-1-26. Main gear door latch mechanisms

latches and locks to assure proper operation of the landing gears and doors. When landing gear actuating cylinders are replaced, or when length adjustments are made, overtravel must be checked.

Overtravel is that action of the cylinder piston beyond the movement necessary for landing gear extension and retraction. The additional action operates the landing gear latch mechanism. Should overtravel be out of limits, substantial excess pressure will be applied to the entire gear assembly.

Because of the wide variety of aircraft types and designs, procedures for rigging and adjusting landing gear will vary. Uplock and downlock clearances, linkage adjustments, limit switch adjustments and other landing gear adjustments vary widely with landing gear design. For this reason, always consult the applicable manufacturers maintenance or service manual before performing any phase of landing gear rigging or installation.

Adjusting landing gear latches. The adjustment of latches is of prime concern to the airframe technician. A latch is used in landing gear systems to hold a unit in a certain position after the unit has traveled through a part of, or all, of its cycle.

For example, on some aircraft, when the landing gear is retracted, each gear is held in the up position by a latch. The same holds true when the landing gear is extended. Latches are also used to hold the landing gear doors in either the open or closed positions.

There are many variations in latch design. However, all latches are designed to accomplish the same thing. They must operate automatically, at the proper time and hold the unit in the desired position. A typical landing gear door latch is described in the following paragraphs.

The landing gear door that is detailed in Figure 3-1-26 is held closed by two door latches. As shown, one is installed near the rear of the door. To have the door locked securely, both locks must grip and hold the door tightly against the aircraft structure. The principal components of each latch mechanism, shown in Figure 3-1-26, are a hydraulic latch cylinder, a latch hook, a spring loaded crank-and-lever linkage and a sector.

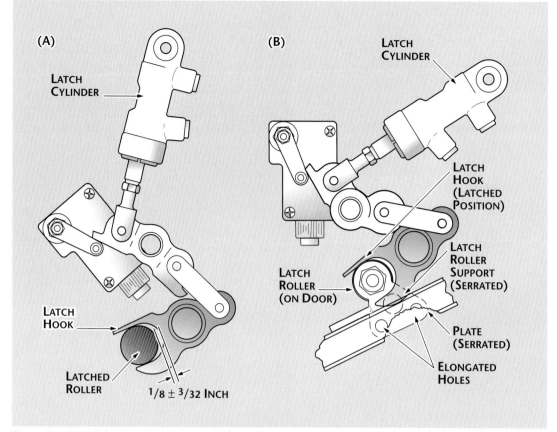

Figure 3-1-27. (A) Latch roller clearance (B) Latch roller support adjustment

The latch cylinder is hydraulically connected with the landing gear control system and mechanically connected, through linkage, with the latch hook. When hydraulic pressure is applied, the cylinder operates the linkage to engage (or disengage) the hook with (or from) the latch roller on the door. In the gear-down sequence, the spring load on the linkage disen-gages the hook. In the gear-up sequence, spring action is reversed when the closing door is in contact with the latch hook and the cylinder operates the linkage to engage the hook with the latch roller.

Cables on the landing gear emergency exten-sion system are connected to the sector to per-mit emergency release of the latch rollers. An uplock switch is installed on, and actuated by, each latch to provide a gear-up indication in the cockpit.

With the gear up and the door latched, inspect the latch roller for proper clearance as shown in Figure 3-1-27, view A. On this installation, the required clearance is 1/8 ±3/32 inch. If the roller is not within tolerance, loosening its mounting bolts and raising or lowering the latch roller support may adjust it. This may be done due to the elongated holes and serrated locking surfaces of the latch roller support and serrated plate (Figure 3-1-27, view B).

Landing gear door clearances. Landing gear doors have specific allowable clearances, which must be maintained between doors and the aircraft structure or other landing gear doors. Adjusting the door hinges and connect-ing links and trimming excess material from the door if necessary could maintain these required clearances.

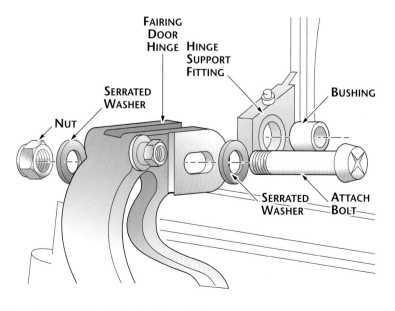

Figure 3-1-28. Adjustable door installation

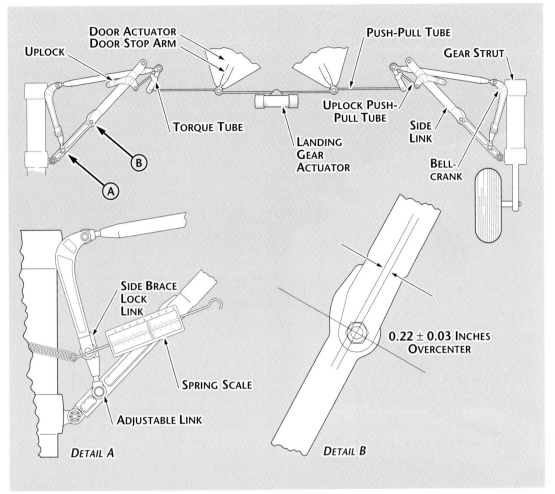

Figure 3-1-29. Landing gear schematic showing overcenter adjustments

On some installations, placing the serrated hinge and serrated washers in the proper position and torquing the mounting bolts adjusts the door hinges. Figure 3-1-28 illustrates this type of mounting, which allows linear adjustments. The length of the elongated bolt hole in the door hinge controls the amount of linear adjustment.

The distance the landing gear doors open or close depends upon the length of door linkage and the adjustment of the doorstops. The manufacturer's maintenance manuals specify the length of door linkages and adjustment of stops or other procedures whereby correct adjustments may be made.

Landing Gear Drag and Side Brace Adjustment

The landing gear side brace illustrated in Figure 3-1-29 consists of an upper and lower link, hinged at the center to permit the brace to jackknife during retraction of the landing gear. The upper end pivots on a trunnion attached to the wheel well overhead. The lower end is connected to the shock strut.

On the side brace illustrated, a locking link is incorporated between the upper end of the shock strut and the lower drag link. Usually in this type installation, the locking mechanism is adjusted so that it is positioned slightly overcenter. This provides positive locking of the side brace and the locking mechanism, and as an added safety feature, prevents inadvertent gear collapse caused by the side brace folding.

To adjust the overcenter position of the side brace locking link illustrated in Figure 3-1-29, place the landing gear in the down position and adjust the lock link end fitting so that the side brace lock link is held firmly overcenter. Manually break the lock link and move the landing gear to a position five or six inches inboard from the down locked position, and then release the gear. The landing gear must free fall and lock down when released from this position.

In addition to adjusting overcenter travel, the down lock spring tension must be checked using a spring scale. The tension should be between 40 to 60 lbs. for the lock link illustrated. The specific tension and procedure for checking will vary from system to system.

Landing Gear Retraction Check

There are several occasions when a retraction check should be performed. First, it should be done during annual inspection of the landing gear system (Figure 3-1-30). Second, when performing maintenance that might affect the landing gear linkage or adjustment, such as changing an actuator, make a retraction check to see whether everything is connected and adjusted properly (Figure 3-1-31). Third, it may be necessary to make a retraction check after a hard or overweight landing has been made which may have damaged the landing gear. Closely inspect the gear for obvious damage and then make the retraction check. And finally, one method of locating malfunctions in the landing gear system is to perform a gear retraction check.

Figure 3-1-30. This shows a Beechcraft Bonanza during a retraction check. Notice that the Bonanza uses a special jacking system.

There are a number of specific inspections to perform when making a retraction check of landing gear. Included are:

- Landing gears for proper retraction and extension
- Switches, lights and warning horn or buzzer for proper operation
- Landing gear doors for clearance and freedom from binding
- Landing gear linkage for proper operation, adjustment and general condition
- Latches and locks for proper operation and adjustment
- Alternate extension or retraction systems for proper operation
- Any unusual sounds such as those caused by rubbing, binding, chafing, or vibration

The procedures and information presented herein were intended to provide familiarization with some of the details involved in landing gear rigging, adjustments and retraction checks and do not have general application. For exact information regarding a specific aircraft landing gear system, consult the applicable manufacturer's instructions.

Alignment

Airplanes do not normally require realignment of the landing gear, even after repairs. That is if the repairs were done correctly and the structure is not misaligned. It is an item on a hard landing inspection list (or should be) and must be accomplished from time to time.

Unlike an automobile where front end alignment is a complex operation and requires special equipment, checking landing gear alignment is not complex. It requires that you know how to make accurate measurements. Like a symmetry check, alignment checks begin with an empty airplane and a straight line down the middle of the floor that is exactly on the center line. Do not level the airplane because all measurements will then be in relation to level, and

Figure 3-1-31. This shows a twin engined airplane undergoing an annual inspection, which includes a retraction check.

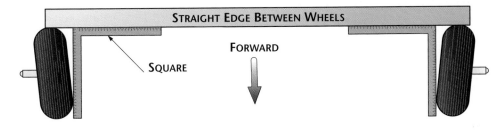

Figure 3-1-32. Measuring toe in

not to the floor. Using a carpenter's Framing Square you can place the short section on the floor and the long section against the side of a tire and wheel. This will show you if the wheel has any camber (the top leans in or out).

Spring steel landing gear may have positive (leaning in) or negative (leaning out) camber. Camber on a spring steel gear is constantly changing as the aircraft speed and weight change during a takeoff or landing. On an oleo gear, angles generally remain the same during operation. Camber is not normally adjustable. It is more of a check on structural alignment.

Next comes checking for toe in or toe out. If toed in the wheel center lines will cross the airplane center line somewhere ahead of the airplane. If toed out they will cross behind the airplane. The manufacturers maintenance manual will normally give you any tolerances established for the airplane you are checking. A landing gear that is toed in will try to follow the center line of the wheel and travel towards the center line of the airplane. It can place a large combined torsion and side load on the entire gear assembly and its attach points. In order to relieve the pressure the tires will slip on the runway. In essence scrubbing along and dragging the tire tread sideways somewhat.

Toe in (or toe out) has a very detrimental effect on tire wear. You can actually see the increased wear by looking at the tread. With toe in each tire rib will have a feathered edge towards the center line; with toe out it will go away from the center line.

To actually measure toe in is relatively easy. It requires two straight boards that are longer than the distance between the outside of the wheels from side to side (Figure 3-1-32). Mark a center line crossway on each board and center them over the center line of the airplane. Block up each end one-half the height of the tires and allow them to touch the tire tread. Place a straight edge along the outside of the wheels and draw a line across the board. Now measure the distance along the board from the center line to the outside line for each wheel, front and rear. You will now have toe in (or out)

measurements for each tire, in reference to the airplane center line.

Depending on how the axle is attached to the strut, you can use one of two ways to adjust for toe in. If the axle is bolted on, there are tapered shims that are available to make the correction. For axles that are not bolted, there is a limited amount of adjustable available by shimming between the landing gear torque links (Figure 3-1-33).

Shock Struts

Shock struts are self-contained hydraulic units that support an aircraft on the ground and protect the aircraft structure by absorbing

Figure 3-1-33. Landing gear torque link

and dissipating the tremendous shock loads of landing. Shock struts must be inspected and serviced regularly to function efficiently.

Since there are many different designs of shock struts, only information of a general nature is included in this section. For specific information about a particular installation, consult the applicable manufacturer's instructions.

The compression stroke of the shock strut begins as the aircraft wheels touch the ground; the center of mass of the aircraft continues to move downward, compressing the strut and sliding the inner cylinder into the outer cylinder. The metering pin is forced through the orifice and, by its variable shape, controls the rate of fluid flow at all points of the compression stroke. In this manner, the greatest possible amount of heat is dissipated through the walls of the shock strut. At the end of the downward stroke, the compressed air is further compressed, limiting the compression stroke of the strut.

If there was an insufficient amount of fluid and/or air in the strut, the compression stroke would not be limited and the strut would bottom.

The extension stroke occurs at the end of the compression stroke as the energy stored in the compressed air causes the aircraft to start moving upward in relation to the ground and wheels. At this instant, the compressed air acts as a spring to return the strut to normal. It is at this point that forcing the fluid to return through the restrictions of the snubbing device produces a snubbing or damping effect. If this extension were not snubbed, the aircraft would rebound rapidly and tend to oscillate up and down, due to the action of the compressed air. Figure 3-1-34 shows a shock strut at normal inflation and at full extension. A sleeve, spacer, or bumper ring incorporated in the strut limits the extension stroke.

Servicing Shock Struts

For efficient operation of shock struts, the proper fluid level and air pressure must be maintained. To check the fluid level, the shock strut must be deflated and in the fully compressed position. Deflating a shock strut can be a dangerous operation unless servicing personnel are thoroughly familiar with high-pressure air valves. Observe all the necessary safety precautions. Refer to manufacturer's instructions for proper deflating technique.

The following procedures are typical of those used in deflating a shock strut, servicing with hydraulic fluid and re-inflating:

(A)   (B)

Figure 3-1-34. Illustration (A) shows a shock strut at normal inflation, while (B) shows a similar strut at full extension.

1. Position the aircraft so the shock struts are in the normal ground operating position. Make certain that personnel, workstands and other obstacles are clear of the aircraft. Some aircraft must be placed on jacks to service the shock struts.

2. Remove the cap from the air valve.

3. Check the swivel hex nut for tightness with a wrench.

4. If the air valve is equipped with a valve core, release any air pressure that may be trapped between the valve core and the valve seat by depressing the valve core. Always stand to one side of the valve, since high-pressure air can cause serious injury, e.g., loss of eyesight.

5. Remove the valve cap.

6. Release the air pressure in the strut by slowly turning the swivel nut counter-clockwise.

7. Ensure that the shock strut compresses as the air pressure is released. In some cases, it may be necessary to rock the aircraft after deflating to ensure compression of the strut.

8. When the strut is fully compressed, the air valve assembly may be removed.

9. Fill the strut to the level of the air valve opening with an approved type of hydraulic fluid.

10. Re-install the air valve assembly, using new O-ring packing. Torque the air valve assembly to the values recommended in the applicable manufacturer's instructions.

11. Install the air valve core.

12. Inflate the strut, using a high-pressure source of dry air or nitrogen. Bottled gas should not be used to inflate shock struts. On some shock struts, the correct amount of inflation is determined by using a high-pressure air gauge. On others, it is determined by measuring the amount of extension (in inches) between two given points on the strut. The proper procedure can usually be found on the instruction plate attached to the shock strut. Shock struts should always be inflated slowly to avoid excessive heating and overinflation.

13. Tighten the swivel hex nut, using the torque values specified in the applicable manufacturer's instructions.

14. Remove the high-pressure air line chuck and install the valve cap. Tighten the valve cap finger tight.

Bleeding Shock Struts

If the fluid level of a shock strut has become extremely low, or if for any other reason air is trapped in the strut cylinder, it may be necessary to bleed the strut during the servicing operation. Bleeding is usually performed with the aircraft placed on jacks. In this position the shock, struts can be extended and compressed during the filling operation, thus expelling all the entrapped air. The following is a typical bleeding procedure:

1. Construct a bleed hose containing a fitting suitable for making an airtight connection to the shock strut filler opening. The base should be long enough to reach from the shock strut filler opening to the ground when the aircraft is on jacks.

2. Jack the aircraft until all shock struts are fully extended.

3. Release the air pressure in the strut to be bled.

4. Remove the air valve assembly.

5. Fill the strut to the level of the filler port with an approved type hydraulic fluid.

6. Attach the bleed hose to the filler port and insert the free end of the hose into a container of clean hydraulic fluid, making

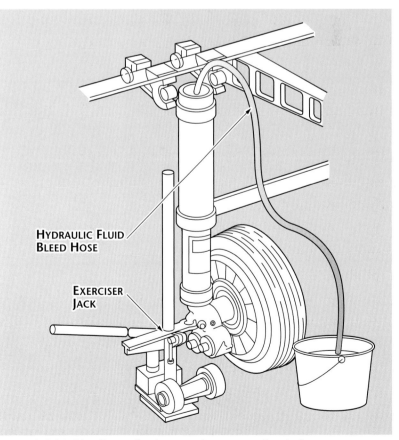

HYDRAULIC FLUID BLEED HOSE

EXERCISER JACK

Figure 3-1-35. Bleeding a shock strut using an exerciser jack

sure that this end of the hose is below the surface of the hydraulic fluid.

7. Place an exerciser jack (Figure 3-1-35) or other suitable single-base jack under the shock strut jacking point. Compress and extend the strut fully by raising and lowering the jack until the flow of air bubbles from the strut has completely stopped. Compress the strut slowly and allow it to extend by its own weight.

8. Remove the exerciser jack, and then lower and remove all other jacks.

9. Remove the bleed hose from the shock strut.

10. Install the air valve and inflate the strut.

Shock struts should be inspected regularly for leakage of fluid and for proper extension. Exposed portions of the strut pistons should be wiped clean daily and inspected closely for scoring or corrosion.

Landing Gear Alignment, Support and Retraction

The landing gear consists of several components that enable it to function. Typical of these are the torque links, trunnion and bracket arrangements, drag strut linkages, electri-

Figure 3-1-36. Torque links

cal and hydraulic gear-retraction devices and gear indicators.

Alignment. Torque links (Figure 3-1-36) keep the landing gear pointed in a straight-ahead direction; one torque link connects to the shock strut cylinder, while the other connects to the

Figure 3-1-37. Trunnion and bracket arrangement. Notice ground lock pin with streamer attached. It must be removed before flight.

piston. The links are hinged at the center so that the piston can move up or down in the strut.

Support. To anchor the gear to the aircraft structure, a trunnion and bracket arrangement (Figure 3-1-37) is usually employed. This arrangement is constructed to enable the strut to pivot or swing forward or backward as necessary when the aircraft is being steered or the gear is being retracted. To restrain this action during ground movement of the aircraft, various types of linkages are used, one being the drag strut.

The upper end of the drag strut (Figure 3-1-37) connects to the aircraft structure, while the lower end connects to the shock strut. The drag strut is hinged so that the landing gear can be retracted.

Emergency extension systems. The purpose of the emergency extension system is to lower the landing gear if the main system fails. Many types of emergency gear retraction systems are in use on modern aircraft. Some of the most common types include a free fall system in which the gear is allowed to fall into position under its own weight. Another type is a manually operated hydraulic pump or a manual crank down type.

Landing gear safety devices. Such safety devices as mechanical downlocks, safety switches and ground locks, may prevent accidental retraction of a landing gear. Mechanical downlocks are built-in parts of a gear-retraction system and are operated automatically by the gear-retraction system. To prevent accidental operation of the downlocks, electrically operated safety switches are installed.

Ground locks. Most aircraft, particularly large ones, are equipped with additional safety devices to prevent collapse of the gear when the aircraft is on the ground. These devices are called *ground locks*. One common type is a pin installed in aligned holes drilled in two or more units of the landing gear support structure. Another type is a spring-loaded clip designed to fit around and hold two or more units of the support structure together. All types of ground locks usually have red streamers permanently attached to them to readily indicate whether or not they are installed (See Figure 3-1-37).

Gear indicators. To provide a visual indication of landing gear position, indicators are installed in the cockpit or flight compartment.

Gear warning devices are incorporated on all retractable gear aircraft and usually consist of a horn or some other aural device and a red warning light. The horn blows and the light comes on when one or more throttles are

retarded and the landing gear is in any position other than down and locked.

Several designs of gear position indicators are used. One type displays movable miniature landing gears, which are electrically positioned by movement of the aircraft gear. Another type consists of two or three green lights, which burn when the aircraft gear is down and locked (Figure 3-1-38). A third type consists of tab-type indicators with markings *up* to indicate that the gear is up and locked, a display of red and white diagonal stripes to show when the gear is unlocked, or a silhouette of each gear to indicate when it locks in the *down* position.

Nosewheel centering. Centering devices include such units as internal centering cams

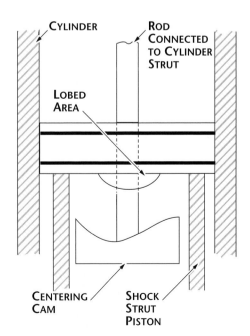

Figure 3-1-38. A typical gear position indicator

Figure 3-1-39. Cutaway view of a nose gear internal centering cam

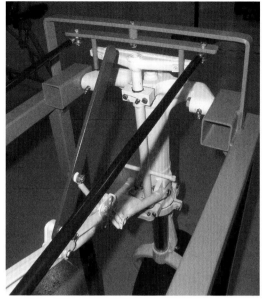

Figure 3-1-40. Nosewheel steering system on a nose gear trainer

(Figure 3-1-39) to center the nose wheel as it retracts into the wheel well. If a centering unit were not included in the system, the fuselage wheel well and nearby units could be damaged.

During retraction of the nose gear, the strut does not support the weight of the aircraft. The strut is extended by means of gravity and air pressure within the strut. As the strut extends, the raised area of the piston strut contacts the sloping area of the fixed centering cam and slides along it. In so doing, it aligns itself with the centering cam and rotates the nose gear piston into a straight-ahead direction.

The internal centering cam is a feature common to most large aircraft. However, other centering devices are commonly found on small aircraft. Small aircraft characteristically incorporate an external roller or guide pin on the strut. As the strut is folded into the wheel well on retraction, the roller or guide pin will engage a ramp or track mounted to the wheel well structure. The ramp/track guides the roller or pin in such a manner that the nose wheel is straightened as it enters its well.

In either the internal cam or external track arrangement, once the gear is extended and the weight of the aircraft is on the strut, the nose wheel may be turned for steering.

Nosewheel Steering System

Light aircraft. Light aircraft are commonly provided nosewheel steering capabilities through a simple system of mechanical linkage

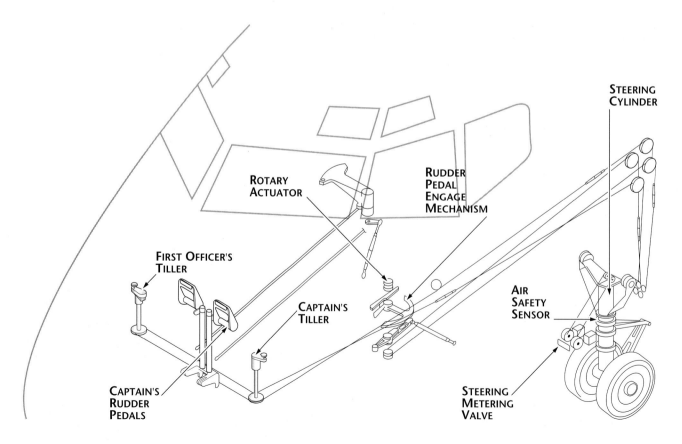

Figure 3-1-41. In this system, interior cables control the steering metering valve. Both the rudder pedals and Captain's tiller can control the aircraft.

hooked to the rudder pedals. Most common applications utilize push-pull rods to connect the pedals to horns located on the pivotal portion of the nosewheel strut. Figure 3-1-40 is an example of a Piper nose wheel steering linkage on a trainer.

Heavy aircraft. Large aircraft, with their larger mass and a need for positive control, utilize a power source for nosewheel steering. Even though large aircraft nosewheel steering system units differ in their construction features, basically all of these systems work in approximately the same manner and require the same sort of units (Figure 3-1-41).

Shimmy Dampers

The three types of shimmy dampers include the piston type, the vane type or one built into the steering system of the aircraft. Shimmy dampers and how they work internally are covered in the hydraulics chapter of this book.

Steering shimmy damper. A steer damper is hydraulically operated and accomplishes two separate functions of steering and/or eliminating shimmying. The type discussed here is designed for installation on nose gear struts and is connected into the aircraft hydraulic system. They are found on older transport aircraft. The steer damper accomplishes two separate functions: one is steering the nosewheel and the other is shimmy damping.

Daily inspection of a steer damper should include a check for leakage and a complete inspection of all hydraulic connections, steer damper mounting bolts for tightness and all fittings and connections between the moving parts of the shock strut and the steer damper wing shaft.

Piston-type shimmy damper. The piston-type shimmy damper is mounted on a bracket at the lower end of the nose gear shock strut outer cylinder. See Figure 3-1-42.

The piston-type shimmy damper requires very little maintenance, however it should be inspected at regular intervals for evidence of leakage. The hydraulic fluid reservoir may require addition of fluid to maintain the proper fluid level. The cam assembly should also be inspected for wear and loose fasteners.

There is generally a minimum of servicing and maintenance; however, it should be checked periodically for evidence of hydraulic leaks around the damper assembly, and the reservoir fluid level must be properly maintained at all times. The cam assembly should be checked for evidence of binding and for worn, loose, or broken parts.

Vane-type shimmy damper. The vane-type shimmy damper is located on the nosewheel shock strut just above the nosewheel fork and may be mounted either internally or externally. If mounted internally, the housing of the shimmy damper is fitted and secured inside the shock strut, and the shaft is splined to the nosewheel fork. If mounted externally, the housing of the shimmy damper is bolted to the side of the shock strut, and the shaft is connected by mechanical linkage to the nosewheel fork.

Maintaining the proper fluid level is necessary to the continued functioning of a vane-type shimmy damper. If a shimmy damper is not operating properly, the fluid level is the first item that should be checked by measuring the protrusion of the indicator rod from the center of the reservoir cover. Inspection of a shimmy damper should include a check for evidence of leakage and a complete examination of all fittings and connections between the moving parts of the shock strut and the damper shaft for loose connections. Fluid should be added only when the indicator rod protrudes less than the required amount. The distance the rod should protrude varies with different models. A shimmy damper should not be over-filled. If the indicator rod is above the height specified on the nameplate, fluid should be bled out of the damper.

Figure 3-1-42. The gear on this Hawker 800XP is a good example of a piston type shimmy damper.

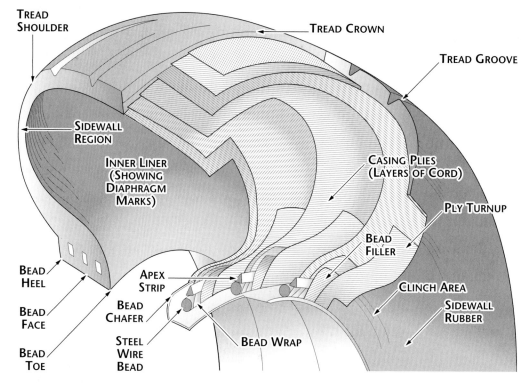

TREAD SHOULDER

TREAD CROWN

TREAD GROOVE

SIDEWALL REGION

INNER LINER (SHOWING DIAPHRAGM MARKS)

CASING PLIES (LAYERS OF CORD)

PLY TURNUP

BEAD FILLER

BEAD HEEL

APEX STRIP

CLINCH AREA

SIDEWALL RUBBER

BEAD FACE

BEAD CHAFER

BEAD TOE

STEEL WIRE BEAD

BEAD WRAP

Figure 3-2-1. Ply layout for a bias tire

Landing Gear System Maintenance

Because of the stresses and pressures acting on the landing gear, inspection, servicing and other maintenance becomes a continuous process. The most important job in the maintenance of the aircraft landing gear system is thorough, accurate inspections. To properly perform the inspections, all surfaces should be cleaned to ensure that no trouble spots go undetected.

Periodically, it will be necessary to inspect shock struts, shimmy dampers, wheels, wheel bearings, tires and brakes. During this inspection:

- Check for the presence of installed ground safety locks.

- Check landing gear position indicators, lights and warning horns for operation.

- Check emergency control handles and systems for proper position and condition.

- Inspect landing gear wheels for cleanliness, corrosion and cracks.

- Check wheel tie bolts for looseness.

- Examine anti-skid wiring for deterioration.

- Check tires for wear, cuts, deterioration, presence of grease or oil, alignment of slippage marks and proper inflation.

- Inspect the landing gear mechanism for condition, operation and proper adjustment.

- Lubricate the landing gear, including the nose wheel steering.

- Check steering system cables for wear, broken strands, alignment and safetying.

- Inspect landing gear shuck struts for such conditions as cracks, corrosion, breaks and security.

- Where applicable, check the brake clearances.

Lubrication. Various types of lubricants are required to lubricate points of friction and wear on the landing gear. These lubricants are applied by hand, an oil can, or a pressure-type grease gun. Before using the pressure-type grease gun, wipe the lubrication fittings clean of old grease and dust accumulations, because dust and sand mixed with a lubricant produce a very destructive abrasive compound. As each fitting is lubricated, the excess lubricant on the fitting and any that is squeezed out of the assembly should be wiped off. Wipe the piston rods of all exposed actuating cylinders; clean them frequently, particularly prior to operation, to prevent damage to seals and polished surfaces.

Section 2

Aircraft Tires and Tubes

Proper care and maintenance of tires have always been important in aircraft maintenance. Because of the modern fast-landing aircraft, careful tire maintenance has become increasingly important.

The following designations refer to construction features and the types of tire casings with which they are used. Tire size designations are discussed later.

Tire Designs

Bias tires. Until the advent of radial tires, all tires were bias tires. They are constructed with plies laid at angles between 30° to 60° to the direction of rotation. Each succeeding ply is laid with the cord angle opposite to the last one. This provides a balanced carcass that is very strong. Bias tires have some problems with high speed inasmuch as they will not hold their round shape well. The tread coming up off of the runway will be wavy, setting up considerable heat and vibration.

Figure 3-2-1 shows the construction details of a bias tube-type aircraft tire. Tubeless tires are similar to tube tires except they have a rubber inner liner that is mated to the inside surface of the tire. The rubber liner helps retain air in the tire. The beaded area of a tubeless tire is designed to form a seal with the wheel flange. Wear indicators have been built into some tires as an aid in measuring tread wear. These indicators are holes in the tread area or lands in the bottom of the tread grooves.

Radial tires. The main difference in a radial tire is the direction the plies are laid. They are laid roughly 90° to the circumference of the tire. This makes the tire much stronger in the direction it is traveling than a bias tire. So much so in fact, that this is the main reason for the *ply rating* as opposed to the actual number of plies. Few radial tires actually have as many plies as the rating indicates. See Figure 3-2-2.

Radial tire sidewalls are more flexible. This is why they look underinflated when compared to a bias tire. It is also the key to their better high-speed performance. It lays a better footprint on the runway, and does not distort as much when lifting the tread up behind the tire during rotation.

Tire Construction

Cord body. The *cord body* consists of multiple layers of nylon with individual cords arranged parallel to each other and completely encased in rubber. The cord fabric has its strength in only one direction. Each layer of coated fabric constitutes one ply of the cord body. Adjacent cord plies in the body are assembled with the cords crossing at nearly right angles to each other.

This arrangement provides a strong and flexible tire that distributes impact shocks over a wide area. The functions of the cord body are to give the tire tensile strength, to resist internal pressures and to maintain tire shape.

Tread. The *tread* is a layer of rubber on the outer surface of the tire. It protects the cord body from abrasion, cuts, bruises and moisture. It is the surface that contacts the ground.

Sidewall. The *sidewall is* an outer layer of rubber adjoining the tread and extending to the beads. Like the tread, it protects the cord body from abrasion, cuts, bruises and moisture.

Bead. The *beads* are multiple strands of high-tensile strength steel wire imbedded in rubber and wrapped in strips of open weave fabric. The beads hold the tire firmly on the rims and serve as an anchor for the fabric plies that are turned up around the bead wires.

Chafing strip. The *chafing strips* are one or more plies of rubber-impregnated woven fabric wrapped around the outside of the beads. They provide additional rigidity to the bead and prevent the metal wheel rim from chafing the tire. Tubeless tires have an additional ply of rubber over the chafing strips to function as an air seal.

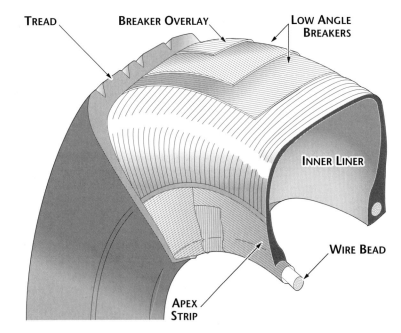

Figure 3-2-2. Ply layout for a radial tire

Figure 3-2-3. A nosewheel tire with a double chine

Breaker. The *breakers* are one or more plies of cord or woven fabric impregnated with rubber. They are used between the tread rubber and the cord body to provide extra reinforcement to prevent bruise damage to the tire. Breakers are not part of the cord body.

Tread patterns. There are many different tread patterns or tread designs. They are plain, ribbed and nonskid. A plain tread has a smooth, uninterrupted surface and commonly called an SC tire (for smooth contour). A ribbed tread has three or more continuous circumferential ribs separated by grooves. A nonskid tread is any grooved or ribbed tread. The most common aircraft tire tread is the ribbed pattern.

Ply rating. The term *ply rating* has superseded reference to the number of cord fabric plies in a tire. This term is used to identify a tire's maximum recommended load for specific types of service. It does not necessarily represent the number of cord fabric plies in a tire. Most nylon cord tires have ply ratings greater than the actual number of fabric plies in the cord body.

Chine. The *chine* is a flared ring that protrudes around one or both sides of a tire. Its purpose is to deflect water to the outside, thus keeping it away from the engine intakes or propellers. It is a term borrowed from the ship building industry. A nose wheel tire will normally have a chine on both sides, (Figure 3-2-3) while main gear normally have a chine on the outside only.

Size Designation

Figure 3-2-4 shows the points of measurement used to designate the size of a tire. For example, a tire with a size designation of 26 X 6.6 would have an outside diameter (measurement A) of 26 inches and a cross-sectional width (measurement B) of 6.6 inches.

The letter X merely separates the two measurements. If the tire's size designation were 26 inches X 6.6 inches -10 inches, then the tire would have a rim diameter (measurement C) of 10 inches. If only one numerical designation is used for a tire, you should assume that it is the outside diameter (measurement A).

Standard identification markings. You should be familiar with the markings on the sidewall of a tire. The markings engraved or embossed on a sidewall are shown in Figure 3-2-5.

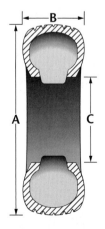

Tire and Tube Sizes

- *This typical diagram indicates the possible dimensions shown on the tire sidewalls.*

- *The symbol (X) is used between dimensions "A" and "B"; hence 26 x 6.6" means the overall outside diameter is 26" and the cross section width is 6.6".*

- *The symbol (-) is used between dimensions "B" and "C"; hence 7.50-10 means that the cross section width is 7.50" and the rim ledge diameter is 10".*

- *The symbol 'R' replaces the (-) for radial tires.*

- *Hence A x B - C, or A x B R C*

(A) OUTSIDE DIAMETER   (B) CROSS SECTIONAL WIDTH   (C) RIM DIAMETER

Figure 3-2-4. Size designation of tires

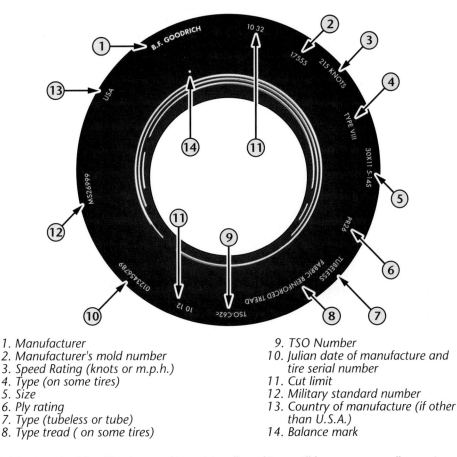

1. Manufacturer
2. Manufacturer's mold number
3. Speed Rating (knots or m.p.h.)
4. Type (on some tires)
5. Size
6. Ply rating
7. Type (tubeless or tube)
8. Type tread (on some tires)
9. TSO Number
10. Julian date of manufacture and tire serial number
11. Cut limit
12. Military standard number
13. Country of manufacture (if other than U.S.A.)
14. Balance mark

Figure 3-2-5. New tire identification markings. Not all markings will be present on all new tires.

Most of the markings are self-explanatory; item 10 has a maximum of 10 characters. The first four positions show the date of manufacture in the form of a Julian date (last digit of the year followed by the day of the year, or 17 Oct 2011 = 1290). The next positions are completed by the manufacturer and are either numbers or letters. They are used to create a unique serial number for a particular tire. The cut limit (11) is expressed in thirty-seconds of an inch and is used to evaluate the depth of cuts in the thread area. Tires are marked with a red dot (14) on the sidewall to indicate the lightweight (balance) point of the tire.

Vent Markings

Tube tires with inflation pressures greater than 100 p.s.i. and all tubeless tires must be suitably vented to relieve trapped air. Tube tires are vented in one of two ways. The first method uses air bleed ridges on the inside tire surface and grooves on the bead faces. The ridges and grooves channel the air trapped between the inner tube and the tire to the outside. The second method uses four or more vent holes that extend completely through each tire sidewall. They relieve both pocketed air and air that accumulates in the cord body by normal diffu-sion through the inner tube and tire. Tube tire vent holes are marked with an aluminum- or white-colored dot (Figure 3-2-6).

Tubeless tires have vent holes that penetrate from the outside of the tire sidewall to the outer plies of the cord body. They relieve air that

Figure 3-2-6. Vent hole markings (white dots) on a new tube type tire

accumulates in the cord body by normal diffusion through the tubeless tire liner and the tire carcass. Vent holes in tubeless tires are marked with a bright green dot.

NOTE: *Retreaded tires may not have the vent holes clearly marked.*

Tire Storage

The life of a tire, whether mounted or dismounted, is directly affected by storage conditions. Tires should always be stored indoors in a dark, cool, dry room. It is necessary to protect them from light, especially sunlight. Light causes ultraviolet (UV) damage by breaking down the rubber compounds. The elements, such as wind, rain and temperature changes, also break down the rubber compounds. Damage from the elements is visible in the form of surface cracking or weather check-

Figure 3-2-7. Incorrect tire storage. Tires, mounted and dismounted, should be stored vertically in a rack and protected from dirt, water, and sunlight.

Figure 3-2-8. This picture shows three common problems. The tire is under-inflated, the nose wheel is extremely dirty, and there is no protection for the wheel bearings.

ing. UV damage may not be visible. Tires must not be allowed to come in contact with oils, greases, solvents, or other petroleum products that cause rubber to soften or deteriorate.

The storeroom should not contain fluorescent lights or sparking electrical equipment that could produce ozone. Tires should be stored vertically in racks and according to size. The edges of the racks must be smooth so the tire tread does not rest on a sharp edge. Tires must never be stacked in horizontal piles (Figure 3-2-7). Tires removed from the storeroom should be based on age from the date of manufacture so the older tires will be used first. This procedure helps to prevent the chance of deterioration of the older tires in stock.

Tire Inspection

There are two types of inspections conducted on tires. One is conducted with the tire mounted on the wheel. The other inspection is conducted with the tire dismounted.

Mounted inspection. During each daily or special inspection, tires must be inspected for correct pressure, tire slippage on the wheel (tube tires), cuts, wear and general condition. Tires must also be inspected before each flight for obvious damage that may have been caused during or after the previous flight. Figure 3-2-8 is a typical underinflated nose gear on a Cessna single-engine airplane sitting in the tiedowns with no chocks.

Maintaining the correct inflation pressure in an aircraft tire is essential to safety and to obtain its maximum service life. Most aircraft inner tubes and tubeless tire liners are made of natural rubber to satisfy extreme low-temperature performance requirements. Natural rubber is a relatively poor air retainer. This accounts for the daily inflation pressure loss and the need for frequent pressure checks.

If this check discloses more than a normal loss of pressure, you should check the valve core for leakage by putting a small amount of suitable leak detection solution or soapy water on the end of the valve and watch for bubbles. Replace the valve core if it is leaking. If no bubbles appear, it is an indication that the inner tube (or tire) has a leak. When the tire and wheel assembly shows repeated pressure loss exceeding five percent of the correct operating inflation pressure, it should be removed from the aircraft.

WARNING: *Overinflation or underinflation can cause catastrophic failure of aircraft tire and wheel assemblies, particularly at the wheel flanges. This could result in injury, death and/ or damage to aircraft or other equipment.*

After making a pressure check, you should always replace the valve cap. Be sure that it is screwed on finger tight. The cap prevents moisture, oil and dirt from entering the valve stem and damaging the valve core. It also acts as a secondary seal if a leak develops in the valve core.

Some aircraft manufacturers use tire slippage marks. The slippage mark is a red paint strip one inch wide and two inches long. It extends equally across the tire sidewall and the wheel rim, as shown in Figure 3-2-9. Tires should be inspected for slippage on the rim as part of a post flight inspection. Failure to correct tire slippage may cause the valve stem to be ripped from the tube.

Tire treads should be inspected to determine the extent of wear. The maximum allowable thread wear for tires without wear depth indicators is when the tread pattern is worn to the bottom of the tread groove at any spot on the tire. The maximum allowable tread wear for tires with tread wear indicators is when the tread pattern is worn either to the bottom of the wear depth indicator or the bottom of the tread groove. These limits should apply regardless of whether the wear is the result of skidding or normal use.

The tread and sidewall should be examined for cuts and embedded foreign objects. Glass, stones, metal and other materials embedded in the tread should be removed to prevent cut growth and eventual carcass damage. A blunt awl or screwdriver maybe used for this purpose. You should be careful to avoid enlarging the hole or damaging the cord body fabric.

WARNING: *When you are probing for foreign objects, be sure you keep the probe from penetrating deeper into the tire. Objects being pried from the tire frequently are ejected suddenly and with considerable force. To avoid eye injury, safety glasses or a face shield should be worn. A gloved hand over the object may be used to deflect it.*

Aircraft should not be parked in areas where the tires may stand in spilled hydraulic fluids, lubricating oils, fuel, or organic solvents (Figure 3-2-10). If any of these materials is accidentally spilled on a tire, it should be immediately wiped with a clean, absorbent cloth. The tires should then be washed with soap and thoroughly rinsed with water.

Dismounted inspection. Whenever a tire has been subjected to a hard landing or has hit an obstacle, it should be removed in accordance with the applicable MMs and dismounted for a complete inspection to determine if any internal damage has occurred. The tire beads should

be spread, and the inside of the tire inspected with the aid of a light. If the lining has been damaged or there are other internal injuries, the tire should be removed from service. You should check the entire bead area and the area just above the bead for evidence of rim chafing and damage. Check the wheel for damage that may damage the tire after it is mounted.

Aircraft Tire Maintenance

In many cases, tire failures are attributed to material failures and/or manufacturing defects when actually improper maintenance was the underlying cause. Poor inspection, improper

Figure 3-2-9. Tire slippage mark on a removed wheel and tire

Figure 3-2-10. This nosewheel tire has no protection from the engine oil dripping on it. Too many tires are left sitting in the tiedowns this way for weeks.

buildup and operation of tires in an underinflated or overinflated condition, are common causes for tire failure.

During the mounting, dismounting and inflating of tires, safety is paramount. Compressed air and nitrogen present a safety hazard if the operator is not aware of the proper operation of the inflation equipment and the characteristics of the inflation medium. It is also very important to know the wheel type and be familiar with the manufacturer's recommended procedure before you attempt to dismount a tire. For specific precautions concerning a particular installation, you should always consult the applicable MM.

Dismounting

In the tire shop, you should recheck tires for complete deflation before disassembling the wheel and *breaking the bead* of the tire. Breaking the bead means separating the bead of the tire from the wheel flange.

Breaking the bead. The use of proper equipment for breaking the bead of the tire away from the wheel flange will save materials and labor. Aircraft tires, inner tubes and wheels can be damaged beyond repair by improper mounting and dismounting equipment and procedures.

Commercially available or locally fabricated equipment that uses either a hydraulically actuated cylinder or a mechanically actuated device may be used, provided the equipment

will not damage the tires or wheels. Figure 3-2-11 shows a bead-breaking tool.

Dismounting divided (split) wheels. The tire bead should be broken away from the wheel and the nuts and bolts removed. If the tire has a tube, push the valve away from the seated position. This will prevent damage to the inner tube valve attachment when you break the tire bead loose. Then, remove the wheel assembly from the tire.

If the tire is tubeless, remove the wheel seal carefully from the wheel half and place it on a clean surface. Wheel seals in good condition may be reused if replacement seals are not available. If the tire has a tube, remove it. Inner tubes can be reused if they are in good condition.

Dismounting remountable flange wheels. The tire bead should be broken away from the wheel according to the bead-breaking procedure. If the tire has a tube, you should push the valve away from the seated position. Again, this will prevent damage to the inner tube valve attachment when you break the bead. If you have trouble removing the flange while the wheel is mounted on the bead-breaking machine, remove the tire from the machine.

Lay the tire and wheel assembly flat with the demountable flange side up. Drive the remountable flange down by tapping it with a rubber, plastic, or rawhide-faced mallet. This should enable you to remove the locking ring.

> **CAUTION:** *Extreme care must be taken when you break the beads loose and remove the lockring on some remountable flange wheels. The toe of the remountable flange may extend very close to the tube valve stem. Excessive travel of the remountable flange or of the tire bead may cut the rubber base of the inner tube valve.*

If the tire is tubeless, remove the wheel seal carefully and place it on a clean surface. Wheel seals in satisfactory condition may be reused if replacement seals are not available. Turn the tire and wheel assembly over and lift the wheel out of the tire. Remember to keep the wheel flange and locking ring together as a unit to avoid mismatch during remounting.

Mounting

Prior to mounting a tire on a wheel, you should inspect the tire and ensure the inside of the tire is free of foreign materials. The inner tube must be inspected for bead chafing, thinning, folding, surface checking, heat damage, fabric liner separation, valve pad separation, damaged valves, leaks and other signs of deterioration.

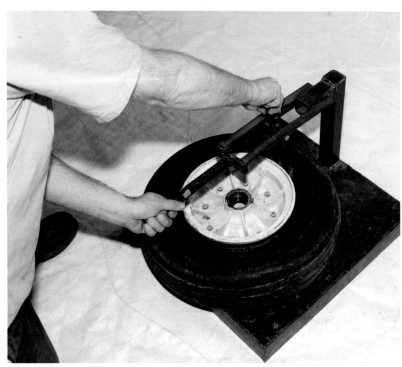

Figure 3-2-11. Using a bead-breaking tool

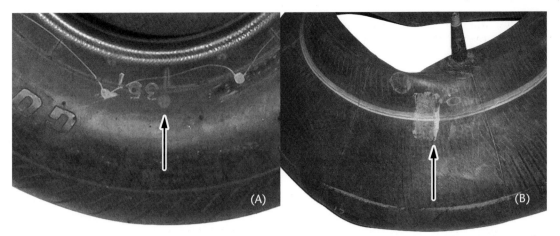

Figure 3-2-12. (A) Red dot balance marker on a tube type tire, and (B) a balance mark on an inner tube

Mounting divided (split) wheels. When you mount a tube tire, dust the tube with talcum powder and insert it in the tire. The tire should be positioned so the balance marker on the tube is located next to the balance marker on the tire.

> **NOTE:** *The balance marker on an inner tube is a stripe of contrasting colors approximately 1/2-inch wide and 2 inches long. It is located on the valve side of the tube. The balance mark on a tire is a red dot approximately one-half inch in diameter. It is located on the sidewall near the bead. (See Figure 3-2-12).*

You should inflate the tube until it is round, and then place the valve-hole half of the wheel into position in the tire. Push the valve stem through the hole. Finally, insert the other half of the wheel and align the bolt holes.

Install four bolts, nuts and washers 90 degrees apart. Start the bolts by hand, and tighten them evenly until the wheel halves seat. Install the remaining bolts, nuts and washers. Tighten the bolts in a crisscross order to prevent distorting the wheel or damaging the inserts. Use a calibrated torque wrench, and tighten each bolt in increments of 25 percent of the specified torque value in a crisscross order until the total torque value required for each bolt in the wheel has been reached.

> **NOTE:** *When Lubtork is specified on the wheel half, coat all the threads and bearing surfaces of the bolt heads. Lubtork must not be used on magnesium wheels. For magnesium wheels, you should use the lubricant specified or MIL-G-21164 lubricant. All excess should be removed.*

Before mounting tubeless tires, check the tire sidewall for the word *tubeless*. Tires without this marking should be treated as tube tires. When you mount tubeless tires, install the valve stem (valve core removed) in the wheel assembly. Removing the valve core prevents unseating of the wheel seal by the pressure built up when the tire is installed.

Insert one wheel half in the tire, and position the tire so the balance marker on the tire is located at the valve stem. Install the wheel seal. Install the other wheel half and align the bolt holes. Install the bolts, washers and nuts in the same manner used for the wheel assembly containing inner tubes.

Mounting remountable flange wheels. When you mount a tube tire on a remountable flange wheel, the inner tube should be prepared and inserted in the tire in the same manner used on a split or divided wheel. The wheel is then positioned on a flat surface with the fixed flange down. Push the tire on the wheel assembly as far as it will go, and guide the valve stem into the valve slot with the fingers. Install the remountable flange on the wheel. Secure the locking ring according to the assembly instructions required by the applicable wheel manual.

When you mount a tubeless tire on a demountable flange wheel, install the valve stem (valve core removed) in the wheel assembly. Removing the valve core prevents unseating the wheel seal by the pressure built up when the tire is installed. The wheel seal should be lubricated with the same lubricant and in the same manner as previously mentioned for split or divided wheel assemblies using tubeless tires. Install the wheel seal on the flange. Secure the locking ring according to the assembly instructions required by the applicable wheel manual.

Tire Inflating

The gas used to inflate a tire depends more on the airplane than on the actual tire. While it

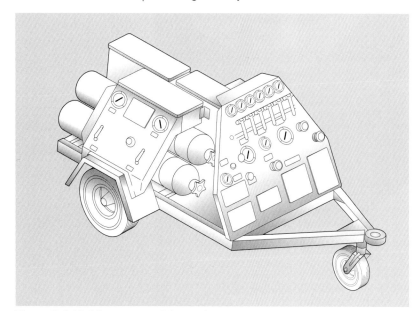

Figure 3-2-13. Nitrogen servicing units

Figure 3-2-14. Inflation safety cage with aircraft tire

exceed the required tire inflation pressure by more than 50 percent.

The tire inflator assembly consists of a remote regulator with low- and high-pressure gauges and a 10-foot service hose. The inflator assembly should be calibrated upon initial receipt, before being placed in service and every 6 months thereafter. The unit should be equipped with a built-in relief valve to prevent over pressurization of a tire during inflation. The relief valve should be set at 20 p.s.i. above the maximum pressure required.

After the buildup of a new tire, it should be placed in a safety cage for inflation. A typical safety cage is shown in Figure 3-2-14. The method of inflation used depends on whether a tube or tubeless tire is being inflated.

To inflate tube tires, you should remove the valve core and place the wheel assembly in the safety cage. Remember that there is more than one style of valve core (see Figure 3-2-15). A "H" on the core means high pressure. Attach a remote tire inflation gauge assembly to the valve stem. Be sure the inner tube is not being pinched between the tire bead and the wheel flange.

On remountable flange wheels, be sure the remountable flange and locking ring are seated properly. Secure the safety cage door and inflate the tire to its maximum operating pressure. This will seat the tire beads against the rim flanges. Deflate the tire and install the valve core. Then, inflate the tire to its maximum operation pressure. You should allow the tire to remain at this pressure for a minimum of ten minutes. At the end of this 10-minute period, there should be no detectable pressure loss.

If no pressure loss is detected, the tire pressure should be reduced to 50 percent of the maximum operating pressure and the tire and wheel assembly stored in a rack, ready for issue.

If there is a significant pressure loss, the tire pressure is reduced to 50 percent of the maximum operating pressure for safety, then the assembly should be removed from the safety cage and the cause of the leak determined. If a slow leak is detected, the air retention test should be extended to 24 hours. If the leakage exceeds five percent, the tire should not be used until remedial action is taken.

A loss of pressure less than five percent may be experienced during the first 24 hours after initial inflation of a new tire. This is attributed to normal tire stretch. The tire pressure should be adjusted accordingly. Tubeless tires are inflated in the same manner as tube tires except the valve core is not removed.

would be nice to inflate all aircraft tires with nitrogen, it simply is not practical to do so. Many light aircraft are simply inflated with shop air. However, shop air should be clean and dry (frequently it is not). Filter the shop air, if necessary.

Nitrogen is used in tires that will see high speed and high altitude. Thus, a Piper J-3 Cub might well use air, while anything with a turbocharger might use nitrogen. All turbine airplanes should use nitrogen.

All high-pressure inflation sources should be equipped with a regulator that limits the line pressure to the inflator (see Figure 3-2-13). The regulator should be set to provide a controlled inlet pressure to the inflator. It should not

Tire Preventative Maintenance

Debris on runways and in parking areas causes tire failures, and results in many tires being removed long before they reach full service life. It is important that those areas be kept clean at all times. When you ground handle an aircraft, do not pivot with one wheel locked or turn sharply at slow speeds.

This not only scuffs off the thread, but also causes internal separation of the cords. Always be sure the aircraft is moving before you attempt a turn. This allows the tire to roll instead of scrape.

You should make every effort to prevent oil, grease, hydraulic fluid, or other harmful materials from coming in contact with the tires. When there is a chance that harmful materials may come in contact with the tires during maintenance, they should be protected by covers. To clean tires that have come in contact with oil, grease, or other harmful material, you should use a brush or cloth saturated in a soap and water solution. Rinse well with tap water.

A holder of a pilot's certificate can replace or repair a tire on an aircraft owned or operated by the pilot. The operation is considered preventative maintenance and is allowed under 14 CFR part 43.

Uneven Tread Wear

If a tire shows signs of uneven or excessive tread wear, the cause should be investigated and the condition remedied before the tire is ruined. Some of the common causes of uneven tread wear are underinflation, overinflation, misalignment and incorrect balance.

Underinflation. Underinflation causes the tire to wear rapidly and unevenly at the outer edges of the tread. An underinflated tire develops higher temperatures during use than a properly inflated tire. This can result in tread separation or blowout failure.

Overinflation. Overinflation reduces the tread contact area, causing the tire to wear faster in the center. Overinflation increases the possibility of damage to the cord on impact with foreign objects.

Misalignment. Figure 3-2-16 shows rapid and uneven tire wear caused by incorrect camber or toe-in. The wheel alignment should be corrected to avoid further wear and mechanical problems.

Balance. Correct balance of the tire, tube and wheel assembly is important. A heavy spot on

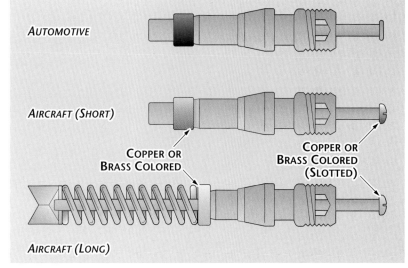

Figure 3-2-15. Valve core identification

Figure 3-2-16. Rapid tread wear caused by misalignment

an aircraft tire causes that spot to always hit the ground first upon landing. This results in excessive wear at the one spot and an early failure at that part of the tire. A severe case of imbalance may cause excessive vibration during takeoff and landing. This makes handling of the aircraft difficult. Figure 3-2-17 shows a tire that is actually out of round and more likely was difficult to balance.

Tire cuts. Tire cuts on the sidewall of a tire are generally more serious than a tread cut. In general, it depends on how deep the cut is. When the cut is deep enough that the cord

Figure 3-2-17. This tire was worn excessively from misalignment and was also out of round, making balance difficult. The high spot shows more wear than the rest of the tire.

Figure 3-2-18. This is a bad sidewall cut from a large piece of debris on the runway. If the cord is cut or damaged, the tire should be replaced.

Figure 3-2-19. The result of being outside and unprotected for too long is severe weather checking

has been cut the tire is due for replacement. Figure 3-2-18 shows a bad sidewall cut from a large piece of debris on the runway.

Weather checking. Figure 3-2-19 shows a mounted tire that has been outside for way too long. The carcass is severely weather checked. Water, dirt and more sunlight can penetrate the cracks and attack the underlying cord. A tire this bad needs replacing.

Nylon flat spotting

If the aircraft stands in one place under a heavy static load for several days, local stretching may cause an out-of-round condition with a resultant thumping during takeoff and landing. This is the so-called *nylon sickness.*

Dual Installations

On dual-wheel installations, tires should be matched according to specific dimensions. Table 3-2-1 is an example only. Do not use it as a service instruction. Tires vary somewhat in size between manufacturers and can vary a great deal after being used. When two tires are not matched, the larger tire supports most, or all, of the load. Since one tire is not designed to carry this increase in load, a failure may result.

Many aircraft manuals will specify a maximum size difference for dual tire installations.

Dual tire inflation. Dual tires must be inflated evenly. One may not be inflated more (or less) than the other. To do so will move a tremendous load to the tire with the higher pressure. So much so in fact, that it may fail with catastrophic consequences. Never push the pressure limits. They are there for a reason.

Matching Criteria

With the availability of several different tire manufacturers products in both bias and radial types, comes the problem of mixability. Some questions must be asked. Which tire can be used with another brand of tire? Who can use them? Can bias and radial tires be mixed on the same truck?

The answer to these questions is maybe. Each aircraft manufacturer will define tire-matching criteria in their aircraft maintenance manual. In most cases, the answer will be no. However, as always the manufacturer is the last word.

Aircraft Tubes

The purpose of the inner tube is to hold the air in the tire. Tubes are identified by the type and size of the tire in which they are to be used.

Identification. Tubes are designated for the tires in which they are to be used. For example, a type I tube is designed for use in a type I tire. The size of the tube is the size of the tire in which it is designed to fit.

Inner tubes required to operate at 100 p.s.i. or higher inflation pressure, are usually reinforced with a ply of nylon cord fabric around the inside circumference. The reinforcement extends a minimum of one-half inch beyond that portion of the tube that contacts the rim.

Type III and type VII inner tubes have radial vent ridges molded on the surface, as shown in Figure 3-2-20. These vent ridges relieve air trapped between the casings and the inner tube during inflation. Inner tube valves are designed to fit specific wheel rims.

Tube storage. Tubes should be stored under the same conditions as new tires. New tubes should be stored in their original containers. Used tubes should be partially inflated (to avoid creasing), dusted with talc (to prevent sticking) and stored in the same manner as tires. Under no circumstances should inner tubes be hung over nails or hooks.

Inspection. Inner tubes should be inspected and classified as serviceable or nonserviceable. Usually, the eye can detect leaks due to punctures, breaks in the tire and cuts. Small leaks may require a soapy water check. Complete submersion in water is the best way to locate small leaks. If the tube is too large to be submerged, spread soapy water over the entire surface and examine carefully for air bubbles. The valve stem and valve base should be swished around to break any temporary seals. The tube should be checked for bent or broken valve stems and stems with damaged threads.

Serviceable tubes. Inner tubes should be classified as serviceable if they are found to be free of leaks and other defects when they are inflated with the minimum amount of air required to round out the tube and water checked.

Nonserviceable tubes. Nonserviceable tubes may be repairable or nonrepairable. Nonserviceable tubes with the following defects should be classified as repairable:

- Bent, chafed, or damaged metal valve threads

- Replaceable leaking valve cores

TIRE OUTSIDE DIAMETER	MAXIMUM DIFFERENCE IN OUTSIDE DIAMETERS
Less than 18 inches	1/8-inch
18 to 24 inches	1/4-inch
25 to 32 inches	5/16-inch
33 to 40 inches	3/8-inch
41 to 48 inches	7/16-inch
49 to 55 inches	1/2-inch
56 to 65 inches	9/16-inch
More than 65 inches	5/8-inch

Table 3-2-1. Tolerances for diameters of paired tires in dual installations. Example chart only. Refer to airframe manufacturers tolerances for specific measurements.

Nonserviceable tubes with the following defects should be classified as nonrepairable:

- Any tear, cut, or puncture that completely penetrates the tube

- Fabric-reinforced tubes with blisters greater than one-half inch in diameter in the reinforced area

- Chafed or pinched areas caused by beads or tire breaks

- Valve stems pulled out of fabric-base tubes

- Deterioration or thinning due to brake heat

- Folds or creases

- Severe surface cracking

- No balance marker

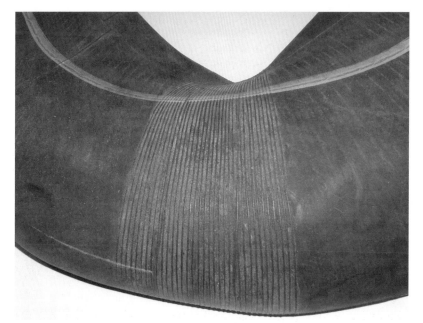

Figure 3-2-20. Inner tube vent ridges

Tire Rebuilding/Retreading

The rebuilding of aircraft tires has been practiced for many years. A rebuilt tire is one that has a new tread section attached to a good carcass of a worn tire. The process is called *recapping*. Each recapped tire saves aircraft operators approximately 75 percent of the cost of a new tire. History shows that a rebuilt tire gives service comparable to a new tire.

Tire Repairs

The following guidelines represent present acceptable industry practices regarding aircraft tire repairs:

For tires operated above 120 m.p.h.

- **Tread area.** Cuts, cracks, or other tread injuries 1 1/2-inches in length and 1/4-inch in width or less on the first cord body ply and which do not extend through more than 40 percent of the actual tire cord body plies are repairable. However, any tread injury repaired by skiving or rasping methods should not have the final repair greater than two inches in length.

- **Tread injuries.** Tread injuries that penetrate a distance equal to 40 percent of actual tire cord body plies and are 1 1/2-inches in length or less on the first cord body should be limited to six per tire, and should not be less than 60 degrees apart along the tire circumference.

- **Sidewall rubber.** Surface defects of any degree (checking radial and circumferential cracks, cuts and snags) may be repaired, provided the injuries do not penetrate into the cord body fabric plies.

- **Bead area.** Minor injuries to the bead area may be repaired, provided the plies are not damaged.

- **Bead seal.** The bead seal should not be affected or intersected by impressions or depressions.

- **Bead face and bead heel.** Those areas should be smooth.

- **Bead toe.** The bead toe should be trimmed so that no edges are exposed above the bead face and so that any bead toe flash remaining does not protrude more than 1/8-inch from the face contour of the bead. If trimming of the bead toe is necessary, the trimming shall not cut or expose the tire cord material or more than one layer of chafer fabric.

- **Chafer strip.** Minor injuries in the chafer strip or slight tire tool injuries in the general bead area are repairable, if they do not extend into the plies of the tire and there is no sign of separation in the bead area. Loose or blistered chafer strips can be repaired or replaced.

- **Inner liner.** Inner liner surface damage and defects other than liner splices that are less than two inches in length, may be repaired. A maximum of ten of these repairs are acceptable with no more than three repairs in any one quadrant. Liner splice damage defects may be repaired if it is less than ten inches in length.

- **Exposed cord.** Exposed cord, either in the breaker or carcass ply, should not exceed one percent of the buffed total tread area on one spot or more than two percent for the entire tire. Exposed fabric should not exceed one carcass ply in depth.

For tires operated below 120 m.p.h.

- **Bead injuries.** Repairs may be made where only the chafe resistant material is damaged or loose, or where minor injuries do not penetrate into more than 25 percent of the tire plies up to three damaged plies.

- **Tread or sidewalls.** Injuries may be repaired by the spot method. This includes cuts in the tread area that are smaller than 1/2-inch in length and do not penetrate more than the following number of plies into the cord body listed in Table 3-2-2.

NUMBER OF PLIES	MAX CUT DEPTH
Less than 8	None
8 through 16	2 plies
More than 16	4 plies

Table 3-2-2. Allowable cut depth for some low speed tires.

Retreadable Tires (Recaps)

Generally, tires with sound cord bodies and beads and those with flat spots that do not extend into more than one carcass ply are retreadable. However, each tire manufacturer has established repair data for their tires. Additionally, each retreader approved by the FAA also has a process specification that must meet both the manufacturers data and FAA AC 145-4.

What this means is that one retreader may not approve a tire that is repairable by another retreader. Once you get accustomed to working with a specific supplier, acceptance or rejection criteria is easier to recognize.

Marking of retreaded tires. The following is a direct quote from FAA AC 145-4:

A. Whenever it is necessary to replace the area containing the required original markings (Ref: TSO-C62c as applicable), or if those markings are damaged, such markings can be replaced by the retreader, except for the TSO identification, which can only be replaced at the direction of the manufacturer. When retreading tires for certain air carriers, they may require additional tire markings. In addition, each retreaded tire shoulder should be permanently embossed with at least the retread information as follows:

(1) The letter R followed by a number 1, 2, etc., to signify the sequential number of retreads applied thereon.

(2) The month and year of the retread application.

(3) The name or identifying letters of the retreader who retreaded the tire.

(4) The plant location of the retreader.

B. Balance marker. A balance marker, consisting of a red dot, is required to be affixed on the sidewall of the tire immediately above the bead to indicate the lightweight point of the tire. The dot is required to remain for any period of storage plus the retread life of the tire.

C. Tread design. All tires should have a full circumferential groove or other tread design, which will provide adequate traction for all operational maneuvers as specified by the manufacturer.

D. Tread reinforcement fabric. The tread reinforcing fabric should not end directly under an outer tread groove.

E. Balance tolerance. All tires should be balanced in accordance with the schedule set forth in TSO-C62c.

F. Tire weight. The weight of a tire should not be greater than the applicable aircraft type certificate limitations unless specifically approved.

Figure 3-2-21 is an example of all the markings on a recapped tire.

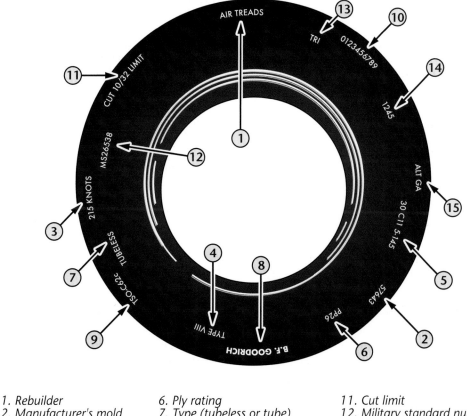

1. Rebuilder
2. Manufacturer's mold number
3. Speed Rating (knots or m.p.h.)
4. Type (on some tires)
5. Size
6. Ply rating
7. Type (tubeless or tube)
8. Original manufacturer
9. TSO Number
10. Original date of manufacture or casing serial number
11. Cut limit
12. Military standard number
13. Number of times rebuilt
14. Julian date of rebuild or month and year of rebuild
15. Rebuilder's plant location

Figure 3-2-21. Markings on a recapped tire. Not all markings will be present on each recapped tire.

Nonrepairable Aircraft Tires

If any of the following conditions exist, repair of the tire is not recommended:

- Injuries to the bead or bead area (except that repairs may be made where only the chafe resistant material is damaged, or loose, or if damage does not extend into the plies of the tire and if there is no sign of separation in the bead area).

- Bead injuries that affect the seal of the bead on tubeless tires.

- Evidence of separation exceeding process specification limits between plies or around bead wire.

- Injuries requiring reinforcement and all injuries requiring sectional repair.

- Kinked or broken bead.

- Weather checking or radial cracks that penetrate into body cords.

- Evidence of flex breaks.

- Loose internally damaged or broken cords.

- Broken or cut cords in the outside sidewall, or shoulder area.

- Evidence of blisters or heat damage to the bead seat where reversion scorching, or rubber flaking has occurred.

- Cracked, deteriorated, or damaged inner liners.

Figure 3-3-2. Typical divided (split) wheel assembly

- Flat spots and skid burns that have penetrated more than one carcass

- Tires that have been saturated with fuel, grease, or oil, to the point where tread adhesion or tire integrity could be adversely affected.

- Tires when sidewalls have been buffed and veneered three times

- Punctures that penetrate through the cord body are not repairable.

Section 3

Aircraft Wheels

Aircraft wheels are removed frequently for tire changes, inspections and lubrication. Wheels are among the most highly stressed parts of an aircraft. High tire pressures, cyclic loadings, corrosion and physical damage contribute to failure of aircraft wheels. Complete failure of an aircraft wheel can be catastrophic.

The dimensions used to identify wheels are not necessarily the dimensions of the wheels themselves. Instead, they refer to dimensions of the tire.

Aircraft wheels are made from either aluminum or magnesium alloys. This provides a strong, lightweight wheel that requires very little maintenance. Wheels are divided into three general types—drop center, split and remountable flange.

The wheels have knurled flanges to prevent the tires from slipping on the wheel. Wheels

Figure 3-3-1. An example of a drop center wheel

Figure 3-3-3. Remountable flange wheel

used with tubeless tires have the wheel halves sealed by an O-ring, and they use special valve stems that are a part of the wheel.

Drop center wheels. Not common in today's environment, drop center wheels were manufactured just like automobile wheels, only many times stronger (Figure 3-3-1). They are one piece, with the brake drum shrunk and bolted to the inside of the wheel, just as an automobile brake system. The brake systems were typically either single circular shoe or expander tube type.

Mostly on main gears, drop center wheels mount tires that are 14 or more plies. They must be mounted and dismounted the same as automobile tires, i.e. using a wheel holder, tire irons and a big tire hammer. The lever type automobile tire changers used by today's tire shops will not do the job. If you have to dismount and mount tires on drop center wheels, find a tire professional who will do the job for you.

Split wheels. Figure 3-3-2 shows a typical *split (divided) wheel*. This type of wheel is divided into two halves. The two halves are sealed by an O-ring and held together with nuts and bolts. Each wheel half is statically balanced. This type of wheel is universally used on nose, main and tail landing gears of modern aircraft.

Remountable flange wheel. The remountable flange wheel is made so one flange of the wheel can be removed to change the tire. How the flange is attached depends on how large the wheel is. If it is a small wheel, the flange is held in place by a lockring. Larger wheel flanges, or rings, are attached with a series of studs and nuts. The wheel is balanced with the flange mounted on the wheel. Then, both the wheel and flange are marked. To ensure proper balance of the wheel during assembly,

the two marks should be lined up. Figure 3-3-3 shows a typical remountable flange wheel. This type of wheel is commonly used on the main landing gear of older aircraft. They are also common as tail or nose wheels on these same aircraft.

Typical Wheel Assembly

The bearing cups are shrink-fitted into the hub of the wheel casting, and are the parts on which the bearings ride. The bearings are tapered roller bearings. Each bearing is made of a cone, retainer and rollers. This type of bearing absorbs side thrust as well as radial loads and landing shocks. These bearings must be cleaned and lubricated in accordance with the appropriate MM.

A three-piece grease retainer keeps the grease in the inboard bearing and keeps out dirt and moisture. It is composed of a felt seal and inner and outer closure rings. A lockring secures the assembly inside the wheel hub. See Figure 3-3-4.

The hubcap seals the outboard side of the hub. It is secured with a lockring. On some aircraft, the hubcap is secured with screws.

All wheels designed for use on the main landing gear are equipped with brake components. These components are attached to the wheel casting. They may consist of either a brake drum or brake drive keys.

The trend tends to be faster and more powerful aircraft with increased load carrying capabilities. This means heavier loads and higher landing speeds. The friction of long landing rollouts and taxiing causes heat to be absorbed by the wheel. Because of the heat, possible wheel failure may occur.

Figure 3-3-4. A typical felt grease retainer; two thin flat washers, a felt ring and a snap ring

This may damage equipment and injure personnel. To prevent this situation, aircraft manufacturers have developed a safety device called a *fusible plug* shown in Figure 3-3-5. The fusible plug contains an alloy that will melt at a specific temperature and permit the tire to deflate, rather than explode. This action occurs if the wheel is exposed to excessive heat.

Wheel Maintenance

Corrosion and loss of bearing lubrication are two of the major causes of failure or rejection of aircraft wheels. It is extremely important to protect aircraft wheels and bearings from water, particularly salt water or runway deicer. Wheel bearing lubrication gets contaminated, or breaks down, from excessive heat and water.

Grease seals in aircraft wheels are not designed to withstand a direct stream of water. When wheels are exposed to a stream of water (such as a hose), it will usually penetrate the hub area, contaminating the bearing lubricant. This contributes to corrosion in the bearing area. All wheel bearings should be lubricated at every tire change, and as required by the applicable maintenance schedule. It is a requirement at each 100-hour and annual inspection.

Cleaning. You should clean bearings, bearing cups, wheel bores and grease retainers with solvent, to remove all traces of the grease, preservative compounds and contamination. Dry the bearings and the hub area with compressed air, but *do not spin the bearings with the air nozzle.* At this point they are no longer lubricated and the r.p.m. will exceed the bearings rating. You should perform a visual inspection of the bearings, bearing retainers and bearing cups with a 10X magnifier. Replace all excessively worn,

dented, scored, or pitted bearing cups. Most bearing cups will display some wear.

Any obvious defects on bearing cone and roller assemblies, including cracks in the bearing retainer, are cause for replacement.

Lubrication. You should repack the bearings with the manufacturers specified grease. They can be repacked by hand or by using a bearing packer. Spread a thin layer of grease on the bearing cups. Inspect the felt grease retainers for deterioration, contamination, or water saturation. Replace them if necessary. You should presoak felt retainers with light oil prior to their installation. Reinstall the wheel on the aircraft according to the applicable maintenance manual (MM).

Installation. When you reinstall the wheel on the aircraft, the proper adjustment of the bearings is extremely important. The following general rules apply to wheel installation:

- Tighten the axle nut while you spin the wheel with your hand.

- When the wheel no longer spins freely, back off the axle nut one castellation (one-sixth turn).

When properly installed and adjusted, the wheel will turn freely, but will not move sidewise.

NOTE: *This procedure may vary from one aircraft to another. Some aircraft require a specific torque to be applied to the axle nut. In these cases, you should refer to the applicable MM.*

- Install the appropriate axle nut safety device (usually a cotter pin).

- Install and lock the hubcap in place.

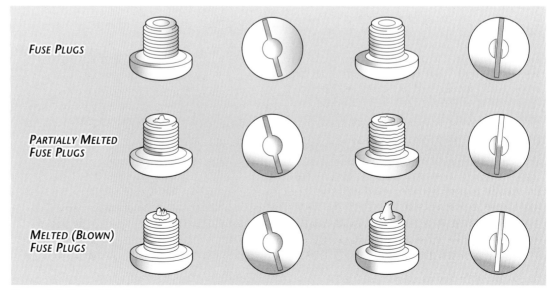

Figure 3-3-5. Fusible plugs

Safety. When you perform tire and wheel maintenance, you should handle inflated and partially inflated wheel assemblies with the same respect because of the destructive potential of a gas under pressure.

Wheel Overhaul

It is possible for you to overhaul an item as uncomplicated as a wheel. It is just another part of the maintenance process. Other responsibilities include painting, cleaning, inspection (lubrication), blending maintenance and corrosion and physical damage blendout.

Painting. When the wheel paint has deteriorated to the extent that touch-up is not feasible, wheels should be stripped and repainted.

Cleaning. To inspect aircraft wheels for cracks, physical damage and corrosion, they must be clean. All dirt, rubber and grease deposits must be completely removed. Cleaning for appearance sake is generally a requirement. Removing stains without damaging the finish is a necessity.

Many wheels will be discolored after the rubber deposits have been removed from the tire bead areas. This discoloration is acceptable, and further cleaning is not necessary. Often there are discolored areas around brake keys that are difficult to remove without damaging the paint.

The following steps describe the wheel cleaning procedures. The process described is for example only. You must check the appropriate manufacturers maintenance manual for recommendations. Not all wheels are the same.

Clean the wheels as follows:

1. Prepare one tank (solution A) of cleaning solution consisting of 4 to 9 parts cleaning solvent and 1 part solvent emulsion cleaner.

2. Prepare another tank (solution B) of cleaning solution consisting of 4 to 9 parts of clean water and 1 part emulsion cleaner.

3. Place the wheel portion to be cleaned on a grill over solution A, and spray it thoroughly with solution A to remove all loose grease and soil.

4. Immerse the wheel portion in solution A, and allow it to soak for 20 minutes.

5. Repeat step 3, and then scrub the tire bead areas with bristle brushes to remove the rubber deposits. Do not use wire brushes.

6. Thoroughly dry the wheel with compressed air.

7. Immerse the wheel portion in solution B, and allow it to soak for 20 minutes.

8. Place the wheel portion on a grill over solution B, and spray it thoroughly with solution B. Remove any remaining soil or grease deposits with liberal amounts of solution B and bristle brushes.

9. Thoroughly wash the wheel portion with a high-pressure stream of clean water to remove all solvents. Compressed air may be used to dry the wheel.

Inspection. You should perform a visual inspection of the wheel for cracks, loose bearing cups, corrosion, physical damage and melted fusible plugs. See Figure 3-3-5.

A plug may not need to be replaced. If the eutectic material appears to be filed, sanded, or broken, you shoulvd assume the serviceable limits have been exceeded and reject the plug.

You should perform the eddy current and dye penetrant inspections on wheels. Inspect all tie bolts for corrosion, elongation, bending, stripped threads, or deformed shanks. You should also perform a magnetic particle inspection for cracks on all steel parts. Any of the listed defects is cause for rejection of the tie bolt. Self-locking tie bolt nuts may be reused provided that the nut cannot be turned onto the tie bolt by hand. On disc wheels, you should inspect brake keys or gears for wear and looseness. Replace worn brake keys and gears, or reattach loose brake keys and gears. Corrosion or rust on brake keys and gears is common, and is not cause for rejection.

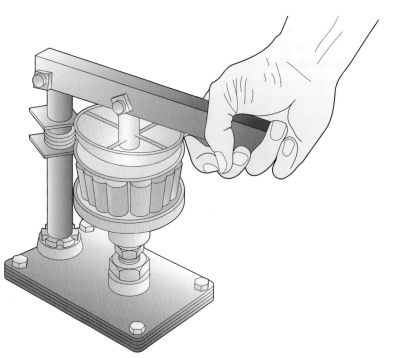

Figure 3-3-6. Pressure repacking of wheel bearings

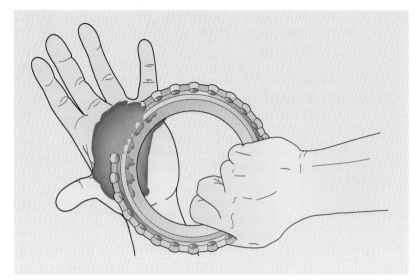

Figure 3-3-7. Hand repacking of wheel bearings

Bearing maintenance. You should remove and inspect the bearing cone and roller assemblies according to the applicable MM. Thoroughly clean the bearings, bearing cups, wheel bores and grease retainers with solvent to remove the grease, preservative compounds and contamination.

You should repack bearings with the recommended grease. Bearings may be repacked either with pressure equipment or by hand. See Figures 3-3-6 and 3-3-7. The pressure method is recommended because it is easier, faster and reduces the possibility of contamination. The pressure method assures a more even distribution of grease within the bearing.

> NOTE: *You should ensure bearings are completely dry before packing them with lubricant. If there is still solvent in the races, it will dilute the grease.*

You should also spread a thin layer of grease on the bearing cups. Inspect the grease retainers for evidence of deterioration, contamination, or water saturation. You should replace them if necessary. Presoak the retainers with light oil prior to installing them.

Corrosion and physical damage blendout. Limited and isolated corrosion and physical damage should be blended. Generally, wheel rims, outside ends of bearing hubs, nicks, gouges and pock marks are not considered significant unless the defect is deeper than 0.020 of an inch. The defect should not be blended out unless there is active corrosion in the defect. However, all burrs must be removed. The maximum depth of blendout will be contained in the component maintenance manual for the product.

The rims, bearing hub ends and tire bead area can be blended out with a medium or fine cut,

half-round or round tile. You should lightly file the damaged area to remove the defects. After the defects have been removed, you should hand polish the areas with 320 or finer grit aluminum oxide paper. All file marks should be removed.

Section 4

Aircraft Brake Systems

For the first 20 years or so of powered flight, airplanes did not need brakes. Landing speeds were slow, airports were not paved and tail skids provided enough drag to slow things down fairly quickly.

The rapid advances in aviation after WWI produced a proliferation of larger, faster and more complicated airplanes. Paved airports and ramps made tail wheels and brakes a necessity.

The coming of the jet age produced airplanes that exceeded anything to date in capacity. They were heavier, faster, carried more weight and landed heavier than ever before. System design and improvements kept pace, and braking systems were in the forefront. Figure 3-4-1 shows the system of an Airbus A300-600. It is typical of large jet transport systems.

Mechanical brakes. The first brake systems were simple cable operated shoe brakes; not more than you would find on a simple go-cart today. Most all were cam operated, with maintenance being nothing more than cable adjustments and replacing brake linings. These systems are still around today in some of the older classical general aviation aircraft. They more or less worked for their intended purposes, but were less than adequate when airplanes got heavier and faster.

Air brakes. Air powered brakes were popular on some of the English WWII military airplanes. Like mechanical brakes, they are still in use today by a small number of classical airplanes. As such, they really demand no knowledge other than the fact that they exist.

Hydraulic brakes. As airplane designs evolved, hydraulic brake systems gained universal acceptance and are used today almost exclusively. From the Piper Cub to the latest jet transport designs, all hydraulic brake systems operate on the same basic principle. They differ greatly in complexity.

Some light aircraft are equipped with a single master cylinder that is operated by a hand-lever

(Johnson bar) and applies brake action to both main wheels simultaneously. Steering on this system is accomplished by nose wheel linkage.

Most, however, use independent brake systems. Each brake system (each side) contains its own fluid reservoir and is completely independent of the aircraft's hydraulic system.

The brakes are installed on each main landing gear wheel and are operated independently of each other. The brakes must supply enough force to provide for holding the aircraft during normal run-up, slowing the aircraft after landing and steering the aircraft. The respective rudder pedal controls the brakes by the use of either a toe pedal or a heel pedal.

Independent-Type Brake System

An independent brake system contains the following components:

- A hydraulic reservoir
- One or two master cylinders
- Mechanical linkage from the rudder pedal to the master cylinder
- Fluid lines and hoses
- A brake assembly for each main gear wheel

In general, the independent-type brake system is used on all small aircraft. This type of brake system is termed *independent* because it has its own reservoir and is entirely independent of the aircraft's main hydraulic system.

Master Cylinders

Master cylinders similar to those used in the conventional automobile brake system power the independent-type brake system However, there is one major difference-the aircraft brake system has two master cylinders while the automobile system has only one.

An installation diagram of a typical independent type brake system is shown in Figure 3-4-2. The system is composed of a reservoir, two master cylinders and mechanical linkage, which connects each master cylinder with its corresponding brake pedal, connecting fluid lines and a brake assembly in each main landing gear wheel.

Each master cylinder is actuated by toe pressure on its related pedal. The master cylinder builds up pressure by the movement of a piston inside a sealed fluid-filled cylinder. The resulting hydraulic pressure is transmitted to the fluid line, which is connected to the brake

Figure 3-4-1. The A300-600 Airbus is a classic example of the brake systems of today.

assembly in the wheel. This action results in the friction necessary to stop the wheel.

When the brake pedal is released, the master cylinder piston is returned to the OFF position by a return spring. Fluid that was moved into the brake assembly is then pushed back to the master cylinder by a piston in the brake assembly. The brake assembly piston is returned to the OFF position by a return spring in the brake.

The typical master cylinder has a compensating port, or valve, that permits fluid to flow from the brake chamber back to the reservoir when excessive pressure is developed in the brake line due to temperature changes. This feature ensures against dragging or locked brakes.

Should a master cylinder break a return spring, one of two things would happen:

- The compensating port would not allow the brake to release
- The brake would drag

Various manufacturers have designed master cylinders for use on aircraft. All are similar in operation, differing only in minor details and construction.

Types of Master Cylinders

There have been several types of master cylinder designs, but they have essentially settled down to one basic type in most general aviation airplanes: the vertical master cylinder and its variations. In this section, we will cover basic vertical cylinders, as well as a common horizontal manual cylinder and a couple of different types of power cylinders.

These units were chosen for their variety of operation methods. In practice, each manufacturer, particularly in larger airplanes, have their own designs. Most all work on one of these principals.

Vertical master cylinders. Each master cylinder is actuated by pressure applied to the toe pedal. The movement of the piston down into the master cylinder provides pressure, which is forced through the fluid lines and against the brake assembly at the wheel. The pressure provides friction in the brake assembly, which results in reduced movement of the wheel (Figure 3-4-3).

When the brake pedal is released, the master cylinder piston is returned to the static position by a return spring. Fluid that was moved into the brake assembly is then pushed back to the master cylinder by a piston in the brake assembly. The brake assembly piston is returned to the *off* position by a return spring in the brake.

The return spring also operates a *compensator valve*. When the master cylinder is returned to the *battery* (normal rest) position the spacer tubing pushes down on the shaft seal washer, allowing a fluid passage between the reservoir and the cylinder to open. This always keeps the cylinder full of fluid.

> **NOTE:** *Any time master cylinder travel is excessive, first check the reservoir fluid level. If it is full, you can be fairly sure that the problem is in the compensator valve. Either the O-ring is damaged or has some form of debris that will not allow it to seal when the pedal is pushed.*

Internal leakage can also allow the pedal to slowly go down with the brakes applied, as the fluid leaks past the seal.

Parking brake lever. Most vertical master cylinders have a built-in parking brake in the form of a wedging type lever. When pressure is applied to the pedal and the parking brake lever raised, it will wedge the shaft so it cannot return to battery, thereby locking the brakes. Pushing on the pedals again and releasing the wedging action, releases the parking brakes, allowing the master cylinder shaft to return to the off position.

> **NOTE:** *Do not set parking brakes on aircraft equipped with hydraulically operated multiple disk-type brake assemblies while the brakes are hot, as this may cause the brake disc to warp.*

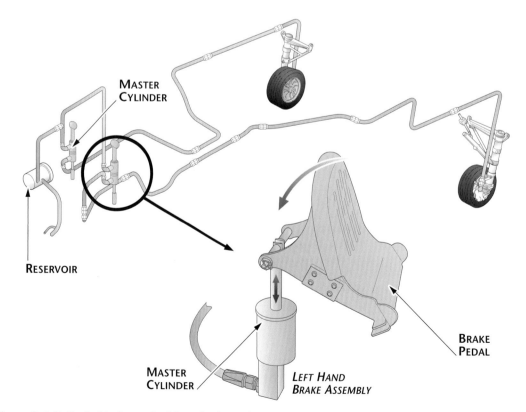

Figure 3-4-2. Typical independent-type brake system

Allow sufficient time for brakes to cool after landing prior to setting parking brake.

These types of parking brakes are very effective, but have three problem areas:

- There is no temperature compensation. When the parking brakes set and ramp temperature raises, the line pressure increases with no relief. This can damage components.

- With continued use, the vertical shaft can become grooved with marks from the parking brake lever. It may not operate properly.

- Debris can get under the lever and not allow it to release properly.

Goodyear master cylinder. Goodyear manufactured one of the original types of master cylinders. A cutaway view of the Goodyear master cylinder is shown in Figure 3-4-4. Fluid is fed by gravity to the master cylinder from an external reservoir. The fluid enters through the cylinder inlet port and compensating port and fills the master

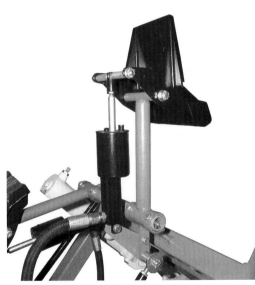

Figure 3-4-3. Vertical master cylinder

cylinder casting ahead of the piston and the fluid line leading to the brake-actuating cylinder.

Brake Assemblies

Brake assemblies commonly used on aircraft are the single-disc, dual-disc, multiple-disc and the segmented rotor types. Automotive shoe type brakes are rare, but are found on some older general aviation airplanes, as well as some ex-WWII trainer types. There are also some older expander tube type brakes that are occasionally found.

The single and dual-disc types are commonly used on small aircraft; the multiple-disc type is normally used on medium-sized aircraft; and the segmented rotor types are commonly found on heavier aircraft.

There are five main manufacturers represented in the brake systems in current use:

- Goodrich
- Goodyear
- Cleveland
- Dunlap
- Messier-Bugatti

Within each manufacturers product line there are enough different sizes and models to confuse anyone. Brake units range from a light airplane two-puck unit that weighs about two pounds each, to airline units that weigh over 250 pounds each.

Pucks are available in a variety of materials, depending on the friction requirements. The materials range from organic to sintered iron, trimetallic, ceramic-metallic hybrid, carbon and carbon-boron.

For each type of puck material, there is a break-in procedure. If relined brakes are put into service without breaking them in properly,

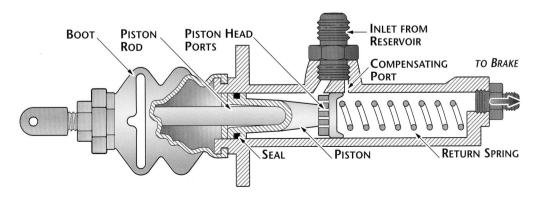

Figure 3-4-4. Goodyear master brake cylinder

their effectiveness and service life will be significantly reduced. Most airplane Maintenance Manuals (MM) will give you a break-in procedure. For Cleveland pads, information on break-in is available in their maintenance manual, or on their web site (www.parker.com/cleveland).

Expander Tube Brakes

The expander tube brake is a low-pressure brake with 360° of braking surface. It is light in weight, has few moving parts and may be found on older aircraft, both large and small.

An exploded view of the expander tube brake is shown in Figure 3-4-5. The main parts of the brake are the frame, expander tube, brake blocks, return springs and clearance adjuster.

The brake frame is the basic unit around which the expander tube brake is built. The main part of the frame is a casting that is bolted to the torque flange of the landing gear shock strut. Detachable metal sides form a groove around the outer circumference into which moving parts of the brake are fitted.

The expander tube is made of neoprene reinforced with fabric, and has a metal nozzle through which fluid enters and leaves the tube.

The brake blocks are made of special brake lining material, the actual braking surface being strengthened by a backing plate of metal. The brake blocks are held in place around the frame and are prevented from circumferential movement by the torque bars.

The brake return springs are semi-elliptical or half-moon in shape. One is fitted between each separation in the brake blocks. The ends of the return springs push outward against the torque bars, while the bowed center section pushes inward, retracting the brake blocks when the brakes are released.

When hydraulic fluid under pressure enters the expander tube, it expands. Since the frame prevents the tube from expanding inward and to the sides, all movement is outward. This forces the brake blocks against the brake drum, creating friction. The tube shields prevent the expander tube from extruding out between the blocks, and the torque bars prevent the blocks from rotating with the drum. Friction created by the brake is directly proportional to brake line pressure.

The clearance adjuster consists of a spring-loaded piston acting behind a neoprene diaphragm. It closes off the fluid passage in the inlet manifold when the spring tension is greater than the fluid pressure in the passage. Tension on the spring may be increased or decreased by turning the adjustment screw. Some of the older models of the expander tube brake are not equipped with clearance adjusters.

For brakes equipped with adjusters, clearance between the brake blocks and drum is usually set to a minimum of 0.002 to 0.015 in., the exact setting depending on the particular aircraft concerned. All brakes on the same aircraft should be set to the same clearance.

To decrease clearance, turn the adjuster knob clockwise; to increase clearance, turn the

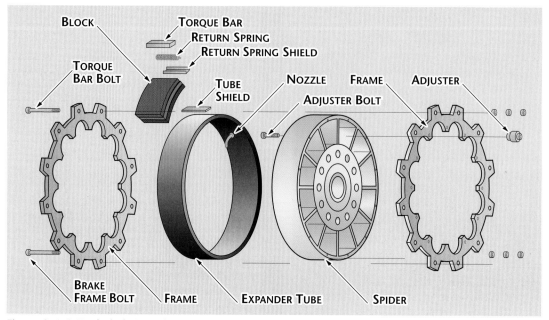

Figure 3-4-5. Exploded view of expander tube brake

adjuster knob counterclockwise. Kept in mind, however, that turning the adjuster knob alone does not change the clearance. The brakes must be applied and released after each setting of the adjuster knob to change the pressure in the brake and thereby change the brake clearance.

Expander tube brakes are difficult to maintain in comparison to disc brakes. Neither will they tolerate the heat output of a disc brake. Expander tube brakes have been replaced by disc brakes on all aircraft built in the last several decades. For the few that still exist, after-market replacement kits are available from STC holders to convert them to modern disc brakes. The FAA still tests on them.

Shoe Brakes

Automotive style two-shoe brakes are not common on airplanes. A few older aircraft, and some crop duster conversions, are about all that exist. There are a few systems that use one large brake shoe, instead of two smaller ones, found on WWII trainers. It is very unlikely you will encounter them. Should you have the occasion to work on any, there is something about their design you should know.

All shoe brakes work on the *servo-shoe* principal. That is, the *leading shoe* (front one on either side) always exerts more pressure than the rear one. In essence, it tries to wrap around the inside of the drum, the shoe anchor point only stops it. In the case of a circular shoe, the anchor point on the trailing end of the shoe stops it. This process greatly increases the pressure that can be applied to the brake.

If you look at a set of automotive brake shoes, you will notice that the leading shoe is shorter than the trailing shoe. It is also made from a different material. This is because the coefficient of friction and the area (square inches of rubbing surface) establish the servo action of the brake. A large-area leading shoe can cause a sudden, or violent, application of brakes. Cutting the shoe back reduces the violence of the brake application to something reasonable.

Disc Brakes

Single disc brakes. The single disc brake is very effective for use on smaller aircraft. Braking is accomplished by applying friction to both sides of a rotating disc, with the disc turning with the landing gear wheel. There are several variations of the single disc brake. All operate on the same principle and differ mainly in the manner in which the disc is attached to the wheel.

Figure 3-4-6 shows two different types of single disc brakes installed on light aircraft. Illustration A is a Cleveland brake assembly. In the Cleveland assemblies, the disc is mounted solid to the wheel. A caliper adaptor is bolted solid to the axle. The caliper carries both brake pucks and floats on pins that can adjust in and out of the adapter to compensate for wear. Changing brake pads (pucks) is a five-minute operation and the wheel never leaves the ground.

Illustration B is a Goodyear brake assembly on an identical landing gear leg. The brake housing is attached solidly to the landing gear axle flange with mounting bolts. The disc slides in and out of the wheel slots to compensate for wear. To replace pucks the airplane must be jacked up and the wheel removed. Then the caliper must be unbolted from the axle so the disc can be separated from the caliper. Next, the pads can be changed, then everything put back together in reverse order.

Figure 3-4-6. Typical single disc brake installations: (A) a Cleveland brake system and (B) a Goodyear brake system

Figure 3-4-7. Dual disc brake

Figure 3-4-8. Edge view of a multiple disc brake showing the disc stackup

Maintenance of the single disc brake may include bleeding, performing operational checks, checking lining wear, checking disc wear and replacing worn linings and discs.

A bleeder valve is provided on the brake housing for bleeding all disc brakes. Bleeding should be performed according to the instructions contained in the aircraft MM.

Operational checks are made during taxiing. Braking action for each main landing gear wheel should be equal, with equal application of pedal pressure and without any evidence of soft or spongy action. When pedal pressure is released, the brakes should release without

any evidence of drag. All disc type brakes must be checked periodically for lining wear. Excessively worn linings must be replaced.

The method used to check lining (puck) wear depends upon the model of the brake assembly. Different methods are described later in this chapter. Before checking the brakes on any aircraft, always refer to the applicable Maintenance Manual (MM) and use the method recommended by the aircraft manufacturer.

Dual disc brakes. Dual disc brakes are used on aircraft where more braking friction is desired with lower pressures.

The dual disc brake is very similar to the single disc type, except that two rotating discs, instead of one, are used. One model of this brake is shown in Figure 3-4-7.

The unit consists of a housing assembly, a center carrier assembly and two rotating discs. The housing assembly contains eight cylinders, each of which contains a piston, a return spring and a self-adjusting pin. Brake linings (pucks) are attached to each piston, to both sides of the center carrier and to the housing assembly, which makes a total of 24 pucks.

When hydraulic pressure is applied to the pistons, the pucks are forced against the first disc, which contacts the pucks in the center carrier. This force moves the center carrier and its pucks move against the second disc, forcing it in contact with the pucks in the housing. In this manner, each disc receives equal braking action on both sides as the pressure is increased. When pressure is released, the return springs force the pistons back to the preset clearance between the pucks and the disc. The self-adjusting feature is identical to that described for the single floating disc brakes. Maintenance of dual disc brakes is the same as for the previous single disc type.

Multiple/trimetallic disc brakes. Multiple disc brakes are heavy-duty brakes designed for use with power brake control valves or power boost master cylinders. The brake assembly consists of a bearing carrier, bearings, the retaining nut, the annular actuating piston and the heat stack. Which is composed of a pressure plate, rotating discs (rotors), stationary discs (stators) and backup plate, an automatic adjuster, retracting springs and various other components.

Regulated hydraulic pressure is applied through the automatic adjuster to a chamber in the bearing carrier.

The bearing carrier is bolted to the shock strut axle flange and serves as a housing for the annular actuating piston. Hydraulic pressure forces the annular piston to move outward, compressing the rotating discs, which are keyed to the landing wheel and the stationary discs, which are keyed to the bearing carrier. The resulting friction causes a braking action on the wheel and tire assembly.

When the hydraulic pressure is relieved, the retracting springs force the actuating piston to retract into the housing chamber in the bearing carrier. The hydraulic fluid in the chamber is forced out by the return of the annular actuating piston, and is bled through the automatic adjuster to the return line. The automatic adjuster traps a predetermined amount of fluid in the brake, an amount just sufficient to give correct clearances between the rotating discs and stationary discs. See Figure 3-4-8.

The trimetallic disc type brakes operate on the same basic principle as the multiple disc brakes and will be discussed in detail later in this chapter.

Carbon brakes. High performance is always expensive, and carbon brake units are no exception. They are the best for very high speed high load stopping systems. Aside from more exotic materials, they do not look much different from other brake systems. The difference is in the performance. Carbon brakes will work when other brakes would melt from the heat. As an example, some carbon/ceramic systems regularly operate at over 2,000°F, a temperature that will melt steel disks.

To perform at this level, carbon brakes do more than simply withstand more heat. The carbon material actually acts as a heat sink, soaking up the heat of friction at a rate that exceeds steel by 2.5 times. The heat is then released more slowly into the rotors. Conventional brake materials work in the other direction; they try to keep the linings cooler for longer lining life. This heats the steel disks more rapidly and reduces their life.

As carbon brakes heat up, they do not lose efficiency, nor does the material wear more rapidly. This makes braking itself much more efficient. It translates directly into greater safety by reducing braking distances.

While on the subject of braking distances, think on this: An aircraft can only brake at a rate that will not exceed the coefficient of friction of its tires against the runway. The power of the brakes and the coefficient of friction cannot exceed the structural strength of the landing gear and its attachment to the aircraft. Any excess brake capacity is useless.

Because of the efficiency of carbon brakes, many brake manufacturers have reduced the size of

Figure 3-4-9. A carbon brake wheel unit. Carbon brakes are designed for maximum friction and thus to handle a very high heat range.

their brake assemblies to four rotor designs. While providing all the braking power that the airframe can absorb, the units weigh almost a ton (2,000 pounds) less than conventional braking systems on a wide body airplane.

Most wide body airplanes, some regular airline airplane types, and almost all military airplanes use carbon brakes.

As an A&P technician, you are not apt to work on carbon brakes until you have acquired enough experience to do so without problems. Besides, if you remember the FARs, you must be under supervision the first time you do anything. Figure 3-4-9 shows a carbon brake assembly.

Power Boost Brake System

As a general rule, the power boost brake system is used on aircraft that land too fast to use the independent-type system, but are too light in weight to require the *power brake control system*. In this type of system, a line is tapped off from the main hydraulic system pressure line, but main hydraulic system pressure does not enter the brakes. Main system pressure is used only to assist pedal movement. This is accomplished through the use of power boost master cylinders.

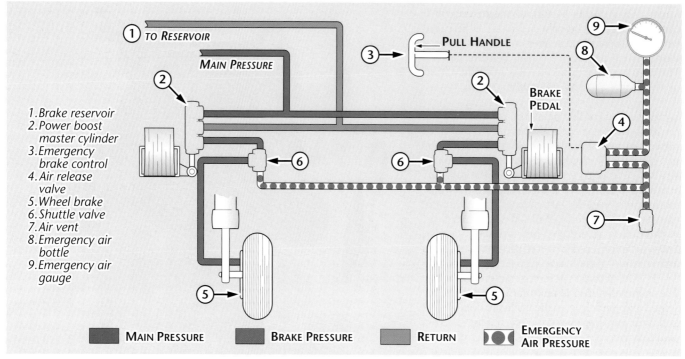

Figure 3-4-10. Power boost brake system

A schematic diagram of a typical power boost brake system is shown in Figure 3-4-10. The normal system consists of a reservoir, two power boost master cylinders, two shuttle valves and the brake assembly in each main landing wheel. A compressed air bottle with a gauge and release valve is installed for emergency operation of the brakes.

In this system (Figure 3-4-10), main hydraulic system pressure is routed from the pressure manifold to the power boost master cylinders. When the brake pedals are depressed, fluid for actuating the brakes is routed from the power boost master cylinders through shuttle valves to the brakes.

When the brake pedals are released, the main system pressure port in the master cylinder is closed off, and fluid is forced out the return port, through the return line to the brake reservoir. The brake reservoir is connected to the main hydraulic system reservoir to assure an adequate supply of fluid to operate the brakes.

When the emergency air system is used, air pressure, directed through a separate set of lines, acts on the shuttle valves, blocking off the hydraulic lines and actuating the brakes.

Power Brake Control Valve System

A power brake control valve system is used on aircraft requiring a large volume of fluid to operate the brakes. Because of the weight and size of the aircraft, large wheels and brakes are required. Larger brakes mean greater fluid displacement and higher pressures. For this reason, independent master cylinder types of systems are not practical on heavy aircraft. A typical power brake control valve system is shown in Figure 3-4-11.

In this system, a line is tapped off from the main hydraulic system pressure line. The first unit in this line is a check valve, which prevents loss of brake system pressure in case of main system failure.

The next unit is the accumulator, the main purpose of which is to store a reserve supply of fluid under pressure. When the brakes are applied and pressure drops in the accumulator, more fluid enters from the main system and is trapped by the check valve. The accumulator also acts as a surge chamber for excessive loads imposed upon the brake hydraulic system.

Following the accumulator are the pilot and copilot's brake valves. The purpose of a brake valve is to regulate and control the volume and pressure of the fluid that actuates the brake.

Four check valves and two one-way restrictors, sometimes referred to as orifice check valves, are installed in the pilot and copilot's brake actuating lines.

The check valves allow the flow of fluid in one direction only. The orifice check valves allow unrestricted flow of fluid in one direction, from the pilot's brake valve; an orifice in the poppet restricts flow in the opposite direction. The purpose of the orifice check valves is to help prevent chatter.

Temperature Compensation

To compensate for increases in pressure due to thermal expansion the next unit in the brake actuating lines is the pressure relief valve. In this particular system, the pressure relief valve is preset to open at 825 p.s.i. to discharge fluid into the return line. The valve closes at 760 p.s.i. minimum.

Each brake actuating line incorporates a shuttle valve for the purpose of isolating the emergency brake system from the normal brake system. When brake-actuating pressure enters the shuttle valve, the shuttle is automatically moved to the opposite end of the valve. This action closes off the inoperative brake system actuating line. Fluid returning from the brakes travels back into the system to which the shuttle was last opened.

Types of Brake Control Valves

Pressure ball check type. A power brake control valve of the pressure ball check type is shown in Figure 3-4-12. The valve is designed to release and regulate main system pressure to the brakes and to relieve thermal expansion when the brakes are not being used. The main parts of the valve are the housing, piston assembly and tuning fork.

The housing contains three chambers and three ports. They are the pressure inlet, brake and return ports.

The piston assembly is made up of a piston head, piston shaft, pilot pin and cross pin. The piston head separates the brake and return chambers. A cup seal prevents fluid from escaping to the return chamber when the brakes are applied. The seal is held in place by a retainer and piston return spring. The piston head has a hole drilled through its center for the flow of fluid to the return port. This hole is opened and closed by the pilot pin. The pilot pin also opens the pressure port. The flange of the pilot pin and the hole in the piston head are lapped together. The piston shaft connects the piston head with the tuning fork. The shaft is slotted, and the cross pin prevents it from turning.

The tuning fork connects the brake pedal linkage with the control valve. It swivels on the housing and limits the maximum pressure directed to the brake. The upper arm of the tuning fork is a bar spring that bends from the point of the fulcrum when hydraulic pressure overcomes toe force.

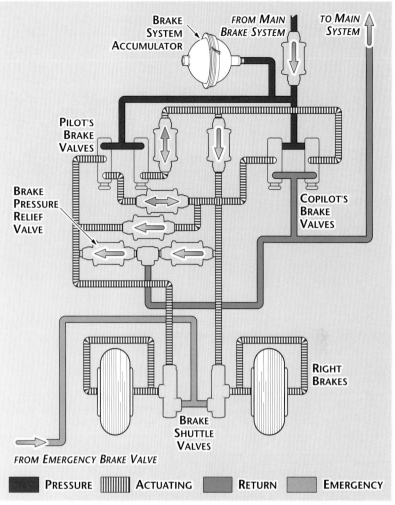

Figure 3-4-11. Typical power brake control valve system

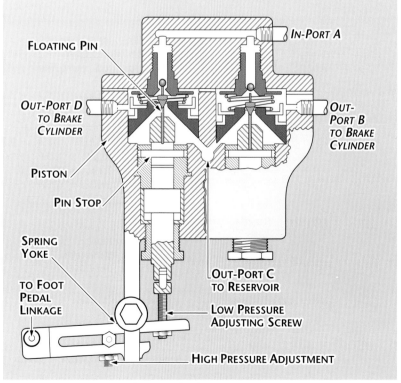

Figure 3-4-12. Power brake control valve (pressure ball check type)

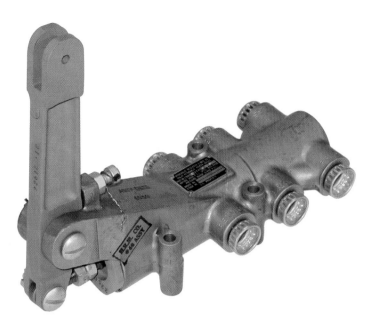

Figure 3-4-13. A contemporary power brake control valve

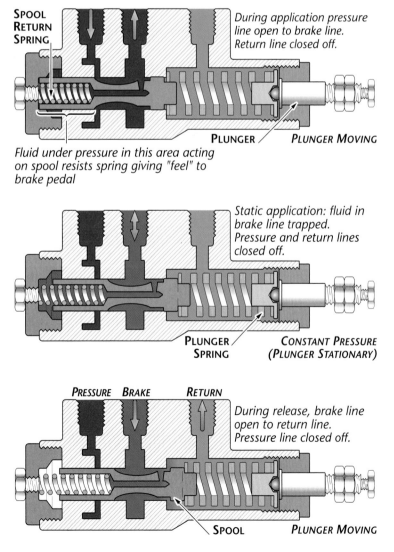

Figure 3-4-14. Power brake control valve (sliding spool type)

Sliding spool type. A sliding spool-type power brake control valve is shown in Figure 3-4-13. This valve is basically a sleeve and a spool installed in a housing. The spool moves inside the sleeve, opening or closing either the pressure or return port of the brake line. Two springs are provided. The large spring, referred to in the illustration as the plunger spring, provides feel to the brake pedal. The small spring returns the spool to the OFF position.

When the plunger is depressed, the large spring moves the spool, which closes off the return port and opens the pressure port to the brake line. When the pressure enters the valve, fluid flows to the opposite end of the spool through a hole. The pressure pushes the spool back far enough toward the large spring to close the pressure port, but not open the return port. The valve is then in the static condition. This movement partially compresses the large spring, giving feel to the brake pedal. When the brake pedal is released, the small spring moves the spool back, opening the return port. This action allows fluid pressure in the brake line to flow out through the return port.

Maintenance of the sliding spool brake control valve is limited to checking the action of the plunger. This is done by manually depressing the plunger until it bottoms, and then releasing it suddenly. If the plunger remains depressed (does not snap out), the valve is binding at the spool and sleeve. If binding occurs, the valve should be replaced. Overhaul of these types of master cylinders is not normally a maintenance function. They are ordinarily the responsibility of a certified repair station. Figure 3-4-13 is a photo of a contemporary power brake control valve. It operates the same as the diagrams in Figure 3-4-14, except the ports are in a different place.

Brake Debooster Cylinder

In some power brake control valve systems, debooster cylinders are used in conjunction with the power brake control valves. These units are generally used on aircraft equipped with a high-pressure hydraulic system and low-pressure brakes. The purpose of the brake debooster cylinder is to reduce the pressure to the brake and increase the volume of fluid flow. Figure 3-4-15 shows a typical debooster cylinder installation.

The unit is mounted on the landing gear shock strut in the line between the control valve and the brake. The schematic diagram in the illustration shows the internal parts of the cylinder.

When the brake is applied, fluid under pressure enters the inlet port to act on the small end of the piston. The ball check prevents the fluid from passing through the shaft. Force is transmitted through the small end of the piston to the large end of the piston. As the piston moves downward in the housing, a new flow of fluid is created from the large end of the housing, through the outlet port, to the brake. Because the force from the small piston head is distributed over the greater area of the large piston head, pressure at the outlet port is reduced. At the same time, the small piston head used to move the large piston head displaces a greater volume of fluid.

Normally, the brake will be fully applied before the piston has reached the lower end of its travel. However, if the piston fails to meet sufficient resistance to stop it (due to a loss of fluid from the brake unit or connecting lines), the piston will continue to move downward until the riser unseats the ball check valve in the hollow shaft. With the ball check valve unseated, fluid from the power control valve will pass through the piston shaft to replace the lost fluid. Since the fluid passing through the piston shaft acts on the large piston head, the piston will move up, allowing the ball check valve to seat when pressure in the brake assembly becomes normal.

When the brake pedal is released, pressure is removed from the inlet port and the piston return spring moves the piston rapidly back to the top of the debooster. This rapid movement causes suction in the line to the brake assembly, resulting in faster release of the brake. Deboosters, in one form or another, are used on most transport type aircraft.

Lockout deboosters. A lockout debooster works like a regular debooster, except it does not automatically refill the (larger) cylinder. There is a valve at the top of the debooster that has an outside handle. When the handle is raised, fluid may transfer from the high pressure to the low pressure side. Thus, the low pressure side is constantly kept filled. A lockout debooster lever check was part of a walk-around inspection.

The thought was that between flights the system could only need fluid added if it acquired a leak. If it acquires a leak, it should be shut off until the leak is fixed.

Lockout deboosters are old technology and are not normally used on current systems.

Emergency Brake System

On all aircraft, except those equipped with independent-type brake systems, an emergency brake system is provided. On some aircraft, a pneumatically operated emergency system is provided. Others have a reserve hydraulic system; an emergency hydraulic reservoir retains a sufficient supply of hydraulic fluid for manual operation of the brakes in case no hydraulic power is available.

The power boost brake system, described earlier, is equipped with a pneumatically operated emergency system. The emergency system consists of a T-handle, compressed air bottle, air release valve and pressure gauge.

Pulling the T-handle operates the system. This releases the compressed air stored in the air bottle. Air pressure unseats the shuttle valves at the air inlet ports and seats the hydraulic pressure ports. Air pressure is then applied directly to the brakes.

Once air pressure has been applied, the brake can be released only by depressing a button on the air release valve. Brake systems must be bled after using the emergency pneumatic systems, and the air storage bottle must be serviced with the specified amount of dry compressed air or nitrogen. A pressure gauge indicates the amount of air in the bottle, in pounds per square inch (p.s.i.).

Large Aircraft Brake Systems

Larger corporate and airline airplanes have a different set of design requirements for braking systems. Basically, they have to have a backup

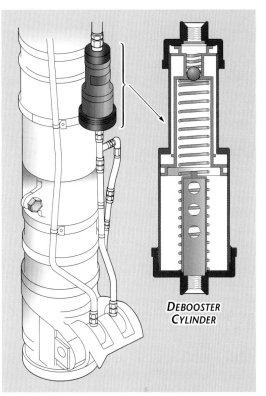

Figure 3-4-15. Brake debooster cylinder

DEBOOSTER CYLINDER

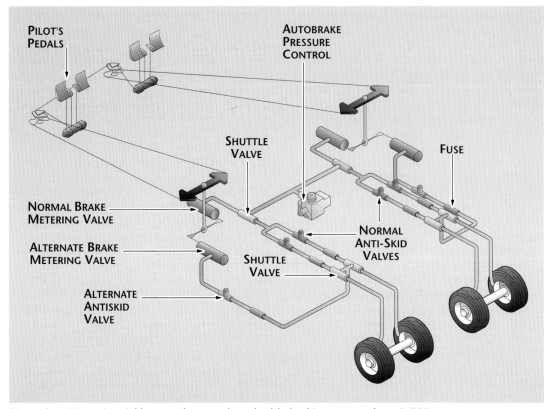

Figure 3-4-16. A pictorial layout of a complete double braking system for a B-727

system for everything, and brake systems are no exception. Figure 3-4-16 is a pictogram of how the brake system lays out. Follow along while reading the following operational description, and you will see that everything has a backup.

All of the brake system components (except the brake pedals, cable runs, swivel valves and the brakes themselves) are located in the main gear wheel well.

Subsystem features. A Boeing 727 has two operating hydraulic systems: A system and B system. The brakes are normally operated by system B hydraulic pressure. Alternate brake hydraulic operation is provided by system A, through alternate brake metering valves, when system B is not available.

One normal and one alternate anti-skid valve for each brake (and for each gear) is installed in the lines from the brake metering valve to the brakes. It modulates brake pressure as required. A hydraulic fuse is downstream of each anti-skid valve to prevent the complete loss of the hydraulic system fluid if a brake line fails.

One shuttle valve for each gear, between the normal brake metering valve and the auto-brake module, passes the output pressure to the brakes. Another set of shuttle valves, one for each brake, located between the normal and alternate brake systems, passes the pressures to the brakes.

System operation. Depressing the brake pedals rotates cable quadrants that operate both the normal and alternate brake metering valves simultaneously. With normal B pressure available, the normal brake metering valve applies hydraulic pressure, through shuttle valves, to the normal anti-skid valves. The anti-skid valves modulate pressure through on and off action to the brakes. When the autobrake system is on, the higher pressure from the auto-

Figure 3-4-17. A complete brake unit from a B-727

brake module is applied through the shuttle valves to the normal anti-skid valves.

When normal hydraulic pressure fails, an alternate brake selector valve opens and system A pressure goes to the alternate brake metering valve. Upon brake application, system A pressure is applied through the alternate anti-skid valves and shuttle valves to the brake units.

Physical description. The main landing gear brakes are multi-disc units consisting of four rotating and five stationary elements. Two of the stationary discs, the pressure plate and the backing plate, have lining on one surface. The other stationary discs, called stators, are lined on both sides and are notched on the inside, enabling them to slide on over the torque tube splines.

The linings on the rotors are made from a heat stable ceramic-metallic material that will maintain its original strength at high temperatures, and are keyed to the wheel. The brake carrier houses six pistons and four automatic adjuster assemblies. Two bleeder valves are provided at the top of the housing and two wear indicator pins are installed on the sides of the housing. Figure 3-4-17 is a complete brake unit from a B-727. It is very large and extremely heavy.

Operation. When the brakes are applied, hydraulic fluid under pressure enters the inlet port of the brake and goes to the pistons. The pistons push against the pressure plate, which presses the rotors and stators together against the torque tube backing plate. Movement of the pressure plate also compresses the return springs in the piston housing by means of the automatic adjuster assemblies.

Each automatic adjuster assembly consists of an adjuster pin spherical ball, metal tube and an adjuster housing. The pin is attached to the pressure plate. The ball is larger in diameter than the tube. Pressure plate movement beyond the compressed spring travel forces the ball deeper into the tube. When pressure is released, return spring force retracts the pressure plate, permitting the separation of the rotors and stators and allowing the wheel to rotate freely. The ball cannot move back in the tube, thus proper brake running clearance is maintained.

Maintenance practices. Airline inspection procedures are more than a bit different from general aviation aircraft. Figure 3-4-18 shows one end of a dual wheel bogie system being jacked up. As you can see, there is simply more to inspect.

Figure 3-4-18. This is a dual-wheel bogie system of an airliner. Notice the torque arms between the brake assemblies and the gear leg itself.

Brake wear is a preflight inspection item. A brake wear indicator is used to determine the wear condition of the brake stack. It consists of a pin fastened to the pressure plate and a bracket fastened to the housing. When the brakes are applied, the distance the pin protrudes through the bracket indicates the amount of wear remaining in the brake. At minimum protrusion, the brake must be changed.

Operation control sequence. The main wheel brakes are normally supplied pressure from the B system. Except for parking, all brake operation is normally controlled by the anti-skid system operation. Shuttle valves are used to isolate manual and autobrake pressure, and to isolate alternate brake pressure from normal sources. Hydraulic fuses are used to protect against complete system fluid loss in the normal and alternate systems.

Major subsystem sequence. Auto braking uses B system pressure to supply inboard and outboard brakes. The normal anti-skid valves are used to adjust pressure to the individual brakes as required.

Normal sequence. Manual braking uses the normal brake metering valves to control brake pressure. When the anti-skid is operating, it functions to control the wheel brake pressure from the source supplying the pressure through the normal or alternate anti-skid valves, as applicable.

Backup operation. Alternate brake pressure is selected automatically when B system pressure drops below approximately 1,500 p.s.i. When this occurs the A system pressure is supplied to the alternate brake metering valves through the alternate brake selector valve. The alternate brake metering valves are operated in parallel with the normal brake metering valves. Fluid from each alternate brake metering valve is routed to the brakes through an alternate anti-skid valve to control pressure to brake pairs.

A brake feel actuator causes increased feel force at the brake pedals during normal brake operation. These forces are at maximum at 600 p.s.i.

A description of the anti-skid system is included in Section 5 of this chapter.

Electric Brakes

As aircraft manufacturers progress toward an "all-electric" aircraft, many of the systems are changing. An all-electric aircraft simply means one that uses electricity to replace all other forms of onboard energy, especially hydraulic. The advantages can be seen in the decreased weight and safety inherent in the electronic systems. These systems are more reliable, with multi-channel redundancy, as well as safer, by eliminating the risk of hydraulic leaks and fire hazards.

Electric brake technology, shown in the block diagram Figure 3-4-19, enhances the efficiency of braking in general and of each individual brake, through faster response, simpler installations and easier diagnostics and maintenance.

Electronic Control. In an electric braking system, electronic control units and electrical wiring replace hydraulic lines and equipment, and electromechanical actuators replace hydraulic pistons. When the pilot steps on the brake pedal, a computer sends information to a control box, which converts these electrical signals into an electromechanical command; the actuators on the brake ring, replacing the hydraulic pistons, press the carbon disks, or heat stacks, against each other as in a conventional hydraulic system.

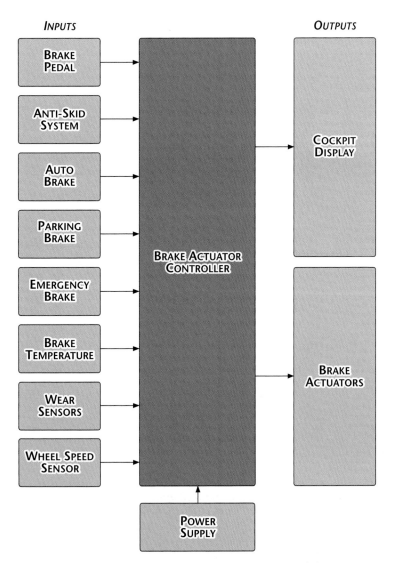

Figure 3-4-19. An electric system reduces the complexity of the system compared to a hydraulic system.

There are three key components to an electric brake system: electric brake actuator controller, electric brake actuators and the brake assembly, which is made up of carbon disks in both the stationary and rotating groups.

Brake Actuator Controller. On a commercial aircraft, a typical installation has four brake actuator controllers mounted in the electrical electronic (EE) bay. Upon receiving a brake command signal from the brake system control unit, or brake pedal, the actuator controllers feed power to four separate electromechanical brake actuators on each brake assembly.

The electromechanical brake actuators provide the clamping force, instead of the hydraulic pistons. The actuators clamp standard carbon heat stacks installed on each of the main landing gear wheels, seen in Figure 3-4-20. The actuators are all line replaceable units (LRUs) that can be replaced without removing the wheel and tire assembly or the brake. This means less maintenance costs because they have fewer parts and a less complicated system.

Brake Actuator. The brake actuator controllers replace a large number of valves and accumulators that store hydraulic fluid, under pressure, to power a hydraulic braking system.

Hydraulic system are more complicated, involving a far more intricate, time consuming removal process when the brakes need to be serviced. With electric brakes, you can just disconnect and replace a few components, and you have no concerns about hydraulic leaks.

When a hydraulic system develops a leak in one of the piston actuators, the entire brake – including the wheel and tire – must be removed, disassembled and completely overhauled. Fluid leakage occurs not only around the heat stacks, but also at valve connections within the hydraulic lines. The connections are bound by hydraulic seals, which don't care for temperature extremes. For this reason, they are prone to develop leakage, and must be constantly inspected to make sure that no leaks are taking place. Most airlines have a daily line level inspection of the landing gear brakes.

With a conventional hydraulic brake system, one leaking hydraulic piston means the entire brake must be removed or disabled and the flight cancelled or delayed. If one of the electromechanical brake actuators is inoperative, the minimum equipment list (MEL) will still permit aircraft dispatch since the other three actuators can adjust for the inoperative actuator.

Built-in Test. Electric brakes also have the advantage of a built-in testing and diagnostic feature that reports the maintenance condi-

Figure 3-4-20. On an electric brake, the hydraulic pistons are replaced with electromechanical actuators.

tion of the brakes, including heat stack wear, on a cockpit digital display. The display uses text and graphics to report the condition of the brakes and the time remaining on the heat stack. In contrast, hydraulic braking systems are equipped with a wear pin within the wheel that needs to be visually inspected prior to each days flights. Wear pins are used on electric brakes, but only for redundancy purposes.

The digital information enables the airlines to forecast, more efficiently, the maintenance activity that needs to be done on the brakes.

General Brake System Maintenance

Proper functioning of the brake system is of the utmost importance. Inspections must be performed at frequent intervals, and maintenance work must be performed promptly and carefully.

Operational checks. There is no one operational test that will fulfill the requirements of all airplanes. Normally the manufacturer will have a procedure spelled out in the MM.

At least there should be a formal system test procedure for complex systems. Simple light aircraft ops tests may be nothing more than

Figure 3-4-21. Wheel brake

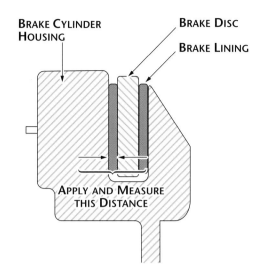

Figure 3-4-22. Wear check method (No. 1)

stepping on the brakes while having the airplane pushed out of the hangar. The brake unit shown in Figure 3-4-21 should be tested without being installed in the wheel.

Spongy brakes. When a brake system has been serviced or repaired, all the air needs to be removed by bleeding the system. A major cause of spongy brakes is the presence of air in the system. When the brake pedal must be pumped several times before full braking is possible; the most common reason is air in the system.

Early brake systems that used an alcohol and castor oil based fluid; the hoses would soften and expand under pressure. The effect was much the same as spongy brakes due to air.

Brake Wear Check

Goodrich and Goodyear brakes. Lining wear for Goodrich and Goodyear systems may be checked by one of two methods. Before checking the brakes on any aircraft, always refer to the applicable MM and use the method recommended by the aircraft manufacturer.

Wear check method (No. 1). Have a person in the cockpit apply the brake, and with the brake applied, measure the distance between the face of the brake disc and the brake housing, as shown in Figure 3-4-22. If this distance has progressed to the maximum specified measurement given in the MM, the brake should be removed and disassembled and the lining pucks inspected for wear.

> **NOTE:** *Linings can only be measured by removing and disassembling the brake. If any puck has worn to a thickness of less than one-sixteenth inch, the entire set must be replaced. NEVER MIX NEW AND USED LININGS.*

Wear check method (No. 2). In using this method, have a person in the cockpit apply the brake. With the brake applied, check the position of the automatic adjusting pins (Figure 3-4-23). If any adjusting pin recedes inside the adjusting pin nut (one-sixteenth to three-eighth inch, the exact amount depending on the brake model), the brake must be removed and disassembled, and the lining thickness checked.

If any lining is worn to a thickness of one-sixteenth inch or less, the entire set of linings must be replaced. Figure 3-4-23 illustrates the normal position of the automatic adjusting pin (protruding out of the adjusting pin nut).

Cleveland brakes. Wear checks for pad thickness and disc condition is more easily accomplished on a Cleveland brake caliper. Look at Figure 3-4-24 and follow along as you read the following paragraphs. The pad thickness and disc thickness can be measured with a ruler by simply doing it. There is no need to disassemble anything. Disc condition can also be readily observed. If grooved too much to continue in service it will be readily apparent. Only then will the wheel need to be removed and the disc replaced.

Any fluid leaks or seeps can also be observed with no disassembly of the unit(s). Should you find any, the caliper can be removed and disassembled, repaired, reinstalled and the system bled, all without removing the wheel.

Removing the caliper and sliding the pressure plate off the torque pins allows the installation of new linings. The old brass rivets would then need to be removed, the plate checked for flatness and new pads riveted back on. Slide the caliper pins into the bushings, slip the inside pressure plates behind the drum and install the bolts.

There are two induced malfunctions possible with Cleveland brakes that do not happen on the other systems:

- If the caliper hose is misaligned, (installed under side pressure), it will try to bind the torque pins as the pads wear and the caliper adjusts. This will result in the pads wearing crooked, i.e. more on one end of the pressure plate that the other. The result is a tapered pad that must be replaced prematurely. The fix is to always check that the hose is properly aligned and the caliper is free to move without binding.

- There is a requirement to clean and lubricate the torque pins so they will remain free floating in their bushings. This will allow for correct wear adjustment. If the pins are lubricated with engine oil, the oil will pick up dirt. The resulting goo will bind the pins as solid as if they were glued.

- This will not allow the automatic adjusters to work. Instead, a brake application will bend the axle bracket to allow the pads to contact the disc. At some point the axle bracket will break. Always be sure the pins are free floating.

Never mix new and used linings on a caliper. They will bind. Also, never use old linings with a new disc.

Emergency System Contamination Check

Check the emergency system for contamination. Remove the plug from the unused pneumatic pressure port on the brake assembly. Position a clean white cloth adjacent to the opening, and slowly pull the emergency brake control handle. Allow airflow through the system for approximately five seconds. There should be no evidence of combustible contaminants on the cloth. If the system is contaminated, the emergency brake pneumatic lines from the brake control valve to the brake assembly must be flushed per the MM.

Bleeding Procedures

There are two general methods of bleeding brake systems – bleeding from top downward (top-down method) and bleeding from the bottom upward (bottom-up method). The method used generally depends on the type and design of the brake system to be bled. In some instances, it may depend on the bleeding equipment available. A general description of each method is presented in the following paragraphs.

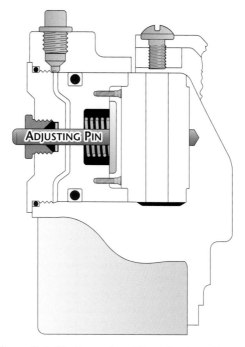

Figure 3-4-23. Normal position of automatic adjusting pin

Top-down method. In using the top-down method, the air is expelled from the system through one of the bleeder valves provided on the brake assembly. A bleeder hose is attached to the bleeder valve, and the free end of the hose is placed in a container that has enough hydraulic fluid to cover the end of the hose. The air-laden fluid is then forced from the system by applying the brakes. If the brake system is a part of the main hydraulic system, a portable hydraulic test stand may be used to supply the pressure.

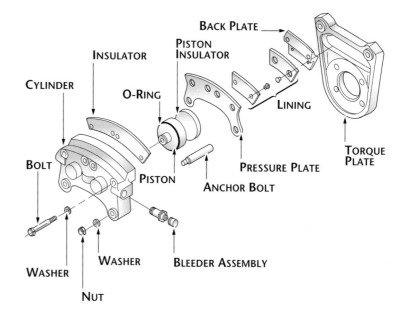

Figure 3-4-24. An exploded view of a Cleveland brake caliper with the axle mounted torque plate

If the system is an independent master cylinder system, the master cylinder will supply the necessary pressure.

In either case, each time the brake pedal is released, the bleeder valve must be closed or the bleeder hose pinched off; otherwise, more air will be drawn back into the system. Bleeding should continue until no more air bubbles come through the bleeder hose into the bleeder container.

Bottom-up method. In the bottom-up method, the air is expelled through the brake system reservoir or other specially provided location. Some aircraft have a bleeder valve located in the upper brake line. In this method of bleeding, a bleeder bomb supplies pressure.

A *bleeder bomb* is a portable tank in which hydraulic fluid is placed, and then put under pressure with compressed air. The bleeder bomb is equipped with an air valve, air gauge and a connector hose. It should also have a safety pop-off valve. The connector hose, which attaches to the bleeder valve on the brake assembly, is provided with a shutoff valve. Another type of bottom bleeder is a pump and reservoir system shown in Figure 3-4-25. These can be shop fabricated easily.

A bleeder valve attach fitting is available from aircraft supply houses, or parts distributors (Figure 3-4-26). It allows a leak-free connection to the bleeder valve. When not in use a 5/32-inch rivet can be inserted into the fitting, head first. This will stop fluid from leaking out of, and dirt from getting into, the supply hose.

Normally, the hose is connected to the lowest bleed fitting on the brake assembly. With the brake bleed fitting opened, opening the bleeder bomb shutoff valve allows pressurized fluid to flow from the bleeder bomb through the brake system until all the trapped air is expelled. The brake bleeder valve is then secured, and the bleeder bomb hose is disconnected.

This method of bleeding should be performed strictly in accordance with specific instructions for the aircraft concerned. Although the bleeding of individual systems presents individual problems, the following precautions should be observed in all bleeding operations:

1. Ensure that the bleeding equipment is absolutely clean and filled with the proper type of hydraulic fluid.

2. Maintain an adequate supply of fluid during the entire operation. A low fluid supply will allow more air to be drawn into the system.

3. Continue bleeding until no more air bubbles are expelled from the system and a firm brake pedal is obtained.

4. Check the reservoir fluid level after the bleeding operation is completed. With brake pressure on, check the entire system for leaks.

Overheated Wheel Brakes

When brakes are used at all, they produce heat. Disc brakes are designed to withstand a considerable amount of heat. After use, a brake system can hold its heat for some time.

CAUTION: *Do not touch the disc on a brake that has just been used. It is hot and will burn you severely.*

Figure 3-4-25. Bleeder pump system for bottom bleeding

Figure 3-4-26. A bleeder valve attach fitting will allow for leak-free operation.

Larger aircraft have more severe problems when they overheat. In the event a large aircraft has been subjected to excessive braking, the wheels may be heated to the point where there is danger of a blowout or fire.

Excessive brake heating weakens tire and wheel structures, increases tire pressure and creates the possibility of fire in the magnesium wheels.

When the brakes on a large aircraft have been used excessively, the fire department should be notified immediately, and all unnecessary personnel should be advised to leave the immediate area. There should be an emergency plan in place where you work to handle these types of occurrences. Even though overheated brakes are in the realm of the line crews, you should still know how to handle the situation.

Brake System Component Maintenance

Components of brake systems are not peculiar to any one system. A given component will vary in shape, size, capacity and manner of operation (depending upon the manufacturer), but the function remains the same. In this section, we will discuss some of the more common brake system component maintenance practices.

Independent system reservoirs. Reservoirs for independent brake systems generally take on one of three forms:

• A cup built into the upper part of the master cylinder.

• A small pint-sized can mounted to the firewall and connected with hoses to gravity feed the master cylinders

• A pressurized reservoir with sight glass and automatic filling from the hydraulic system.

Clean the reservoir inside and out with cleaning solvent. Use a fiber brush on threads. Dry the interior with clean, dry compressed air from a regulated low-pressure source.

After the reservoir is cleaned and the cure-date repair kit parts have been installed, conduct a leakage test. This is accomplished by connecting a source of 25-p.s.i air to the filler port and applying pressure. The reservoir should then be submerged in a tank of water for a minimum of two minutes. No leakage should be seen.

Power brake valves. Maintenance of these valves is limited to removal and replacement. After installation, rig the valves. Make an operational check of the brake system in accordance with the MMs.

Brake shuttle valves. Shuttle valve maintenance is generally limited to repairing leakage. Tightening the end caps may usually repair external leakage. If this does not stop the leakage, the end cap O-ring should be replaced.

Removing and flushing the unit with clean hydraulic fluid can usually repair internal leakage. Excessive heating is a good indication of internal leakage through a shuttle valve. Excessive cycling of the emergency system pump is also an indication of a leaky shuttle valve.

After an emergency system has been operated, all emergency system pressure should be bled off as soon as possible and the normal system restored to operation.

Brake Assembly Maintenance

Single and Dual Disc Brakes

Goodrich and Goodyear brakes. Automatically adjusted single and dual disc brakes are designed to provide a satisfactory running clearance between the brake disc and the brake linings. The self-adjusting feature of the brake maintains the desired lining and puck-to-disc clearance, regardless of lining or puck wear. When you apply the brakes, hydraulic pressure moves each piston and its pucks or linings against the disc or discs as applicable.

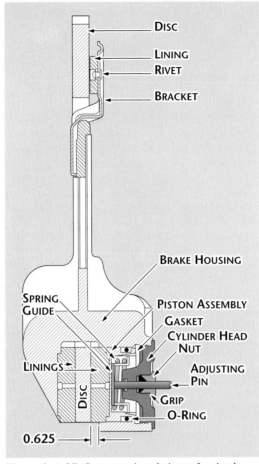

Figure 3-4-27. Cross-sectional view of a single disc brake assembly with captured torquing-type automatic adjuster

As the linings wear, the piston pushes against the adjusting pin (through the spring guide) and moves the pin against the friction of the adjusting pin grip. When you release the brake pressure, the force of the return spring moves the piston away from the brake disc, but it does not move the adjusting pin, which is held by the friction of the pin grip. The piston moves away from the disc until it stops against the head of the adjusting pin.

Thus, regardless of the amount of wear, the same travel of the piston will be required to apply the brake, and the running clearance will be maintained.

The automatic adjusting feature may be referred to as a captured torquing type or captured nontorquing-type. Figure 3-4-27 shows a typical captured torquing-type automatic adjuster. It is mandatory that clearance be established between the linings and the discs before torquing the automatic adjusting nut to the amount specified for the brake involved. Otherwise, the brake will drag until an amount equal to the built-in clearance is worn from the face of the linings. With the adjusting nuts properly torqued, the friction between the grip and

the adjusting pin is great enough to overcome the compression of the return spring, and the adjusting pin will be pulled through the grip only to compensate for lining wear.

After torquing the automatic adjusting nuts to the specified value, back them off and retorque several times. This procedure will ensure proper mating of all parts and the correct torque on the final assembly.

Figure 3-4-28 shows the captured nontorquing-type automatic adjuster used on some single and dual disc brake assemblies.

Brakes that contain nontorquing adjusters can be identified by the locknut and threaded bushing over each adjusting pin. The only difference between the torquing and nontorquing-type automatic adjustment is the method used to restrict the movement of the adjusting pin. The torquing-type adjustment uses a tapered grip, and the nontorquing uses one or more 1/4-inch wide grips composed of brass liners.

Spare grips are shipped with pilot pins installed to open the grip to the approximate diameter of the adjusting pin, thus preventing damage to the grip during installation. The pilot pin is expelled as the grip is forced over the adjusting pin. If grips are to be reused when a brake is disassembled, they should have the pilot pins reinstalled before assembly in the brake.

Brake repairs on the single disc brake consist of replacing linings, worn or damaged sealing devices, brake release units, or brake discs (Figure 3-4-29). Lining replacement and cure-date kit installation contain the following steps:

1. Remove the lockwire and unscrew the cylinder heads (brake release units); remove the release units from the housing.

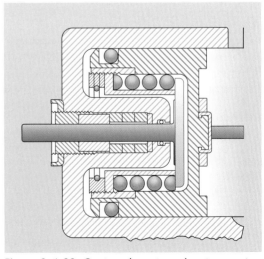

Figure 3-4-28. Captured nontorquing-type automatic adjuster

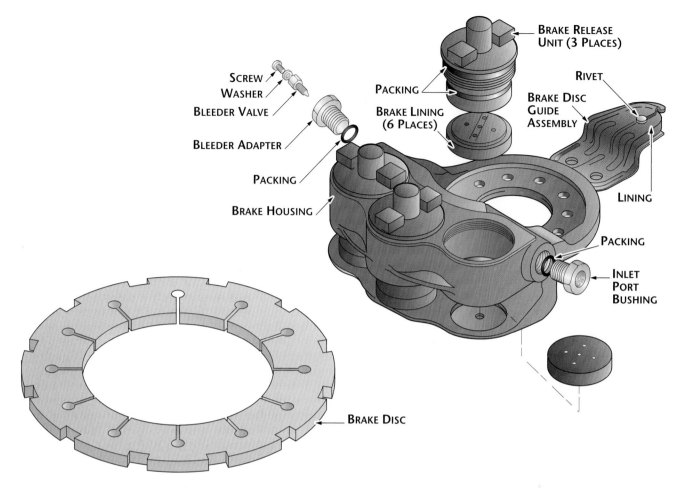

Figure 3-4-29. Single disc brake-repair and parts replacement diagram

2. Remove the disc from the brake housing.

3. Remove the inlet plug and bushing, the bleeder adapter and O-ring packings.

4. Remove the brake linings from the pistons, the brake housing and the disc guide.

5. Clean the brake assembly components with low-pressure compressed air. Wash all metal parts in cleaning solvent. Dry with compressed air.

6. Check the release units for damages, nicks and gouges. If damaged, replace the complete release unit.

7. Check the brake housing for cracks and cylinder walls for nicks or other visible damage. Damage will necessitate turning in the complete brake assembly to supply for disposition.

8. Install new linings in the housing cavities and rivet on the disc guide lining. Friction fit will hold the linings in the housing cavities. Do NOT use cement.

9. Install new linings in the piston cavities using brake lining adhesive specified for such use (for example, Pliogrip No. 3).

10. Install brake disc into the brake housing

11. Dip brake release unit packings from the cure-date kit into the hydraulic fluid and install on the brake release units.

12. Coat the piston of the release unit with a light coating of hydraulic fluid and install in the housing. Tighten the cylinder heads against the housing as specified in the MM.

13. Reinstall the inlet plug, bushing and bleeder adapter into the housing. Use new packings that have been dipped in hydraulic fluid.

14. Lockwire the cylinder heads, bleed the brake and test the brake for leakage and proper operation. Test pressure for this brake assembly is 1,100 p.s.i. Hold the pressure for two minutes and check the assembly for leaks.

15. Release and reapply the pressure 10 times, and check for proper brake operation and release of the discs.

16. Allow the brake to stand two minutes with pressure relieved to check for static fluid leakage.

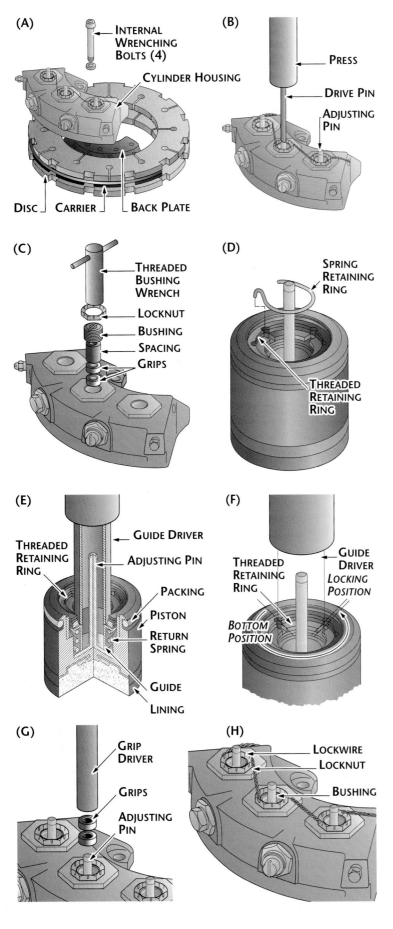

Figure 3-4-30. Seal replacement and piston return adjustment

On dual disc brakes, as well as some single disc brakes, the linings may be replaced without disturbing the brake hydraulic system. The shock strut must be raised with a wheel jack until the wheel is clear of the ground.

The wheel is removed, and the four internal wrenching bolts that attach the brake housing to the backplate are removed. The two setscrews located at each side of the brake housing are unscrewed enough to allow removal of the axle flange attaching bolts. Make certain the brake assembly is supported before you remove the bolts, or damage to the brake hose could result. Remove the brake linings from the pistons, center carrier, backplate and disc guide.

Riveted linings must be drilled. Snap-on or friction-fit linings can be easily pried off with a common screwdriver. Remove dirt and other foreign particles from the brake assembly components by the use of low-pressure compressed air. Wear safety eye protection during this operation.

Clean the external surfaces of the brake parts with a cloth dampened with cleaning solvent.

Replace any brake lining attaching buttons that are damaged. The housing, backplate, center carrier and all bolts should be inspected for damage, cracks, or leakage, as applicable. If the brake has hydraulic leakage or if the housing, backplate, or center carrier is damaged or cracked, the complete brake assembly should be replaced.

Inspect the disc for minimum thickness, maximum width of the keyways and warping. Check the disc for warping by using a straightedge across the face of the disc. Replace a brake disc that is worn excessively.

When a brake disc keyway is worn excessively or elongated, inspect the brake disc drive keys within the wheel assembly for damage and security. Replace the drive keys or the wheel if the damage exceeds the limitations specified in the applicable MM.

The new linings are installed in the brake pistons, the center carrier and the backplate. The disc guide lining is riveted to the disc guide. The pistons are pushed back into the piston housing until a maximum of 1/8-inch of lining is protruding beyond the housing. Assemble the brake on the axle flange, and torque all attaching bolts as well as the four internal wrenching bolts to the specifications provided in the MM.

The fore and aft axle attaching nuts on the brake housing must have their flat surface toward the setscrew on the final torque. The setscrews are tightened against the flat surfaces to safety the nuts. Secure the four internal wrenching bolts with lockwire. The wheel is installed and the shock strut lowered. Perform an operational check to verify proper operation. Specified steps throughout the lining and disc replacement procedures and the final security of all attachments require quality assurance verification as indicated in the MM.

Figure 3-4-30 shows the various steps involved in replacing the piston seals and adjusting the return mechanism. The internal wrenching bolts holding the cylinder housing to the carrier and backplate are removed (view A).

The cylinder housing is placed under a press, as illustrated in view B. Use the press and the drive pin to force the adjusting pins through their grips and remove the pistons from the housing. Make sure that the drive pin is centered on the adjusting pin to prevent damaging the adjusting packings and grips.

Next, cut the lockwire on the locknut. Use the threaded bushing wrench, illustrated in view C, to remove the locknuts, bushings, spacers and grips from the housing. Remove the spring retaining ring from within the piston, as shown in view D.

With the linings still attached to the pistons, support the pistons in a press. Use a 3 inch length of 7/8-inch diameter steel tubing to force the guides to the bottom on the adjusting pins, as shown in view E.

Hold the guides in the bottomed position and turn the threaded retaining rings clockwise until the rings are snug against the bottom guides. Back off the threaded retaining rings

3/4 of a turn counterclockwise from the bottomed positions, if necessary, continue turning counterclockwise to the next locking position, as shown in view F. Secure the threaded retainer with the wire retaining ring.

Replace the piston packings that have been dipped in hydraulic fluid, and ensure that the packings and adjusting pin stems are lubricated with hydraulic fluid.

The piston assemblies are then installed in the cylinder housing and forced to the complete brake-off positions – bottomed in the housing cavities. The pistons are supported against their linings to the brake-off position. Use the press and the grip driver, as illustrated in view G, to force the grips, one at a time, over the adjusting pins until they are bottomed. The pistons must remain in the complete brake-off position when the grips are installed. Place the spacers over the adjusting pins and install the bushings finger tight. Hold the bushings in finger tight positions and install and tighten the locknuts. Safety wire the locknuts, as shown in view H.

> **NOTE:** *On some brake assemblies, the adjusting pin bushing (adjusting pin nut) is torqued to a specified value.*

The brake assembly must be tested following reassembly. Connect the brake assembly to a hydraulic supply source. Bleed the brake assembly and apply 600 p.s.i.

> **CAUTION:** *Before applying pressure, make sure that the brake is assembled properly with all bolts torqued and brake discs in position. Failure to do so could result in injury to personnel.*

Hold the test pressure for two minutes while you are checking the brake assembly for leaks. Release and apply the pressure 10 times to be sure that the brake functions properly. The brake discs should be free when hydraulic pressure is released. Allow the brake to stand for two minutes with pressure released and check for static fluid leakage.

If the brake assembly is not to be installed immediately, install any attaching hardware that is part of the assembly, fill with preservative hydraulic fluid and cap or plug all openings to prevent contamination.

Cleveland Brakes

These brake systems are much simpler to disassemble and reassemble because the work can be performed without jacking the airplane or removing the wheel. An example is shown in Figure 3-4-31.

Figure 3-4-31. A dual caliper Cleveland brake on a King Air 90. Notice the easy access to the pressure plate attachment bolts.

Figure 3-4-32. Riveting new brake pads onto the backing plate is a simple operation.

- Insert new brass brake rivets in the pads with the heads in the countersunk holes.

- Place one rivet head over the anvil of the riveting tool, and the rivet header into the recess of the rivet. Hold the other rivet in place to keep the pad aligned. See Figure 3-4-32.

- Generally, a single blow on the rivet header will form the rivet head.

- Now set the other rivet(s) in the pads

The inner pressure plate with its' new linings can be slid over the anchor pins. The outer pressure plate(s) can be inserted behind the brake disk and the attach bolts installed. Once everything is installed the attach bolts must be torqued to specifications. Some bolts are torqued dry, while others require *Lubtork* (a thread lubricant). The Cleveland Maintenance manual lists different procedures for different torquing operation. The caliper should be checked for binding and adjustment, i.e. smooth sliding on the anchor pins. Figure 3-4-33 is a rebuilt caliper.

Initial adjustment. Once relined the brakes will be seriously out of adjustment and the system will need to be bled before the units can be initially adjusted. Bottom up bleeding generally works well. Be careful not to overflow the master cylinder and cause brake fluid to pool in the cabin floor.

Once bled, several applications of the brakes will normally adjust the caliper to the operating clearance. Once the operating clearance is set and the fluid levels rechecked, the process must be completed by *conditioning the pads*. The conditioning-in process is thoroughly covered in the Cleveland Maintenance Manual. A personal copy of the manual can be downloaded from www.parker.com/cleveland.

Removing the caliper. To remove a caliper it is only necessary to remove the bolts that fasten it to the outer pressure plate. Be sure to catch the outer pressure plate(s) before removing the last bolt so they do not fall to the floor. The caliper can then be withdrawn by sliding it off the anchor pins.

Once removed, the inner pressure plate, and any thermal insulating plates, can be slid off the anchor pins. Note that smaller units do not have thermal insulator plates. The pistons will normally be held in the caliper bore(s) by the friction of the O-ring seal(s) and will not fall out. The thickness of the brake disk should now be measured and compared to specifications to determine airworthiness.

If the seals are to be replaced, place a catch pan under the unit and remove the pistons.

Reassembly. After cleaning the piston(s) and cylinder bore(s) lubricate new seals with the correct brake fluid and reinstall them in the cylinder(s), pushing them completely to the rear of the cylinders. Replace any damaged insulators by simply sliding them over the anchor pins. When the brake pads are to be changed, also do the following:

- Using a brake pad riveting tool in a shop vise, drive out the old rivets.

- Clean the pressure plate so no debris will be under the new pad.

Figure 3-4-33. A Cleveland caliper with dual pistons and new brake pads.

Automatic adjusting. In view of the complexity of other adjusting systems, the Cleveland system is simplistic, but works very well. When the brakes are applied and the clearance is normal, the O-ring rolls somewhat in the cylinder bore, instead of sliding. As the pads wear and the clearance increases, the distance is greater than the amount the O-ring can roll. As a result, it rolls, and then slides to take up the space. When the brakes are released, the O-ring rolls back an infinitesimal amount, providing running clearance for the pads.

Troubleshooting. To keep the automatic adjustment feature working, it is only necessary to keep the caliper clean. If oil collects on the caliper and pistons, dirt will collect and build up. This buildup is enough to freeze the caliper in place and cause the brakes not to adjust.

Each application of brakes will try to bend the torque plate bolted to the axle. It will not take too much flexing and will crack and break. Another way the adjuster can be defeated by oil and grime is by allowing oil to collect on the torque pins. They will glue themselves to the bushing in the torque plate. Instead of bending the torque plate with each brake application, the action will cause the bushing to pull out.

There will be no loss of brake clearance as the bushing works its way out. The only way to notice if it happens is to actually take a second and look at the bushings location. Make it part of a pre-flight inspection.

Trimetallic Disc Brakes

Figure 3-4-34 shows a typical trimetallic brake assembly. The trimetallic brake assembly consists of a brake housing subassembly, a keyed torque tube and torque tube spacer, a housing backplate, stationary and rotating discs and a pressure plate subassembly.

Description. The brake housing subassembly, keyed torque tube and spacer and the housing backplate are bolted together to form the basic brake assembly. The remaining components of the brake assembly are mounted over the keyed torque tube and between the brake housing and the housing backplate. The metallic-faced rotating discs have keyways that engage drive keys in the wheel so that they rotate with the wheel.

The rotating discs are separated by the stationary discs, which are keyed to the torque tube. The mating surfaces of these rotating and stationary discs constitute the major friction-braking surfaces of the brake.

Additional friction surfaces exist between the outer face of one rotating disc and the hous-

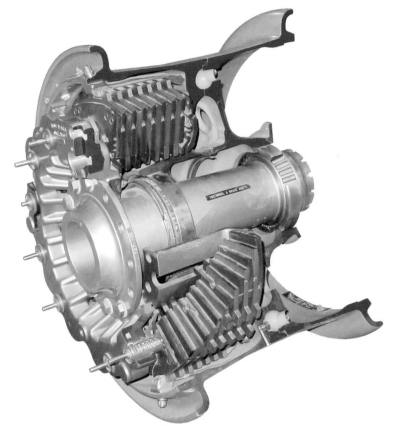

Figure 3-4-34. A multiple disc metallic lined brake, installed in a cutaway wheel, for a Boeing jet airliner

ing backplate and between the outer face of the rotating disc at the opposite end and the pressure plate subassembly.

The pressure plate subassembly consists of the pressure plate, replaceable wear plate and wear plate insulator. These three parts are riveted together. The pressure plate serves as a seat for the self-adjusting pins of the self-adjusting mechanism, and rests against the insulators installed in the outer ends of the brake pistons.

The pressure plate is the component through which force is directly transmitted during application and release of the brakes. The wear plate is keyed to the torque tube to prevent rotation of the complete subassembly, and serves as the friction surface for the outer face of the adjacent rotating disc. The wear plate insulator prevents brake heat from being transferred to the pressure plate and the brake pistons.

The brake pistons transmit hydraulic pressure through the pressure plate subassembly to the brake discs. Standard O-rings and backup rings around each piston prevent hydraulic fluid leakage and entry of contaminants. The pistons are further protected against heat transfer from the pressure plate subassembly by individual insulators installed in the ends of each piston where it contacts the pressure plate.

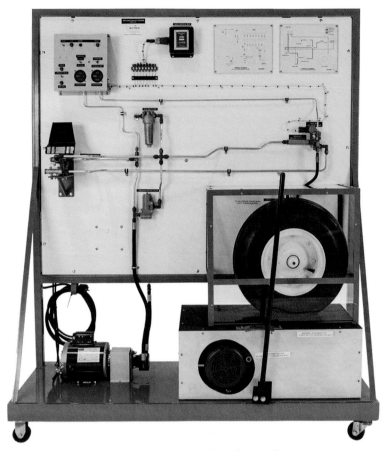

Figure 3-5-1. An anti-skid system trainer that shows all components at one time.

tons is controlled by the aircraft's brake metering system in response to the operating of the brake pedals. Braking action applied to the wheel brake is proportional to the pressure exerted on the brake pedal.

Pressure applied to the brake actuates all of the pistons within the brake housing. These pistons, in turn, force the pressure plate subassembly laterally against the discs and against the housing backplate. As the pressure is applied and the brake starts to actuate, the lateral movement of the pressure plate subassembly pulls the self-adjusting pins, the split collar grip and tube subassemblies and the return spring guides against the return springs, compressing them until the spring guides bottom in the housings. When the hydraulic pressure is relieved, the return spring mechanisms, acting through the heads of the self-adjusting pins, pull the pressure plate subassembly back to the released position. The pistons also return to their deactuated positions. The extent of the return motion is limited by engagement of the spring guides with the retaining ring stops inside the spring housing.

As the discs wear, self-adjusting pins and tubes are pulled through the split collar grips by the force exerted on the pressure plate by the pistons. This small movement of the adjusting pins and tubes, relative to the grips, is equivalent to the combined wear of all the discs. When pressure is removed from the brake, the return springs return the pressure plate and the brake pistons to the designed reset clearance and maintain a constant displacement.

Maintenance. Maintenance of a multi disc metallic brake assembly consists of cleaning and inspection. Normally, brake overhaul is accomplished by either a special department with the proper tooling and training, or by a certified repair station.

Self-adjusting mechanisms are located around the brake housing. They accomplish normal release of the brake and provide a continuing adjustment action to compensate for brake wear. Each mechanism consists of a self-adjusting pin, a spring housing and bushing, a return spring guide, a retaining ring, a grip and tube subassembly and a self-locking nut. The grip and tube subassembly mounts over the self-locking pin, with the grips being installed firmly on the tube.

As disc wear occurs, movement of the adjusting pins through the split collar grips provides automatic adjustment. The retaining ring inside the spring housing serves as a stop and retainer for the spring guide, which, in turn, holds the return spring in position. The head of the self-adjusting pin engages the pressure plate subassembly to allow brake release when pressure is removed.

Operation. When the landing gear wheel is rotating, the metallic–faced rotating discs of the brake assembly rotate freely between the stationary steel discs. When pressure is applied to the brake assembly pistons, the rotating and stationary discs are forced together, creating friction between their surfaces. The amount of hydraulic pressure applied to the brake pis-

Section 5
Anti-skid System

A feature found in high performance aircraft braking systems is skid control, or *anti-skid* protection. This is an important system because if a wheel goes into a skid, its braking value is greatly reduced.

The skid control system performs four functions:

1. Normal skid control
2. Locked wheel skid control

3. Touchdown protection

4. Fail-safe protection

The main components of the system consist of a skid control generator for each braked wheel, a skid control box, skid control valves for each wheel brake, a skid control switch, a warning lamp and an electrical control harness with a connection to the landing gear squat switch. The way that the system is installed in an airplane makes it impossible to see the components all at the same time. However, by looking at the system trainer in Figure 3-5-1 it is possible to see the entire system.

Normal skid control. Normal skid control comes into play when wheel rotation slows down but has not come to a stop. When this slowing down happens, the wheel sliding action has just begun, but has not yet reached a full scale slide. In this situation, the skid control valve removes some of the hydraulic pressure to the wheel. This permits the wheel to rotate a little faster and stop its sliding. The more intense the skid is, the more braking pressure is removed. The skid detection and control of each wheel is completely independent of the others. The wheel skid intensity is measured by the amount of wheel slow-down.

Skid control generator. The *skid control generator* is the unit that measures the wheel rotational speed (Figure 3-5-2). It also senses any changes in the speed, as well as the rate of change. It is a small electrical generator, one for each braked wheel, mounted in the wheel axle. The generator armature is coupled to, and driven by, the main wheel through the drive cap in the wheel. As it rotates, the generator develops a voltage and current signal. The signal strength indicates the wheel rotational speed. This signal is fed to the *skid control box* through the harness.

Skid control box. The box reads the signal from the generator and senses change in signal strength. It can interpret these as developing skids, locked wheels, brake applications and brake releases. It analyzes all it reads and then sends appropriate signals to solenoids in the skid control valves. Figure 3-5-3 is a skid control box. In practice, it would be mounted in the fuselage.

Skid control valves. The *skid control valves* mounted after the brake control valve are solenoid operated. Electric signals from the skid control box actuate the solenoids. If there is no signal (because there is no wheel skidding), the skid control valve will have no effect on brake operation. But, if a skid develops, either slight or serious, a signal is sent to the *skid control valve solenoid*. This solenoid's action lowers the metered pressure in the line between the meter-

ing valve and the brake cylinders. It does so by dumping fluid into the reservoir return line whenever the solenoid is energized. Naturally, this immediately relaxes the brake application. The pressure flow into the brake lines from the metering valves continues as long as the pilot depresses the brake pedals. But the flow and pressure is rerouted to the reservoir instead of to the wheel brakes.

The system pressure enters the brake control valve where it is metered to the wheel brakes in proportion to the force applied on the pilot's foot pedal. However, before it can go to the brakes, it must pass through a skid control valve. There, if the solenoid is actuated, a port is opened in the line between the brake control valve and the brake. This port vents the brake application pressure to the utility system return line. This reduces the brake application, and the wheel rotates faster again. The system is designed to apply enough force to operate just below the skid point. This gives the most effective braking.

Pilot control. The pilot can turn off the operation of the anti-skid system by a switch in the cockpit. A warning lamp lights when the system is turned off or if there is a system failure.

Locked wheel skid control. The *locked wheel skid control* causes the brake to be fully released when its wheel locks. A locked wheel easily occurs on a patch of ice due to lack of tire friction with the surface. It will occur if the normal skid control does not prevent the wheel from reaching a full skid. To relieve a locked wheel skid, the pressure is bled off longer than in normal skid function. This is to give the wheel time to regain speed. The locked wheel skid control is out of action during aircraft speeds of less than 15-20 mph.

Touchdown protection. The *touchdown protection circuit* prevents the brakes from being

Figure 3-5-2. Skid control generator that mounts inside the wheel axle.

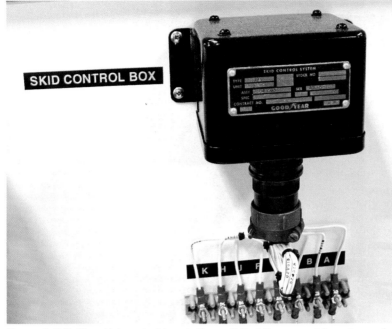

Figure 3-5-3. The skid control box is the unit that figures out how to apply the brakes so the wheels will not skid.

applied during the landing approach even if the brake pedals are depressed. This prevents the wheels from being locked when they contact the runway. The wheels have a chance to begin rotating before they carry the full weight of the aircraft. Two conditions must exist before the skid control valves permit brake application. Without them, the skid control box will not send the proper signal to the valve solenoids. The first is that the squat switch must signal that the weight of the aircraft is on the wheels. The second is that the wheel generators sense a wheel speed of over 15-20 mph.

Fail-safe protection. The *fail-safe protection circuit* monitors operation of the skid control system. It automatically returns the brake system to full manual in case of system failure. It also turns on a warning light.

Anti-skid System (Boeing 727)

The Boeing 727 anti-skid system description is typical of a large aircraft installation.

Anti-skid System Operation

The anti-skid system is designed to provide maximum effective braking for any runway condition without skidding. Individual transducers for each main gear wheel provide wheel speed information to a control unit located in the aircraft's electronics bay. This information is processed by the control unit, which detects impending skids and produces correction signals to the brake units.

Anti-skid valves in both the normal and alternate brake system are controlled by these signals to reduce brake pressure and prevent skids and brake lockups.

General component locations. All anti-skid systems start with a wheel speed transducer that is installed in each main gear wheel axle. The anti-skid control unit is located in the electronic equipment bay. Four normal and two alternate anti-skid valves are mounted in the left and right main gear wheel wells.

Subsystem features. The anti-skid system is programmed to provide locked wheel protection during touchdown, and wheel speed sensing and skid protection during braking. A guarded on-off switch provides control of the anti-skid system. A warning light provides notice of defective systems. A built-in-test capability (BITE) provides continuous self test and fault warning.

Anti-skid protection is provided during both normal and alternate manual braking and during autobrake operation. An operational anti-skid system is required to use autobrakes.

General operation. The anti-skid system operates by overriding the pilot's metered pressure or autobrake commands. The anti-skid system controls hydraulic pressure to each brake until optimum wheel braking is obtained. Regardless of prevailing weather conditions, airplane stopping distances are minimized and directional control is maintained. In addition to individual wheel skid protection, wheels are compared inboard to inboard and outboard to outboard for locked wheel protection. Touchdown protection prevents brake pressure application before wheel touchdown and spinup.

Anti-skid Control Valve

The anti-skid control valve is an electrically controlled hydraulic valve which meters hydraulic pressure applied to a brake in accordance with signals received from the anti-skid control unit.

Six anti-skid control valves are used. One for each main gear brake for normal operation and two are used for alternate operation. Three valves are located in each of the main gear wheel wells. The valves are used to modulate brake pressure to the brake assemblies in either the manual or automatic braking modes.

The anti-skid valves consist of an electrically controlled first stage valve and a hydraulically controlled second stage valve. The first stage valve contains the torque motor, which positions a flapper between two hydraulic ports (return and pressure). The second stage valve is spring offset to the left and pressure bias controlled by the drilled passageway in the spool. With no control signal to the torque motor, the flapper is towards the return nozzle. The spool is spring offset to the left.

When braking occurs, hydraulic pressure is ported through the second stage valve directly to the brake assemblies. The pressure is also acting around the annular recess, through the nozzle into the first stage valve. As pressure in the first stage chamber increases, the increased pressure also acts on the spring offset side of the spool keeping the valve displaced to the left.

Operation. If the anti-skid system senses a wheel slowing down too fast, a signal is sent to the torque motor windings. This causes the flapper to move towards the pressure nozzle, restricting fluid into the chamber and letting more fluid escape to the return line.

Pressure drops in the first stage chamber and acts on the spring offset side of the spool. The pressure on the left side of the second stage valve (through the drilled passage way) forces it to the right, closing the brake pressure in-line and porting some pressure to return. The amount of second stage valve movement depends upon the torque motor current, which in turn depends upon the amount of brake pressure reduction required for wheel spin up. Conversely, a decrease in torque motor current allows more brake pressure to be applied.

A check valve is installed between the pressure line and the brake line to prevent a hydraulic lock in the brake line when applied pressure is removed.

4

Aircraft Electrical Systems

Aircraft electrical system design has evolved from simple systems consisting of magnetos, a battery, alternator and power distribution circuitry for lights and radios into highly complex systems such as those found on the new generation Boeing and Airbus air transport category aircraft. Almost all the other aircraft systems such as flight controls, hydraulic systems, fuel systems, ice and rain protection systems, engine overheat and fire detection systems to name a few, depend on an electrical power generation and distribution system. New aircraft designs with "fly-by-wire" technology have even abandoned traditional hydraulic actuator methods for making flight control inputs in favor of lightweight wiring and electrically powered actuators.

Aircraft electronic systems, known as avionics, are also an integral part of the electrical system in almost all aircraft. Avionics depend on electrical power to control and monitor functions such as communications, navigation, engine control and instrumentation, flight deck and cabin environmental controls and flight management computers. A typical aircraft electrical system functions much like the human body's central nervous system, monitoring, regulating and controlling everything on board the aircraft necessary for manned flight.

State-of-the-art avionics systems usually consist of an interrelated group of smaller electronic systems; each dedicated to a single function, but linked through an onboard computer. Large transport category aircraft are generally built with three or more such avionics systems, providing redundancy so that one failed system need not result in an in-flight emergency. An excellent source of information on the avionics in new generation aircraft is the book, *Avionics: Systems and Troubleshooting*, by T.K. Eismin, which is also published by Avotek®.

Learning Objectives:

REVIEW
- basic principles of electrical theory
- types of wiring diagrams and wiring symbols

DESCRIBE
- properties of generators and alternators
- characteristics and functions of switches, relays and current limiters

EXPLAIN
- how to identify, size, and install aircraft wiring
- importance of monitoring and controlling electrical load

Left. A Boeing 747 has 171 miles of wire and 971 assorted lights, switches and gauges. This is more material than a person can learn at one time. These complicated systems can be learned one wire at a time through careful study of electrical diagrams, logic charts and trouble-shooting charts.

Photo courtesy of Lufthansa

Regardless of how simple or complex the electrical system is, the fundamental of physics surrounding electrical power generation, distribution and usage are the same. For that reason, a review of electrical terms and principles is useful before getting into the practical applications of electrical theory and the study of electrical circuits.

Section 1

Electrical Theory Overview

Ohm's Law and Power

Ohm's Law is the basis for the relationship between voltage, resistance and current. This law applies to all *direct current (DC)* circuits. In a modified form, it also applies to *alternating current (AC)* circuits. Discussed later in this chapter is the difference between DC circuits and AC circuits, the merits of each, and typical uses of each type of circuit.

Ohm's experiments showed that current flow in an electrical circuit is directly proportional to the amount of voltage applied to the circuit. In other words, as the voltage increases, the current increases; and when the voltage decreases, the current flow decreases. This relationship is true as long as the resistance in the circuit remains constant. When the resistance changes, current will also change. This relationship is illustrated by the Ohm's law formula and may be expressed in the following equation:

$$I = E/R$$
$$\text{or, } E = I \times R$$

"I" is current in amperes, "E" is the potential difference measured in volts, and "R" is the resistance measured in ohms (designated by the Greek letter omega, whose symbol is Ω). If given any two of these circuit quantities, the third is found by simple algebraic transposition.

Electric current is usually referred to as current or current flow. As covered in the electricity chapter of *Introduction to Aircraft Maintenance*, current flow in only one direction is called direct current, or DC. Current that reverses itself periodically at a certain frequency is called alternating current, or AC.

Electrical power is usually measured in watts. The watt is a relatively small unit of power considering that a motor rated at one horsepower

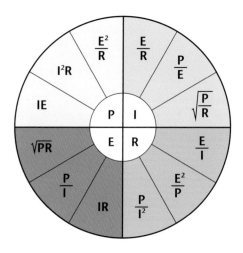

Figure 4-1-1. Summary of basic equations using the volt, ampere, ohm and watt

consumes 746 watts of electrical energy when running. A much more common unit of measure for electrical power is the kilowatt (1,000 watts). In measuring amounts of electrical energy consumed, the kilowatt-hour is used. For example, if that one horsepower motor consumes electrical energy for 10 hours, it has used 7,460 watt - hours or 7.46 kilowatt hours of electrical energy.

The basic formula for finding electrical power in a circuit is:

$$P = I \times E$$

"P" is the power in watts, "I" is current in amperes, and "E" is emf (electromagnetic force) in volts. It is possible to substitute the Ohm's law values for E in the power formula to obtain the following:

$$P = I^2 \times R$$

From this equation, the power in watts in a circuit varies as the square of the circuit current in amperes and varies directly with the circuit resistance in ohms.

Figure 4-1-1 is a summary of all the possible equations using the electrical units volt (E), ampere (I), ohm (R) and watt (P).

Electromagnetism

Knowledge of magnetism has existed for many centuries, but it was not until the eighteenth century that this body of knowledge was combined with research and experimentation of the properties and characteristics of electrical phenomena. Magnetism is in fact so integral to the generation and utilization of electricity that aircraft systems designed to produce, control and utilize electrical power could not function without it.

In 1820, it was demonstrated that an electric current produces magnetic effects. Conversely, it was later discovered by Faraday that a magnet in motion has the ability to generate electricity. This physical motion is the basis for the most common means of generating aircraft electrical power, whether it be a generator, alternator, or one of the newer designs in power genera-tion components known as an *integrated drive generator (IDG)*. The chemical action that occurs in several different types of aircraft batteries is another method used for producing electrical current, however, use of batteries as a power source is usually reserved for emergency or backup electrical power.

Generators and alternators produce electrical current flow when mechanical energy is used to rotate their conductive wire coils through a magnetic field. To show how an electric current can be created by a magnetic field, see Figure 4-1-2 and recall the demonstration conducted with a permanent magnet, coil of wire and a galvanometer.

Several turns of a conductor are wrapped around a cylindrical form, and the ends of the conductor are connected together to form a complete circuit, which includes a galvanom-eter. If a simple bar magnet is plunged into the cylinder, the galvanometer can be observed to deflect in one direction from its zero (center) position (A). When the magnet is at rest inside the cylinder, the galvanometer shows a reading of zero, indicating that no current is flowing (B). The galvanometer in Figure (C) indicates a current flow in the opposite direction when the magnet is pulled from the cylinder.

The same results may be obtained by holding the magnet stationary and moving the cylinder over the magnet, indicating that a current flows when there is relative motion between the wire coil and the magnetic field. The induced cur-rent caused by the relative motion of a conduc-tor and a magnetic field always flows in such a direction that its magnetic field opposes the motion. When a conductor is moved through a magnetic field, as shown in Figure 4-1-3, an *elec-tromotive force (emf)* is induced in the conductor. The direction (polarity) of the induced emf is determined by the magnetic lines of force and the direction the conductor is moved through the magnetic field.

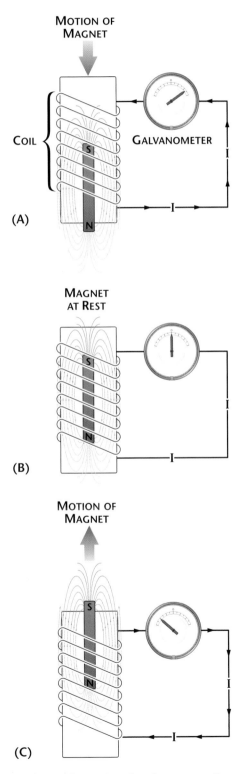

Figure 4-1-2. Illustration showing current flow when a magnet is moved through a coil.

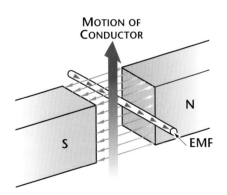

Figure 4-1-3. Conductor moving through a mag-netic field creating a current flow

The generator left-hand rule (not to be confused with the left-hand rules used with a coil) can be used to determine the direction of the induced emf, as shown in Figure 4-1-4. The index finger of the left hand is pointed in the direction of the magnetic lines of force (north to south), the thumb is pointed in the direction of movement of the conductor through the magnetic field, and the second finger points in the direction of the electrical current flow. When two of

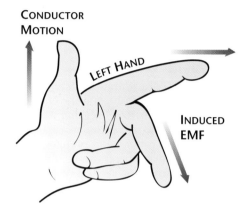

Figure 4-1-4. The generator left-hand rule showing the direction of induced emf

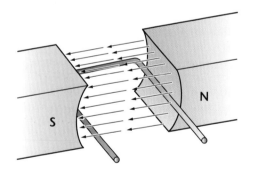

Figure 4-1-5. When a looped conductor is rotated through the magnetic field a voltage is induced in both sides.

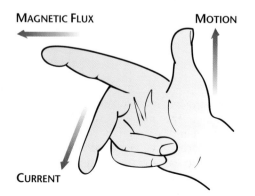

Figure 4-1-6. The right hand rule for electric motors

these three factors are known, the third may be determined by the use of this rule.

When a looped conductor is rotated in a magnetic field (Figure 4-1-5), a voltage is induced in each side of the looped conductor. Since each side of the looped conductor cuts through the magnetic field in opposite directions as physical rotation occurs (one side moving up while the other is moving down), current flow will be continuous, moving in on one side of the loop and out the other side.

The same electrodynamics that make a generator or alternator produce electricity are in play when electrical current is utilized to do work, such as raising or lowering flaps. Whereas generators convert mechanical motion into electrical current, electric motors convert electrical current back into mechanical motion to accomplish work.

In the same way that the generator left-hand rule may be used to explain the relationship between direction of the magnetic lines of force, direction of current flow and direction of loop rotation, the right-hand rule for electric motors may be used to determine the same parameters (Figure 4-1-6). The index finger of the right hand is pointed in the direction of the magnetic lines of force (north to south), the thumb is pointed in the direction of movement of the conductor through the magnetic field, and the second finger points in the direction of the electrical current flow. When two of these three factors are known, the third may be determined by the use of this rule.

Alternating Current

Figure 4-1-7 shows a simple generator loop. If the loop is rotated half a turn, loop side A and loop side B will have exchanged positions and the induced emf in each half of the loop wire will have reversed its original direction. This happens because the wire formerly cutting the lines of force in an upward direction is now moving downward. The value of an induced emf depends on three factors:

1. The number of wires moving through the magnetic field

2. The strength of the magnetic field

3. The speed of rotation

A generator used to produce alternating current is usually known as an *AC generator* or an *alternator*. The simple generator shown in Figure 4-1-7 illustrates one of the methods of generating an alternating voltage (emf). The rotating loop (marked A and B) is placed between two magnetic poles (N and S), and the

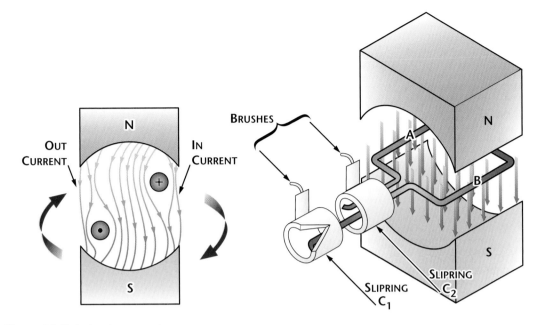

Figure 4-1-7. A simple generator

ends of the loop are connected to two metal slip or collector rings (C1 and C2). Current is transferred from the collector rings to the wiring by brushes for distribution of the electrical power to electrical loads such as lighting circuits, hydraulic system motors, or anti icing heating elements.

As wire B moves down across the field, a voltage is induced which causes the current to flow outward. When the wires are formed into a loop, the voltages induced in the two sides of the loop are combined. Therefore, for explanatory purposes, the action of either conductor, A or B, while rotating in the magnetic field is similar to the action of the loop.

An alternating voltage creates the potential for alternating current and if the circuit is closed as in Figure 4-1-2, current will flow. The magnitude of that current depends on several factors such as:

- The resistance of the load that is using the current

- Size of the wire that carries the electrical current

- The voltage produced by the generator or alternator

The rotating loop is also known as the rotor. For a more detailed explanation of what happens as the generator/alternator rotor spins, see Figure 4-1-8.

At position 1, conductor A moves parallel to the lines of force. Since it cuts no lines of force, the induced voltage is zero. As the conductor advances from position 1 to position 2, the voltage induced gradually increases.

At position 2, the conductor moves perpendicular to the flux and cuts a maximum number of lines of force; therefore, a maximum voltage is induced. As the conductor moves beyond two, it cuts a decreasing amount of flux at each instant, and the induced voltage decreases.

At position 3, the conductor has made one-half of a revolution and again moves parallel to the lines of force, and no voltage is induced in the conductor. As the A conductor passes position 3, the direction of induced voltage reverses since the A conductor now moves downward, cutting flux in the opposite direction.

As the A conductor moves across the south pole, the induced voltage gradually increases in a negative direction, until at position 4, the conductor again moves perpendicular to the flux and generates a maximum negative voltage. From position 4 to 5, the induced voltage gradually decreases until the voltage is zero and the conductor and wave are ready to start another cycle.

The curve shown at position 5 is called a sine wave. It represents the polarity and magnitude of the instantaneous values of the voltage generated. The horizontal baseline is divided into degrees of rotation, or time, and the vertical distance above or below the baseline represents the value of voltage at each particular point in the rotation of the loop

In addition to generators, *DC to AC static inverters* are also commonly used as a source of AC

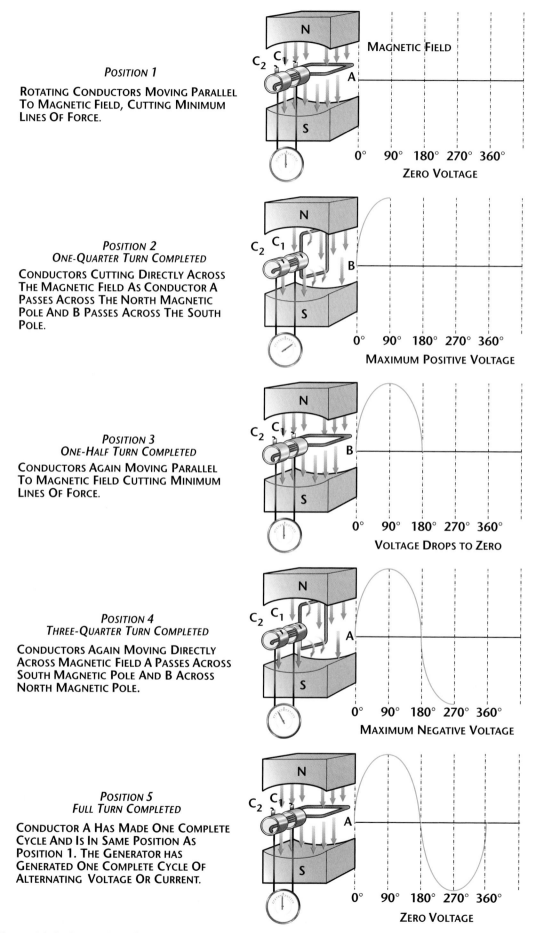

POSITION 1

ROTATING CONDUCTORS MOVING PARALLEL TO MAGNETIC FIELD, CUTTING MINIMUM LINES OF FORCE.

MAGNETIC FIELD

ZERO VOLTAGE

POSITION 2
ONE-QUARTER TURN COMPLETED

CONDUCTORS CUTTING DIRECTLY ACROSS THE MAGNETIC FIELD AS CONDUCTOR A PASSES ACROSS THE NORTH MAGNETIC POLE AND B PASSES ACROSS THE SOUTH POLE.

MAXIMUM POSITIVE VOLTAGE

POSITION 3
ONE-HALF TURN COMPLETED

CONDUCTORS AGAIN MOVING PARALLEL TO MAGNETIC FIELD CUTTING MINIMUM LINES OF FORCE.

VOLTAGE DROPS TO ZERO

POSITION 4
THREE-QUARTER TURN COMPLETED

CONDUCTORS AGAIN MOVING DIRECTLY ACROSS MAGNETIC FIELD A PASSES ACROSS SOUTH MAGNETIC POLE AND B ACROSS NORTH MAGNETIC POLE.

MAXIMUM NEGATIVE VOLTAGE

POSITION 5
FULL TURN COMPLETED

CONDUCTOR A HAS MADE ONE COMPLETE CYCLE AND IS IN SAME POSITION AS POSITION 1. THE GENERATOR HAS GENERATED ONE COMPLETE CYCLE OF ALTERNATING VOLTAGE OR CURRENT.

ZERO VOLTAGE

Figure 4-1-8. Generation of a sine wave

power and are functionally unique in that they have the capability to convert DC power into AC power. Static inverters use a two-stage power conversion topology that provides regulated output AC voltage. A high frequency switching DC/DC converter isolates and scales the 28 VDC input to a high voltage internal DC bus.

Power is then transferred to a high frequency DC/AC bridge inverter circuit to generate an output of 115 VAC or 230 VAC at 50 Hz, 60 Hz or 400 Hz. Some static inverters can be connected in parallel or three phase for higher power ratings. They are designed with control technology that ensures proportional current sharing without the need for output voltage matching. Newer static inverter design features include high reliability, low weight and high efficiency.

Frequency. Regardless of which device is used to generate AC voltage and AC current, certain characteristics of the produced voltage and current are the same. *Frequency* is the term used to describe the number of cycle events that occur in one second. See Figure 4-1-9. With the unit of frequency being the *hertz*, 400 cycles per second is referred to as a frequency of 400 hertz, and is written 400 Hz, the commonly used frequency in aircraft AC electrical systems.

In a generator, the voltage and current pass through a complete cycle of values each time a conductor (coil) passes under a north and south pole of the magnet. The number of cycles for each revolution of the coil or conductor is equal to the number of pole pairs. The frequency is therefore equal to the number of cycles in one revolution multiplied by the number of revolutions per second. Expressed in equation form:

$$F = \frac{\text{Number of poles}}{2} \times \frac{\text{r.p.m.}}{60}$$

Where P/2 is the number of pole pairs, and r.p.m./60 the number of revolutions per second. If in a 2-pole generator, the conductor is turning at 3,600 r.p.m., the revolutions per second (r/s) are:

$$r/s = \frac{3600}{60} = 60 \text{ revolutions}$$

Since there are two poles, P/2 is one, and the frequency is 60 Hz. In a 16-pole generator with an armature speed of 3,000 r.p.m., substitute in the equation,

$$F = \frac{16}{2} \times \frac{3600}{60}$$
$$F = 400 \text{ Hz}$$

Phase. In addition to frequency and cycle characteristics, alternating voltage and current also have a relationship called *phase*. In a circuit

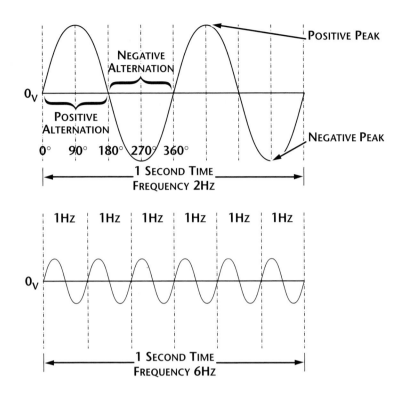

Figure 4-1-9. Frequency in cycles per second

that is fed (supplied) by one electrical power source, there must be a certain phase relationship between voltage and current if the circuit is to function efficiently. In a system fed by two or more electrical power sources, not only must there be a certain phase relationship between voltage and current of one source, but there must be a phase relationship between the voltages and the other sources. In addition, the relationship between two or more circuits may be established comparing the phase characteristics of one to the phase characteristics of the others.

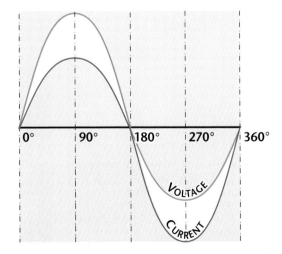

Figure 4-1-10. In-phase condition of current and voltage

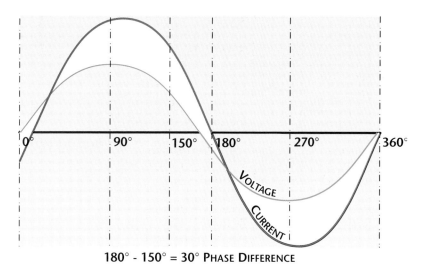

180° - 150° = 30° Phase Difference

Figure 4-1-11. Voltage and current 30° out of phase

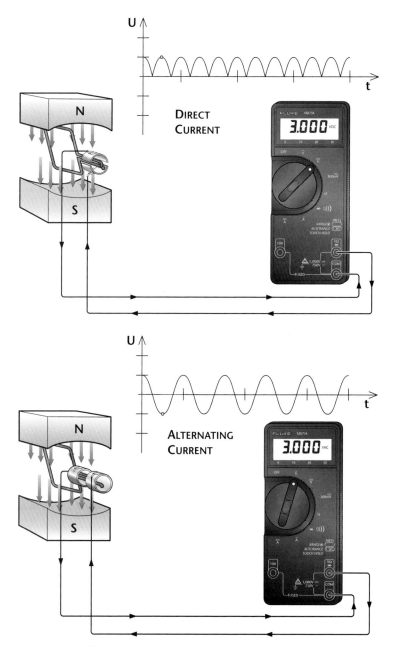

Figure 4-1-12. Different waveforms of DC generators and AC alternators

When two or more sine waves pass through 0° and 180° at the same time and reach their peaks at the same time, an in-phase condition exists, as shown in Figure 4-1-10. The peak values (magnitudes) do not have to be the same for the in-phase condition to exist.

When the sine waves pass through 0 and 180° at different times and reach their peaks at different times, an out-of-phase condition exists, as shown in Figure 4-1-11. The amount that the two sine waves are out of phase is indicated by the number of electrical degrees between corresponding peaks on the sine waves. In Figure 4-1-11, the current and voltage are 30° out of phase.

Impedance. AC circuits contain resistance and may contain *inductive reactance, capacitive reactance* or both. The combination of resistance, inductive reactance and capacitive reactance in a circuit determines that circuit's total opposition to the flow of current, which is known as *impedance (Z). Resistance (R)* cannot be added directly with inductive reactance or capacitive reactance. Instead, the relationship between these types of opposition to current flow may be represented by a right triangle. The law of right triangles or the Pythagorean theorem is the formula used for determining a circuit's impedance.

Power. Power, when used as a term to describe the potential for an electrical circuit to do work, is obtained by the equation $P = E \times I$, as stated earlier in this chapter. Therefore, if two amps of current are flowing through a circuit load at a pressure (electromotive force) of 220 volts, the *true power* consumed is 440 watts.

Apparent power, on the other hand, is the product of the effective voltage as indicated by a voltmeter and the effective current, as indicated by an ammeter. Apparent power equals true power only when the circuit's total opposition to the flow of current is purely resistance (no inductive or capacitive reactance).

The *power factor* is the ratio of *true power* to the *apparent power* and is usually expressed as a percentage. When there is capacitance or inductance in the circuit, the current and voltage will not be exactly in phase and the *true power* (as read on a wattmeter) will be less than the apparent power.

Direct Current

DC Generators. Direct current (DC) power generation components and DC circuits are used on all aircraft for certain applications such as low-current lighting, battery powered engine starting systems and emergency backup power for radios, to name a few. Many single-engine and light twin-engine general aviation aircraft use 12-volt DC systems, but it is more common

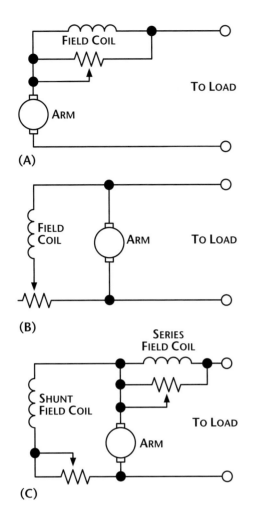

Figure 4-1-13. There are three ways of connecting a generator field windings to the load: (A) in series, (B) as a shunt, (C) in series-shunt.

to find 28-volt DC systems on turbine-powered aircraft and on all transport category aircraft. The most commonly used sources of power for DC circuits are batteries and DC generators.

A generator is rated in power output. Since a generator is designed to operate at a specific voltage, the rating is usually given as the number of amperes the generator can safely supply at its rated voltage.

Generator rating and performance data are stamped on the name plate attached to the generator. When replacing a generator it is important to choose one of the proper rating.

DC generators function in much the same way as AC generators except for the manner in which current generated by the rotating loops (rotor coils) is transferred to the power distribution wiring. Instead of slip rings, a commutator is mounted on the end of the rotor and electrical current is transferred to a set of brushes in such a way that current flows in only one direction through the brushes. Figure 4-1-12 illustrates both principles.

Types of DC generators

There are three types of DC generators:

- Series wound
- Shunt wound
- Series-shunt or compound-wound

The difference in type depends on the relationship of the field windings to the external circuit, and can be seen in Figure 4-1-13 A-C.

Series wound. The field winding of a series generator is connected in series with the external circuit or load. Series generators have very poor voltage regulation under changing load, since the greater the current through the field coil to the external circuit, the greater the induced e.m.f. and the greater the output voltage. Therefore, when the load is increased, the voltage increases, when the load decreases the voltage decreases. Regulation is provided by a rheostat in the in parallel with the field windings, shown in A of Figure 4-1-13. Since the regulation is so poor in the series wound generator, it is never used in aircraft.

Shunt wound. A shunt wound generator has its field windings connected in parallel with the output circuit. If a constant voltage is desired, the shunt wound generator is not suitable for rapidly fluctuating loads. Any increase in load causes a decrease in output voltage and decrease in load causes an increase in voltage. The armature and the load are connected in series, and all of current flowing in the external circuit passes through the armature windings. Voltage regulation can be controlled by a inserting a rheostat in series with the field windings, as shown in B of Figure 4-1-13. For a given setting of the field rheostat the output voltage will drop as the load is applied, and the rheostat must be reset for the new values.

Compound-wound. A compound-wound generator combines a series winding and a shunt winding in such a way that the characteristics of each are used to advantage. The series field coils are connected in series with the armature circuit, and are mounted on the same poles as the shunt field coils. The series fields then contribute a magneto-motive force, which influences the main field of the generator. This compounding creates a generator whose output voltage may increase or decrease with changes in the load. Regulation is achieved by the use of rheostats in both the series and shunt fields, and is shown in C of Figure 4-1-13.

Batteries

Various types of rechargeable batteries are in use in aircraft electrical systems, including *lead-acid*

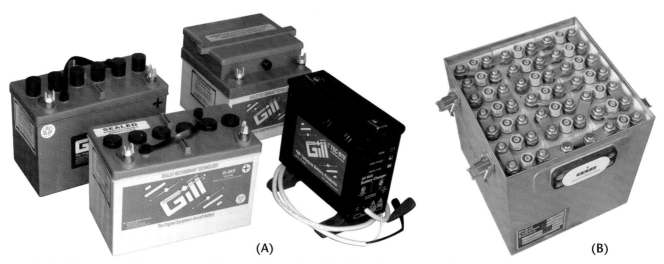

Figure 4-1-14. (A) Several different types of lead acid batteries, (B) a 20-cell nickel-cadmium battery

and *nickel-cadmium (NiCad)*. Figure 4-1-14 shows different sizes and types of batteries that are in current use. Although available for quite some time, nickel cadmium batteries did not come into extensive use in aviation until increases in the number of commercial and executive jet aircraft made them economically practicable. The many advantages of the nickel-cadmium battery were well known, but its initial cost was several times that of the lead-acid battery.

The increasing use of the nickel-cadmium battery (referred to by a variety of trade names) stems largely from the low maintenance cost derived from the long service life of the battery. Additionally, the nickel-cadmium battery has a short recharge time, excellent reliability and good starting capability.

However, a nickel-cadmium battery does suffer from a process that is not found in lead-acid batteries. It is called thermal runaway.

Thermal runaway can result in a chemical fire and/or explosion of the nickel-cadmium battery under recharge by a constant-voltage source, and is due to cyclical ever-increasing temperature and charging current.

Figure 4-1-15. A battery quick disconnect plug

One or more shorted cells or an existing high temperature and low charge can produce the cyclical sequence of events:

1. Excessive current

2. Increased temperature

3. Decreased cell(s) resistance

4. Further increased current

5. Further increased temperature

This will not become a self-sustaining thermal-chemical action if the constant-voltage charging source is removed before the battery temperature exceeds 160°F.

Battery maintenance. Batteries are generally mounted inside a box, which is placed in a mount or on a platform that has been coated, to protect it from the corrosive effects of electrolyte spills and other corrosive by-products of the battery. The battery is held securely in the mount or on the platform using hold down brackets and other common hardware.

The battery box has a removable top with a vent tube nipple at each end. When the battery is installed in an airplane, a vent-tube is attached to each nipple. One tube is the intake tube and is exposed to the slipstream, and the other is the exhaust vent tube.

On some older airplanes the exhaust vent tube is attached to the battery drain sump, which is a glass jar containing a felt pad moistened with a concentrated solution of sodium bicarbonate (baking soda).

Most newer airplanes have the airstream directed through the battery case where battery gases are picked up and then expelled overboard without damage to the airplane.

To facilitate installation and removal of the battery in some aircraft, a quick disconnect assembly is used to connect the power leads to the battery. This assembly, which is shown in Figure 4-1-15, attaches the battery leads in the aircraft to a receptacle mounted on the side of the battery. The receptacle covers the battery terminal posts and prevents accidental shorting during the installation and removal of the battery. The plug consists of a socket and a handwheel with a coarse-pitch thread. It can be readily connected to the receptacle by the handwheel. Another advantage of this assembly is that the plug can be installed in only one position, eliminating the possibility of reversing the battery leads.

Section 2

Generators and Alternators

Description of Generator Systems

For the purpose of this discussion, a Piper Aircraft Comanche Delco Remy generator system has been chosen. The generator is of the two brush, shunt type and is controlled by a regulator operating on the principal of inserting resistance into the generator field circuit to cause a reduction of generator voltage and current output.

With each generator is a regulator assembly, composed of a voltage regulator and current regulator, to prevent overloading of the battery and electrical circuits. Also with the regulator is a reverse current cutout to prevent the generator from being motorized by the battery when the generator output drops below the battery voltage.

Field current between the regulator and generator may be broken through the use of the master switch. The generator is located on the front lower right side of the engine and utilizes a belt drive from the engine crankshaft. The generator voltage regulator is located on the engine firewall.

Checking Generator System

Analyzing complaints of generator regulator operation, any of several basic conditions may be found:

- **Fully charged battery and low charging rate:** indicates normal generator-regulator operation

- **Fully charged battery and a high charging rate:** indicates that the voltage regulator is not reducing the generator output as it should. A high charging rate to a fully charged battery will damage the battery and the accompanying high voltage is very injurious to all electrical units.

This operating condition may result from:

- Improper voltage regulator setting

- Defective voltage regulator unit

- Grounded generator field circuit (in either generator, regulator or wiring)

- Poor ground connection at regulator

- High temperature, which reduces the resistance of the battery to charge so that it will accept a high charging rate even though the voltage regulator setting is normal

If the trouble is not due to high temperature, determine the cause of trouble by disconnecting the lead from the regulator "F" (field) terminal with the generator operating at medium speed. If the output remains high, the generator field is grounded either in the generator or in the wiring harness. If the output drops off, the regulator is at fault, and it should be checked for a high voltage setting or grounds.

Several scenarios are possible:

1. **Low battery and high charging rate:** This is normal generator-regulator action.

2. **Low battery and low or no charging rate:** This condition could be due to:

   - Loose connections frayed or damaged wires

   - Defective battery

   - High circuit resistance

   - Low regulator setting

   - Oxidized regulator contact points

   - Defects within the generator

If the condition is not caused by loose connections, frayed or damaged wires, proceed as follows to locate cause of trouble.

To determine whether the generator or regulator is at fault, shortly ground the "F" (field) terminal of the regulator and increase generator speed. If the output does not increase, the generator is probably at fault and it should be checked. If the generator output increases, the trouble is due to:

- A low voltage (or current) regulator setting

- Oxidized regulator contact points which insert excessive resistance into the generator field circuit so that output remains low.

- Generator field circuit open within the regulator at the connections or in the regulator wiring

Burned resistances, windings or contacts. These result from open circuit operation or high resistance in the charging circuit. Where burned resistances, windings or contacts are found, always check wiring before installing a new regulator. Otherwise, the new regulator may also fail in the same way.

Burned relay contact points. This is due to reversed generator polarity. Generator polarity must be corrected after any checks of the regulator or generator or after disconnecting and reconnecting leads.

Adjustments, Tests and Maintenance of Generator System

The best assurance of obtaining maximum service from generators with minimum trouble is to follow a regular inspection and maintenance procedure. Periodic lubrication where required,

Figure 4-2-1. Sectional view of generator

inspection of the brushes and commutator and checking of the brush spring tension are essentials in the inspection procedure. In addition, disassembly and thorough overhauling of the generator at periodic intervals are desirable as a safeguard against failures from accumulations of dust and grease and normal wear of parts. This is particularly desirable on installations where maintenance of operating schedules is of special importance.

In addition to the generator itself, the external circuits between the generator, regulator and battery must be kept in good condition since defective wiring or loose or corroded connections will prevent normal generator and regulator action. At times, it maybe found necessary to adjust the voltage regulator. Figure 4-2-1 shows a cutaway generator, allowing the internal wiring and brush holders to be seen.

Test and Maintenance of Generator

Inspection of generator. At periodic intervals, the generator should be inspected to determine its condition. The frequency required is determined by the type of service in which it is used. High speed operation, excessive dust or dirt, high temperatures and operating the generator at or near full output most of the time are all factors which increase bearing, commutator and brush wear. The units should be inspected at approximately 100 hour intervals.

The inspection procedure is as follows:

1. First inspect the terminals, external connections and wiring, mounting, pulley and belt. Then remove the cover band so that the commutator, brushes and internal connections can be inspected. If the commutator is dirty, it may be cleaned with a strip of No. 00 sandpaper. Never use emery cloth to clean the commutator.

2. The sandpaper maybe used by holding it against the commutator with a wood stick while the generator is rotated, moving it back and forth across the commutator. Gum and dirt will be sanded off in a few seconds. All dust should be blown from the generator after the commutator has been cleaned. A brush seating stone can also be used to clean the commutator.

3. If the commutator is rough, out of round or has high mica, the generator must be removed and disassembled so that the armature can be turned down in a lathe and the mica undercut.

4. If the brushes are worn down to less than half their original length, replace them. Compare the old brush with a new one

to determine how much it is worn. New brushes should be seated to make sure that they are in good contact with the commutator. A convenient tool for seating brushes is a brush seating or bedding stone. This is a soft abrasive material which, when held against a revolving commutator, disintegrates so that particles are carried under the brushes and wear their contacting faces to the contour of the commutator in a few seconds. All dust should be blown from the generator after the brushes are seated. See Figure 4-2-2.

5. The brush spring tension must be correct since excessive tension will cause rapid brush and commutator wear, while low tension causes arcing and burning of the brushes and commutator. Brush spring tension can be checked with a spring gauge hooked on the brush armor brush attaching screw. Bending the brush spring as required, can correct it. If the brush spring shows evidence of overheating (blued or burned), do not attempt to readjust it, but install a new spring. Overheating will cause a spring to lose its temper.

6. Always inspect the braided copper conductor connection, commonly called a pigtail, to the brush. Check for fraying or other signs of wear. The pigtails eliminate possible sparking to the brush guides caused by the movement of the brushes within the holder, thus minimizing side wear of the brush.

7. The belt should be checked to make sure that it is in good condition and has correct tension. Low belt tension will permit belt slippage with a resulting rapid belt wear and low or erratic generator output. Excessive belt tension will cause rapid belt and bearing wear. Check the tension of a new belt 25 hours after installation.

Shunt generator output. The current setting of the current regulator with which the shunt generator is used determines the maximum output of shunt generators.

Checking Defective Generators

If the generator-regulator system does not perform according to specifications (generator does not produce rated output or produces excessive output), and the trouble has been isolated in the generator itself, the generator may be checked further as follows to determine the location of trouble in the generator.

No output. If the generator will not produce any output, remove the cover band and check the commutator, brushes and internal connections. Sticking brushes, a dirty or gummy commuta-

Figure 4-2-2. Most generator maintenance is in the area of brush and brush holders

tor or poor connections may prevent the generator from producing any output. Thrown solder on the cover band indicates that the generator has been overloaded (allowed to produce excessive output) so it has overheated and melted the solder at the commutator riser bars. Solder thrown out often leads to an open circuit and burned commutator bars. If the brushes are satisfactorily seated and have good contact with the commutator, and the cause of trouble is not apparent, use a set of test points and a test lamp as follows to locate the trouble (leads must be disconnected from generator terminals).

Raise the grounded brush from the commutator and insulate with a piece of cardboard. Check for grounds with test points from the generator main brush to the generator frame. If the lamp lights, it indicates that the generator is internally grounded. Location of the ground can be found by raising and insulating all brushes from the commutator and checking the brush holders, armature, commutator and field separately. Repair or replace defective parts as required.

NOTE: *If a grounded field is found, check the regulator contact points, since a grounded field may have permitted an excessive field current which will have burned the regulator contact points. Burned regulator points should be cleaned or replaced as required.*

If the generator is not grounded, check the field for an open circuit with a test lamp. The lamp should light when one test point is placed on the field terminal or grounded field lead and the other is placed on the brush holder to which the field is connected. If it does not light, the circuit is open. If the open is due to a bro-

Figure 4-2-3. A growler is a standard fixture for checking armatures for grounds.

ken lead or bad connection, it can be repaired, but if the open is inside one of the field coils, it must be replaced.

If the field is not open, check for a short circuit in the field by connecting a battery of the specified voltage and an ammeter in series with the field circuit. Proceed with care, since a shorted field may draw excessive current, which might damage the ammeter. If the field is not within specification, new field coils will be required.

> **NOTE:** *If a shorted field is found, check the regulator contact points, since a shorted field may have permitted excessive field current which would have caused the regulator contact points to burn. Clean or replace points as required.*

If the trouble has not yet been located, check the armature for open and short circuits. Open circuits in the armature are usually obvious; since the open circuited commutator bars will arc every time they pass under the generator brushes so that they will soon burn. If the bars are not burned badly and the open circuit can be repaired, the armature can usually be saved. In addition to repairing the armature, generator output must be brought down to specifications to prevent overloading by readjustment of the regulator.

Short circuits in the armature are located by use of a growler (Figure 4-2-3). The armature is placed in the growler and slowly rotated (while a thin strip of steel such as a hacksaw blade is held above the armature core). The steel strip will vibrate above the area of the armature

core in which short circuited armature coils are located. If the short circuit is obvious, it can often be repaired so that the armature can be saved.

Unsteady or low output. If the generator produces a low or unsteady output, the following factors should be considered:

- A loose drive belt will slip and cause a low or unsteady output.

- Brushes which stick in their holders, or low brush spring tension, will prevent good contact between the brushes and commutator so that output will be low and unsteady. This will also cause arcing and burning of the brushes and commutator.

- If the commutator is dirty, out of round or has high mica, generator output is apt to be low and unsteady. The remedy here is to turn the commutator down in a lathe and undercut the mica. Burned commutator bars may indicate an open circuit condition in the armature as already stated above.

Excessive output. When a generator produces excessive output, first determine if the problem is in the generator or some other part of the system. If the generator output remains high, even with the "F" (field) terminal lead disconnected, then the trouble is in the generator itself, and it must be further analyzed to locate the source of trouble.

In the system which has the generator field circuit grounded externally, accidental internal grounding of the field circuit would prevent normal regulation so that excessive output might be produced by the generator. On this type of unit, an internally grounded field which would cause excessive output may be located by use of test points connected between the "F" terminal and the generator frame. Leads should be disconnected from the "F" terminal and the brush to which the field lead is connected inside the generator should be raised from the commutator before this test is made.

If the lamp lights, the field is internally grounded. If the field has become grounded because the insulation on a field lead has worn away, re-insulating the lead can repair it. It is also possible to make repair where the ground has occurred at the pole shoes by removing the field coils, re-insulating and reinstalling them. Installing new insulating washers or bushings can repair a ground at the "F" terminal stud.

> **NOTE:** *If battery temperature is excessive, battery overcharge is apt to occur, even though regulator settings are normal. Consult the applicable maintenance manual trouble shooting chart for a procedure.*

Noisy generator. A loose mounting, drive pulley or gear, worn or dirty bearings or improperly seated brushes may cause noise emanating from a generator. Cleaning and relubrication may sometimes save dirty bearings, but replace worn bearings. Brushes can be seated as explained later. If the brush holder is bent, it may be difficult to reseat the brush so that it will function properly without excessive noise. Such a brush holder will require replacement.

Armature service. The armature should be checked for opens, shorts and grounds as explained in following paragraphs. If the armature commutator is worn, dirty, out of round or has high mica, the armature should be put in a lathe so the commutator can be turned down and the mica undercut. Undercut the mica and clean the slots carefully to remove any trace of dirt or copper dust. As a final step in this procedure, sand lightly with No. 00 sandpaper the commutator to remove any slight burrs left because of the undercutting procedure.

Open-circuited armatures can often be saved when the open is obvious and repairable. The most likely place an open will occur is at the commutator riser bars. This usually results from overloading of the generator which causes overheating and melting of the solder. Repair can be effected by resoldering the leads in the riser bars (using rosin flux) and turning down the commutator in a lathe to remove the burned spot and then undercutting the mica as explained in the previous paragraph. In some heavy-duty armatures, the leads are welded into the riser bars and resoldering cannot repair these.

Short circuits in the armature are located by use of a growler, seen in Figure 4-2-3. The armature is revolved in the growler with a steel strip, such as a hacksaw blade, held above it the blade will vibrate above the area of the armature core in which the short is located. Copper or brush dust in the slots between the commutator bars sometimes causes shorts between bars, which can be eliminated by cleaning out the slots. Bending wires slightly and re-insulating the exposed bare wire can often eliminate shorts at crossovers of the coils at the core end.

Use of a test lamp and test points detects grounds in the armature. If the lamp lights when one test point is placed on the commutator with the other point on the core or shaft, the armature is grounded. Grounds occur because of insulation failure, which is often brought on by overloading and consequent overheating of the generator. Repairs can sometimes be made if grounds are at core ends (where coils come out of slots) by placing insulating strips between core and coil which has been grounded.

Polarizing generator. After a generator has been repaired and reinstalled, or at any time after a generator has been tested, it must be repolarized to make sure that it has the correct polarity with respect to the battery it is charging. Failure to repolarize the generator may result in burned relay contact points, a run-down battery and possibly serious damage to the generator itself. The procedure to follow in correcting generator polarity depends upon the generator-regulator wiring circuits; that is, whether the generator field is internally grounded or is grounded through the regulator. Polarizing or flashing the field also produces a residual magnetism into the field shoes, which induces a small voltage into the field windings. This small voltage is what starts the generation process when the generator is driven by the aircraft engine.

> **CAUTION:** *Not all generators are flashed the same, so always consult the maintenance manual first.*

Radio interference. The output of a DC generator normally has some AC noise superimposed on the DC signal. The noise, or hash, created by brush arcing, causes interference in the aircraft avionics. The noise can be removed from the DC signal by connecting a capacitor between the armature and ground in parallel with the armature windings. The AC passes to ground through the low impedance path provided by the capacitor.

Voltage Regulator Operation

The only practical method of maintaining a constant voltage output from an aircraft generator under varying conditions of speed and load is to vary the strength of the magnetic field. The strength of the magnetic field is varied through the use of increasing or decreasing the resistance in the field circuit.

Description of regulator. The regulator shown in Figure 4-2-4 consists of a cutout relay, a voltage regulator and a current regulator unit. The cutout relay closes the generator to battery circuit when the generator voltage is sufficient to charge the battery, and it opens the circuit when the generator slows down or stops. The voltage regulator unit is a voltage-limiting device that prevents the system voltage from exceeding a specified maximum and thus protects the battery and other voltage-sensitive equipment. The current regulator unit is a current-limiting device that limits the generator output so as not to exceed its rated maximum.

Cutout relay. The cutout relay (Refer to Figure 4-2-4), has two windings, a series winding of a few turns of heavy wire and a shunt winding

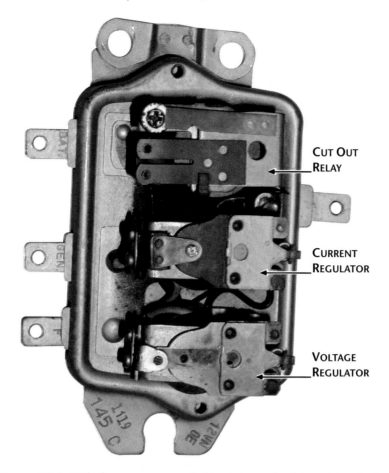

CUT OUT RELAY

CURRENT REGULATOR

VOLTAGE REGULATOR

Figure 4-2-4. With the cover removed, it is obvious why this part is called a three unit regulator.

This reverse flow of current through the series winding causes a reversal of the series winding magnetic field. The magnetic field of the shunt winding does not reverse. Therefore, instead of helping each other, the two windings now oppose so that the resultant magnetic field becomes insufficient to hold the armature down. The flat spring pulls the armature away from the core so that the points separate; this opens the circuit between the generator and battery.

Current regulator. The current regulator (Refer to Figure 4-2-4) has a series winding of a few turns of heavy wire which carries all generator output. The winding core is assembled into a frame. A flat steel armature is attached to the frame by a flexible hinge so that it is just above the core. The armature has a contact point that is just below a stationary contact point. When the current regulator is not operating, the tension of a spiral spring holds the armature away from the core so that the points are in contact. In this position, the generator field circuit is completed to ground through the current regulator contact points in series with the voltage regulator contact points.

When the load demands are heavy, as for example, when electrical devices are turned on and the battery is in a discharged condition, the voltage may not increase to a value sufficient to cause the voltage regulator to operate. Consequently, generator output will continue to increase until the generator reaches rated maximum current. The current regulator is set for this current value. Therefore, when the generator reaches rated output, this output, flowing through the current regulator winding, creates sufficient magnetism to pull the current regulator armature down and open the contact points. With the points open, resistance is inserted into the generator field circuit so that the generator output is reduced.

As soon as the generator output starts to fall off, the magnetic field of the current regulator winding is reduced, the spiral spring tension pulls the armature up, the contact points close and directly connect the generator field to ground. Output increases and the above cycle is repeated. The cycle continues to take place while the current regulator is in operation 50 to 200 times a second, preventing the generator from exceeding its rated maximum. When the electrical load is reduced (electrical devices turned off or battery comes up to charge), then the voltage increases so that the voltage regulator begins to operate and tapers the generator output down. This prevents the current regulator from operating.

Either the voltage regulator or the current regulator operates at any one time - the two do not operate at the same time.

of many turns of fine wire. The shunt winding is connected across the generator so that generator voltage is impressed upon it at all times. The series winding is connected in series with the charging circuit so that all generator output passes through it. The relay core and windings are assembled into a frame. A flat steel armature is attached to the frame by a flexible hinge so that it is centered just above the stationary contact points. When the generator is not operating, the armature contact points are held away from the stationary points by the tension of a flat spring riveted on the side of the armature.

When the generator voltage builds up a voltage great enough to charge the battery, the magnetism induced by the relay windings is sufficient to pull the armature toward the core so that the contact points close. This completes the circuit between the generator and battery. The current, which flows from the generator to the battery, passes through the series winding in a direction to add to the magnetism holding the armature down and the contact points closed.

When the generator slows down or stops, current begins to flow from the battery to the generator.

Resistances. The current and voltage regulator circuits use a common resistor that is inserted in the field circuit when either the current or voltage regulator operates. A second resistor is connected between the regulator field terminal and the cutout relay frame, which places it in parallel with the generator field coils. The sudden reduction in field current occurring when the current or voltage regulator contact points open, is accompanied by a surge of induced voltage in the field coils as the strength of the magnetic field changes. These surges are partially dissipated by the two resistors, thus preventing excessive arcing at the contact points. The second resistor is not present on all regulators. Many aircraft regulators have this resistor omitted.

Temperature compensation. Voltage regulators are compensated for temperature by means of a bimetal thermostatic hinge on the armature. This causes the regulator to regulate at a higher voltage when cold, which partly compensates for the fact that a higher voltage is required to charge a cold battery. Many current regulators also have a bimetal thermostatic hinge on the armature. This permits a somewhat higher generator output when the unit is cold, but causes the output to drop off as temperature increases.

Regulator polarity. Some regulators are designed for use with negative grounded systems. Other regulators are designed for use with positive grounded systems. Using the wrong polarity regulator on an installation will cause the regulator contact points to pit badly and give short life. As a safeguard against installation of the wrong polarity regulator, all regulators of this type have the model number and the polarity clearly stamped on the end of the regulator base.

Regulator Maintenance

Mechanical checks and adjustments (air gaps, point opening) must be made with battery disconnected and regulator preferably off the aircraft.

> **CAUTION:** *The cutout relay contact points must never he closed by hand with the battery connected to the regulator. This would cause a high current to flow through the units that could seriously damage them.*

Electrical checks and adjustments may be made either on or off the airplane. The regulator must always be operated with the type of generator for which it is designed.

The regulator must be mounted in the operating position when electrical settings are checked

and adjusted and it must be at operating temperature. Figure 4-2-5 shows an air gap adjustment.

Specified generator speeds for testing and adjusting. All generators must be operated at a speed sufficient to produce current in excess of the specified setting. Voltage of the generator must be kept high enough to insure sufficient current output, but below the maximum operating voltage of the voltage regulator unit.

Polarizing the generator. After any tests or adjustments the generator on the airplane must be polarized after leads are connected, but before the engine is started, as follows:

• After reconnecting leads, momentarily connect a jumper lead between the "GEN" and "BAT" terminals of the regulator. This allows a momentary surge of current to flow through the generator, which correctly polarizes it. Failure to do this may result in severe damage to the equipment, since reversed polarity causes vibration, arcing and burning of the relay contact points.

Cleaning contact points. The contact points of a regulator will not operate indefinitely without some attention. It has been found that a great majority of all regulator trouble can be eliminated by a simple cleaning of the contact points, plus some possible readjustment. The flat points should be cleaned with a spoon or riffler file. On negative grounded regulators that have the flat contact point on the regulator armatures, loosen the contact bracket mount-

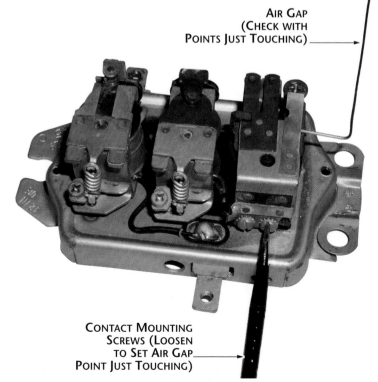

Figure 4-2-5. Three unit voltage regulator air gap adjustment

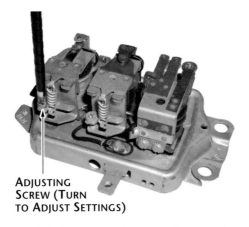

ADJUSTING
SCREW (TURN
TO ADJUST SETTINGS)

Figure 4-2-6. Adjusting voltage regulator setting

ing screws so that the bracket can be tilted to one side. A flat file cannot be used successfully to clean the flat contact points since it will not touch the center of the flat point where point wear is most apt to occur. Never use emery cloth or sandpaper to clean the contact points. Remove all the oxides from the contact points but note that it is not necessary to remove any cavity that may have developed.

Regulator Checks and Adjustments

Voltage regulator. Two checks and adjustments are required on the voltage regulator, air gap and voltage setting. Voltage regulator setting is shown in Figure 4-2-6.

Air gap. To check air gap, push armature down until the contact points are just touching and then measure air gap. Adjust by loosening the contact mounting screws and raising or lowering contact bracket as required. Be sure the points are lined up and after adjustment tighten the screws.

Voltage setting. There are two ways to check the voltage setting, the fixed resistance method and the variable resistance method.

Fixed resistance method:

- Connect a fixed resistance between the battery terminal and ground as shown in Figure 4-2-7 after disconnecting the battery lead from the battery terminal of the regulator. The resistance should be 1 1/2-ohms for a 14 volt unit. It must be capable of carrying 10 amperes without any change of resistance with temperature changes.

- Connect a voltmeter from regulator "BAT" (battery) terminal to ground.

- Place the thermometer within 1/4-inch of regulator cover to measure regulator ambient temperature.

- Operate generator at specified speed for 15 minutes with regulator cover in place to bring the voltage regulator to operating temperature.

- Cycle the generator.

Method 1. Move voltmeter lead from "BAT" to "GEN" terminal of regulator. Retard generator speed until generator voltage is reduced to 4 volts. Move voltmeter lead back to "BAT" terminal of regulator. Bring generator back to specified speed and note voltage setting.

Method 2. Connect a variable resistance into the field circuit as in Figure 4-2-7. Turn out all resistance. Operate generator at specified speed. Slowly increase (turn in) resistance until generator voltage is reduced to 4-volts. Turn out all resistance again and note voltage setting (with voltmeter connected as in Figure 4-2-7).

- Regulator cover must be in place.

- Note the thermometer reading and select the "Normal Range of Voltage" for this temperature as listed in the specifications.

- Note the voltmeter reading with regulator cover in place.

- To adjust voltage setting, turn adjusting screw. Refer to Figure 4-2-7. Turn clockwise to increase setting and counterclockwise to decrease setting.

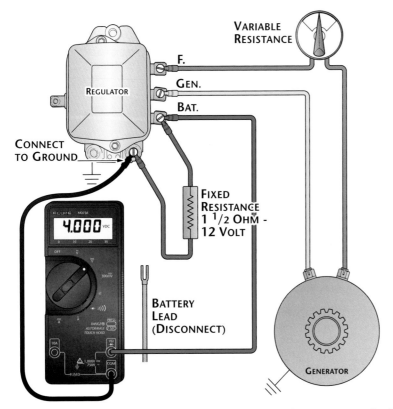

Figure 4-2-7. Checking voltage setting using the fixed resistance method

CAUTION: *If adjusting screw is turned down (clockwise) beyond the range, the spring support may not return when screw is backed off. In such case, turn screw counterclockwise until there is ample clearance between screw head and spring support. Then bend spring support up carefully until it touches the screw head. Final setting of the unit should always be made by increasing spring tension, never by reducing it. If setting is too high, adjust unit below required value and then raise to exact setting by increasing the spring tension. After each adjustment and before taking reading, replace the regulator cover and cycle the generator.*

Variable resistance method:

- Connect ammeter and 1/4-ohm variable resistor in series with the battery as shown in Figure 4-2-8.

NOTE: *It is very important that the variable resistance be connected at the "BAT" terminal as shown in Figure 4-2-7 rather than at the "GEN" terminal even though these terminals are in the same circuit. Any small resistance added to the circuit between the generator and* this point will simply be offset by a rise in generator voltage without affecting the output shown at the ammeter.

- Connect voltmeter between "BAT" terminal and ground

- Place thermometer within 1/4-inch of regulator cover to measure regulator ambient temperature

- Operate generator at specified speed. Adjust variable resistor until current flow is 8 to 10 amperes. If less current than is required above is flowing, it will be necessary to turn on airplane lights to permit increased generator output. Variable resistance can then be used to decrease current flow to the required amount.

- Allow generator to operate at this speed and current flow for 15 minutes with regulator cover in place in order to bring the voltage regulator to operating temperature.

- Cycle the generator by either method listed in "Fixed Resistance Method" of "Voltage Setting" procedure.

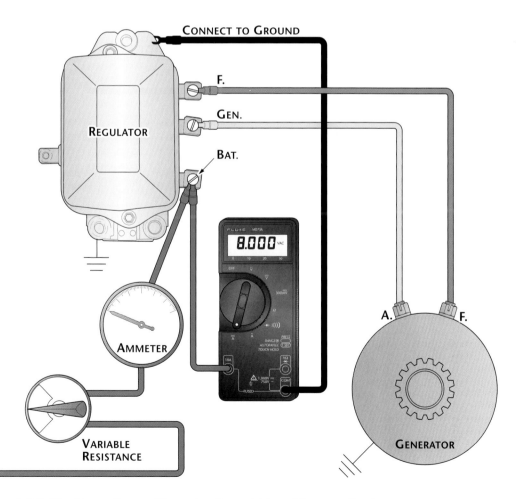

Figure 4-2-8. Checking voltage setting using the variable resistance method.

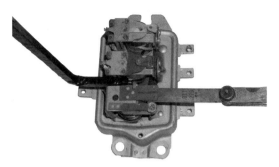

Figure 4-2-9. Cutout relay point opening check and adjustment

- Note the thermometer reading and select the "Normal Range" of voltage for this temperature as listed in the specifications.

- Note the voltmeter reading with regulator cover in place.

- Adjust voltage regulator as required as described in "Fixed Resistance Method of Voltage Setting Procedure." In using the variable resistance method, it is necessary to readjust the variable resistance after each voltage adjustment to assure that 8 to

10 amperes are flowing. Cycle generator after each adjustment before reading voltage regulator setting with cover in place.

Cutout relay. The cutout relay requires three checks and adjustments: air gap, point opening and closing voltage. The air gap and point opening adjustments must be made with the battery disconnected. Check the MM for settings. See Figure 4-2-9.

Closing voltage. First connect regulator to proper generator and battery. Connect voltmeter between the regulator "GEN" terminal and ground. Then use either of the following methods:

- **Method 1.** Slowly increase generator speed and note relay closing voltage. Decrease generator speed and make sure the cutout relay points open.

- **Method 2.** In addition, add a variable resistor connected into the field circuit. See Figure 4-2-10. Use a 25-ohm 25-watt resistor. Operate generator at medium speed with variable resistance turned all in. Slowly decrease (turnout) the resis-

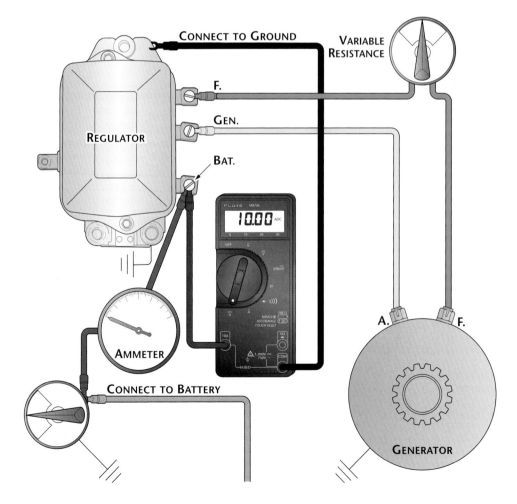

Figure 4-2-10. Checking current regulator, load method

tance until cutout relay points close. Note closing voltage. With cover in place, slowly increase (turn in) resistance to make sure points open.

Adjust closing voltage by turning adjusting screw. Turn the screw clockwise to increase the setting and counterclockwise to decrease the setting.

Current regulator. Two checks and adjustments are required on the current regulator, air gap and current setting.

Air gap. Check and adjust in exactly the same manner as for the voltage regulator.

Current setting. Current regulator setting on current regulators having temperature compensation should be checked by the following method:

Load method:

- Connect ammeter into charging circuit as in Figure 4-2-10.

- Turn on all accessory load (lights, radio, etc.) and connect an additional load across the battery (such as a carbon pile or band of lights) to drop the system voltage approximately one volt below the voltage regulator setting.

- Operate generator at specified speed for 15 minutes with cover in place. (This establishes operating temperature). If current regulator is not temperature-compensated, disregard 15 minute warm-up period.

- Cycle generator and note current setting

- Adjust in same manner as described for voltage regulator. Refer to Figure 4-2-6.

Jumper lead method: Use only for current regulators without temperature compensation.

- Connect ammeter into charging circuit as in Figure 4-2-11.

- Connect jumper lead across voltage regulator points as in Figure 4-2-11.

- Turn on all lights and accessories or load battery as under Load Method.

- Operate generator at specified speed and note current setting.

- Adjust in same manner as described for the voltage regulator. Refer to Figure 4-2-6.

Radio By-Pass Condensers

The installation of radio by-pass condensers on the field terminal of the regulator or gen-

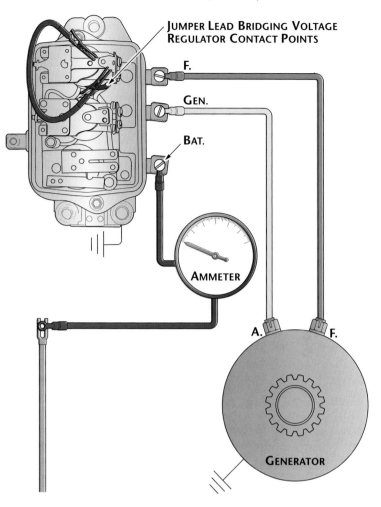

Figure 4-2-11. Checking current regulator, jumper lead method

erator can cause the regulator contact points to burn and oxidize so that generator output will be reduced and a run down battery will result. If a condenser is found connected to either of these terminals, replace the condenser and clean the regulator contact points as previously explained.

High Points on Regulator Performance and Checks

- The voltage regulator unit limits the voltage of the circuit, thus protecting the battery, distributor points, lights and other accessories from high voltage.

- The current regulator unit provides protection to the generator, preventing it from exceeding its maximum rated output.

- Never set the current regulator above the maximum specified output of the generator.

- Many of the regulators are designed to be used with a positive grounded battery while others are designed to be used with a negative grounded battery only. Never attempt to use the wrong polarity regulator on an application.

- The majority of reported regulator troubles arise from dirty or oxidized contact points which cause a reduced generator output. Clean the contact points with a spoon or riffler file. Never use emery cloth or sandpaper to clean points.

- Always make sure that the rubber gasket is in place between the cover and base before

(A)

(B)

(C)

Figure 4-2-12. Three of the most common types of alternator mountings (A) Chrysler type (B) Delco Remy type and (C) Prestolite type

replacing the cover. The gasket prevents entrance of moisture, dust and oil vapors, which might damage the regulator.

- The proper testing equipment in the hands of a qualified mechanic is necessary to assure proper and accurate regulator settings. Any attempt on the part of untrained personnel to adjust regulators is apt to lead to serious damage to the electrical equipment and should therefore be discouraged.

- After any generator or regulator tests or adjustments, the generator must be polarized in order to avoid damage to the equipment.

- It is recommended that following replacement or repair of a generator or regulator may be adjusted on a test bench as a matched unit.

Alternator System

Different kinds of alternators. Basically, there are three different kinds of alternators used on light aircraft, including small piston engine twins. One, used by Piper Aircraft, is a Chrysler product. Another, manufactured by Delco Remy, is normally a belt drive model. Yet a third kind is normally used on larger Continental engines and is driven by a gear drive. The gear drive can be any of three different places on an engine. See Figure 4-2-12.

For the purpose of this discussion, a Piper Aircraft Comanche alternator system has been chosen. It is the aviation equivalent of a Chrysler automotive system.

Description of Alternator System

One 12-volt battery and a 14-volt direct current 37 or 60 ampere alternator supplies the electrical power. The alternator is located on the front lower right side of the engine and utilizes a belt drive from the engine crankshaft. Many advantages in both operation and maintenance are derived from this system. The main advantage is that full electrical power output is available regardless of engine r.p.m.

The alternator has no armature or commutator and only a small pair of carbon brushes, which make contact with a pair of copper slip rings. The rotating member of the alternator, known as the rotor, is actually the field windings. The rotor draws only 1/13th (37 amp) or 1/20th (60 amp) of the current output. Therefore, there is very little friction and negligible wear and heat in this area. The alternating current is converted to direct current by diodes pressed into the end bell housing of the alternator. The diodes are highly reliable solid state devices, but are easily damaged if current flow is reversed through them.

The alternator system does not require a reverse current relay, because of the high back resistance of the diodes and the inability of the alternator to draw current or motorize. A current regulator is unnecessary because the windings have been designed to limit the maximum current available. Therefore, the voltage regulator is the only control needed.

An additional latching circuit is used to help keep the master solenoid closed when the battery voltage is low and the engine starter is being operated. This circuit routes voltage from the alternator to the master solenoid coil, thus holding the master solenoid in the closed position and allowing the starter to function properly. This circuit will also supply some voltage to the battery. A diode is placed into this circuit to prevent the reverse flow of current from the battery to the alternator.

The circuit breaker panel contains a circuit breaker marked ALT FIELD. If the field circuit breaker trips, it will result in a complete shutdown of power from the generating system. After a one or two minute cool-down period, the breaker can be reset manually. If tripping reoccurs and holding the breaker down will not prevent continual tripping, then a short exists in the alternator field.

Unlike previous systems, the ammeter does not indicate battery discharge, but displays the load in amperes placed on the generating system. With all electrical equipment off (except master) the ammeter will indicate the amount of charging current demanded by the battery. This amount will vary, depending on the percentage of charge in the battery at the time. As the battery becomes charged, the amount of current displayed on the ammeter will reduce to approximately two amperes. The amount of current shown on the ammeter will tell immediately whether the alternator system is operating normally.

It is important to note that on installations where the ammeter is in the generator or alternator lead, and the regulator system does not limit the maximum current that the generator or alternator can deliver, the ammeter can be red-lined at 100 percent of the generator or alternator rating. The generators rating is usually given as the number of amps the generator can safely supply at its rated voltage.

The amount of current shown on the ammeter is the load in amperes that is demanded by the electrical system from the alternator. As a check, take for example a condition where the battery is demanding 10 amperes charging current, and then switch on the landing light. Note the value in amperes placarded on the panel for the landing light fuse (10-amps) and multiply this by 80 per cent, the total is a current of 8 amperes. This is the approximate current drawn by the light. Therefore, when the light is switched on, there will be an increase of current from 10 to 18 amperes displayed on the ammeter. As each unit of electrical equipment is switched on, the currents will add up and the total, including the battery, will appear on the ammeter.

Using the example that the airplane's maximum continuous load with all equipment on is approximately 30-amperes for the 37-ampere alternator or 48-amperes for the 60-ampere alternator, this approximate 30- or 48-ampere value, plus approximately two amperes for a fully charged battery, will appear continuously under these flight conditions. If the ammeter reading were to go much below this value, under the aforementioned conditions, trouble with the generating system would be indicated and corrective action should be taken by switching off the least essential equipment.

Alternator System Test Procedure

Start engine and set throttle for 1,000 to 1,200 r.p.m.

Switch on the following loads and observe the ammeter output increase as indicated:

- Rotating beacon – 3 to 6 amps
- Navigation and instrument lights (bright position) - 4 to 6 amps.
- Landing light - 7 to 9 amps
- Radio - 4 to 6 amps each

If alternator does not meet above indications, refer to troubleshooting chart. Follow troubleshooting procedure outlined on a chart in a systematic fashion checking each cause and isolation procedure under a given trouble before proceeding with the following cause and isolation procedure.

Bench Testing an Alternator

Field current draw. Connect a test ammeter in series between a 12-volt battery positive post and the alternator field terminal. Connect a jumper wire to a machined surface on one of the alternator end shields (ground) and to the negative battery post. The reason for connecting to the machined surface is to ensure a good electrical connection. The end shields are treated to

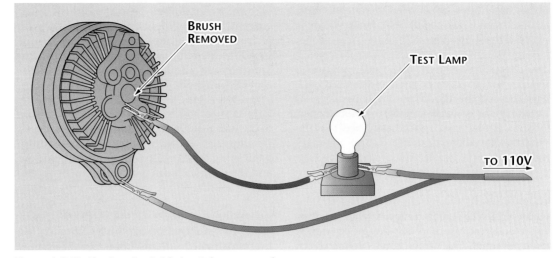

Figure 4-2-13. Testing the field circuit for a ground

oppose corrosion. The material used to treat the end shields is not a good electrical conductor.

Observe the ammeter to determine the current flowing through the rotor coil and connected circuit and record the amount. Slowly rotate the rotor with the pulley while watching the meter. The current will be a little less while rotating the rotor than when it is stationary. However, if the slip rings are clean and the brushes are making good contact, the reading should be fairly even. A slight fluctuation will be normal due to variation in turning speed when operated by hand.

The current draw should not be less than 2.3 amperes nor more than 2.7 amperes. A reading of less than 2.3 amperes indicates high resistance due to poorly soldered coil leads at the slip rings, dirty, oily slip rings or poor brush contact. A reading higher than 2.7 indicates shorted coil windings.

Testing alternator internal field circuit for a ground. To test the alternator internal field circuit for a short circuit to ground, proceed as follows:

- Remove the ground brush and using a 110-volt test lamp, place one test probe to a machined surface at one of the alternator end shields. Refer to Figure 4-2-13. The test lamp should not light.

- If the test lamp lights, carefully observe the order in which the parts were installed during removal of the insulated brush assembly. Remove the three through bolts. Then, separate the two end shield assemblies. Touch one of the test lamp probes to one of the slip rings and the remaining test probe to the rotor shaft. The lamp should not light. If the lamp lights, the rotor assembly is grounded and requires replacement.

- If the test lamp does not light, the ground condition was in the insulated brush assembly and the parts were either assembled wrong or damaged and short circuiting through to ground. Inspect the brush holder and insulated washer. Replace if damaged. The stack of parts attaching the insulated brush holder assembly to the end shield must always be installed in the proper sequence as follows: Insulated brush holder, "FLD" (FIELD) terminal, insulating washer, lock washer and attaching screw.

Inspection. Inspect the condition of the alternator components paying special attention to the condition of the slip rings for indications of oil, burns or worn areas. Inspect brushes for signs of sticking in holder or shield and for wear.

Inspect the bearing surface of the rotor shaft and the roller bearings at the rectifier end. Rotate the rotor in the drive end shield to feel for roughness in the drive end bearing. Inspect the grease retainer, if so equipped, on late alternators. Inspect the rectifier leads especially at connections for a good solder joint also inspect insulation. Rectifier/stator lead must be pushed down into the slots that are cast into the end shield and cemented with special cement as specified by the maintenance manual.

Testing rectifiers. Test the rectifiers as per the maintenance manual. Piper Aircraft has a special tool to make the process less painful.

Voltage Regulator Servicing (Chrysler Type)

Air gap adjustment. (Regulator removed). The most accurate method of measuring the air gap between the lower side of the armature and the top of the core is to use a volt test lamp.

- Connect one lead from the test lamp to a 12-volt battery positive post and the remaining lead to the regulator IGN terminal.

- Connect a jumper wire from the battery negative post to the regulator "Field" terminal.

- Insert a 0.048 inch gauge between the armature and the core at the hinge side of the stop.

- Press the armature (not the contact reed) down against the gauge. The test lamp light should dim.

- Insert a 0.052 inch gauge and when the armature is pressed down the lamp should not dim.

- If an adjustment is required, loosen the adjustable bracket retaining screw and raise or lower the support as required to bring the air gap to specifications and retighten retaining screw.

The regulator resistance units can be checked by simply inverting the regulator (Figure 4-2-14) and looking at the bottom.

NOTE: *The base air gap is 0.048 to 0.052 inch. However, the transfer voltage determines the final air gap. The transfer voltage is the rise in voltage from the reading taken while operating on the upper contact at 1,250 r.p.m. with a load of 15-amps and the voltage taken at 2,200 r.p.m. with a load of 7-amps or less. This difference should not be less than 0.2-volt nor more than 0.7-volt. If the transfer voltage is less than 0.2-volt, it is permissible to increase the air gap but not to exceed 0.005 inch (0.057).*

- If the transfer voltage is greater than 0.7-volt, the air gap can be decreased not to exceed 0.005 inch (0.043). These adjustments should only be made following the tests and then retested after adjustment.

Figure 4-2-14. Voltage regulator resistance units

Contact clearance adjustment. The distance between the upper and lower contacts is preset at the factory and the contact clearance between the movable contact and the lower contact should be correct. Even though the air gap is readjusted, the contact clearance should remain the same.

Should the regulator be misadjusted, the contact clearance can be returned to the specified 0.014 inch ± 0.002 inch by bending the lower contact bracket.

Voltage Regulator Fusible Wire Replacement

- Cut fuse wire above solder connection at the base and unwind wire at top bracket.

CAUTION: *If an attempt is made to unsolder the old fuse, the very small wire from voltage coil may be damaged.*

- Tin end of fuse wire (use resin core solder only).

- Holding tinned end of new fuse wire into recessed rivet at base of regulator and against old piece of fuse wire that remains, cause a drop of solder from soldering iron to fall on these, parts. Allow solder to cool sufficiently for fuse wires to make a good solder joint.

- Pull new fuse wire up enough to remove slack and wrap it around the bracket, then solder the coiled wire to the bracket and cut off the surplus fuse wire.

- The original fuse wire is machine wound on the upper bracket. Replacement fuse should be soldered to the bracket to ensure a good electrical contact.

Solid State Voltage Regulator

Checking voltage regulator. The regulator is a fully transistorized unit in which all of the components are encapsulated in epoxy, which makes field repair of the unit impractical, and if it does not meet the specifications, it must be replaced. The regulator may be tested by the following procedure:

1. Be sure that the battery is fully charged and in good condition.

2. Check the alternator according to the manufacturer's instructions, to determine if it is functioning properly. This test must be done with the regulator out of the circuit. After completing this test, reconnect the regulator into the circuit.

3. Use a good quality accurate voltmeter with at least a 15-volt scale.

4. Connect the positive voltmeter lead to the red wire at the regulator harness connector or terminal block. Connect the negative voltmeter lead to the regulator housing.

NOTE: *Do not connect the voltmeter across the battery, because the regulator is designed to compensate for resistance contained within the wiring harness.*

5. With the alternator turning at sufficient r.p.m. to produce a half load condition, or approximately 25-amperes output, the voltmeter should read between 13.6 and 14.3 volts. The ambient temperatures surrounding the voltage regulator should be between 50°F to 100°F while this test is being made.

6. The voltage regulator heat sink, or case, is the ground connection for the electronic circuit. Therefore, if this unit is tested on the bench, it is most important that a wire, No. 14, be connected between the regulator case and the alternator. If the regulator does not regulate between 13.6 and 14.4-volts, one of the following conditions may exist:

CONDITION	CAUSE	SOLUTION
Regulates, but out of specification	The regulator is out of calibration	The regulator must be replaced
The voltmeter continues to read battery voltage	Poor or open connections within the wiring harness	Clean/tighten connection
	The regulator is open	Replace the regulator
Voltage continues to rise	Regulator housing not grounded, or shorted	Regulator must be replaced

These are some of the things to look for in case of failure:

- Poor or loose connections.
- Poor ground on the regulator housing.
- Shorted alternator windings.
- A grounded yellow wire (This will cause instantaneous failure).
- Disconnecting the regulator while the circuit energized.
- Open circuit operation of the alternator (The battery disconnected).

Over-Voltage Relay

Checking over-voltage relay. The relay may be tested with the use of a good quality, accurate voltmeter, with a scale of at least 20-volts and a suitable power supply, with an output of at least 20-volts or sufficient batteries with a voltage divider to regulate voltage. The test equipment may be connected by the following procedure:

1. B+ is connected to "Bat" of the over-voltage control.

2. B- is connected to the frame of the over-voltage control.

3. Be sure both connections are secure, and connected to a clean, bright surface.

4. Connect the positive lead of the voltmeter to the "Bat" terminal of the over-voltage control.

5. Connect the negative lead of the voltmeter to the frame of the over-voltage control.

6. The over-voltage control is set to operate between 16.2-volts to 17.3 volts. By adjusting the voltage, an audible "click" may be heard when the relay operates.

7. If the over-voltage control does not operate between 16.2 and 17.3-volts it must be replaced.

Checking Generator or Alternator Belt Tension

If properly installed and checked periodically, the generator or alternator drive belt will give very satisfactory service. However, an improperly tensioned belt will wear rapidly and may slip and reduce unit output. Consequently, a belt should be checked for proper tension at the time of installation, again after 25 hours of operation and each 100 hours thereafter.

The method of checking belt tension is simple and requires little time for accomplishment. This method of checking belt tension consists of measuring torque required to slip the belt at the pulley on the generator or alternator, and is accomplished as follows:

- Apply a torque indicating wrench to the nut that secures the pulley to the generator or alternator and turn the pulley in a clockwise direction. Observe the torque shown on the wrench at the instant the pulley slips.

- Check the torque indicated above with the torque specified in the maintenance manual chart. Adjust the belt tension accordingly.

DC Starter-Generators

While all reciprocating engine installations use separate electric starters and generators or alternators, most turboprop installations do not. Instead, they use a single unit that combines the two functions. It is called a starter-generator (Figure 4-2-15). Starter-generators are constructed with two sets of field coils and a common armature. The result is a compact unit that is very powerful in either function and only takes up the space of one unit.

When activated as a starter, the field leads are shunted together and disabled, allowing the unit to run as an electric starter. Once the engine is started and the starter switch inactivated, the field coils are energized and the generator function begins.

The following discussion is a description of the power system operation of a King Air 200. The airplane is equipped with two 28 volt units rated at 250 amps each. With the start mode selected, they function as a starter. With the start mode deselected and the generator mode selected, the starter-generator then generates regulated voltage to the generator bus.

Generator Operation

The generators are brought on line by the generator switches located with the battery switch.

Two generator control panels control generator operation. The generator control panels provide voltage regulation, generator paralleling, load sharing, reverse current sensing and control, overvoltage protection and overexcitation protection.

Voltage regulation. The generator output voltage is sensed at the generator side of the generator bus contactor. The control panel supplies the generator field excitation current required to supply the electrical load and maintain a bus voltage of 28 VDC. Figure 4-2-16 is an illustration of a starter-generator voltage regulator.

The generator voltage sense input to the generator control panel is at pin "B", voltage regulator power input is at pin "J" and the regulator output to the generator field is from pin "M" of the control panel connector.

Generator control. When the engine start switch is activated in either the ignition and engine start or the starter only position, the field sense relay is activated, shunting the field leads together and rendering the generator inoperative.

Figure 4-2-15. A combination starter-generator

CAUTION: *Do not exceed the starter motor operating time limits of 30 seconds on, 5 minutes off, 30 seconds on, 5 minutes off, 30 seconds on then 30 minutes off.*

The equalization of the generators is accomplished by utilizing the voltage developed in the generator compensator windings. This voltage is sensed at the interpole terminal of the winding, terminal "D" of the generator. The paralleling circuit includes the LH and RH field grounding relays and the LH and RH control panels. Equal load sharing is dependent upon equalizing the resistance of the generator windings and the external circuitry. The generator control panels

Figure 4-2-16. A voltage regulator in a King Air starter-generator system

are designed to control the generators and the load shared within 2 1/2 percent.

Reverse current protection. The control panel provides the reverse current protection and the forward current control. Bus voltage is sensed at the bus side of the line contactor while generator output voltage is sensed at the generator side of the line contactor. Whenever the generator is operating and the control switch is placed in the RESET position, the generator output voltage will rise to the regulated voltage. When the generator switch is placed in the ON position, voltage output from pin "H" of the control panel will close the line contactor to connect the generator to the generator bus.

The same voltage will also be routed through a Zener diode to the coil of a control relay to apply bus voltage to the coil of the appropriate generator bus tie relay. The Zener diode will ensure that the generator bus tie relay will be opened if generator output voltage drops below the regulated voltage. The output of either generator is connected to the center bus and to the battery.

When the generator slows down to the point where it can no longer maintain a positive load, the generator will begin to draw current from the airplane bus. This reverse current passes through the compensator windings of the generator and the resultant voltage is sensed at the interpole terminal of the generator. The generator control panel then removes the voltage from the coil of the bus contactor, permitting the contactor to open and remove the generator from the airplane bus.

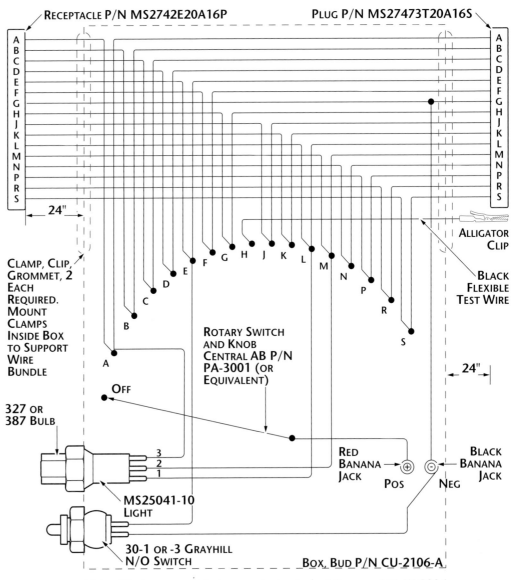

Figure 4-2-17. A wiring schematic for a starter–generator test unit

Should a circuit fault supply generator output (or bus voltage) to the field of a generator, the affected unit will attempt to assume the full electrical load. If bus voltage rises above 28 VDC, the generator is temporarily removed from the bus by the reverse current limiter. Failure of the reverse current limiter could cause the affected generator to assume the entire electrical load. The resultant bus voltage would depend upon the generator speed, the electrical load and the nature of the fault.

Reverse current protection circuits may be checked for operation as follows:

- Reset the generators to bring generator output up to regulated voltage.

- Shut down both engines, leaving the battery and both generators ON.

NOTE: *As the engines slow down and reverse current begins to flow into the generators, the generators will be removed from the line.*

- Normal indications are: Loadmeter = zero; generator-out lights = ON.

- Check for abnormal indications, showing that reverse current protection is not being accomplished.

Overvoltage protection. The generator control panel provides overvoltage protection. Should the output voltage reach 32 VDC, the overvoltage protection portion of the generator control panel will open the coil circuit of the bus contactor, isolating the overvoltage generator from the airplane bus. The normally regulated generator will again come on line to supply the system.

Over excitation protection. The generator control panel provides over excitation protection. This portion of the control panel will activate in the event the generator load and speed conditions are such, or the nature of the fault is such that the generator voltage increases without control, but does not reach an overvoltage condition. Should the generator field reach the designed limitation value, the circuitry providing over excitation protection will be activated to remove the affected generator from the bus.

Undervoltage protection. Undervoltage detection and protection is provided when the generator control panel compares line voltage with an internal reference. If line voltage drops below 18 VDC for more than 4 seconds, the under voltage circuit then uses the over-voltage circuit as a switch to trip the field relay.

Starter/Generator and Controls Troubleshooting

Generator control panel test unit. A Generator Control Panel Test Unit should be used to gain access to the individual inputs and outputs of the generator control panel. A wiring schematic (Figure 4-2-17) provides adequate information for building the test unit.

CAUTION: *Connect and disconnect test unit only when power is OFF.*

The wiring harness plug is disconnected from the generator control panel and connected to the receptacle of the test unit. The plug from the test unit is inserted into the receptacle of the generator control panel, thereby connecting the test unit in series with the control panel. Do not connect the control panel when checking continuity.

A sensitive multimeter, capable of measuring voltages accurate to within one percent, is connected to the test unit by way of the banana jacks on the face of the unit; proper polarity of these connections must be closely observed as both positive and negative values will be measured. The alligator clip from pin "G" should be positively grounded to the airplane structure.

All voltage measurements are made with one generator on and a 50 percent load. Maximum generator output is 300 amperes; so enough electrical equipment should be turned on to establish about 150 amperes of load on the electrical system. Refer to the electrical load utilization chart in the maintenance manual for load requirements of the various electrical systems and components.

When checking the generator equalizer circuit for operation, a voltage drop of two or three volts at pin "B" is an indication that the generator equalizer circuit is operating properly.

The following items should be observed when using the test unit:

- Failure of the generator control relay to properly short the shunt field of the generator may produce transient voltage spikes at pin "M" of the control panel during start. The magnitude of the spikes may be sufficient at times to permanently damage the internal circuitry of the control panel. A dimly illuminated lamp on the test unit during start is a normal indication. Should the lamp flash brightly when starts are initiated and the start relay closes, a voltage transient has been sensed at pin "M" of the control panel. Another bright flash of the lamp when the start is terminated and the start relay opens may occur during this check.

- If pin "H" is shorted to ground, correct the fault before replacing the control panel.

- All resistance and continuity checks are made with the engines off, the battery off and the control panel disconnected.

- The test switch on the test unit applies a ground signal to the equalizer circuit of the control panel (pin "E").

- All checks are made with only one generator on the line at a time.

CAUTION: *Connect and disconnect the test unit only when the system is off.*

Carbon Pile Voltage Regulator

Vibrating-type regulators cannot be used with generators which require a high field current since the contacts will pit or burn. Heavy duty generator systems require a different type of regulator, such as the carbon pile voltage regulator.

The carbon pile voltage regulator depends on the resistance of a number of carbon disks arranged in a pile, or stack. The resistance of the carbon stack varies inversely with the pressure applied. When the stack is compressed under appreciable pressure, the resistance in the stack is less. When the pressure is reduced, the resistance of the carbon stack increases, because there is more air space between the disks, and air has high resistance.

Pressure on the carbon pile depends upon two opposing forces: a spring and an electromagnet. The spring compresses the carbon pile, and the electromagnet exerts a pull that decreases the pressure. The coil of the electromagnet is connected across the generator terminal B and through a rheostat (adjustable knob) and resistor (carbon disks) to ground.

When the generator voltage varies, the pull of the electromagnet varies. If the generator voltage rises above a specific amount, the pull of the electromagnet increases, decreasing

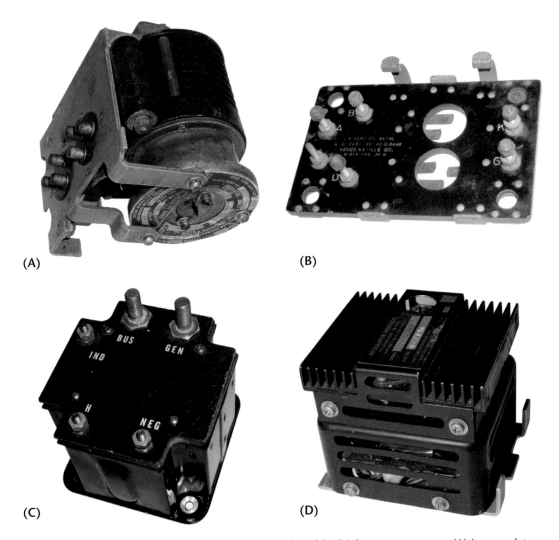

(A)

(B)

(C)

(D)

Figure 4-2-18. Carbon pile voltage regulators are used in older high output systems. (A) is a regulator, (B) is the regular mounting base, (C) is a reverse current cutout, while (D) is a solid state replacement for a carbon pile regulator.

the pressure exerted on the carbon pile and increasing its resistance. Since this resistance is in series with the field, less current flows through the field winding, there is a corresponding decrease in field strength, and the generator voltage drops.

On the other hand, if the generator output drops below the specified value, the pull of the electromagnet is decreased and the carbon pile places less resistance in the field winding circuit. In addition, the field strength increases and the generator output increases. A small rheostat provides a means of adjusting the current flow through the electromagnet coil. Figure a shows a typical 24-volt voltage regulator with its mount base and reverse current relay.

A carbon pile regulator and the necessary components to install one are shown in Figure 4-2-18. Carbon pile regulator systems are not normally used since the advent of solid state electronics.

Large Aircraft Generators

Extreme electrical demands can be placed on transport airplanes. None of the electrical generation systems discussed so far can come close to supplying the demand. The discussion that follows should help you understand just how different they really are.

The Boeing-727 is a good example of an airline type generating system. The electrical system's primary source of power is AC generators, and most components are solid state.

The AC bus system is designed so that individual generators can supply individual load buses, or the engine-driven generators can be paralleled through a synchronizing bus. This means that any one or certain combinations of, generators can supply the load buses. See Figure 4-2-19.

KW/KVAR Explanation

Before we get into the discussion, we need to do some more review of a few terms common on large AC systems. It is necessary to consider three types of power in assessing the amount of work an AC generator is doing:

- KW (real power) represents the energy the generator is supplying to accomplish work such as operating the radios, galley, lights, etc.

- KVAR (reactive power) is the amount of energy that is used in magnetizing the

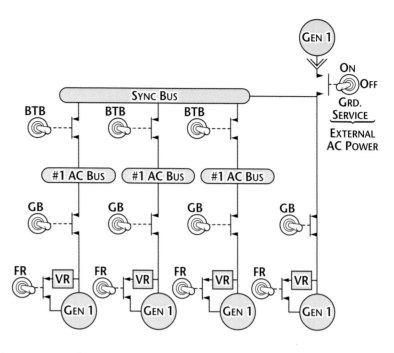

Figure 4-2-19. The basic generator system for a Boeing 727

iron cores of relays, transformers, motors and solenoids, as well as the energy used in charging any capacitors (condensers) in the circuit.

- KVA (apparent power) is the amount of work the generator senses that it is doing. Manufacturer's limitations are given in apparent power because of the heating effect on the generator. It is calculated by measuring the current and voltage of a generator and multiplying them together. This unit is the kilovolt ampere.

KW meter. The KW meters indicate the rate at which energy is being delivered by the generators to the airplane systems. This energy leaves the electrical system and does not return. It follows that a continuous torque is required on the generator shafts to supply this energy. Parallel operation of generators requires that the generators share the KW load. If one generator were to assume more than its share of the load, its temperature would increase and the temperature of its CSD would also increase. Often, the total KW load during parallel operation is larger than a single generator can tolerate, so it follows that load sharing is imperative. Load sharing is accomplished by a combined action of the load controllers and the CSD. The load controllers compare the KW load of the individual generators to the average KW load of the system and provide electrical signals to the CSD's to affect KW load balance.

The CSD's must respond properly to these sharing signals and develop the correct torque on the generator shafts. When the KW instrument readings show a KW balance among the generators operating in parallel, the load controllers and CSD's are working properly. The operating control panel for a Boeing 727 is shown in Figure 4-2-20.

KVAR meter. The KVAR meter indicates the rate at which energy is being delivered by the generators to the magnetic fields in motor coils and other magnetic devices. This energy does not leave the electrical systems. In fact, it returns to the generators and is not lost. No torque is required of the CSD to provide these KVAR loads. Therefore, the CSD's and load controllers are not involved in KVAR load balancing.

Unfortunately, high KVAR loads have a very bad effect on the magnetic conditions inside the generators. These KVAR loads tend to decrease the main magnetic field. The generator voltage would be unacceptably low if it were not for the ability of the voltage regulators to adjust the field current upward to compensate for the demagnetizing effect of the KVAR loads.

In parallel operation, if one generator were to carry more than its share of KVAR load, its field current would be very high. The heating effect inside the generator field windings would cause damage.

The voltage regulators must compare the KVAR load on the individual generator to the average KVAR load to determine the field current necessary to affect KVAR load balance. If the KVAR instruments show a KVAR balance, the voltage regulators are functioning well.

Generators

Generator and equipment location. Each engine-driven generator is located on its respective engine. The APU driven generator is on the APU that is permanently mounted in the wheel well. Electrical leads from all four of these generators go to the electric equipment racks in the electronics compartment in the forward belly of the aircraft. Generator control units are located in the electronics compartment. They are remotely operated. Distribution

ESSENTIAL POWER FAILURE LIGHT
Illuminates any time essential AC bus is not energized and the battery bus is energized. Also illuminates master warning light on center instrument panel.

RESIDUAL VOLTS SWITCH
Use requires that generator field relay be tripped. When depressed AC voltmeter indication is shown on inner (0-30) scale, and indicates residual voltage of selected generator.

KVARS SWITCH
When pressed KW/KVAR meters will indicate KVARs.

SYNCHRONIZING LIGHTS
Illuminated - Generator not in phase with synchronous bus. Extinguished - Generator synchronized and condition acceptable to parallel operation.

AC METERS SELECTOR

ESSENTIAL POWER SOURCE SELECTOR
Selects power source for essential AC bus standby power.

GENERATOR POSITION- NO.1, NO. 2 OR NO. 3
If no fault exists and the field relay is closed the generator selected will supply AC power to the essential AC bus.

APU OR EXT PWR POSITION
If no fault exists and the field relay is closed, the generator selected will supply AC power to the essential AC bus.

STANDBY
Provides powering of essential flight instruments and radios from battery in event of complete AC failure. It must be pushed in to rotate to this position

AC VOLTMETER
Indicates generator, bus tie, APU, or external power voltage as selected by AC meters selector (outer scale). Indicates generator residual voltage on inner scale.

FREQUENCY METER
Indicates generator, bus tie, APU, or external power frequency as selected by AC meters selector.

Figure 4-2-20. The cockpit generator control panel for a Boeing 727

of electrical power is from the electronics compartment to the main AC load buses, which are located on a panel in the cockpit.

The generators produce 115 Volt, 3 phase, 400 cycle AC Power and are rated at 40 KVA. The APU generator is identical to the engine driven generators but depends on the APU to maintain a constant r.p.m. The APU generator is rated at 60 KVA because much better cooling is provided for it.

A Boeing 727 has an APU located in the wing root. It is similar to the APU in Figure 4-2-21, which shows an APU in the tail of a B757.

Each generator consists of an *exciter section* and a *main generator section*. The exciter armature rotates within a magnetic field created initially by permanent magnets. With the field relay closed, the voltage regulator will control the strength of the exciter's magnetic field. This strength determines the output of the main generator section since it is excitation for the *main generator rotating field*. The exciter section contains a *stability winding*, which stabilizes the voltage output of the generator under varying electrical loads.

Since the exciter field contains permanent magnets, a small amount of voltage (15±) is generated with the generator breaker tripped. This is known as residual voltage and can be read by pushing the *residual volts* button adjacent to the *AC volts* meter. This button changes the scale from 100-130 volts to 0-30 Volts. It is used in troubleshooting to ascertain whether the generator is turning.

Generator operational load limits, like many electrical component limitations, are predicated on temperature protection. Each engine driven generator continuous load limit is 36 KW. The maximum continuous load on the APU generator is 54 KW (read as 165 AMPS on the APU/EXT load meter).

Generator control panels. Each generator is controlled and protected by its own control panel located in the equipment compartment. The panels provide automatic parallel protection in addition to field relay, generator breaker and bus tie breaker control.

The Fault Protection feature provides automatic clearing of the following faults:

- Over/under excitation
- Over/under voltage
- Exciter ceiling
- Instability
- Over/Underspeed

A red warning light is located near each generator control panel in the electronics compartment. Illumination of the light indicates that the associated components are powered and lethal voltage is present.

Generator controls and indicators. There are nine generator switches on the electrical panel. They control the generator *field relays (FR)*, *generator breakers (GB)* and the *bus tie breakers (BTB)*, which are located in the electronic compartment. Each switch is spring loaded to the neutral position and actuates a relay by momentary movement to the close or trip position. The relays are of the mechanical latch type and will remain in the last selected position. They require electrical power to operate.

The amber lights adjacent to the switches indicate circuit open. Although these relays can be tripped by moving the switch or by the protective circuit sensing a fault, they can only be closed by moving the switch.

The schematic in 4-2-19 illustrates the function of each switch as follows:

- Closing the field relay (FR) completes the exciter circuit through the voltage regulator to the generator.

Figure 4-2-21. The APU in a Boeing 727 is similar to this unit installed in the tail of a Boeing 757.

- Closing the generator breaker (GB) connects individual generator output to its AC load bus.

- Closing the bus tie breaker (BTB) ties the load bus to the SYNC bus and puts that generator in parallel operation with other operating generators. This also provides for power to all load buses from any operating generator.

- On the ground, the load buses can be powered from the SYNC bus by external AC power or the APU generator. To do this the bus tie breakers must be closed.

The system is protected by interlocks, which prevent simultaneous use of any two power sources, other than engine driven generators. Closing a GB will automatically trip the external power switch or the APU GB, if either of them is supplying the SYNC bus. With the APU generator operating, closing its GB will trip all three main GB's or the external power switch.

With external power plugged into the airplane, turning the external power switch on will trip the GB's or the APU GB. The F/E determines the source of power to supply the SYNC bus.

Speed switch. When the CSD attains a predetermined r.p.m. (5,300±) the limit governor ports oil pressure to close a speed switch, making it possible to use electrical power from the generator. At about 42-45% N_2 engine r.p.m.,

the frequency and voltage indications should stabilize.

If the CSD overspeeds or underspeeds beyond predetermined values, a drop in oil pressure will allow the speed switch to open and generator output is no longer available. If the speed switch opens as a result of an underspeed and r.p.m. subsequently returns to normal, the speed switch will close, making generator output available again. However, if the speed switch opens because of an overspeed, oil pressure will be blocked. This condition cannot be corrected until CSD rotation ceases.

The speed switch, when closed, permits closing of the GB and selection of essential power to that generator. If the speed switch opens, the GB will trip and essential power will fail (if selected to that generator).

Frequency control. The frequency of the AC power produced by the generator depends upon the r.p.m. at which the generator is driven by its CSD. On engine start, the CSD will stabilize at a speed preset within the speed governor. If this speed is not satisfactory, as read on the FREQ meter, it could be adjusted through a range of about 10 CPS by turning the FREQ knob located next to the KW/KVAR meter. This speed adjustment is effective only during isolated operation. It is made by sending a signal through the load controller, which "biases" the speed governor. See Figure 4-2-22.

Figure 4-2-22. Frequency adjusting portion of the generator controls

In parallel operation, the FREQ knob will have little effect on the frequency. Frequency, in this case, is controlled automatically by the load controller, which adjusts the torque of its CSD to share equally the total electrical load with the other generators. A generator, which is carrying a disproportionate share of the load in parallel, may respond to frequency control input to more closely share an equal load.

CSD Disconnect Mechanism

A guarded momentary switch on the F/E's panel, when actuated, causes the CSD to disconnect from the engine accessory drive. The drive can be reconnected only on the ground, after engine rotation stops. A "T" handle on the CSD is used to reconnect or disconnect the drive on the ground.

Generator Paralleling

Synchronizing lights. The synchronizing lights on the AC meters selector panel indicate the phase difference between power on the SYNC bus and the selected generator. The left SYNC light monitors phase A of the selected generator. The frequency meter indicates CPS of phase B and the right SYNC light monitors phase C.

Automatic paralleling. This term, though commonly used, is a misnomer. There is no automatic paralleling of generators on this airplane. The term "autoparallel" refers to the automatic protection afforded the system when the generators are paralleled using the GB's. After GB #1 is closed (normal operation) and generator #1 is supplying the SYNC bus, this protection will prevent GB #3 from closing (regardless of GB #3 switch position) if the difference in frequency and/or phase relationship is not within preset limits. Generator #3 can be adjusted into compatibility by adjusting its frequency.

After engine start, when the generators are being paralleled, the essential power selector is positioned to GEN 1, GEN 2 and GEN 3 before any generator breaker is closed. This is done to ensure that each will sustain an electrical load, that each CSD speed switch is closed and that there will be no power interruption to essential AC.

Manual paralleling. This term is used to describe paralleling generators using the BTB's. There is no protective circuit to prevent paralleling generators. Therefore, caution must be used to assure that the generators are properly synchronized and the SYNC lights are out when using the BTB's.

New Generation Aircraft

In the digital age, most new generation aircraft have all of the various functions performed automatically. This makes the cockpit crew a lower workload and helps reduce the possibilities for human error.

Section 3

Aircraft Wiring

The aircraft components that generate electrical power and all the various systems that place a load on that electrical power collectively make up the more complex portion of an aircraft electrical system. Ironically, all the complex electrical components are completely reliant on the simplest portion of an aircraft's electrical system - the wiring. Wire bundles carry all the power and electrical signals necessary for the satisfactory performance of any modern aircraft and are absolutely a critical component of aircraft design.

The reliability of electrical systems and subsystems depends, in part, directly on properly installed wiring. Improperly or carelessly maintained wiring not only can cause system failure but can also be a source of both immediate and potential danger. Therefore, an aircraft maintenance technician must be equipped with the knowledge and techniques for installing wiring properly, inspection of existing wiring installations and recommended maintenance of electrical system wire and cable.

The procedures and practices presented in this chapter are representative of the type generally seen in the aircraft maintenance industry, however, they are general in nature and do not take the place of the individual aircraft manufacturer's instructions and approved practices.

Wire and Cable

To clarify the meanings of the terms *wire* and *cable,* it is generally accepted terminology to describe a *wire* as a single, solid conductor or as a stranded conductor covered with an insulating material. *Cable,* as used in aircraft electrical installations, includes:

- Two or more separately insulated conductors in the same jacket (multi-conductor cable)

- Two or more separately insulated conductors twisted together (twisted pair)

Figure 4-3-1. All wiring systems use a wide variety of wire and cable sizes and types.

- One or more insulated conductors, covered with a metallic braided shield (shielded cable)

- A single insulated center conductor with a metallic braided outer conductor (radio frequency cable). The concentricity of the center conductor and the outer conductor is carefully controlled during manufacture to ensure that they are coaxial. Figure 4-3-1 shows a typical aircraft wire harness containing both wires and cables.

Wire size and type. Wire is manufactured in sizes according to a standard known as the AWG (American Wire Gauge). As shown in Table 4-3-1, the wire diameters become smaller as the gauge numbers become larger. The largest wire size shown in Table 4-3-1 is number 0000, and the smallest is number 50. Sizes larger than 0000 and smaller than 50 are manufactured but are not commonly used in aircraft electrical systems.

Gauge numbers are useful in comparing the diameter of wires, but some types of wire or cable cannot be accurately measured with a gauge. Large wires are usually stranded to increase their flexibility for installations that require forming and bending, which is usually the case in airframes. The total area of a cross section of stranded wire can be determined by multiplying the area of one strand (usually computed in circular mils when diameter or gauge number is known) by the number of strands in the wire or cable.

Many factors affect the selection of a wire size in aircraft electrical system design:

- *Power loss (IR loss)* in the line represents electrical energy converted into heat.

Use of large conductors reduces resistance and therefore the IR loss, but large conductors are heavier, more expensive and require more substantial airframe mounting supports.

- *Voltage drop (IR drop)* variations occur in the line when loads demand more or less current or load resistance changes. This results in variations in the IR drop in the line if the source maintains a constant line voltage. Wide variations in the IR drop in the line cause poor voltage regulation at the load.

Reducing load current lowers the amount of power transmitted, whereas a reduction in line resistance increases the size and weight of conductors required. A compromise must be reached whereby the voltage variation at the load is within tolerable limits and the weight of line conductors does not exceed aircraft design limits.

Another factor to be considered is the *current-carrying ability* of the conductor. When current is drawn through the conductor, heat is generated. The temperature of the wire will rise until the heat radiated or otherwise dissipated, is equal to the heat generated by the passage of current through the line. The heat generated in the conductor cannot be readily dissipated into air when covered by insulation. To prevent burning of the insulation from excessive heat (caused by current that exceeds the wire's current rating), conductor size must be matched with the expected load current.

A general recommendation is that the voltage drop in the main power cables from the aircraft generation source or the battery to the bus should not exceed 2% of the regulated voltage when the generator is carrying rated current or the battery is being discharged at a five minute rate.

The resistance of the current return path through the aircraft structure is always considered negligible. However, this is based on the assumption that adequate bonding of the structure or a special electric current return path has been provided and is capable of carrying the required electric current with a negligible voltage drop. A resistance measurement of 0.005 ohm from the ground point of the generator or battery to the ground terminal of any electrical device is considered satisfactory. Another satisfactory method of determining circuit resistance is to check the voltage drop across the circuit. If the voltage drop does not exceed the limit established by the aircraft or product manufacturer, the resistance value for the circuit is considered satisfactory. When using the voltage drop method of checking a circuit, the input voltage must be maintained at a constant value.

		CROSS SECTION							
GAUGE	DIAMETER IN MILS	CIRCULAR MILS	SQUARE INCH	OHMS PER 1,000 FT	FEET PER OHM	POUNDS PER 1,000 FT	FEET PER POUND	OHMS PER POUND	POUNDS PER OHM
0000	460.0	211,600	0.1662	0.04901	20,400	640.5	1.561	0.00007652	13,070
000	409.6	167,800	0.1318	0.06182	16,180	507.8	1.969	0.0001217	8,215
00	364.8	133,100	0.1045	0.07793	12,830	402.8	2.482	0.0001935	5,169
0	324.9	105,600	0.08291	0.09825	10,180	319.5	3.130	0.0003075	3,252
1	289.3	83,690	0.06573	0.1239	8,070	253.3	3.947	0.0004891	2,044
2	257.6	66,360	0.05212	0.1563	6,398	200.9	4.978	0.0007781	1,285
3	229.4	52,620	0.04133	0.1971	5,074	159.3	6.278	0.001237	808.3
4	204.3	41,740	0.03278	0.2485	4,024	126.3	7.915	0.001967	508.5
5	181.9	33,090	0.02599	0.3134	3,190	100.2	9.984	0.003130	319.5
6	162.0	26,240	0.02061	0.3952	2,530	79.44	12.59	0.004975	201.0
7	144.3	20,820	0.01635	0.4981	2,008	63.03	15.87	0.007902	126.5
8	128.5	16,510	0.01297	0.6281	1,592	49.98	20.01	0.01257	79.58
9	114.4	13,090	0.01028	0.7925	1,262	39.62	25.24	0.02000	49.99
10	101.9	10,380	0.008155	0.9988	1,001	31.43	31.82	0.03178	31.47
11	90.7	8,230	0.00646	1.26	793	24.9	40.2	0.0506	19.8
12	80.8	6,530	0.00513	1.59	629	19.8	50.6	0.0804	12.4
13	72.0	5,180	0.00407	2.00	500	15.7	63.7	0.127	7.84
14	64.1	4,110	0.00323	2.52	396	12.4	80.4	0.203	4.93
15	57.1	3,260	0.00256	3.18	314	9.87	101	0.322	3.10
16	50.8	2,580	0.00203	4.02	249	7.81	128	0.514	1.94
17	45.3	2,050	0.00161	5.05	198	6.21	161	0.814	1.23
18	40.3	1,620	0.00128	6.39	157	4.92	203	1.30	0.770
19	35.9	1,200	0.00101	8.05	124	3.90	256	2.06	0.485
20	32.0	1,020	0.000804	10.1	98.7	3.10	323	3.27	0.306
21	28.5	812	0.000638	12.8	78.3	2.46	407	5.19	0.193
22	25.3	640	0.000503	16.2	61.7	1.94	516	8.36	0.120
23	22.6	511	0.000401	20.3	49.2	1.55	647	13.1	0.0761
24	20.1	404	0.000317	25.7	39.0	1.22	818	21.0	0.0476
25	17.9	320	0.000252	32.4	30.9	0.970	1,030	33.4	0.0300
26	15.9	253	0.000199	41.0	24.4	0.7692	1,310	53.6	0.0187
27	14.2	202	0.000158	51.4	19.4	0.610	1,640	84.3	0.0119
28	12.6	159	0.000125	65.3	15.3	0.481	2,080	136	0.00736
29	11.3	128	0.000100	81.2	12.3	0.387	2,590	210	0.00476
30	10.0	100	0.0000785	104	9.64	0.303	3,300	343	0.00292
31	8.9	79.2	0.0000622	131	7.64	0.240	4,170	546	0.00183
32	8.0	64.0	0.0000503	162	6.17	0.194	5,160	836	0.00120
33	7.1	50.4	0.0000396	206	4.86	0.153	6,550	1,350	0.000742
34	6.3	39.7	0.0000312	261	3.83	0.120	8,320	2,170	0.000460
35	5.6	31.4	0.0000246	331	3.02	0.0949	10,500	3,480	0.000287
36	5.0	25.0	0.00000196	415	2.41	0.0757	13,200	5,480	0.000182
37	4.5	20.2	0.00000159	512	1.95	0.0613	16,300	8,360	0.000120
38	4.0	16.0	0.00000126	648	1.54	0.0484	20,600	13,400	0.0000747
39	3.5	12.2	0.00000962	847	1.18	0.0371	27,000	22,800	0.0000438
40	3.1	9.61	0.00000755	1,080	0.927	0.0291	34,400	37,100	0.0000270
41	2.8	7.84	0.00000616	1,320	0.756	0.0237	42,100	55,700	0.0000179
42	2.5	6.25	0.00000491	1,660	0.603	0.0189	52,900	87,700	0.0000114
43	2.2	4.84	0.00000380	2,140	0.467	0.0147	68,300	146,000	0.00000684
44	2.0	4.00	0.00000314	2,590	0.386	0.0121	82,600	214,000	0.00000467
45	1.76	3.10	0.00000243	3,350	0.299	0.00938	107,000	357,000	0.00000280
46	1.57	2.46	0.00000194	4,210	0.238	0.00746	134,000	564,000	0.00000177
47	1.40	1.96	0.00000154	5,290	0.189	0.00593	169,000	892,000	0.00000112
48	1.24	1.54	0.00000121	6,750	0.148	0.00465	215,000	1,450,000	0.000000690
49	1.11	1.23	0.000000968	8,420	0.119	0.00373	268,000	2,260,000	0.000000443
50	0.99	0.980	0.000000770	10,600	0.0945	0.00297	337,000	3,570,000	0.000000280

AMERICAN WIRE GAUGE ENGLISH UNITS VALUES AT 20°C

Note 1 – The fundamental resistivity used in calculating the tables is the International Annealed Copper Standard size, 0.153 28 ohm-g/m^2 at 20°C. The temperature coefficient, for this particular resistivity, is a_{20}=0.00393 per °C, or a_0=0.00427. However, the temperature coefficient is proportional to the conductivity, hence the change of resistivity per °C is a constant, 0.0000597 ohm-g/m^2. The "constant mass" temperature coefficient of any sample is $\alpha t = \frac{0.000597 + 0.000005}{\text{resistivity in ohm - g/m}^2 \text{ at t deg. C}}$ The density is 8.89 g/cm^3 at 20°C.

Note 2 – The values given in the table are only for annealed copper of the standard resistivity. The user of the table must apply the proper correction for copper of any other resistivity. Hard-drawn copper may be taken as about 2.5 percent higher resistivity than annealed copper.

Table 4-3-1. An American Wire Gauge wire chart

NOMINAL SYSTEM VOLTAGE	ALLOWABLE VOLTAGE DROP CONTINUOUS OPERATION	INTERMITTENT OPERATION
14	0.5	1
28	1	2
115	4	8
200	7	14

Table 4-3-2. Allowable voltage drop between bus and equipment

Insulation. Insulation resistance and dielectric strength are two fundamental properties of any insulation material such as rubber, Teflon, vinyl, glass or plastic. These properties determine their suitability for certain uses. Insulation resistance and dielectric strength are entirely different and distinct properties.

Insulation resistance is the resistance to current leakage through and over the surface of the insulation material. Insulation resistance can be measured with a megger without damaging the insulation to determine the general condition of the insulation. This test however, does not provide all the information needed to ascertain whether the insulation is serviceable. Clean, dry insulation having cracks or other faults might show a high value of insulation resistance but would not be suitable for use.

Dielectric strength is the ability of an insulator to withstand electrical potential differences and is usually expressed in terms of the voltage at which the insulation begins to break down and fail due to the electrostatic stress. Maximum dielectric strength values can be measured by raising the voltage of a test sample until the insulation breaks down.

Only the minimum amount of insulation is applied to any particular type of wire or cable, based on the specific job for which it is designed. This is necessary to keep the cost of wiring lower and to minimize the stiffening effect insulation has on wire or cable. In addition, great varieties of environmental and electrical conditions under which conductors are operated are determining factors.

Various types of conductor insulation materials are used for maximum effectiveness in a specific application or type of installation. Earlier insulation materials such as rubber, silk and paper are no longer used in most aircraft systems. Superior insulation materials such as polyvinyl, nylon, Teflon and fiberglass have been introduced and are more commonly used now. Another example is Teflon-Kapton-Teflon, a highly efficient wire insulation material that Boeing currently uses in the B737 and B757 aircraft electrical wiring.

Wires must be sized so that they:

- Have sufficient mechanical strength to allow for service conditions
- Do not exceed allowable voltage drop levels; protected by system circuit protection devices
- Meet circuit current carrying requirements

Mechanical strength of wires. If it is desirable to use wire sizes smaller than No. 20, particular attention should be given to the mechanical strength and installation handling of these wires, e.g., vibration, flexing and termination. Do not use wire containing less than 19 strands. Consideration should be given to the use of high-strength alloy conductors in small gauge wires to increase mechanical strength. As a general practice, wires smaller than size No. 20 should be provided with additional clamps and be grouped with at least three other wires. They should also have additional support at terminations, such as connector grommets, strain relief clamps, shrinkable sleeving or telescoping bushings. They should not be used in applications where they will be subjected to excessive vibration, repeated bending or frequent disconnection from screw termination.

Voltage drop in wires. The voltage drop in the main power wires from the generation source

VOLTAGE DROP	RUN LENGTHS (FEET)	CIRCUIT CURRENT (AMPS)	WIRE SIZE FROM CHART	CHECK CALCULATED VOLTAGE DROP (VD)= (RESISTANCE/FT) (LENGTH) (CURRENT)
1	107	20	No. 6	VD= (0.00044 ohms/ft) (107) (20)= 0.942
0.5	90	20	No. 4	VD= (0.00028 ohms/ft) (90) (20)= 0.504
4	88	20	No. 12	VD= (0.00202 ohms/ft) (88) (20)= 3.60
7	100	20	No. 14	VD= (0.00306 ohms/ft) (100) (20)= 6.12

Table 4-3-3. Examples of determining required tin plated copper wire size and checking voltage drop using Figure 4-3-2

or the battery to the bus should not exceed 2 percent of the regulated voltage when the generator is carrying rated current or the battery is being discharged at the 5-minute rate. The tabulation shown in Table 4-3-2 defines the maximum acceptable voltage drop in the load circuits between the bus and the utilization equipment ground.

Resistance. The resistance of the current return path through the aircraft structure is generally considered negligible. However, this is based on the assumption that adequate bonding to the structure or a special electric current return path has been provided that is capable of carrying the required electric current with a negligible voltage drop. To determine circuit resistance check the voltage drop across the circuit. If the voltage drop does not exceed the limit established by the aircraft or product manufacturer, the resistance value for the circuit may be considered satisfactory. When checking a circuit, the input voltage should be maintained at a constant value. Tables 4-3-3 and 4-3-4 show formulas that may be used to determine electrical resistance in wires and some typical examples.

Duty cycle. The duty cycle is the time interval that the wiring is carrying current and voltage. There are two ratings, intermittent and continuous. An intermittent duty cycle can be in operation for a maximum of two minutes, all other circuits are considered continuous duty cycles.

Resistance Calculation Methods

Figures 4-3-2 and 4-3-3 provide a convenient means of calculating maximum wire length for the given circuit current.

Values in Tables 4-3-3 and 4-3-4 are for tin-plated copper conductor wires. Because the resistance of tin-plated wire is slightly higher than that of nickel- or silver-plated wire, maximum run lengths determined from these charts will be slightly less than the allow-able limits for nickel- or silver-plated copper wire and are therefore safe to use. Figures 4-3-2 and 4-3-3 can be used to derive slightly longer maximum run lengths for silver- or nickel-plated wires by multiplying the maximum run length by the ratio of resistance of tin-plated wire, divided by the resistance of silver- or nickel-plated wire.

As an alternative method or a means of checking results from Figure 4-3-2, continuous flow resistance for a given wire size can be read from Table 4-3-5 and multiplied by the wire run length and the circuit current. Intermittent current flow is no more than two minutes. For intermittent flow, use Figure 4-3-3.

Voltage drop calculations for aluminum wires can be accomplished by multiplying the resistance for a given wire size, defined in Table 4-3-6, by the wire run length and circuit current.

When the estimated or measured conductor temperature (T_2) exceeds 20°C, such as in areas having elevated ambient temperatures or in fully loaded power-feed wires, the maximum allowable run length (L_2), must be shortened from L_1 (the 20°C value) using the following formula for copper conductor wire:

$$L_2 = \frac{(254.5°C)\,(L_1)}{(234.5°C) + (T_2)}$$

For aluminum conductor wire, the formula is:

$$\frac{(258.1°C)\,(L_1)}{(238.1°C) + (T_2)}$$

These formulas use the reciprocal of each material's resistively temperature coefficient to take into account increased conductor resistance resulting from operation at elevated temperatures.

To determine T_2 for wires carrying a high percentage of their current carrying capability at elevated temperatures, laboratory testing using a load bank and a high temperature chamber is recommended. Such tests should be run at

MAXIMUM VOLTAGE DROP	WIRE SIZE	CIRCUIT CURRENT (AMPS)	MAXIMUM WIRE RUN LENGTH (FEET)	CHECK CALCULATED VOLTAGE DROP (VD)= (RESISTANCE/FT) (LENGTH) (CURRENT)
1	No. 10	20	39	VD= (0.00126 ohms/ft) (39) (20)= 0.98
0.5	--		19.5	VD= (0.00126 ohms/ft) (19.5) (20)=0.366
4	--		156	VD= (0.00126 ohms/ft) (156) (20)= 3.93
7	--		No. 14	VD= (0.00126 ohms/ft) (273) (20)= 6.88

Table 4-3-4. Examples of determining maximum tin plated copper wire size and checking voltage drop using Figure 4-3-3

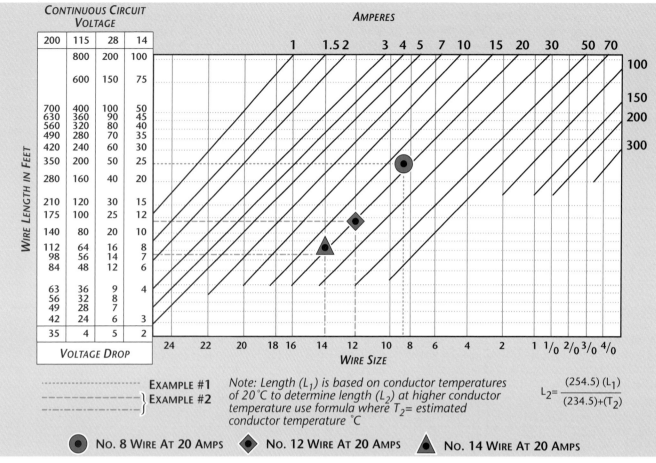

Figure 4-3-2. Conductor chart for single copper wire with continuous flow

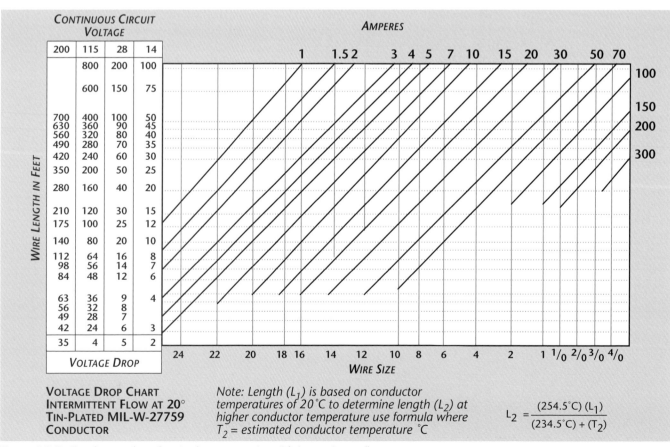

Figure 4-3-3. Conductor chart for single copper wire with intermittent flow

WIRE SIZE	CONTINUOUS DUTY CURRENT (AMPS) - WIRES IN BUNDLES, GROUPS, HARNESSES, OR CONDUITS. (SEE NOTE #1)			MAX. RESISTANCE OHMS/ 1,000FT @20°C TIN PLATED CONDUCTOR (SEE NOTE 2)	NOMINAL CONDUCTOR AREA CIRC. MILS
	WIRE CONDUCTOR TEMPERATURE RATING				
	105 °C	150 °C	200 °C		
24	2.5	4	5	28.40	475
22	3	5	6	16.20	755
20	4	7	9	9.88	1,216
18	6	14	12	6.23	1,900
16	7	11	14	4.81	2,426
14	10	14	18	3.06	3,831
12	13	19	25	2.02	5,874
10	17	26	32	1.26	9,354
8	38	57	71	0.70	16,983
6	50	76	97	0.44	26,818
4	68	103	133	0.28	42,615
2	95	141	179	0.18	66,500
1	113	166	210	0.15	81,700
0	128	192	243	0.12	104,500
00	147	222	285	0.09	133,000
000	172	262	335	0.07	166,500
0000	204	310	395	0.06	210,900

Note 1 - Rating is for 70°C ambient, 33 or more wires in the bundle for sizes 24 through 10, and 9 wires for size 8 and larger, with no more than 20 percent of harness current carry capacity being used, at an operating altitude of 60,000 feet.
Note 2 - For resistance of silver or nickel-plated conductors see wire specifications.

Table 4-3-5. Current carrying capacity and resistance of copper wire

WIRE SIZE	CONTINUOUS DUTY CURRENT (AMPS) - WIRES IN BUNDLES, GROUPS, HARNESSES, OR CONDUITS.		MAX. RESISTANCE OHMS/1,000FT
	WIRE CONDUCTOR TEMPERATURE RATING		
	105°C	150°C	20°C
8	30	45	1.093
6	40	61	0.641
4	54	82	0.427
2	76	113	0.268
1	90	133	0.214
0	102	153	0.169
00	117	178	0.133
000	138	209	0.109
0000	163	248	0.085

Table 4-3-6. Current carrying and resistance of aluminum wire

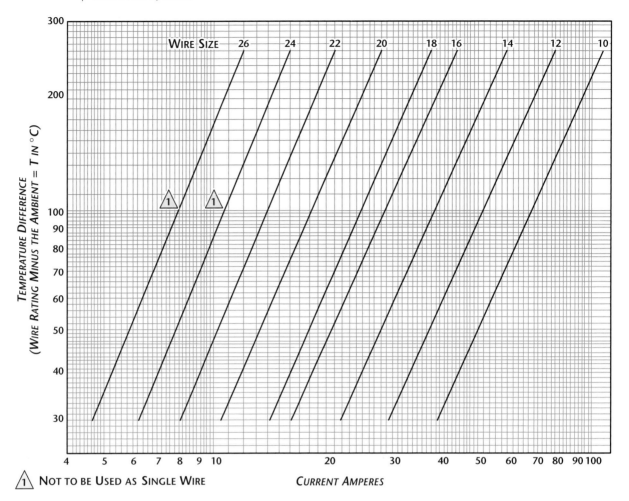

Figure 4-3-4. Single copper wire in free air for 26 to 10 gauge wire

anticipated worse case ambient temperature and maximum current loading combinations.

Approximate T_2 can be estimated using the following formula:

$$T_2 = T_1 + (T_R - T_1)(\sqrt{I_2/I_{max}})$$

Where:

T_1 = Ambient Temperature

T_2 = Estimated Conductor Temperature

T_R = Conductor Temperature Rating

I_2 = Circuit Current (A=amps)

I_{max} = Maximum Allowable Current (A=amps) at T_R

This formula is quite conservative and will typically yield somewhat higher estimated temperatures than are likely to be encountered under actual operating conditions.

> **NOTE:** *Table 4-3-5 and 4-3-6 shows the conductor resistance of aluminum wire and that of copper wire (two numbers higher) are similar. Accordingly, the electric wire current*

in Table 4-3-5 can be used when it is desired to substitute aluminum wire and the proper size can be selected by reducing the copper wire size by two numbers and referring to Table 4-3-5. The use of aluminum wire size smaller than No. 8 is not recommended.

Methods for Determining Current Carrying Capacity of Wires

This part contains methods for determining the current carrying capacity of electrical wire, both as a single wire in free air and when bundled into a harness. It presents derating factors for altitude correction and examples showing how to use the graphical and tabular data provided for this purpose. In some instances, the wire may be capable of carrying more current than is recommended for the contacts of the related connector. In this instance, it is the contact rating that dictates the maximum current to be carried by a wire. Wires of larger gauge may need to be used to fit within the crimp range of connector contacts that are adequately rated for the current being carried. Figure 4-3-6

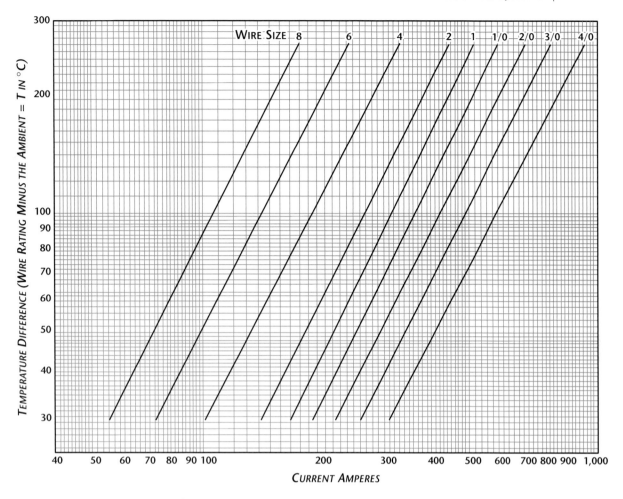

Figure 4-3-5. Single copper wire in free air for 8 to 4/0 gauge wire

gives a family of curves whereby the bundle derating factor may be obtained.

Effects of heat aging on wire insulation. Since electrical wire may be installed in areas where inspection is infrequent over extended periods, it is necessary to consider heat-aging characteristics in the selection of wire. Resistance to heat is of primary importance in the selection of wire for aircraft use, as it is the basic factor in wire rating. Where wire may be required to operate at higher temperatures due either to high ambient temperatures, high-current loading or a combination of the two, selection should be made on the basis of satisfactory performance under the most severe operating conditions.

Maximum Operating Temperature

The current that causes a temperature steady state condition equal to the rated temperature of the wire should not be exceeded for obvious reasons. Rated temperature of the wire may be based upon the ability of either the conductor or the insulation to withstand continuous operation without degradation.

Single wire in free air. Determining a wiring system's current carrying capacity begins with determining the maximum current that a given-sized wire can carry without exceeding the allowable temperature difference (wire rating minus ambient °C). The curves are based upon a single copper wire in free air. See Figures 4-3-4 and 4-3-5.

Wires in a harness. When wires are bundled into harnesses, the current derived for a single wire must be reduced, as shown in Figure 4-3-6 The amount of current derating is a function of the number of wires in the bundle and the percentage of the total wire bundle capacity that is being used.

Harness at altitude. Since heat loss from the bundle is reduced with increased altitude, the amount of current should be derated. Figure 4-3-7 gives a curve whereby the altitude-derating factor may be obtained.

By looking at Figure 4-3-8, it shows why the proper rating of individual wires within a bundle is so important. If one wire becomes overheated, it can damage a significant number of other wires and bundles.

Aluminum conductor wire. When aluminum conductor wire is used, sizes should be selected based on current ratings shown in Table 4-3-6 The use of sizes smaller than #8 is discouraged. Aluminum wire should not be attached to engine mounted accessories or used in areas having corrosive fumes, severe vibration, mechanical stresses or where there is a need for frequent disconnection. Use of aluminum wire is also discouraged for runs of less than 3 feet. Termination hardware should be of the type specifically designed for use with aluminum conductor wiring.

Instructions for Use of Electrical Wire Chart

Correct size. To select the correct size of electrical wire, two major requirements must be met:

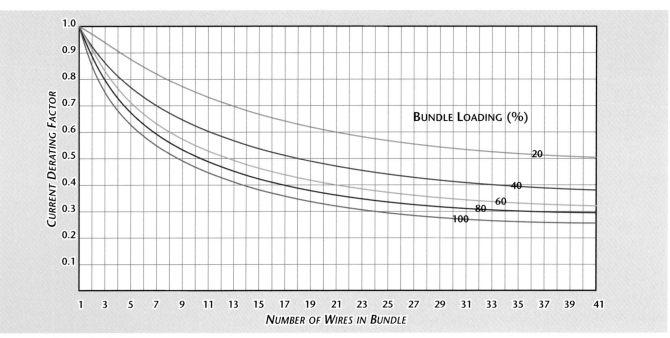

Figure 4-3-6. Bundle derating curves

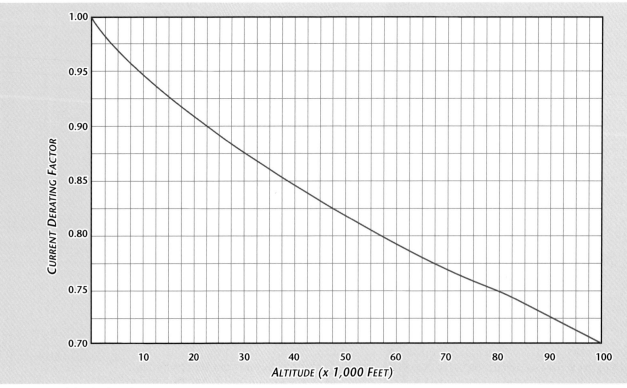

Figure 4-3-7. Altitude derating curve

1. The wire size should be sufficient to prevent an excessive voltage drop while carrying the required current over the required distance. (See Table 4-3-2, Tabulation Chart, for allowable voltage drops).

2. The size should be sufficient to prevent overheating of the wire carrying the required current.

To meet the requirements in selecting the correct wire size using Figure 4-3-2 or Figure 4-3-3, the following information is necessary:

- Wire length in feet

- Number of amperes of current to be carried

- Allowable voltage drop permitted

- Required continuous or intermittent current

- Estimated, or measured, conductor temperature

- If the wire is to be installed in conduit and/or bundle

- If the wire is to be installed as a single wire in free air

Example 1:

Find the wire size in Figure 4-3-2 using the following known information:

- The wire run is 50 feet long, including the ground wire.

- The current load is 20 amps

- The voltage source is 28 volts from the bus to the equipment.

- The circuit has continuous operation.

- Estimated conductor temperature is 20°C or less.

The scale on the left of the chart represents maximum wire length (in feet) to prevent an excessive voltage drop for a specified voltage source system (e.g., 14V, 28V, 115V, 200V). This voltage is identified at the top of scale, and the corresponding voltage drop limit for continuous operation at the bottom. The scale (slant lines) on top of the chart represents amperes. The scale at the bottom of the chart represents wire gauge. The steps for finding the wire size are as follows:

1. From the left scale, find the wire length 50 feet under the 28V source column.

2. Follow the corresponding horizontal line to the right until it intersects the slanted line for the 20-amp load.

3. At this point, drop vertically to the bottom of the chart. The value falls between No. 8 and No. 10. Select the next larger size wire to the right, in this case No. 8. This small-

Figure 4-3-8. This is a run of wire bundles that run the length of the fuselage in a B-757. They are located above the ceiling panels and the gasper. Notice the cable run partially enclosed in a plastic guard above the wiring.

est size wire can be used without exceeding the voltage drop limit expressed at the bottom of the left scale. This example is plotted on the wire chart, Figure 4-3-2. Use Figure 4-3-2 for continuous flow and Figure 4-3-3 for intermittent flow.

Procedures in Example No. 1 can be used to find the wire size for any continuous or intermittent operation (maximum two minutes). Voltage (e.g. 14 volts, 28 volts, 115 volts, 200 volts) as indicated on the left scale of the wire chart in Figure 4-3-2 and 4-3-3.

Example 2:

Using Figure 4-3-2, find the wire size required to meet the allowable voltage drop in Table 4-3-2 for a wire carrying current at an elevated conductor temperature using the following information:

- The wire run is 15.5 feet long, including the ground wire

- The circuit current (I2) is continuously 20 amps

- The voltage source is 28 volts

- The wire type used has a 200°C conductor rating and it is intended to use this thermal rating to minimize the wire gauge. Assume that the method described in Table 4-3-1 was used and the minimum wire size to carry the required current is #14.

- Ambient temperature is 50°C under hottest operating conditions

The procedures for finding wire size in this second example are as follows:

1. The estimated calculation methods outlined in Tables 4-3-1 and 4-3-2 may be used to determine the estimated maximum current

(I_{max}). The #14 gauge wire mentioned above can carry the required current at 50°C ambient (allowing for altitude and bundle derating).

A. Use Figure 4-3-2 to calculate the I_{max} a #14 gauge wire can carry.

Where:

T_2 = estimated conductor temperature

T_1 = 50°C ambient temperature

T_R = 200°C maximum conductor rated temperature

B. Find the temperature differences (T_R-T_1) = (200°C - 50°C) = 150°C.

C. Follow the 150°C corresponding horizontal line to intersect with #14 wire size, drop vertically and read 47 amps at bottom of chart (current amperes).

D. Use Figure 4-3-7 the left side of chart reads 0.91 for 20,000 feet, multiply by derating amps factor:

0.91 x 47 amps = 42.77 amps

Use Figure 4-3-6, find the derate factor for 8 wires in a bundle at 60 percent. First, find the number of wires in the bundle (8) at bottom of graph and intersect with the 60 percent curve meet. Read derating factor, (left side of graph) which is 0.6, derating factor by amps. Caculate I_{max}.

0.6 x 42.77 amps = 26 amps

This is the maximum current the #14 gauge wire could carry at 50°C ambient.

I_{max} = 26 amps

This is the maximum run length for size #14 wire carrying 20 amps from Figure 4-3-2.

L_1=15.5 ft

2. Determine the T_2 and the resultant maximum wire length when the increased resistance of the higher temperature conductor is taken into account.

$T_2 = T_1+(T_R-T_1)(\sqrt{I_2/I_{max}})$

T_2 = 50°C + (200°C - 50°C) ($\sqrt{20A/26A}$)

= 50°C + (150°C) (0.877)

T_2 = 182°C

$L_2=\dfrac{(254.5°C)\ (L_1)}{(234.5°C)\ +\ (T_2)}$

$L_2=\dfrac{(254.5°C)\ (15.5\ ft.)}{(234.5°C)\ +\ 182°C}$

L_2= 9.5 ft

The size #14 wire, selected using the methods outlined, is too small to meet the voltage drop limits from Figure 4-3-2 for a 15.5 feet long wire run.

3. Select the next larger wire (size #12) and repeat the calculations as follows:

This is the maximum run length for 12 gauge wire carrying 20 amps from Figure 4-3-2.

L_1=24 ft

This is the maximum current the size #12 wire can carry at 50°C ambient). Use calculation methods outlined in Table 4-3-4.

I_{max} = 37 amps

T_2 = 50°C + (200°C - 50°C) ($\sqrt{20A/37A}$)

= 50°C + (150°C) (-540) = 131°

$L_2=\dfrac{(254.5°C)\ (L_1)}{(234.5°C)\ +\ (T_2)}$

$L_2=\dfrac{(254.5°C)\ (24\ ft.)}{(234.5°C)\ +\ 131°C} = \dfrac{6,108}{366}$

$L_2=\dfrac{(254.5°C)\ (24\ ft.)}{366} = 16.7\ ft$

The resultant maximum wire length, after adjusting downward for the added resistance associated with running the wire at a higher temperature, is 15.4 feet, which will meet the original 15.5 foot wire run length requirement without exceeding the voltage drop limit expressed in Figure 4-3-2.

Computing Current Carrying Capacity

Example 1:

Assume a harness (open or braided), consisting of 10 wires, size #20, 200°C rated copper and 25 wires, size #22, 200°C rated copper, will be installed in an area where the ambient temperature is 60°C and the airplane is capable of operating at a 60,000-foot altitude. Circuit analysis reveals that seven of the 35 wires in the bundle (7/35 = 20 percent) will be carrying power currents nearly at or up to capacity. Use the following procedure to compute current carrying capacity:

1. Refer to the "single wire in free air" curves in Figure 4-3-2. Determine the change of temperature of the wire to determine free air ratings. Since the wire will be in an ambient of 60°C and rated at 200°C, the change of to temperature is 200°C - 60°C = 140°C. Follow the 140°C temperature difference horizontally until it intersects with wire size line on Figure 4-3-2. The

free air rating for size #20 is 21.5 amps, and the free air rating for size #22 is 16.2 amps.

2. Refer to the "bundle derating curves" in Figure 4-3-6, the 20 percent curve is selected since circuit analysis indicate that 20 percent or less of the wire in the harness would be carrying power currents and less than 20 percent of the bundle capacity would be used. Find 35 (on the abscissa) since there are 35 wires in the bundle and determine a derating factor of 0.52 (on the ordinate) from the 20 percent curve.

3. Derate the size #22 free air rating by multiplying 16.2 by 0.52 to get 8.4 amps in-harness rating. Derate the size #20 free air rating by multiplying 21.5 by 0.52 to get 11.2 amps in-harness rating.

4. Refer to the "altitude derating curve" of Figure 4-3-7, look for 60,000 feet (on the abscissa) since that is the altitude at which the airplane will be operating. Note that the wire must be derated by a factor of 0.79 (found on the ordinate). Derate the size #22 harness rating by multiplying 8.4 amps by 0.79 to get 6.6 amps. Derate the size #20 harness rating by multiplying 11.2 amps by 0.79 to get 8.8 amps.

5. To find the total harness capacity, multiply the total number of size #22 wires by the derated capacity (25 x 6.6 = 165.0 amps) and add to that the number of size #20 wires multiplied by the derated capacity (10 x 8.8 = 88 amps) and multiply the sum by the 20 percent harness capacity factor. Thus, the total harness capacity is (165.0 + 88.0) x 0.20 = 50.6 amps. It has been determined that the total harness current should not exceed 50.6 amps, size #22 wire should not carry more than 6.6 amps and size #20 wire should not carry more than 8.8 amps.

6. Determine the actual circuit current for each wire in the bundle and for the whole bundle. If the values calculated in step #5 are exceeded, select the next larger size wire and repeat the calculations.

Example 2:

Assume a harness (open or braided), consisting of 12, size #12, 200°C rated copper wires, will be operated in an ambient of 25°C at sea level and 60°C at a 20,000-foot altitude. All 12 wires will be operated at or near their maximum capacity. Use the following procedure:

1. Refer to the "single wire in free air" curve in Figure 4-3-2, determine the temperature difference of the wire to determine free air ratings. Since the wire will be in ambient of 25°C and 60°C and is rated at 200°C, the temperature differences are 200°C - 25°C = 175°C and 200°C - 60°C = 140°C respectively. Follow the 175°C and the 140°C temperature difference lines on Figure 4-3-5 until each intersects wire size line, the free air ratings of size #12 are 68 amps and 61 amps, respectively.

2. Refer to the "bundling derating curves" in Figure 4-3-6, the 100 percent curve is selected because we know all 12 wires will be carrying full load. Find 12 (on the abscissa) since there are 12 wires in the bundle and determine a derating factor of 0.43 (on the ordinate) from the 100 percent curve.

3. Derate the size #12 free air ratings by multiplying 68 amps and 61 amps by 0.43 to get 29.2 amps and 26.2 amps, respectively.

4. Refer to the "altitude derating curve" of Figure 4-3-7, look for sea level and 20,000 feet (on the abscissa) since these are the conditions at which the load will be carried. The wire must be derated by a factor of 1.0 and 0.91, respectively.

5. Derate the size #12 in bundle ratings by multiplying 29.2 amps at sea level and 26.6 amps at 20,000 feet by 1.0 and 0.91, respectively, to obtain 29.2 amps and 23.8 amps. The total bundle capacity at sea level and 25°C ambient is 29.2x12=350.4 amps. At 20,000 feet and 60°C ambient the bundle capacity is 23.8x12=285.6 amps. Each size #12 wire can carry 29.2 amps at sea level, 25°C ambient or 23.8 amps at 20,000 feet, and 60°C ambient.

6. Determine the actual circuit current for each wire in the bundle and for the bundle. If the values calculated in Step #5 are exceeded, select the next larger size wire and repeat the calculations.

Identifying Wire and Cable

Aircraft electrical system wiring and cable may be marked with a combination of letters, numbers and colors to identify the wire, the circuit it belongs to, the gauge number and other information necessary to relate the wire or cable to a wiring diagram. Such markings are called the identification code. Unfortunately, aircraft manufacturers do not adhere to a common standard procedure for marking and identifying wiring to simplify the task. Rather, each manufacturer usually develops its own identification code roughly following ATA system naming conventions. Figure 4-3-9 illustrates the backside of an avionics bay in an airliner. It is a classic example of why wire numbering is so important.

Figure 4-3-9. Without accurately numbered wires, it would be all but impossible to maintain these systems.

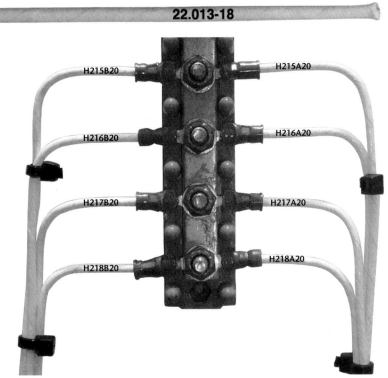

Figure 4-3-10. Typical wire markings in an auto-pilot system

One such identification practice (Figure 4-3-10) shows a marking and spacing that is representative of a typical wire marking in an auto-pilot system. The number 22 in the code refers to the system in which the wire is installed, e.g., the autoflight system. The next set of numbers, 0.013, is the wire number, and the 18 indicates the wire size.

A letter or group of letters and numbers added to the basic identification number typically identifies other system components, such as plugs and jacks. These letters and numbers in some cases indicate the physical location of the component in the system. Interconnected cables are also marked in some systems to indicate location, proper termination and use.

In order to ensure visibility and to be legible, the size and color of the marking stamped on the insulation should contrast with the insulation color. For example, black stamping should be used with light-colored backgrounds, or white stamping on dark-colored backgrounds.

Wires are usually marked at intervals of not more than 15 inches lengthwise and within 3 inches of each junction or terminating point. Figure 4-3-10 shows wire identification at a terminal block.

An alternative means of identification is frequently used with coaxial cable and wires at terminal blocks and junction boxes. The identification consists of a marking or stamping a wiring sleeve rather than the wire itself. A flexible *vinyl sleeving,* either clear or white opaque, is commonly used for general purpose wiring. For high-temperature applications, *silicone rubber* or *silicone fiberglass sleeving* is recommended. Where resistance to synthetic hydraulic fluids or other solvents is necessary, clear or white opaque *nylon sleeving* can be used.

The preferred method is usually an identification marking stamped directly on the wire (insulation) or on the sleeving; several other methods are also used. Figure 4-3-11 shows a piece of white shrink tubing with the wire number stamped on it before being installed.

Wiring Installation and Routing

A number of recommended procedures are used for installing aircraft electrical wiring. It is useful to discuss and define the various types of wire/cable installations and procedures in common terms. The following are common terms and descriptions:

- **Open wiring.** Open wiring is any wire, wire group or wire bundle not enclosed in a conduit.

- **Wire group.** A wire group is two or more wires going to the same location and physically tied together to retain their group identity. See Figure 4-3-12.

- **Wire bundle.** A wire bundle is two or more wire groups tied together because they are going in the same direction at the point where the tie is located.

- **Electrically protected wiring.** Electrically protected wiring are wires that include (in the circuit) protection against overloading, such as fuses, circuit breakers or other limiting devices.

- **Electrically unprotected wiring.** Electrically unprotected wiring are wires or large cables (generally from generators to main bus distribution points) that do not have protection, such as fuses, circuit breakers or other current-limiting devices.

The practice of grouping or bundling some wiring should be avoided in cases such as electrically unprotected power wiring and wiring going to duplicate equipment or components that are vital or related to safety-of-flight. The practice of equipping aircraft with duplicated equipment, such as several electrically powered hydraulic pumps is known as redundancy and is a means of continued safe flight operation if one vital system fails. For example, grouping or bundling VHF #1 and VHF #2 wiring could result in the loss of both radios if enough damage occurred in the bundle carrying wiring for both radios. This, of course, would defeat the purpose of redundancy and should be avoided.

In general, wire bundles should contain less than 75 wires or measure 1 1/2 to 2 inches in diameter where practicable. When several wires are grouped at junction boxes, terminal blocks, panels, etc., identity of the group within a bundle can be retained.

Twisted wire. Several wires running parallel to one another may be twisted together uniformly when specified on the engineering drawing, usually to mitigate electromagnetic interference or distortion of an electrical signal. The most common examples are wiring near a magnetic compass or flux valve, three-phase distribution wiring, and radio wiring as specified on engineering drawings.

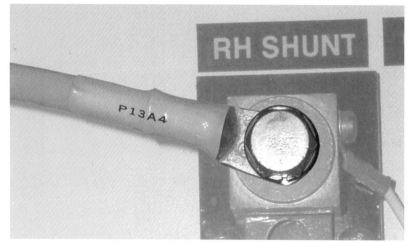

Figure 4-3-11. A numbered sleeve made from shrink tubing

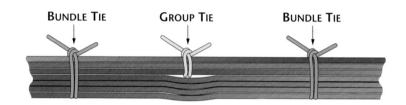

Figure 4-3-12. Groups of wires for specific circuits are frequently bundled, then included in main bundle.

Twist the wires so that they will lie snugly against each other, making approximately the number of twists per foot as shown in Table 4-3-7. Always check wire insulation for damage after twisting. If the insulation is torn or frayed, replace the wire.

Radius of bends. The radius of bends in wire groups or bundles should be not less than 10 times the outside diameter of the wire group or bundle. However, at terminal strips, where wire is suitably supported at each end of the bend, a minimum radius of three times the outside diameter of the wire, or wire bundle, is normally acceptable. There are, of course, exceptions to these guidelines in the case of certain types of cable; for example, coaxial cable should never be bent to a smaller radius than ten times the outside diameter.

Wire splice spacing. Spliced connections in wire groups or bundles should be located so

TWISTS PER INCH									
Wire Size	#22	#20	#18	#16	#14	#12	#10	#8	#6	#4
2 Wires	10	10	9	8	7 1/2	7	6 1/2	6	5	4
3 Wires	10	10	8 1/2	7	6 1/2	6	5 1/2	5	4	3

Table 4-3-7. Twisting wires together helps to reduce the electromagnetic interference

that they can be easily inspected when trouble-shooting a wiring problem. Splices must also be staggered (Figure 4-3-13) to prevent the bundle from becoming excessively enlarged. Splices that are not self-insulated must be covered

Figure 4-3-13. When several splices are necessary at one location stagger them as indicated.

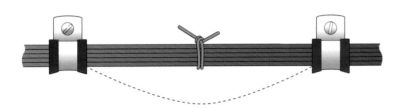

Figure 4-3-14. Some wire slack helps future maintenance tasks.

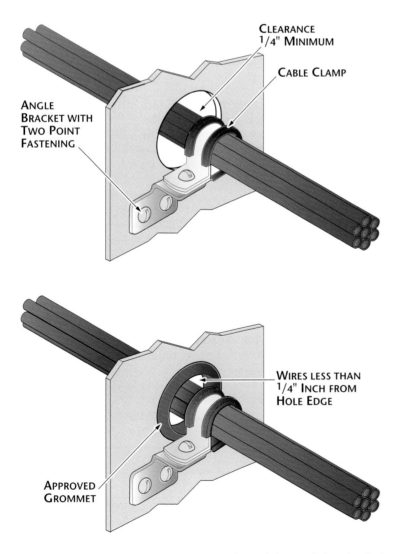

Figure 4-3-15. Two methods of running wire through former lightening holes

with plastic or some other suitable insulation material and securely tied at both ends. There should not be more than one splice between two terminals on any one cable.

Bundle slack (sag). The wire run or bundle should not be unduly taut nor should it be excessively slack. A suitable amount of slack between wire/cable supports normally does not exceed a maximum of 1/2-inch deflection with normal hand force. Slightly more slack may be acceptable if the wire bundle is thin and the clamps are far apart, but should never be so great that the wire bundle could possibly chafe against any surface. The proper amount of slack is important near each end of the wire bundle for a number of reasons. Figure 4-3-14 shows slack in a wire bundle.

Enough slack allows the technician to accomplish other maintenance tasks on equipment near the wire bundle, especially when connectors must be disconnected and reconnected. Slack also makes it easier to replace terminals without resulting in wires too short to be reconnected. Enough slack helps to prevent mechanical strain on the wires, wire junctions and supports permits free movement of shock and vibration-mounted equipment.

Wiring should be routed, either parallel with or at right angles to, the airframe stringers or ribs whenever practicable. An exception to this general rule is coaxial cable, which is generally routed as directly as possible to avoid signal distortion and minimize line loss.

Wire bundles must be adequately supported throughout their length with a sufficient number of supports to keep unsupported lengths short and minimize vibration. Several factors must be considered for the protection of wires, cables, wire groups and wire bundles when deciding where to route them:

- Exposure to chafing or abrasion

- Exposure to high temperatures

- Locations that would encourage their use as handholds or supports for cargo or personal items

- Possible damage caused by personnel moving within the aircraft

- Possible damage from cargo stowage or shifting

- Possible damage from battery acid fumes, spray or spillage

- Possible damage from solvents and fluids

Chafing and abrasion. Wires, cables and bundles must be installed in such a way that possible chafing or abrasion is prevented in locations where contact with sharp surfaces, control cables or other wires would damage the insulation. Insulation damage can lead to short circuits, malfunctions or intermittent operation of electrical loads or inadvertent operation of equipment. Direct short circuits to airframe ground can cause fires on board the aircraft, a critical and potentially deadly problem in flight. Insulated cable clamps such as the popular Adel clamp should be used to support wire bundles, especially where the wiring passes through holes in a rib, stringer or bulkhead. In addition, any wire that lies closer than 1/4-inch to the edge of the hole requires that a suitable grommet be installed to line the edges of the hole. See Figure 4-3-15.

When installing nylon or rubber grommets, the task is sometimes made easier by cutting the grommet first. The cut should be made at an angle of 45° to the axis of the wire bundle hole. Position the cut at the top of the hole and secure the grommet in place with an approved general purpose adhesive.

High temperature environments. Unless necessary, wiring should be routed outside of areas in the airframe which are subjected to high temperatures. Such airframe locations include exhaust stacks and spaces containing engine bleed air ducts. In addition, wiring should not touch or be adjacent to high temperature components such as large power resistors that dissipate a lot of heat. Failure to heed these tips can possibly result in deterioration or melting of certain types of insulation. The distance between wiring and these components is normally specified by engineering drawings.

In cases where wiring must be routed through hot areas, wiring with a special insulating material must be used. Some materials that withstand high temperatures include asbestos, fiberglass and Teflon or Teflon derivatives. Additional protection in the form of conduits may also be required. Common sense mandates that wiring repairs in hot areas must be accomplished with the same or other equivalent high-temperature resistant materials.

If coaxial cables must be run through high temperature areas of the airframe, be certain to use only those types of coax that are approved for such environments. Most types of coax have soft plastic insulation, such as polyethylene, which is especially subject to deformation and deterioration at elevated temperatures. All high-temperature areas should be avoided when installing these cables insulated with plastic or polyethylene.

Protection against solvents and fluids. A good rule of thumb for preventing damage to wiring caused by solvents or other corrosive fluids such as Skydrol is to avoid installations in areas that tend to collect such fluids. One such area is typically the lowest four inches of the aircraft fuselage. Some electrical systems, however, depend on wiring connections or termination of wires in such areas. In such a case, the possibility of any wire becoming soaked with fluids must be considered, and plastic tubing should be used to protect the wire. This tubing should extend past the exposure area in both directions and tied at each end. If the wire has a low point between the tubing ends, provide a 1/8-inch drain hole, as shown in Figure 4-3-16. This hole should be punched into the tubing after the installation is complete and the low point definitely established by using a hole punch to cut a half circle. Care should be taken not to damage any wires inside the tubing when using the punch.

No wiring should ever be routed below an aircraft battery. All wires and cables near an aircraft battery should be inspected frequently. Any evidence of insulation discoloration is reason for wiring replacement.

Wheel well area. The wheel well area in retractable gear aircraft is perhaps the most difficult area for ensuring the safety of wiring installations. Wiring hazards include exposure to hydraulic fluids, possible pinching and severe flexing caused by landing gear retraction and extension, and heat given off from hot wheel brakes. Whenever possible all wiring should be in conduit. For wires that flex the most common form of protection includes sleeves of flexible tubing securely held at each end. There should be no relative movement at points where protective flexible tubing is secured.

Wheel well wiring and its insulating tubing should be carefully inspected at frequent inter-

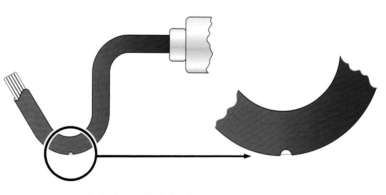

A drainage hole 1/8 inch diameter at lowest point in tubing.

Make the hole after installation is complete and lowest point is firmly established.

Figure 4-3-16. Anytime wire is routed through flexible (or non-flexible) tubing that has a low spot, a water drain hole must be provided.

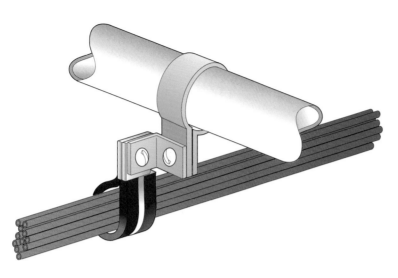

Figure 4-3-17. A method of placing an Adel clamp on tubing

less than two inches but more than 1/2-inch, a polyethylene sleeve may be installed over the wire bundle to give further protection.

Another method that can be used to maintain a rigid separation is depicted in Figure 4-3-17. This clamp arrangement is not considered a wire bundle support, but merely a means for maintaining safe separation. In no case should any wire be routed nearer than 1/2-inch to a plumbing line, and a plumbing line that carries flammable fluids or oxygen cannot ever be used as a support for wires or a wire bundle.

It is good common practice to maintain a minimum clearance of at least three inches between control cables and wiring. In cases where this cannot be accomplished, mechanical guards should be installed to prevent contact between wiring and control cables.

vals (usually determined by the maintenance program under which the aircraft is certified), and any wires or tubing showing signs of wear, deterioration or chafing should be replaced. In addition, ensure that there is no strain on wiring attachments when landing gear strut parts are fully extended, but neither should slack be excessive.

Miscellaneous routing precautions. Installations that require wire bundles to be routed parallel to combustible fluid or oxygen lines should have as much fixed separation as possible designed into the routing. Wire bundles should be on a level with, or above, the plumbing lines. Wiring clamps should be spaced close enough together that a wire broken at a clamp will not contact a plumbing line and create a short circuit and/or fire hazard. In cases where at least six inches of separation is not possible, you may clamp the wire bundle and the plumbing line to the same structure to prevent any relative motion. If the separation is

Cable clamp installation. Figure 4-3-18 shows the correct hardware to use for attaching an Adel clamp. Figure 4-3-19 shows the proper technique for installing cable clamps. The mounting screw should be above the wire bundle whenever possible. It is also desirable that the long straight back portion of the cable clamp rest against a structural member where practicable.

When assembling the wire bundles and installing cable clamps, take care to select the proper size clamp to avoid pinching any wiring and damaging the insulation. A clamp too large will not provide the necessary support and a clamp too small will damage the wiring.

Adel clamps are widely used and are available with a rubber inserts to cushion the wire bundles and prevent damage to tubular structures as shown in Figure 4-3-20. Such clamps must fit tightly, but should not become deformed when the mounting hardware is tightened in place.

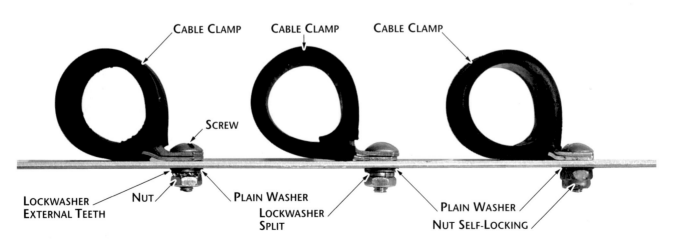

Figure 4-3-18. Although elastic stop nuts are the principal lock, split and star lock washers and plain nuts may also be used

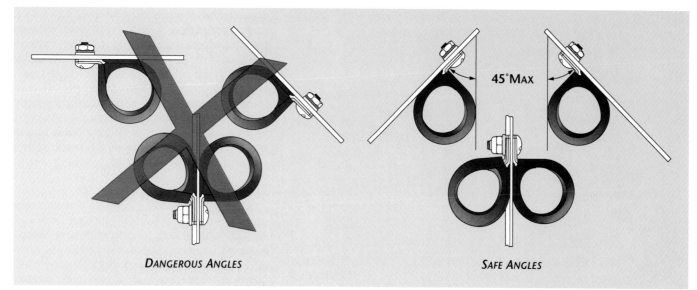

Figure 4-3-19. Do's and don'ts for Adel clamp installation

Conduit. Conduit is the term used to describe a protective tubular covering in which aircraft wiring is routed from one point to another. It provides mechanical protection of wires and cables, may be made from metallic or nonmetallic material, and may be either rigid or flexible, depending on the aircraft application.

Determining the proper conduit size for a specific wire bundle or cable requires several considerations. To allow for ease in maintenance and possible future circuit expansion, the conduit inner diameter should be approximately 25% larger than the anticipated maximum diameter of the conductor bundle. The nominal diameter of a rigid metallic conduit is normally the *outside diameter (O.D.)*, so the *inside diameter (I.D.)* is obtained by subtracting an amount equal to twice the conduit wall thickness.

Whenever conductors are run through conduit, they are susceptible to abrasion damage where they enter and exit the conduit. Therefore, conduit end fittings are attached in such a manner that the conductors contact only smooth surfaces. It is rare, but when fittings are not used, the conduit end should be flared to prevent damage to the conductor insulation or internal wiring.

Conduit is usually installed with clamps along the conduit run to provide support and limit movement. You should be aware of several common conduit installation hazards and take steps to avoid those problems by keeping the following details in mind:

- Before placing conductors in a conduit, drill a few drain holes in the conduit at the lowest point of the conduit run. Carefully remove any drilling burrs from the drain holes.

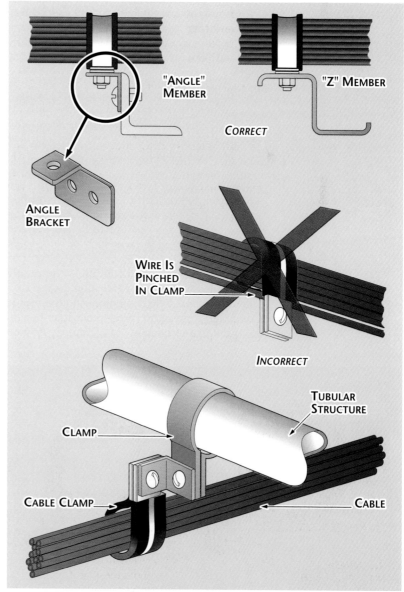

Figure 4-3-20. Good and bad methods of locating and installing Adel clamps and acceptable mounting hardware

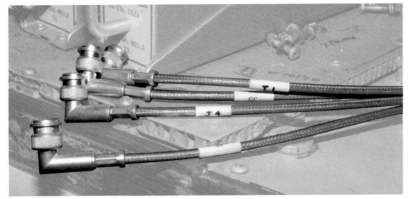

Figure 4-3-21. Shielded RF cable that will connect to ground through the connector

- Do not install conduit in a location where use as a handhold or footstep would be likely.

- Utilize enough clamps of the right size to support the conduit and prevent chafing against the airframe structure. Pay particular attention to the conduit end fittings, ensuring adequate support and making sure that conductor bundle tension does not place excessive stress on the end fittings.

- Repair damaged conduit sections before damage occurs to the conductors within. Adhere to acceptable tube bend radii for rigid conduit as prescribed by the manufacturer's instructions and reject kinked or wrinkled bends in rigid conduit.

When threading wires in conduit it can sometimes be difficult to get a bundle pulled through. Powdered soapstone or tire talcum blown through the conduit first will make it much easier to pull the wires.

Flexible aluminum braid conduit is commonly produced in either a bare flexible design or a rubber-covered style. System wiring that needs protection from radio interference may benefit from the use of flexible brass conduit. In gen-eral, flexible conduit may be used where it is impractical to use rigid conduit, such as areas that have motion between conduit ends or where complex bends are necessary. If flexible conduit must be cut to a certain length, use a technique recommended to minimize fraying of the braid when cutting desired lengths with a hacksaw. One such technique consists of wrapping the flexible conduit with duct tape or some other type of adhesive tape over the area where the cut must be made.

Shielding. The braided shield of a shielded conductor is commonly connected to airframe ground on each end of the shielded cable run. This provides a path to ground for any electromagnetic interference that may possibly be introduced into the center conductor. Figure 4-3-21 shows shielded RF cable. In this manner, extraneous voltages are drained and are not a problem to the center conductor voltage and current. In addition to grounding the shield, it is always advisable to avoid placing low voltage signal wiring in the same conduit or bundle that carries power mains, relay coil drives, relay contact leads or other high level voltages or currents.

Engine ignition noise (static) has traditionally plagued aircraft communication systems in reciprocating engine aircraft and the complete shrouding of the engine in a metal cowl with a metal firewall between the engine and radio installation has not altogether eliminated audible static. By shielding the actual source, most unwanted signals can be suppressed, so high tension wires should be shielded with wire braid, spark plugs jacketed in metal and the magnetos housed in a metal case. The shields on all of these are grounded to the engine block. An additional measure of prevention is typically added by placing the radio antenna some distance from the engine and connecting the antenna to the radio with a shielded coaxial cable, grounding its outer braid at each end.

Bonding and grounding. Bonding is the term used to describe the task of connecting metal objects, wiring or other conductive materials in such a way that electrical energy is easily transferred with a minimum of resistance. Grounding is the term used to describe the task of connecting metal objects, wiring or other conductive materials to the aircraft primary structure in such a way that a return path for electrical current is established, completing the circuit. An aircraft primary structure includes the spars, stringers, ribs and other structural members that form the fuselage or wing structure of the aircraft, and is commonly referred to as airframe ground. See Figure 4-3-22.

A great many bonding and grounding connections are made in all aircraft electrical systems for a number of purposes:

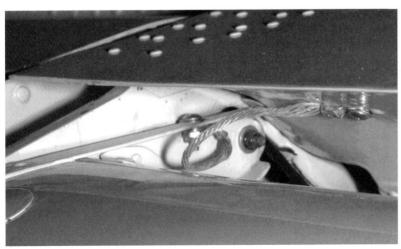

Figure 4-3-22. A bonding strap on an installed flap

- Provide protection to aircraft and personnel against hazards from lightning discharge

- Provide electrical current return paths

- Reduces the generation of radio-frequency electromagnetic voltages

- Provides added protection to personnel against electrical shock hazards

- Improves overall quality and stability of radio transmission and reception

- Prevents accumulations of electrostatic charges between two conductive surfaces

Follow each aircraft manufacturer's standard practices for bonding and grounding techniques where provided. When specific or detailed standard practices are not provided, use the following recommended general procedures and precautions to make bonding or grounding connections:

- Bond or ground equipment cases to the primary aircraft structure where practicable.

- Ensure that bonding or grounding procedures weakens no part of the aircraft structure.

- Bond parts individually where possible.

- Make bonding or grounding connections against smooth, clean surfaces.

- Make bonding or grounding connections in such a way that the connection will not break or loosen because of vibration, expansion and contraction (caused by heating and cooling of metal surfaces), or relative movement in normal service.

- Make bonding and grounding connections in protected areas of the primary airframe structure whenever possible.

- Bonding jumpers should be kept as short as practicable.

- Bonding jumper installations should not interfere with the operation of movable aircraft assemblies or control surfaces. Furthermore, normal movement of aircraft assemblies such as landing gear, control surfaces and doors should not result in damage to the bonding jumper.

Bonding and grounding connections are susceptible to deterioration caused by electrolytic action unless precautionary measures are taken. Corrosion may develop rapidly, especially in operational environments near warm coastal areas. Salt water acts as an electrolyte and accelerates oxidation of the metals at the connection, especially if dissimilar metals are in contact with one another. Left unchecked,

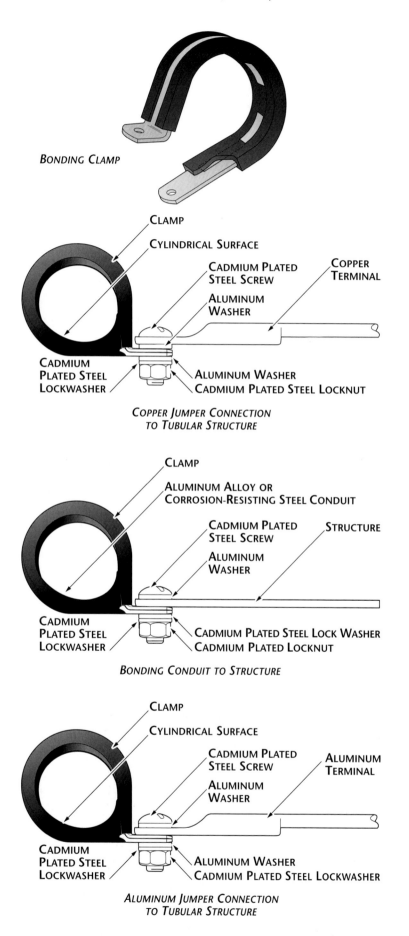

Figure 4-3-23. Three methods of attaching bonding clamps to tubular structure

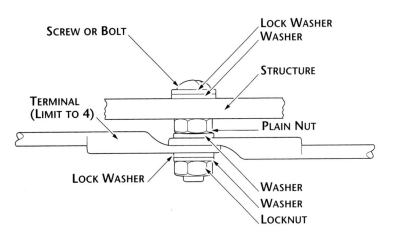

Figure 4-3-24. Grounding a stud to a flat surface

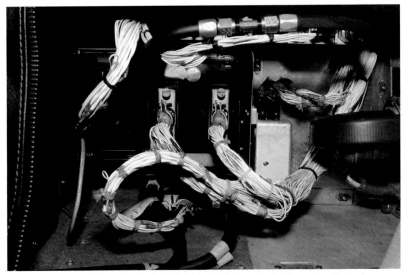

Figure 4-3-25. This avionics bay shows well laced wire bundles that also have enough slack to allow wire maintenance.

corrosion causes the electrical qualities of a bonded or grounded connection to develop high resistance and poor connectivity.

Aluminum alloy jumpers are recommended for most connections; however, copper jumpers can be used to bond together parts made of stainless steel, cadmium-plated steel, copper, brass or bronze. Where contact between dissimilar metals cannot be avoided, jumper and hardware material combinations should be chosen which minimize potential corrosion. In all cases, the part most likely to corrode should be the jumper or associated hardware rather than the airframe. Figure 4-3-23 illustrates a few recommended hardware combinations for making bonding connections.

At locations where finishes are removed, apply a protective finish. One such method is to anodize the area with products such as Alumiprep and Alodyne. This procedure greatly reduces corrosion tendencies at the completed con-

nection. Alodyne is a chromic acid conversion process that leaves a corrosion resistant film on aluminum surfaces.

Use of solder to attach bonding jumpers should be avoided since the soldering process actually creates conditions favorable for the development of corrosion. Where tubular structural members need to be bonded, use clamps to which a jumper is attached. Clamp material should be selected which minimizes the probability of corrosion. In addition, bonding jumpers that carry a substantial amount of ground return current should be large enough to meet the required current rating with a negligible voltage drop.

Several types of bonding and grounding connections are commonly used on flat surfaces, usually by means of through-bolts or screws where there is easy access for installation. One variation on bolted connections involves making a stud connection where a bolt or screw is locked securely to the structure, thus becoming a stud. See Figure 4-3-24. Grounding or bonding jumpers may then be removed or added to the shank of the stud (bolt) without removing the stud from the structure.

In locations where access to the nut for repairs would be difficult, a nut plate may be installed. Nut plates are riveted or welded to a clean area of the structure

If bonding or grounding connections are made to a tab riveted to a structure, it is very important to clean the bonding or grounding surface and make the connection as through it were being made to the structure. In the event that the tab is removed later, the rivets should be replaced with rivets one size larger, and the mating surfaces of the structure and the tab should be cleaned again to ensure removal of all corrosion, dirt and old anodic film.

Bonding or grounding connections can be made to aluminum alloy, magnesium or corrosion-resistant steel tubular structure, but exercise caution to distribute the screw and nut pressure by means of plain washers because of the ease, with which aluminum can be deformed.

Hardware used to make bonding or grounding connections should be selected based on mechanical strength, current to be carried and ease of installation. If aluminum or copper jumpers make connection to the structure of a dissimilar material, a washer of suitable material should be installed between the dissimilar metals so that any corrosion will occur on the washer, which is expendable. When repairing or replacing existing bonding or grounding connections, the same type of hardware used in the original connection should always be used.

The resistance of all bond and ground connections should be tested after connections are made before re-finishing. The resistance of each connection should normally not exceed 0.003 ohm. Resistance measurements need to be of limited nature only for verification of the existence of a bond, but should not be considered as the sole proof of satisfactory bonding. The length of jumpers, methods and materials used and the possibility of loosening the connections in service should be considered.

Lacing, Tying and Bundling

In addition to clamps and conduit, virtually every aircraft wiring installation is secured with cable ties or lacing cord, using techniques known as lacing, tying and bundling.

Lacing. *Lacing* is the securing together of a group or bundle of wires by a continuous piece of cord forming loops at regular intervals around the group or bundle. Only one of the most common methods is shown in Figure 4-3-26B. For extensive information on the cord lacing procedure, go to Chapter 7 of FAA AC43.13-1B.

Tying. *Tying* is the procedure of securing together a group or bundle of wires by individual ties placed around the group or bundle at regular intervals. Today the preferred method is to use nylon wire ties; in the past lacing cord or rib stitching cord was common. Lacing cord is typically made of black or white multi-strand, waxed nylon in a narrow flat ribbon form. Nylon cord is used because it is moisture and fungus resistant.

Wire bundle. A *wire bundle* is composed of two or more wires or groups tied or laced together to form a secure, neat and efficient wiring installation, and to facilitate maintenance. Figure 4-3-25 shows well laced avionics bundles.

Wire groups and bundles are laced or tied with cord to provide ease of installation, maintenance and inspection.

A *wire group* defines two or more wires tied or laced together that are all part of one electrical system, such as strobe lighting, electric flaps or radar altimeter. It is common practice to tie all wire groups or bundles where the clamps/supports are more than 12 inches apart. Figure 4-3-26 illustrates a standard recommended procedure for tying a wire group or bundle, which you will encounter in virtually every aircraft wiring installation. Start the tie by wrapping the lacing cord around the wire group to form a clove hitch knot (Figure 4-3-26A). Next, tie a square knot with an extra loop and trim the free ends of the cord. The lacing is then continued at regular intervals with half hitches along

the wire group or bundle (Figure 4-3-26B) and at each point where a wire or wire group branches off to an electrical device or connection. The half hitches should be spaced to make the bundle neat and secure.

You may on occasion make temporary ties using colored cord to fabricate a wire group or bundle on a workbench or hangar floor for installation in the aircraft later.

A recommended procedure for tying a wire group that branches off the main wire bundle is shown in Figure 4-3-27. The branch-off tie is located on the main bundle just past the

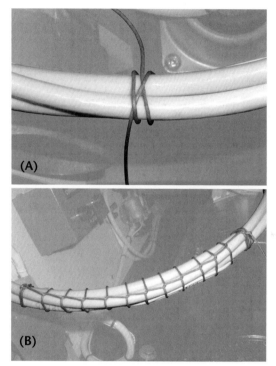

Figure 4-3-26. (A) A clove hitch is used as a start knot, (B) A wiring bundle laced with cord

Figure 4-3-27. Tying off a wire branch

branch-off point. Continue the lacing along the branched-off wire group, using regular spacing. Space the ties so that the bundle is laced up neatly and securely.

Whether laced or tied, wire bundles should be secured to prevent slipping, but not so tightly that the cord cuts into or deforms the insulation. This applies especially to coaxial cable, which has a soft dielectric insulation between the inner and outer conductor. Portions of a wire group or bundle located inside a conduit

are normally not tied or laced, but where wire groups or bundles are routed inside enclosures such as a junction box, use lacing or tying procedures only.

Conductor Termination

Aircraft wiring and cable must be cut to length in order to facilitate easier installation, maintenance and repair. To do this, wire and cable runs in aircraft are broken at specified locations by junctions, including connectors, terminal blocks or buses. All wires and cables should be cut to the lengths specified on aircraft manufacturer or equipment manufacturer drawings and wiring diagrams. Make the cut clean and square, using tools recommended by the wire manufacturer to avoid deformation of the wire or cable. It may be necessary to restore the shape of some large-diameter multi-stranded wires after cutting. Good cuts depend on cutting tool blades that are sharp and free from nicks. A dull blade will deform and extrude wire ends.

Stripping wire insulation. Preparing wire for assembly to connectors, terminals or splices requires that the insulation be removed from a short length of the conductor to expose the bare conductor. This procedure is known as stripping and is accomplished in a number of ways. Copper wire can be stripped with a variety of tools and according to the wire size and insulation type. Care must be exercised to avoid accidentally cutting or nicking the individual copper strands since this compromises

Figure 4-3-28. A common light duty hand operated wire stripper

Figure 4-3-29. Wire stripping with a hand operated tool

the integrity of the connection. Aluminum wire, on the other hand, must be stripped very carefully, using extreme care because individual aluminum wire strands have a tendency to break very easily if they have been nicked or damaged in any way.

Follow these general precautions when stripping any type of wire:

- When using any type of wire stripper, hold the wire so that it is perpendicular to cutting blades.

- Adjust automatic stripping tools carefully and follow the manufacturer's instructions to avoid nicking, cutting or otherwise damaging strands. This is especially important for aluminum wires and for copper wires smaller than No. 10. Examine stripped wires for damage, then cut off and re-strip (if length is sufficient), or reject and replace any wires having more than the allowable number of nicked or broken strands listed in the manufacturer's instructions.

- Make sure insulation is clean-cut with no frayed or ragged edges and trim if necessary.

- Make sure all insulation is removed from stripped area. Some types of wires are supplied with a transparent layer of insulation between the conductor and the primary insulation. Any remaining insulation will cause a defective connection, so ensure all insulation has been removed. If necessary, examine the stripped wire with a magnifying glass.

- When using hand strippers to remove lengths of insulation longer than 3/4-inch, it is easier to accomplish in two or more operations.

- Re-twist copper strands by hand or with pliers, if necessary, to restore natural lay and tightness of strands. Ensure hands and any tools used are clean to avoid contaminating the wiring.

The majority of wire-stripping tasks will normally be done with a pair of hand wire strippers. The tool shown in Figure 4-3-28 is the wire stripping tool most commonly encountered, and may be used on most wiring types. The cutting blades are removable and are replaceable when they become dull. In addition, several choices of cutting blades are available to handle a variety wire sizes.

Safely stripping wires with a hand stripper is a skill made easier by referring to the following suggestions and Figure 4-3-29.

- Insert the wire into exact center of correct cutting slot for wire size to be stripped. Each slot is marked with wire size. Move the wire until the length extending from the blade side of the stripping tool is the recommended length of bare wire required to complete the electrical connection.

- Slowly squeeze the stripper handles together until the clamp side of the tool securely holds the wire by its insulation and watch the blades bite into the insulation while continuing to squeeze the handles. When enough pressure has been applied, the stripping tool jaws will move apart, removing the insulation and revealing bare wire.

- Once the tool's jaws have moved as far apart as possible, quickly release pressure on the handles, allowing the clamping jaw and cutting jaw to fully open. Remove the stripped wire and proceed with the remaining steps to complete the connection.

- There are many different brands and types of commercial wire strippers and crimping pliers, including many that will do both (Figure 4-3-30). In almost all cases they will not be able to do a correct job, because they cannot be calibrated. In the case of commercial strippers, they will invariably nick the wire.

Wire splices and terminal connectors.
Splicing of electrical conductors should be kept

Figure 4-3-30. A selection of commercial wiring tools

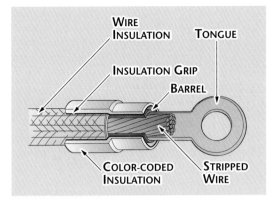

Figure 4-3-31. A wire terminal shown in a sectional view

to a minimum and avoided entirely in locations subject to extreme vibrations. Individual wires within a group or bundle may be spliced, as long as the completed splice is located where it can be visually inspected as needed. Splices should be staggered to prevent the bundle from becoming excessively enlarged or distorted in shape.

Many types of insulated and non-insulated aircraft splice connectors are manufactured by companies such as Amp, and are available for splicing individual wires. Self-insulated splice connectors such as the popular butt-splice are normally the preferred connector, however, a

Figure 4-3-32. Two sizes of ratcheting wire crimping tools

non-insulated splice connector may be specified for a certain application. You would normally cover non-insulated splice connectors with plastic sleeving, secured at both ends with lacing cord or a plastic tie. Solder splices have been used in the past, but they are particularly brittle and not usually recommended.

Whereas *solderless splices* are used to join electric wires to form permanent continuous runs, solderless terminal *lugs* permit easy and efficient connection to and disconnection from terminal blocks, bus bars or other electrical equipment. The great majority of solderless terminal lugs and splices are made of tin-coated copper or aluminum and are either pre-insulated or un-insulated, depending on the desired application. Most manufacturers color code the insulation on a splice or lug to indicate the wire size for which it is made.

Terminal lugs are generally available in three types for use in different space conditions. These are the flag, straight and right-angle lugs. The most commonly used terminal lug is a straight ring lug, designed to slide over the shank of a connecter stud. Terminal lugs are *crimped* (sometimes called *staked* or *swaged*) to the wires by means of hand or power crimping tools.

Copper wiring is terminated with solderless, pre-insulated straight copper terminal lugs. The insulation is part of the terminal lug and extends beyond its barrel so that it will cover a portion of the wire insulation, making the use of an insulation sleeve unnecessary.

The most popular pre-insulated terminal lugs contain an insulation grip (a metal reinforcing sleeve) beneath the insulation for extra gripping strength on the wire insulation. Color-coded insulation is used to identify the wire sizes that can be terminated with each of the terminal lug sizes. See Figure 4-3-31.

In addition to special wire stripping tools, special crimping tools are needed for installing splices and terminals on the ends of wire. While hand-held crimping tools are the most widely used in aircraft maintenance, aircraft manufacturing facilities may use portable power or stationary power tools for making thousands of crimped terminals quickly. These tools crimp the barrel of the typical terminal lug to the conductor and simultaneously crimp the insulation grip to the wire insulation for a very efficient and reliable electrical connection.

Better hand crimping tools all have a self-locking ratchet that prevents opening the tool until the crimp is complete. See Figure 4-3-32. Higher quality hand crimping tools are equipped with a "nest" of various size inserts to fit different size terminal lugs. All types of hand crimping tools need to be calibrated on a regular basis with gauges to ensure the proper adjustment of crimping jaws and a consistently reliable electrical connection.

A properly crimped terminal should provide a joint between the wire and the terminal that is as strong as the tensile strength of the wire itself.

Proper insertion of both the wire and the terminal in the crimping tool is essential for a good crimp. Figure 4-3-33 shows the correct placement of wire and lug in the hand tool. After stripping the wire insulation to proper length, insert the terminal lug tongue first into hand tool crimping jaws until the terminal lug barrel butts flush against the tool stop. Insert the stripped wire into the terminal lug barrel until the wire insulation butts flush against the end of the barrel, and then squeeze the tool handles until the ratchet releases. Remove the completed assembly and examine it for proper crimp. Figure 4-3-34 shows a good crimp on both the terminal and the insulation.

Un-insulated terminal lugs must be insulated after crimping. Before the days of shrink tubing, pieces of transparent flexible tubing called *sleeves* were used. The tubing was also called *spaghetti*. They are stilled used today. The sleeve provides electrical and mechanical protection at the connection. Certain types of pre-insulated lugs known as "hand-shakes" need this sleeve over the exposed portion of the two lugs where they "shake hands". When the size of the sleeving used is such that it will fit tightly over the terminal lug, the sleeving need not be tied; otherwise, it should be tied with lacing cord as illustrated in Figure 4-3-35.

Aluminum terminal lugs must be used with aluminum wire. You may recall that use of aluminum wire has both advantages and disadvantages. It has the advantage of being lighter in weight than copper, however, when aluminum is subjected to bending, it develops *work hardening*, making it brittle. This results in failure or breakage of strands much sooner than in a similar case with copper wire. Aluminum also forms a high-resistance oxide film immediately upon exposure to air. To counteract disadvantages, it is very important to use the most reliable installation procedures.

Use only aluminum terminal lugs to terminate aluminum wires. All aluminum terminal lugs incorporate an inspection hole that permits checking the depth of wire insertion. The barrel of aluminum terminal lugs is filled with a petrolatum-zinc dust compound. This compound removes the oxide film from the

Figure 4-3-33. Using the terminal stop on an AMP ratcheting crimping tool

Figure 4-3-34. A properly crimped ring terminal

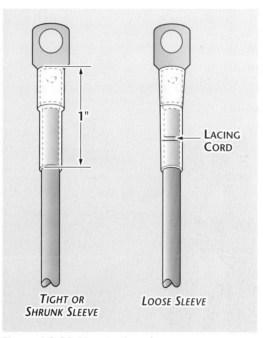

Figure 4-3-35. How to tie a sleeve on a connector

aluminum by a grinding process during the crimping operation. The compound will also minimize later oxidation of the completed connection by excluding moisture and air. The compound is retained inside the terminal lug barrel by a plastic or foil seal at the end of the barrel. See Figure 4-3-36.

As stated earlier, self or pre-insulated splice connectors such as the popular butt-splice are normally the preferred connector. These permanent copper splices join small wires of sizes

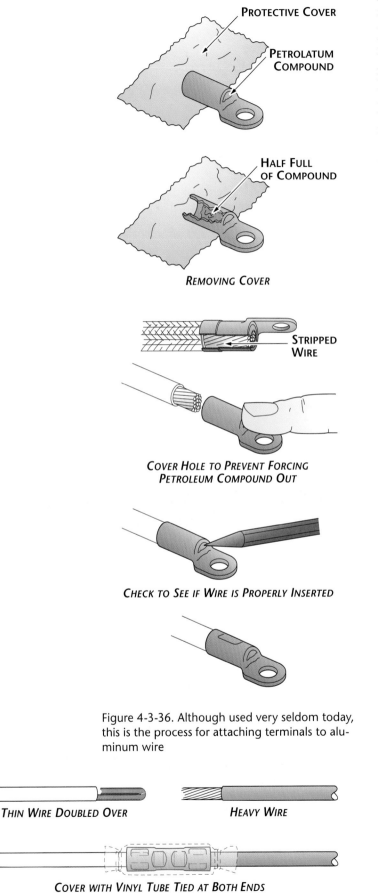

Figure 4-3-36. Although used very seldom today, this is the process for attaching terminals to aluminum wire

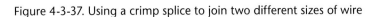

Figure 4-3-37. Using a crimp splice to join two different sizes of wire

22 through 10 and each splice size can be used for more than one wire size. Splices are usually color-coded in the same manner as the pre-insulated small tin-coated copper ring-terminal lugs. Splices are also used to connect one wire size with another, as depicted in Figure 4-3-37. The same crimping tools used for the pre-insulated ring terminal may be used to crimp this type of splice. The crimping procedures are the same except that the crimping operation must be done twice, once for each end of the splice.

Emergency splicing repairs. Emergency repairs for damaged or broken aircraft wiring should always be made in accordance with the aircraft manufacturer's standard practices. This includes crimped splices, use of terminal lugs from which the tongue has been cut off, or by soldering together and potting broken strands (Figure 4-3-38). These repairs are applicable to copper wire, however damaged aluminum wire cannot be temporarily spliced. Bear in mind that the following repairs are for temporary emergency use only and should be replaced as soon as possible with permanent repairs. Some aircraft manufacturers may prohibit emergency repairs, and in those cases, the manufacturer's procedures take precedence and must be followed.

In rare cases when neither a permanent splice nor a terminal lug is available, a broken wire can be repaired with solder and potting compound (if permitted by the aircraft manufacturer):

- Install a piece of plastic sleeving about 3 inches long, and of the proper diameter to fit loosely over the insulation, on one piece of the broken wire.

- Strip approximately 1 1/2-inch from each broken end of the wire.

- Lay the stripped ends side by side and twist one wire around the other with approximately four turns.

- Twist the free end of the second wire around the first wire with approximately four turns. Solder wire turns together, using 60/40 tin-lead resin-core solder.

- When solder is cool, draw the sleeve over the soldered wires and tie at one end. If potting compound is available, fill the sleeve with potting material and tie securely.

- Allow the potting compound to set without touching for the amount of time recommended by the potting compound manufacturer. Check potting compound instructions for the length of time needed to achieve full cure and optimum electrical characteristics.

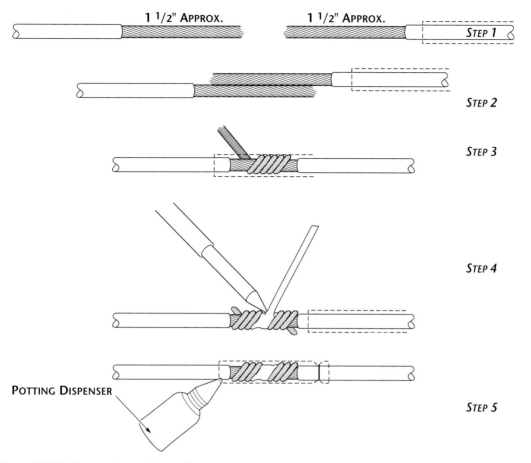

1 1/2" APPROX.　1 1/2" APPROX.

STEP 1

STEP 2

STEP 3

STEP 4

POTTING DISPENSER

STEP 5

Figure 4-3-38. How to do a solder splice

Terminal block and barrier strip connections. Recommended methods should be used when installing terminal lugs on terminal blocks (sometimes known as barrier strips) to enhance security and prevent loosening. This is important for counteracting the loosening effects of vibration and temperature swings that cause expansion and contraction of metal. When the lug terminals are physically "locked" in the correct position, they cannot move in a direction that causes loosening of the hardware. See Figure 4-3-39.

Terminal blocks are normally supplied with studs secured in place by a plain washer, an external tooth lockwasher and a nut. In connecting terminals, a recommended practice is to place copper terminal lugs directly on top of the nut, followed with a plain washer and elastic stop nut, or with a plain washer, split steel lockwasher and plain nut.

Aluminum terminal lugs should be placed over a plated brass plain washer, followed with another plated brass plain washer, split steel lockwasher and plain nut or elastic stop nut. The plated brass washer should have a diameter equal to the tongue width of the aluminum terminal lug. Consult the manufacturer's instructions for recommended dimensions of

Figure 4-3-39. A correctly installed terminal barrier block

these plated brass washers. Do not place any washer in the current path between two aluminum terminal lugs or between two copper terminal lugs. Also, do not place a lockwasher directly against the tongue or pad of the aluminum terminal.

Figure 4-3-40. A junction box with a series of barrier strips.

To join a copper terminal lug to an aluminum terminal lug, place a plated brass plain washer over the nut which holds the stud in place; follow with the aluminum terminal lug, a plated brass plain washer, the copper termi-

nal lug, plain washer, split steel lockwasher and plain nut or self-locking, all-metal nut. As a general rule, use a torque wrench to tighten nuts to ensure sufficient contact pressure. Manufacturer's instructions provide installation torques for all types of terminals.

Junction boxes. In many cases, terminal strips/barrier strips are located in a protective enclosure known as a junction box. A typical junction box is constructed of an aluminum alloy or fiberglass and is mounted in a protected area of the airframe. When junction boxes are located in a fire zone, or engine compartment, they are made of stainless steel. When making connections or troubleshooting an electrical problem in a junction box, exercise caution, especially if any circuitry passing through the junction box must be powered up for troubleshooting. Metal tools, wristwatches and rings are metal, and are excellent conductors which, if accidentally placed between a power source and ground, can become welded in place, causing serious burns or delivering a deadly shock. Figure 4-3-40 is a junction box with six rows of barrier strips.

Connectors

Electrical connectors, consisting of plugs and receptacles, are critical components in an aircraft wiring system that facilitate electrical system maintenance and troubleshooting, especially if frequent disconnection is anticipated. Connectors are available in a large number of types, sizes and styles. Several popular connector manufacturers include Bendix/Amphenol, Deutsche, Burndy and ITT Cannon. At one time Cannon products were so prevalent that multi-pin aircraft electrical connectors were commonly called "cannon plugs". Historically, electrical connectors have been designed and designated by various military specifications, and an astounding number of connector types are available. Figure 4-3-41 shows a collection of various connectors in an engine compartment.

Some connectors have removable pins, into which wires are either crimped or soldered whereas other connectors have captive pins that are permanently potted in place. Connectors have been particularly vulnerable to corrosion in the past due to condensation within the shell. Because of that, special connectors with waterproof features have been developed to replace non-waterproof connectors in areas where moisture causes a problem. When replacing a connector, you must either replace it with another of the same basic type and design, or use a newer style approved by the manufacturer.

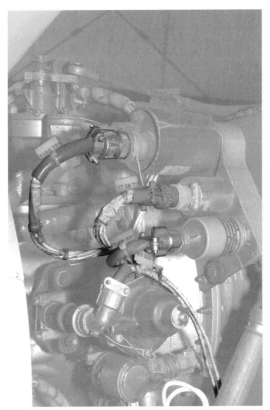

Figure 4-3-41. Wide varieties of connectors are necessary in every wiring installation.

In some cases, aircraft manufacturers may specify a chemically inert waterproof jelly treatment for connectors that are susceptible to corrosion difficulties. As a rule of thumb, the receptacle socket-type insert (female side) should be used for the side of the connection that remains *live* or *hot* after the connector halves are disconnected. This helps to prevent unintentional grounding of a power source should a protruding pin contact airframe ground.

Connector Type Identification

A comprehensive catalog of every mil-spec electrical connector ever made would fill an entire bookcase. Fortunately, the only group of electrical connectors need to be familiar with are the well known 'AN' or 'MS' style cylindrical connectors that are based on one of two military specifications.

MIL-C-5015. *MIL-C-5015* applies mainly to connectors used in aircraft and aeronautical equipment several decades old. This specification covers a series of cylindrical connectors that range from less than one inch in diameter to nearly three inches, and contain from 1 to 100 contacts, or pins. This series has been in use for many years, and the numbering system for these connectors has remained essentially unchanged.

Some distinguishing features of *MIL-C-5015* connectors include large, prominent "male" pins and "female" pins (socket style), collectively referred to as contacts. Screw threads used for connecting the two connector halves are available in either coarse or fine threads.

MIL-C-26482, Series 1 and Series 2. A newer series of connectors, *MIL-C-26482*, Series 1 and Series 2, are physically smaller in size compared to *MIL-C-5015* series, but the largest number of contacts available is 61. They are part of the 'MS' family of part numbers, but also include part number prefixes such as 'PT,' 'KPT,' 'KPSE,' and others. Pin or contact options include both non-removable solder cup and crimp-style removable. Crimp style connectors tend to be more reliable under some conditions, but are more difficult to assemble. They usually require specialized crimping, extraction and insertion tools for assembly and repair, driving up the cost. MIL-C-26482 series pins are usually small with gold plating and use a 'Bayonet' style locking ring instead of screw threads.

AN connectors. Connectors identified by the AN family of part numbers are also divided into classes with the manufacturer's variations in each class. The manufacturer's variations are differences in appearance and in the method of meeting a specification. There are five basic

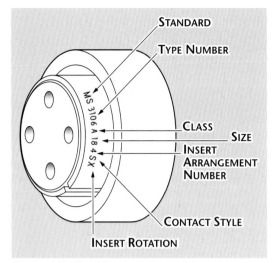

Figure 4-3-42. MS connector marking

classes of AN connectors used in most aircraft. Each class of connector has slightly different construction characteristics. Classes A, B, C and D are made of aluminum, and class K is made of steel. These classes are described as follows:

- **Class A.** Solid, one-piece back shell, general-purpose connector.

- **Class B.** Connector back shell separates into two parts lengthwise. Used primarily where it is important that the soldered connectors be readily accessible. The back shell is held together by a threaded ring or by screws.

- **Class C.** A pressurized connector with inserts that are not removable. Similar to a class A connector in appearance, but the inside sealing arrangement is sometimes different. It is used on walls of bulkheads of pressurized equipment.

- **Class D.** Moisture and vibration resistant connector which has a sealing grommet in the back shell. Wires are threaded through tight-fitting holes in the grommet, thus sealing against moisture.

- **Class K.** A fireproof connector used in areas where it is vital that the electric current is not interrupted, even though the connector may be exposed to continuous open flame. Wires are crimped to the pin or socket contacts and the shells are made of steel. This class of connector is normally longer than other classes of connectors.

Finding Part Numbers

Most connectors have their part numbers stamped or engraved somewhere on the insert or shell. Those on the insert are typically two digits or two digits and a letter, a dash and then two more digits followed by a letter. A few examples are 28-21S and 14S-5P.

The first pair of digits represents the shell size code. The second pair indicates the contact arrangement, and the last letter in the string indicates whether the connector contains pin contacts or socket contacts. A connector with visible male pins will always be designated 'P' in the last position, whereas a connector with female pins will always be designated 'S.' Figure 4-3-42 shows illustrations and part numbers for the most popular connectors.

MS3102A14S-5S. The first pair of characters, MS, indicates the connector series. This can be either a standard designation or something proprietary to a specific manufacturer. The four digits following the series, 3102, indicate the physical connector type. For the MIL-C-5015 spec, these are as follows:

- **3100** - A 3100 is a wall-mounted receptacle, designed to carry wiring through a bulkhead or wall. Provides strain relief for the wire bundle.

- **3101** - A 3101 is a cable-connecting plug/receptacle. These receptacles or plugs do not have mounting flanges. They are designed to be used on extension cables and provide strain relief.

- **3102** - A 3102 is a box-mounting receptacle. These differ from type 3100 in that they have no strain relief, and are designed to bring wiring in and out of a protected chassis. The 3102 series is the most common type of chassis receptacle used on military gear.

- **3106** - The 3106 is a straight plug. This is the most common series of plugs used to mate with the 3100 - 3102 series for control or power cables. These plugs can be equipped with a number of strain relief styles, including clamp, potting cups and watertight styles (MS-E series only).

- **3108** - The 3108 is the same as the 3106, but has a backshell that allows wiring to exit at 90 degrees.

The next single character indicates the connector class:

- **A** - Most common and least expensive, has a solid endbell

- **B** - Same as Class A, but with a split endbell

- **E** - Rated for hostile environments, such as moisture, oils, grease, etc.

- **F** - Replaces Class E in some types of connectors

The next two or three characters, 14S, identify the shell size. This code does not always have a letter following it. The remaining pair of characters, 5S, identifies the contact arrangement. If this code is followed by 'S,' it indicates internal-style pins (socket). If it is followed by P, it indicates external contacts (plug). Examples of MS connectors are shown in Figure 4-3-43.

The numbers for the newer 26482 spec connectors are nearly identical to those described above. The differences are mainly in the four-digit string that identifies the connector type. Here are some examples:

- **3116** - Equivalent to 3106, but sized under MIL-C-26482 instead of MIL-C-5015. Straight plug, solder cup non-removable pins

- **3126** - Same as 3116, but has crimp-style removable pins.

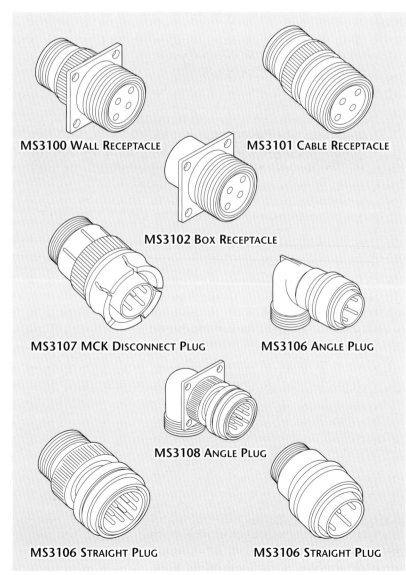

MS3100 WALL RECEPTACLE

MS3101 CABLE RECEPTACLE

MS3102 BOX RECEPTACLE

MS3107 MCK DISCONNECT PLUG

MS3106 ANGLE PLUG

MS3108 ANGLE PLUG

MS3106 STRAIGHT PLUG

MS3106 STRAIGHT PLUG

Figure 4-3-43. Examples of MS connectors

Connector/Cable Fabrication

Constructing any type of wire bundle with connectors is an art which requires the proper tooling to produce a quality product. Solder-type pins require good soldering techniques, skills and equipment, but for assembly of crimp-style connectors, use of recommended tooling is critical. Using incorrect tooling can result in early failure or poor reliability of connectors and equipment, a scenario that could end up being a lot more costly in the long term.

Before fabricating a coax cable, check the equipment manufacturers instructions. Some units require a cable of a specific length, while most are simply routed as direct as possible.

Installation of Connectors

The following procedures outline one recommended method of assembling connectors to receptacles:

1. Locate the proper position of the plug in relation to the receptacle by aligning the key of one part with the groove or keyway of the other part.

2. Start the plug into the receptacle with a light forward pressure and engage the threads of the coupling ring and receptacle.

3. Alternately push in the plug and tighten the coupling ring until the plug is completely seated.

4. Use special connector pliers to tighten coupling rings one sixteenth to one eighth turn beyond finger-tight if space around the connector is too small to obtain a good finger grip.

5. Never use force to mate connectors to receptacles. Do not hammer a plug into its receptacle; and never use a torque wrench or pliers to lock coupling rings.

A connector is generally disassembled from a receptacle in the following manner:

1. Use connector pliers to loosen coupling rings which are too tight to be loosened by hand.

2. Alternately pull on the plug body and unscrew the coupling ring until the connector is separated.

3. Protect disconnected plugs and receptacles with caps or plastic bags to keep debris from entering and causing faults.

4. Do not use excessive force, and do not pull on attached wires.

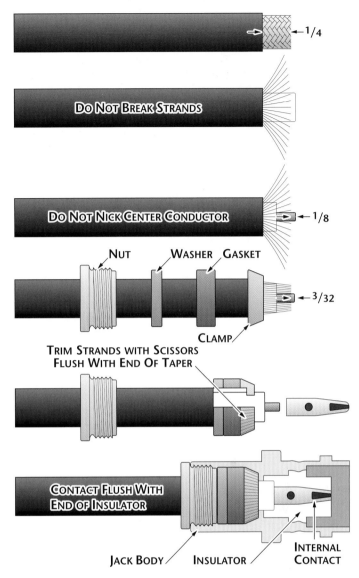

Figure 4-3-44. This figure shows the necessary steps to install the most common type of BNC connector

Coaxial Connectors

A great many connectors have been developed for coaxial and triaxial cables (a center conductor with two separate shields), but the connector most commonly seen terminating coaxial cable is the BNC connector. Developed as a miniature version of the Type C connector just prior to 1950, BNC is an abbreviation for Bayonet, Neill Concelman, named after Amphenol engineer Carl Concelman. The BNC product line is a miniature quick connect/disconnect RF (radio frequency) connector, which makes it extremely useful for removing and installing avionics equipment. BNC connectors feature two bayonet lugs on the female connector that lock the two connector halves with only a quarter turn of the coupling nut. BNC's are ideally suited for cable termination for miniature and subminiature coaxial cable types such as

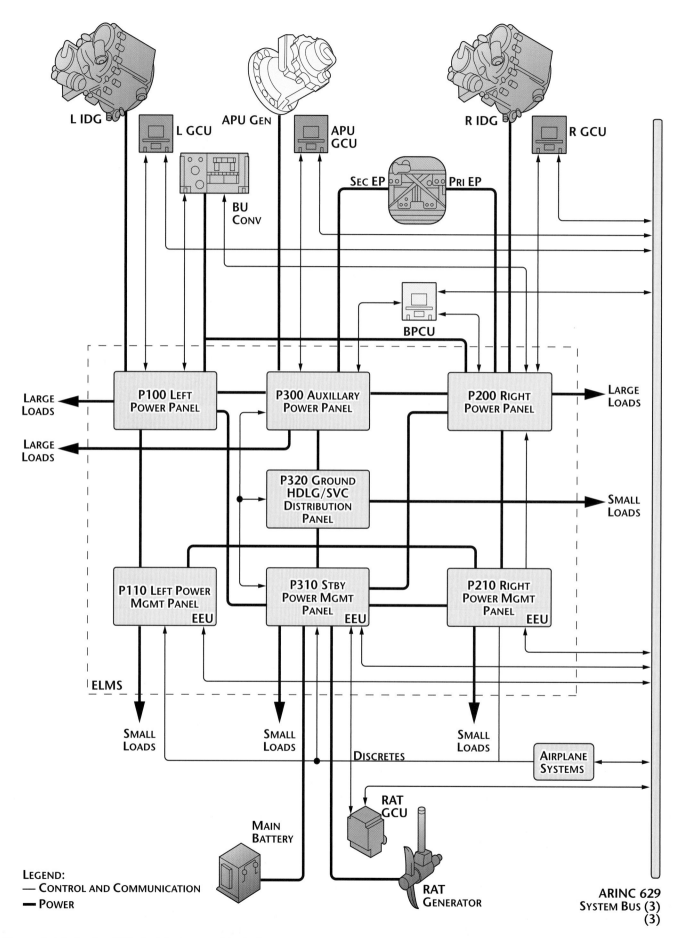

Figure 4-4-1. Boeing 777 electrical power block diagram

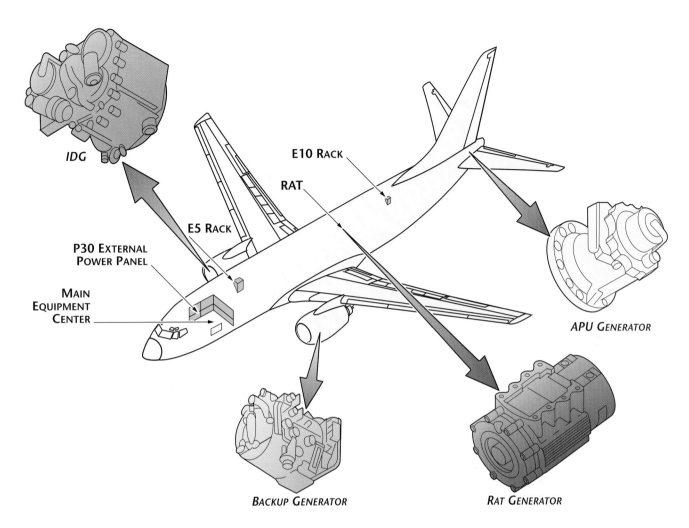

Figure 4-4-2. Boeing 777 electrical power pictorial diagram

RG-58, RG-59, RG-179 and RG-316. Figure 4-3-44 shows the steps necessary to install a removable BNC connector.

Amphenol 50 Ω BNC connectors are miniature, lightweight units designed to operate up to 11 GHz and typically yield low reflection through 4 GHz. Designed to accommodate a large variety of RG and industry standard cables, BNC connectors are available in crimp/crimp, clamp/solder, SURETWIST® and field serviceable termination styles.

Another line of BNC connectors with 75 Ω specifications meets the need for higher performance impedance-matched cable interconnections. These connectors can be used in a variety of applications where 75 Ω impedance connections are needed to insure low signal distortion.

Part numbers that are listed with the appropriate M39012 number are military grade connectors produced in accordance with and actively qualified to the military specification MIL-C-39012.

Section 4
Wiring Diagrams

Wiring diagrams are the *electrical roadmaps* that make it possible to see how all the aircraft electrical wiring and electrical components relate and connect to one another. Several types of wiring diagrams that were designed for specific purposes have been in use for many years.

The simplest of all electrical wiring diagrams is the *block diagram* that typically shows very few component symbols and very little detail, but rather provides a basic understanding and overview of the system design. When a new aircraft electrical system is first learned, the familiarization training usually includes several block diagrams to quickly give a large overview of the system at hand (Figure 4-4-1).

For example, the following Boeing 777 electrical power block diagram clearly shows the three main sources of electrical power, the left

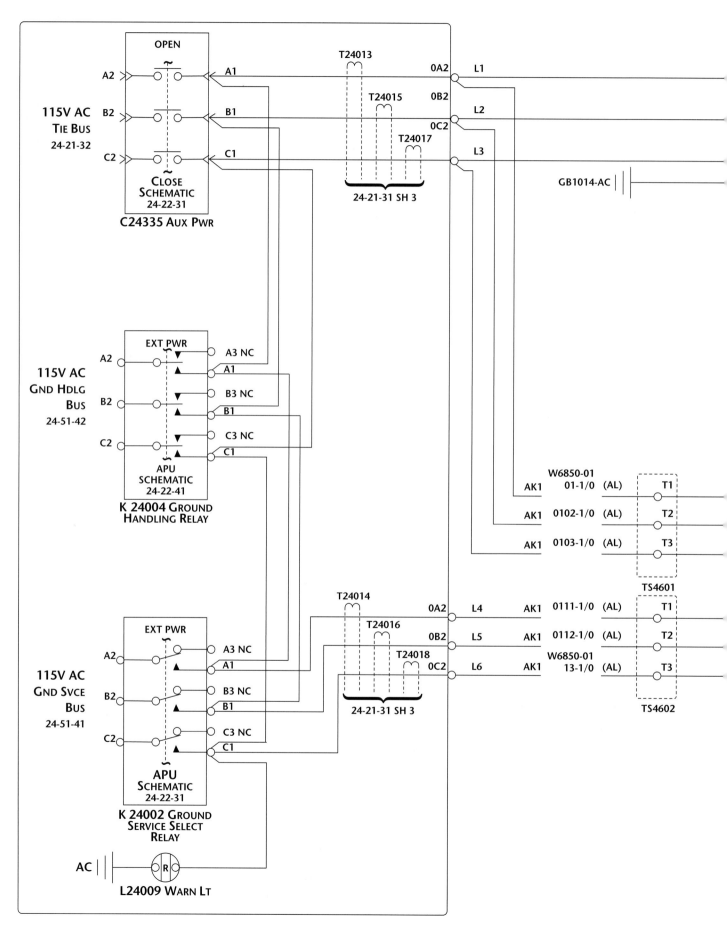

Figure 4-4-3. Boeing 777 APU generator power & regulation schematic diagram

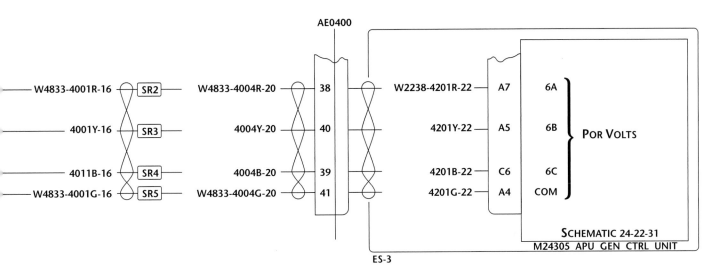

AE0400

W4833-4001R-16 — SR2 — W4833-4004R-20 — 38

4001Y-16 — SR3 — 4004Y-20 — 40

4011B-16 — SR4 — 4004B-20 — 39

W4833-4001G-16 — SR5 — W4833-4004G-20 — 41

W2238-4201R-22 — A7 — 6A

4201Y-22 — A5 — 6B

4201B-22 — C6 — 6C

4201G-22 — A4 — COM

POR VOLTS

SCHEMATIC 24-22-31

M24305 APU GEN CTRL UNIT

ES-3

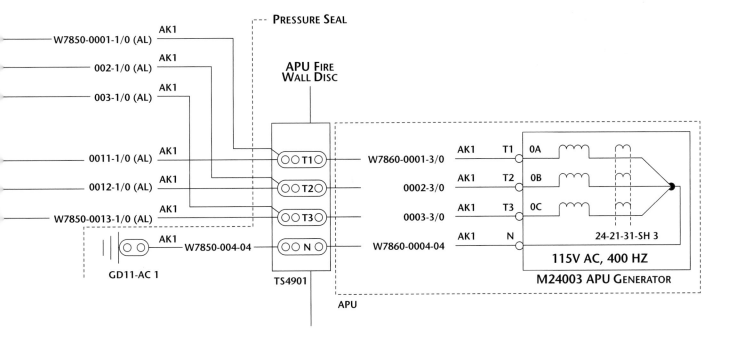

PRESSURE SEAL

APU FIRE WALL DISC

W7850-0001-1/0 (AL) AK1

002-1/0 (AL) AK1

003-1/0 (AL) AK1

0011-1/0 (AL) AK1

0012-1/0 (AL) AK1

W7850-0013-1/0 (AL) AK1

AK1 W7850-004-04

GD11-AC 1

T1 W7860-0001-3/0 AK1 T1 0A

T2 0002-3/0 AK1 T2 0B

T3 0003-3/0 AK1 T3 0C

N W7860-0004-04 AK1 N

24-21-31-SH 3

115V AC, 400 HZ

M24003 APU GENERATOR

TS4901

APU

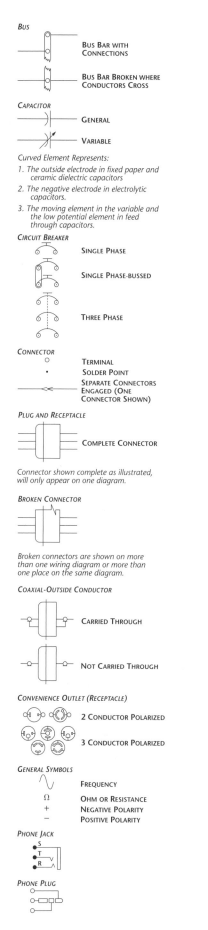

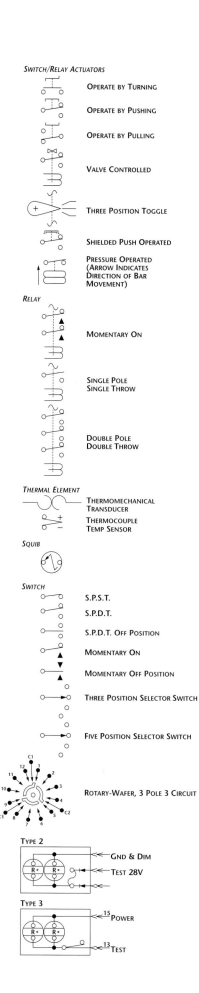

Figure 4-4-4. Boeing 737 schematic symbols

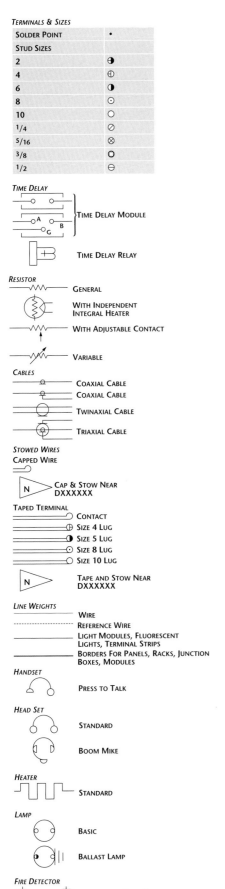

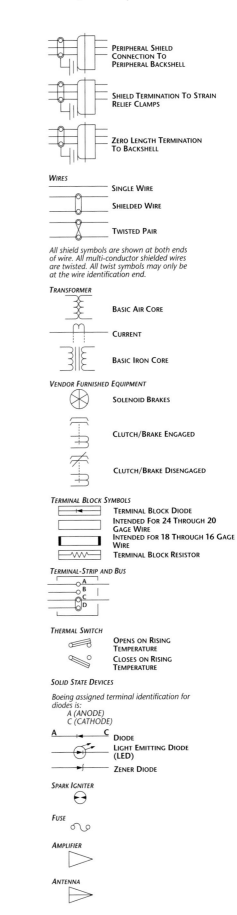

TERMINALS & SIZES

SOLDER POINT	•
STUD SIZES	
2	
4	
6	
8	
10	
1/4	
5/16	
3/8	
1/2	

TIME DELAY

TIME DELAY MODULE

TIME DELAY RELAY

RESISTOR

GENERAL

WITH INDEPENDENT INTEGRAL HEATER

WITH ADJUSTABLE CONTACT

VARIABLE

CABLES

COAXIAL CABLE
COAXIAL CABLE
TWINAXIAL CABLE
TRIAXIAL CABLE

STOWED WIRES
CAPPED WIRE

CAP & STOW NEAR
DXXXXXX

TAPED TERMINAL

CONTACT
SIZE 4 LUG
SIZE 5 LUG
SIZE 8 LUG
SIZE 10 LUG

TAPE AND STOW NEAR
DXXXXXX

LINE WEIGHTS

WIRE
REFERENCE WIRE
LIGHT MODULES, FLUORESCENT LIGHTS, TERMINAL STRIPS
BORDERS FOR PANELS, RACKS, JUNCTION BOXES, MODULES

HANDSET

PRESS TO TALK

HEAD SET

STANDARD

BOOM MIKE

HEATER

STANDARD

LAMP

BASIC

BALLAST LAMP

FIRE DETECTOR

CONTINUOUS LOOP ENGINE FIRE DETECTOR

1 WIRES SHIELDED (A)
2 WIRES SHIELDED (B)
4 WIRES TWISTED (C)

SHIELDED WIRE - SHIELD GROUNDED

OVERBRAID SYMBOL - SHIELDED, TWISTED AND SINGLE WIRES SHOWN

OVERBRAID SYMBOL

ARINC 629

ARINC 629

COUPLER

COUPLER

DATA BUS PANEL- BSYSB CROSS CABINET

BATTERY

SINGLE-CELL

MULTI-CELL

AC THREE PHASE

FLAG NOTES

OR GENERAL

SERVICE BULLETIN FLAG

CUSTOMER ORIGINATED CHANGE FLAG

PERIPHERAL SHIELD CONNECTION TO PERIPHERAL BACKSHELL

SHIELD TERMINATION TO STRAIN RELIEF CLAMPS

ZERO LENGTH TERMINATION TO BACKSHELL

WIRES

SINGLE WIRE

SHIELDED WIRE

TWISTED PAIR

All shield symbols are shown at both ends of wire. All multi-conductor shielded wires are twisted. All twist symbols may only be at the wire identification end.

TRANSFORMER

BASIC AIR CORE

CURRENT

BASIC IRON CORE

VENDOR FURNISHED EQUIPMENT

SOLENOID BRAKES

CLUTCH/BRAKE ENGAGED

CLUTCH/BRAKE DISENGAGED

TERMINAL BLOCK SYMBOLS

TERMINAL BLOCK DIODE
INTENDED FOR 24 THROUGH 20 GAGE WIRE
INTENDED FOR 18 THROUGH 16 GAGE WIRE
TERMINAL BLOCK RESISTOR

TERMINAL-STRIP AND BUS

THERMAL SWITCH

OPENS ON RISING TEMPERATURE
CLOSES ON RISING TEMPERATURE

SOLID STATE DEVICES

Boeing assigned terminal identification for diodes is:
A (ANODE)
C (CATHODE)

DIODE
LIGHT EMITTING DIODE (LED)
ZENER DIODE

SPARK IGNITER

FUSE

AMPLIFIER

ANTENNA

integrated drive generator, Right integrated drive generator and auxiliary power unit (APU) gvenerator. Each generator is controlled by its own Generator Control Unit (GCU) and provides power to its own power panel. In addition, several sources of emergency power are shown, the main aircraft battery and Ram Air Turbine (RAT) generator. It is also shown by the block diagram that there are paths for electrical load-sharing and a means of routing virtually any of the power sources to any of the power panels for redundancy and safety backup power.

Pictorial diagrams provide all the benefits of block diagrams, however, extra detail is added to show some wiring connections and some of the symbology adopted by the particular aircraft manufacturer. Pictorials are the next logical step used in gaining a more thorough understanding of an electrical system. See Figure 4-4-2.

Schematic diagrams are the most detailed of all electrical wiring/system diagrams. They rely heavily on circuitry symbols and require the greatest amount of interpretation in order to be useful. Each aircraft manufacturer provides a key or legend in their wiring manual to explain the meaning of each symbol used in the schematics. Schematic diagrams are the most heavily used of all drawings in troubleshooting electrical circuit problems because they provide the greatest amount of detail.

The 777 APU Generator Power & Regulation schematic, Figure 4-4-3, shows several relay symbols with wiring connections. The dashed lines labeled T24013, T24014, etc. are symbolic of current transformers, used for measuring current flow in each of the three phases of power generated by the APU generator.

Figure 4-4-4 contains the symbols used for the wiring diagrams of a Boeing 737. These are typical of all Boeing airplanes. In fact they are used in a myriad of other manufacturers drawings.

Section 5
Electrical Components

Switches and Relays

The function of any switch in an electrical circuit is to provide a simple and convenient method of connecting or disconnecting electrical loads (lighting, solenoids, motors or any other electrical device that consumes electrical power) from their power source. Switch design is driven by many factors, including voltage, current flow, switch location and ergonomics (for manually operated switches).

Several types of specially designed switches may be used for high current applications in power delivery circuits. One type has a snap-action design that enables rapid opening and closing of contacts (independent of the speed with which its operating toggle or plunger operates). This feature minimizes damage to the switch contacts caused by arcing and prolongs switch life. Another heavy current switch type known as a *contactor, relay or relay-switch* has

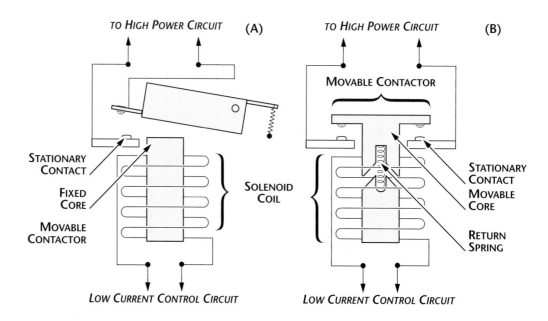

Figure 4-5-1. There are two types of relays: (A) fixed core and (B) moveable core.

very rugged construction and has sufficient contact capacity to make, break and carry continuous load current. Figure 4-5-1 shows both types of relays. These special switches have several unique features, described as follows:

- Ability to control the opening and closing of main switch contacts that carry enormous amounts of current using a very low power switch This isolated low-current control circuit increases operational safety of the switch.

- Remote operation. The low-power control circuit in a typical relay is a solenoid that is operated remotely by a small switch located some distance from the relay, usually somewhere on one of the cockpit instrument panels.

Relays are used as switching devices where a weight reduction can be achieved or electrical controls can be simplified.

A switch's nominal voltage and current rating is customarily stamped on the switch housing by the switch manufacturer. The nominal current rating represents the continuous current rating with the contacts closed that the switch is capable of carrying. When selecting a switch for a circuit based on nominal current ratings, bear in mind that the switch may be required to handle much higher amounts of momentary current when the switch contacts close and begin carrying current. The following circuit types fall into that category:

- **High current rush circuits.** Circuits containing incandescent lamps can draw an initial current that is 15 times greater than the continuous current. Switch contact burning or welding may occur if the switch has a current rating too low to handle this momentary rush of current. As the filament of a typical incandescent lamp heats up and begins producing light, its resistance simultaneously increases, reducing current flow.

- **Inductive circuit.** Magnetic energy stored in a solenoid coil or relay is released as the field around the coils collapses, generating a voltage high enough to cause current to arc across the switch contacts as the control switch is opened.

- **Motor circuits.** Direct-current motors will draw several times their rated current during the first few seconds of motion. As the motor spins up, counter emf is generated which reduces the current draw from the power source. In addition, magnetic energy stored in their armature and field coils induces a voltage spike and surge of current when the control switch is opened.

NORMAL SYSTEM VOLTAGE	TYPE OF LOAD	DERATING FACTOR
28 VDC	Lamp	8
28 VDC	Inductive (relay-solenoid)	4
28 VDC	Resistive (heater)	2
28 VDC	Motor	3
12 VDC	Lamp	5
12 VDC	Inductive (relay-solenoid)	2
12 VDC	Resistive (heater)	1
12 VDC	Motor	2

Notes:
1. To find the nominal rating of a switch required to operate a given device, multiply the continuous current rating of the device by the derating factor corresponding to the voltage and type of load.

2. To find the continuous rating that a switch of a given nominal rating will handle efficiently, divide the switch nominal rating by the derating factor corresponding to the voltage and type of load.

Table 4-5-1. This table provides a means of derating a switch when an additional choice must be made

An aircraft electrical systems engineer will be designing circuitry that would require making switch selections. An aircraft maintenance technician normally will conduct operational checks or troubleshoot an existing circuit. This activity may require switch replacement in accordance with the aircraft manufacturer maintenance manual. Persistent switch malfunctions, however, may indicate that a circuit needs a different switch, rated to carry higher voltages and current, and it is important to know how to select the proper switch.

Tables 4-5-1 and 4-5-2 will assist in the choice of a rating for a replacement switch.

Table 4-5-2 illustrates how to select the proper nominal switch rating when the continuous load current is known. This table provides a means of *derating* a switch to avoid damage from momentary rushes of current and to obtain reasonable switch efficiency and service life. Since a decision on how to position a switch at installation may also be necessary, standard conventions or configurations may apply in the absence of specific maintenance manual instructions.

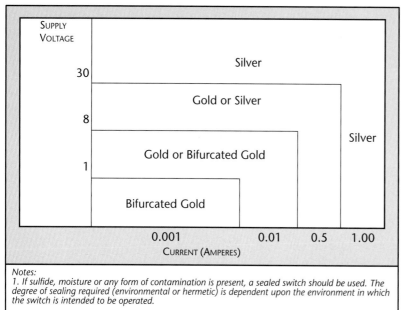

SUPPLY VOLTAGE		
30	Silver	
8	Gold or Silver	Silver
1	Gold or Bifurcated Gold	
	Bifurcated Gold	

CURRENT (AMPERES): 0.001   0.01   0.5   1.00

Notes:
1. If sulfide, moisture or any form of contamination is present, a sealed switch should be used. The degree of sealing required (environmental or hermetic) is dependent upon the environment in which the switch is intended to be operated.

2. If particle contamination in any form is likely to reach the contacts, bifurcated contacts should be used.

3. Low-voltage high-current loads are difficult to predict and may result in a combined tendency of noncontact, sticking, and material transfer.

4. High-voltage high-current applications may require the use of silver nickel contacts.

Table 4-5-2. Selection of contact material for a replacement switch

Standard convention for aircraft switch operation requires two-position *on-off* switches to be mounted so that an upward or forward movement of the switch toggle reaches the on position. When the switch controls movable aircraft elements, such as landing gear or flaps, the toggle should move in the same direction as the desired motion. In addition, a switch guard may be mounted over the switch to prevent inadvertent switch operation. These guards are commonly red in color to indicate that switch operation may result in a hazard to personnel or damage to equipment unless operational precautions are heeded.

As outlined in Chapter 5 of *Introduction to Aircraft Maintenance,* switches are designated by the number of poles, throws and positions they have. A pole of a switch is its movable blade or contactor. The number of poles is equal to the number of circuits, or paths for current flow, that can be completed through the switch at any one time.

The throw of a switch indicates the number of circuits, or paths for current, that it is possible to complete through the switch in each ON position. The number of ON positions a switch has is the number of places at which the operating device (toggle, plunger, etc.) will come to rest and at the same time close one or more circuits.

A *single-pole single-throw (SPST)* switch, as shown in Figure 4-5-2, makes it possible to complete only one circuit through the switch. A single-pole switch through which two circuits can be completed (not at the same time) is a *single-pole double-throw (SPDT)* switch. See Figure 4-5-3.

A switch with two contactors or poles, each of which completes only one circuit, is a *double-pole single-throw (DPST)* switch. A double-pole single-throw toggle-type switch is illustrated in Figure 4-5-4.

A double-pole switch that can complete two circuits, one circuit at a time through each pole

Figure 4-5-2. Single-pole, single throw switch (SPST)

Figure 4-5-3. Single-pole, double throw switch (SPDT)

Figure 4-5-4. Double-pole, single throw switch (DPST)

Figure 4-5-5. Double-pole, double throw switch (DPDT)

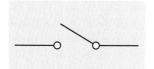

SINGLE - POLE SINGLE THROW

SINGLE - POLE DOUBLE THROW

DOUBLE - POLE SINGLE THROW

DOUBLE - POLE DOUBLE THROW

Figure 4-5-6. How the switches look in a schematic drawing

is a *double-pole double-throw (DPDT)* switch. A toggle-type switch is shown in Figure 4-5-5.

Schematic representations for the most commonly used switches are shown in Figure 4-5-6.

The switches depicted in Figures 4-5-2 through 4-5-5 are simple *toggle switches*; the most commonly used manually operated aircraft switches. Another very similar and very common aircraft switch is a *rocker switch*. They are available in a number of different styles and configurations. The handle of this switch rocks to the ON or OFF position, giving the switch its name.

Rotary switches provide the ability to select many different circuit connections in a small compact space. See Figure 4-5-7.

With the advent of low power control circuitry and computerized aircraft system operation, the *Micro switch* is being used in increasing numbers in new aircraft design. Until recently, micro switches were used mainly as limit switches to provide automatic control and precise positioning of landing gear, flap actuator motors, etc. A microswitch is shown in Figure 4-5-8.

Another type of switch popular with aircraft designers for precise positioning of landing gear, flaps and trim tabs is known as a *proximity switch*. Proximity switches operate using the principles of magnetism to operate an internal switch. When the "target" that is attached to a moving surface is positioned over the switch, the switch closes magnetically, completing the circuit.

Circuit Limiting Devices

Aircraft electrical systems should be protected with *circuit breakers* or *fuses* located as close as possible to the electrical power source bus. Normally, the manufacturer of the electrical equipment specifies the fuse or circuit breaker current rating to be used when installing equipment.

The only circuits that do not require protection are the main circuits of motors and those that do not present a hazard by their omission.

The circuit breaker or fuse should open the circuit before the wiring or equipment overheats to the point of emitting smoke. To accomplish this, the time current characteristic of the protection device must fall below that of the associated conductor. Circuit protector characteristics should be matched to obtain the maximum utilization of the connected equipment.

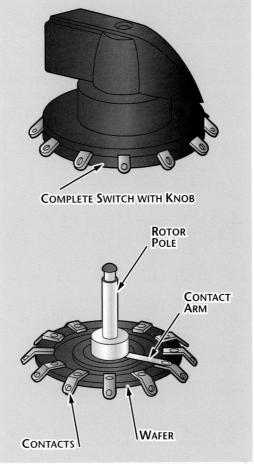

Figure 4-5-7. This illustration shows a complete rotary switch, and its contacts.

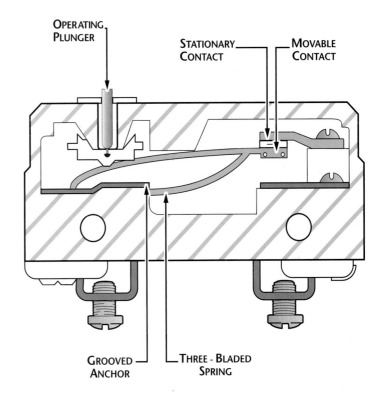

Figure 4-5-8. A microswitch has the ability to control motion to a very close measurement.

An example of a chart used in selecting the proper circuit breaker and fuse protection for copper conductors in a DC circuit is shown in Table 4-5-3.

Table 4-5-3 is a limited chart, applicable to a specific set of ambient temperatures and wire bundle sizes, and is presented as a typical example only. Safe circuit design depends on such guides when selecting the correct conductor and circuit breaker for a specific purpose.

All resettable circuit breakers should be installed at a point in the circuit they are protecting to be able to open the circuit regardless of the position of the operating control when an overload or circuit fault exists. Such circuit breakers are referred to as trip-free. Automatic reset circuit breakers automatically reset themselves after a circuit overload and should not be used as circuit protection devices in aircraft. See Figure 4-5-9.

WIRE AN GAUGE COPPER	CIRCUIT BREAKER AMPERES	FUSE AMPERES
22	5	5
20	7.5	5
18	10	10
16	15	10
14	20	15
12	30	20
10	40	30
8	50	50
6	80	70
4	100	70
2	125	100
1		150
0		150

Basis of chart:
(1) Wire bundles in 135°F ambient and altitudes up to 30,000 feet.
(2) Wire bundles of 15 or more wires, with wires carrying no more than 20 percent of the total current carrying capacity of the bundle as given in Specification MIL-W-5088 (ASG).
(3) Protectors in 75° to 85°F ambient
(4) Copper wire Specification MIL-W-5088
(5) Circuit breakers to Specification MIL-C-5809 or equivalent
(6) Fuses to Specification MIL-F-15160 or equivalent

Table 4-5-3. DC wire and circuit protector chart

Figure 4-5-9. Circuit breakers are available in many styles and capacities to meet a wide variety of electrical conditions.

Section 6

Electrical Power Sources and Monitoring

Proper installation and maintenance of an aircraft electrical system requires knowledge of electrical load limits, acceptable means of controlling or monitoring electrical loads, and circuit protection devices. Aircraft maintenance technicians have the responsibility to ensure that an aircraft electrical system will perform safely after scheduled maintenance, troubleshooting or new equipment installation.

Electrical Load Limits

When installing additional equipment that consumes electrical power in an aircraft, the total electrical load must be safely controlled or managed within the rated limits of the affected components of the aircraft's power-supply system.

Before any aircraft electrical load is increased, the associated wires, cables and circuit protection devices (fuses or circuit breakers) should be checked to determine that the new electrical load (previous maximum load plus added load) does not exceed the rated limits of the existing wires, cables or protection devices.

The generator or alternator output ratings prescribed by the manufacturer should be compared with the electrical loads that may be imposed on the affected generator or alternator by newly installed equipment. If a comparison shows that the probable total connected electrical load may exceed the output load limits of the electrical power generation capabilities of the aircraft generator(s) or alternator(s), the load should be reduced to prevent possible overload. In electrical systems that include

Figure 4-6-1. Beechcraft King Air 200 electrical system is midway up the complexity scale

Photo courtesy of Dynamic Aviation

a storage battery, ensure that the battery is continuously charged in flight, except when short, intermittent loads are connected such as a radio transmitter, a landing-gear motor or other similar devices that may place short-time demand loads on the battery.

Controlling or Monitoring the Electrical Load

Placards are typically installed in the cockpit or flight deck locations in plain view of aircraft crewmembers to provide cautionary information on the various combinations of electrical loads that can safely be connected to the power source.

When an ammeter is electrically located in the battery lead, and the regulator system limits the maximum current that the generator or alternator can deliver, a voltmeter can be installed on the system bus. As long as the ammeter does not read *discharge* (except for short, intermittent loads such as operating the gear and flaps) and the voltmeter remains at *system voltage,* the generator or alternator will not be overloaded.

In installations where the ammeter is electrically located in the generator or alternator lead, and the regulator system does not limit the maximum current that the generator or alternator can deliver, the ammeter should be redlined at 100% of the generator or alternator rating. If the ammeter reading is never allowed to exceed the red line, except for short, intermittent loads, the generator or alternator will not be overloaded.

Where the use of placards or monitoring devices is not practicable or desired, and where assurance is needed that the battery in a typical small aircraft generator/battery power source will be charged in flight, the total continuous connected electrical load may be held to approximately 80% of the total rated generator output capacity. When more than one generator is used in parallel, the total rated output is the combined output of the installed generators.

When two or more generators are operated in parallel and the total connected system load can exceed the rated output of one generator, the electrical system must be designed with the capability to load-shed quickly if a generator or engine failure occurs. A specified procedure requiring pilot intervention may also be provided by the manufacturer whereby the total load can be reduced to a quantity which is within the rated capacity of the remaining operable generator(s).

Electrical loads should be connected to inverters, alternators or similar aircraft electrical power sources in such a manner that the rated limits of the power source are not exceeded, unless some type of effective monitoring means is provided to keep the load within prescribed limits.

Beech King Air Electrical Power Supply and Distribution System

The electrical systems of aircraft range from a system as simple as the venerable Piper Cub with magnetos, a battery, a few lights and pos-

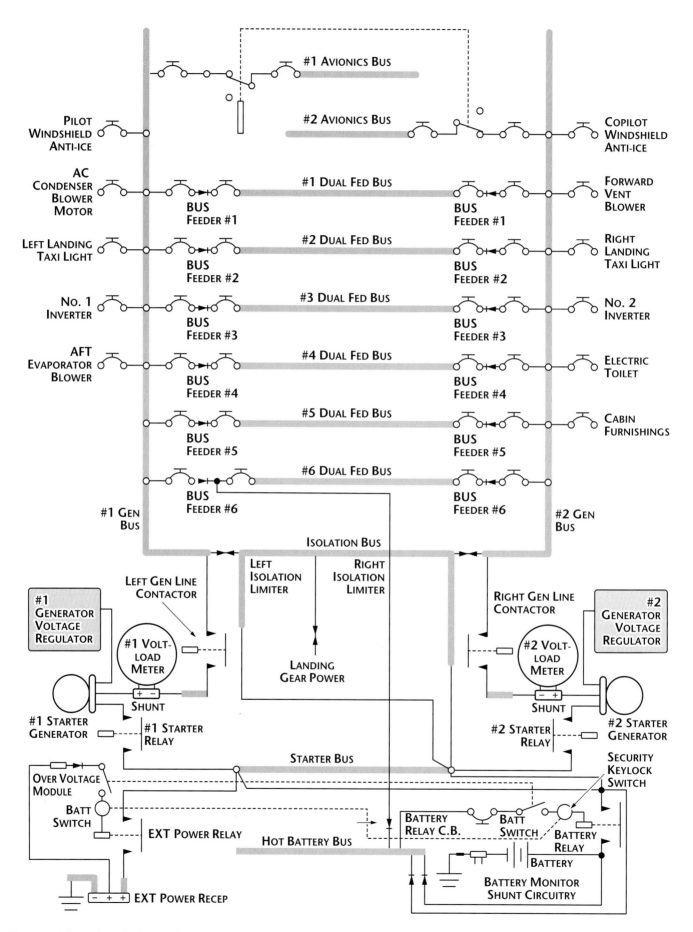

Figure 4-6-2. A King Air electrical generation and major distribution schematic

sibly a generator to incredibly complex systems on aircraft like the Boeing 777 or Airbus A340.

An aircraft electrical system that is approximately halfway up the complexity scale is that of the Beech Model 200 Super King Air (Figure 4-6-1). The prefix "Super" was later dropped by Raytheon (who had acquired Beechcraft) in 1996.

The electrical system utilizes both 28 VDC and 115 VAC electrical power. Most of the electrical system circuitry operates off DC power sources. This is illustrated in Figure 4-6-2.

King Air DC Power

DC power is used for starting the engines, operation of the retractable landing gear, flap motors, standby fuel pump, ventilation blower, lighting and some electronic equipment. Three sources of DC power include one 24-volt (20-cell) 34-ampere hour nickel-cadmium battery and two 250-ampere starter generators. The battery is mounted in the right wing center section, accessible through a panel on the top of the wing. One starter-generator is mounted on each engine accessory drive, and each is fully capable of functioning as a starter or a generator. DC power may also be obtained from an external ground power cart plugged into an external power receptacle on the right engine nacelle.

When used as an engine starter, 28 volts DC is required for starter rotation. Selecting engine start switches routes power through the respective 5-ampere start control circuit breakers on the overhead circuit breaker panel from either the aircraft battery or a ground power cart.

When in use as a generator, the unit is capable of producing 250 amperes at 28 VDC. Each starter-generator is controlled by a generator control unit (GCU). The output of each generator passes through a cable to the respective generator bus (Figure 4-6-2). The left and right generator buses, in turn, supply power to several other DC power buses for distribution to various electrical components. The generators are paralleled to balance the DC loads between the two units. When one of the generating systems is not on-line, and no fault exists, aircraft DC power requirements continue to be supplied, from the other generating source.

Most DC distribution buses are connected to both generator buses but have isolation diodes to prevent power crossfeed between the generating systems, when connection between the generator buses is lost. If either generator is lost because of a ground fault, the remaining generator in operation is still able to supply power to the equipment on that distribution bus. The only exceptions are those DC distribution buses that have dedicated connections to the

Figure 4-6-3. This overhead panel contains most electrical generation and primary distribution controls

Figure 4-6-4. A loadmeter marked in percent of generator output

inoperative generator's bus. When a generator is not operating, reverse current and over-voltage protection is automatically provided. For a detailed review of typical starter-generator unit operation, refer to the turbine engines chapter of *Aircraft Powerplant Maintenance*.

King Air Electrical System Controls

Controls and indicators associated with the DC supply system are located on the overhead control panel (Figure 4-6-3) and consist of a single battery switch (BATT), two generator switches (#1 GEN and #2 GEN) and two volt-loadmeters.

Battery switch. A switch, placarded BATT (Figure 4-6-3), is located on the overhead control panel under the master switch. The BATT switch controls the DC power to the aircraft bus system through the battery relay, and must be on when using external power. The BATT switch is forced OFF when the master switch bar is placed aft.

Generator switches. Two switches (Figure 4-6-3, previous page) placarded #1 GEN and #2 GEN are located on the overhead control panel under the master switch. The three-position toggle switches control electrical power from the designated generator to paralleling circuits and the bus distribution system. Switch positions are placarded reset, on and off. Reset is forward (spring-loaded back to on), on is center and off is aft. When a generator is placed off-line from the aircraft electrical system, due either to fault or from placing the GEN switch in the off position, the affected unit cannot have its output restored to aircraft use until the GEN switch is moved to reset, then on.

Master switch. All electrical current may be shut off using the master switch bar (Figure 4-6-3), that extends above the battery and generator switches. The master switch bar is moved forward when a battery or generator switch is selected on. The bar forces all four switches into the off position when it is moved aft.

Volt-loadmeters. Two meters (Figure 4-6-3), on the overhead control panel display voltage readings and show the rate of current usage from left and right generating systems. Each meter is equipped with a spring-loaded push-button switch that causes the meter to indicate main bus voltage when manually pressed. Each meter normally shows output amperage reading from the respective generator, unless the push-button switch is pressed to obtain bus voltage reading. Current consumption is indicated as a percentage of total output amperage capacity for the generating system monitored (Figure 4-6-4).

Battery monitor. A characteristic of nickel-cadmium batteries is that if they become over-heated, the battery charge current increases and may indicate that thermal runaway is imminent. The aircraft has a charge-current sensor that will detect a charge current. The charge current system senses battery current through a shunt in the negative lead of the battery. Any time the battery charging current exceeds approximately 7-amperes for 6 seconds or longer, the yellow battery charge annunciator light and the master fault caution light will illuminate.

Following a battery engine start, the caution light will illuminate approximately six seconds after the generator switch is placed in the ON position. The light will normally extinguish within two to five minutes, indicating that the battery is approaching a full charge. The time interval will increase if the battery has a low state of charge, the battery temperature is very low, or if the battery has previously been discharged at a very low rate (i.e., battery operation of radios or lights for prolonged periods). The caution light may also illuminate for short intervals after landing gear and/or flap operation. Illumination of the caution light in cruise flight indicates that conditions exist for possible battery thermal runaway. The battery should be turned off. Typically, the battery becomes usable again after a 15 to 20 minute cool down period.

Generator out warning light. Two caution/advisory annunciator panel lights illuminate when either generator is not delivering current to the aircraft DC bus system. These lights are placarded #1 DC GEN and #2 DC GEN. Two flashing *master caution* lights and illumination of either fault light indicates that either the identified generator has failed or voltage is insufficient to keep it connected to the power distribution system.

DC external power source. External DC power can be applied to the aircraft through an external power receptacle on the underside of the right wing leading edge just outboard of the engine nacelle, accessible through a hinged access panel. DC power is supplied through the DC external plug and applied directly to the battery bus after passing through the external power relay. The external power source should be off while connecting the power cable to or removing it from the aircraft's external power supply receptacle. The holding coil circuit of the relay is energized by the external power source when the keylock and BATT switches are in the on position. The *ground power unit (GPU)* must be adjusted to regulate at 28 volts maximum to prevent damage to the aircraft battery and electronics.

King Air AC Power

The AC electrical system is shown in Figure 4-6-5. AC power is produced by inverter units numbered #1 and #2. These inverters take DC power from the aircraft electrical system and convert it to AC power. Both inverters are rated at 750 Volt-Amps and provide single-phase 115 volt and 26 volt 400 Hz AC outputs. The inverters are protected by circuit breakers mounted on the DC power distribution panel mounted beneath the floor. Aircraft equipment operating from single-phase AC power includes several avionics systems and engine instruments for fuel flow and torque meters. Controls and indicators of the AC power system are located on the overhead control panel and on the caution/advisory annunciator panel. AC power is selected ON using the inverter select switches placarded inverter #1 and #2 located on the overhead control panel.

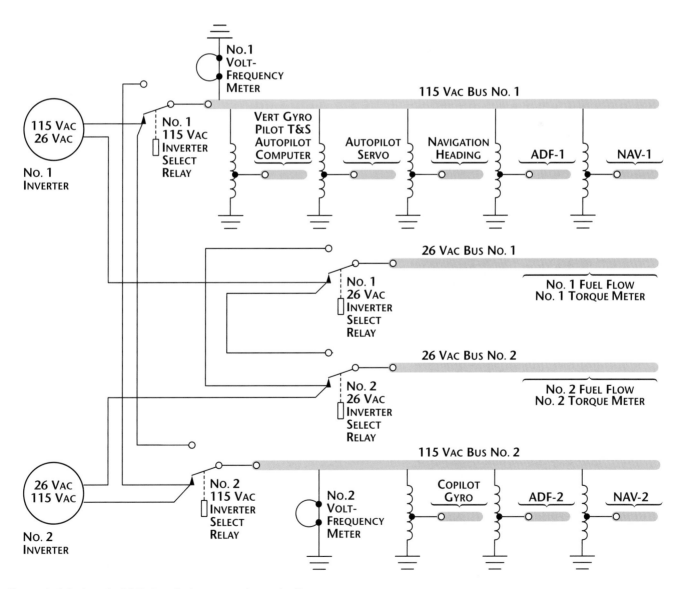

Figure 4-6-5. A typical AC electrical system schematic diagram

Two flashing master caution lights and the illumination of an annunciator caution light #1 inverter or #2 inverter indicate an inverter failure.

Two volt-frequency meters (Figure 4-6-3), are mounted in the overhead control panel to provide monitoring capability for both 115 VAC buses. Normal display on the meter is shown in frequency (Hz). To read voltage, press the button located in the lower left corner of the meter. 115 VAC and 400 Hz on the meters indicates normal inverter output.

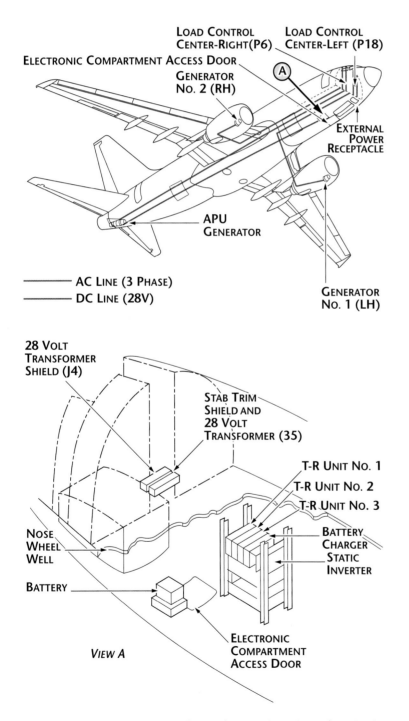

LOAD CONTROL CENTER-RIGHT(P6)
LOAD CONTROL CENTER-LEFT (P18)
ELECTRONIC COMPARTMENT ACCESS DOOR
GENERATOR NO. 2 (RH)
Ⓐ
EXTERNAL POWER RECEPTACLE
APU GENERATOR

——— AC LINE (3 PHASE)
——— DC LINE (28V)

GENERATOR NO. 1 (LH)

28 VOLT TRANSFORMER SHIELD (J4)
STAB TRIM SHIELD AND 28 VOLT TRANSFORMER (35)
T-R UNIT NO. 1
T-R UNIT NO. 2
T-R UNIT NO. 3
NOSE WHEEL WELL
BATTERY CHARGER
STATIC INVERTER
BATTERY
ELECTRONIC COMPARTMENT ACCESS DOOR
VIEW A

Figure 4-6-6. Basic power generation and conversion system for a Boeing 737-300

Boeing 737 Power Supply

AC Electrical Power

The electrical power system of a Boeing 737-300 aircraft consists of 28 VDC, 28 VAC and 115 VAC systems. Three isolated generators that supply 115/200-volt, three-phase, 400-Hz power are primary sources of power for the main AC buses and the whole aircraft electrical system. Each engine drives a generator and the APU drives the third generator. For ground operations, 115/200-volt, three-phase, 400 Hz external power may be connected to the aircraft. Figure 4-6-6 shows the basic electrical distribution.

Transformers that reduce a 115 VAC input down to 28 VAC provide power to the 28 VAC buses. Emergency power is obtained from a static inverter that converts 28 VDC battery power to 115 VAC power. Emergency power supplies critical flight loads on the AC standby bus when the primary power sources have either failed or shut down. The static inverter is located on an electrical equipment shelf in the electronic compartment, or as it is also known, the E&E bay.

DC Electrical Power

Three transformer-rectifier (T-R) units provide 28 VDC power by converting 115 VAC power to 28 VDC power. T-R units are on electrical equipment shelf E3-1. A 36-amp hour nickel-cadmium battery provides 28 VDC power to start the APU. The battery provides power to loads that are connected to the battery bus when other power sources are de-energized and is mounted in the electronic compartment near the electrical equipment shelves.

Each engine-driven generator is driven by a constant-speed drive (CSD) in order to obtain a generator speed of 6,000 r.p.m.. The CSD is a mechanical, differential, hydraulically controlled unit attached to the engine accessory gearbox. Figure 4-6-7 shows a generator drive (disconnected). Manual control, monitoring lights and dial indicators for the drives are located on the forward overhead panel in the control cabin. Warning lights provide indications of low oil pressure and high oil temperature.

Torque is supplied by the engine to the input shaft at various engine speeds. This torque is transmitted to the input end of the planetary differential gear unit in the drive. Depending on the difference between the output speed and 6,000 r.p.m., the variable displacement hydraulic unit will boost or retard the speed of the planetary differential output gear to maintain an output speed of 6,000 r.p.m. as required by the governor.

Aircraft Wiring Diagrams

A complete book of wiring diagrams for a commercial airliner would contain many hundreds of pages. Trying to teach the operation of each and every system is a project way beyond the ability of this chapter. To understand the complex systems of modern airliners the basics must first be learned and then applied to the more complex schematics one wire at a time. While at first this may seem an impossible task, in practice it really isn't. An example of this process is presented in Figure 4-6-8. It is a block diagram of the Boeing 737-300 power generation system. It also includes parts of the distribution system up to the various bus bars.

Start with the generators and follow the system through as it divides and starts supplying the various bus bars. Then follow from each buss bar to its conclusion. You will find it actually proceeds in a very logical manner. That is why binary logic charts can be used to provide current flow charts. See the "Binary Logic" chapter of *Introduction to Aircraft Maintenance* for a review.

Once on the job, the more complex electrical drawings can be followed the same way as the block diagram in Figure 4-6-8.

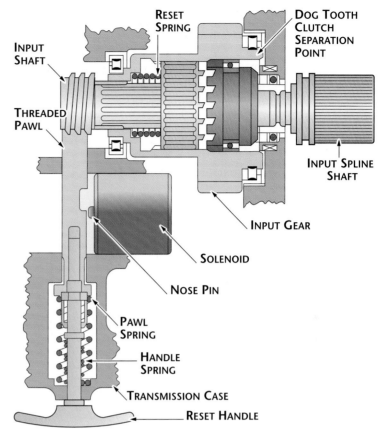

Figure 4-6-7. A generator drive showing the manual disconnect

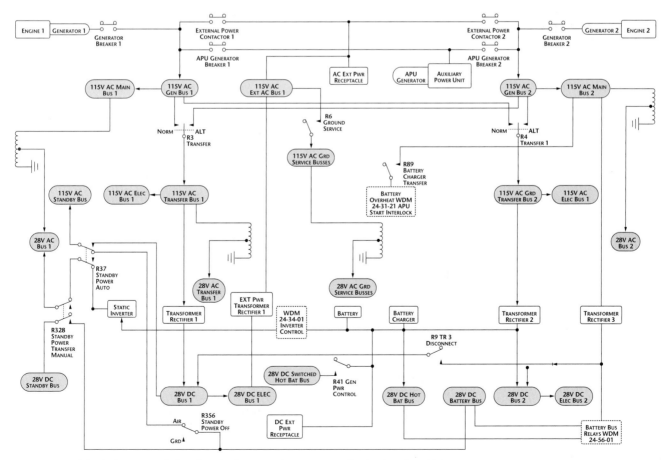

Figure 4-6-8. A block diagram of the electrical generation system of a Boeing 737-300

5

Aircraft Instrument Systems

Section 1

Introduction

Successful flight has always been dependent on the pilot's ability to understand what the aircraft is doing. This was true even with the Wright Flyer in 1903. The earliest aircraft were flown by pilots who sat outside the aircraft, usually on top of the wing, in the airflow. They were able to hear the engine, feel the wind in their face, see the attitude of the aircraft and its relationship to the ground and have some judgment of the height above the ground. As aircraft grew in speed and complexity and pilots moved into enclosed cockpits, they became ever distant from these raw sensory inputs. Additionally, human senses, which are optimized for walking speeds, are insufficient to effectively analyze flight at high speeds.

This brought about the development of instrumentation to aid the pilot in effectively measuring the condition of the aircraft and its systems. The science of instrumentation in aviation dates to the Wright Brothers. The Wright Flyer had a tachometer, a stopwatch and an anemometer (see Figure 5-1-1). The primary purpose of these instruments was to identify engine problems before engine failure occurred so that the airplane could be landed safely.

Safe and reliable operation also requires information about the speed and altitude of the aircraft. The Wright Flyer had a string hanging from the forward canard that the pilot could use to judge the amount of yaw. Instrumentation was soon developed to measure air pressures and changes in pressure. These instruments were calibrated to show altitude, airspeed,

Left. Aircraft instrument systems started as little more than oil pressure gauges to warn the pilot of engine failure. Today's jet aircraft, like this Hawker 4000, use advanced computer systems to monitor many systems on the aircraft.

Photo courtesy of Hawker Beechcraft Corporation

Figure 5-1-1. The instrumentation on the Wright Flyer was very basic.

Figure 5-1-2. The Piper J-3 Cub instrument panel shows how simple the instrumentation was in early aircraft.

rate-of-climb and other information that the pilot needed for safe flight.

The desire to fly in all weather conditions motivated individuals to develop "blind flying" instruments permitting aircraft to safely fly in clouds and other conditions where the pilots visual senses could not be relied upon to maintain level flight. The most successful of the early instruments was the Sperry Gyroscope. Most modern gyroscopic instruments operate on the same principles.

Early instruments relied on simple mechanical mechanisms to display the relevant information to the pilot. Much of the pilot's flight information was obtained visually. Figure 5-1-2 shows a Piper Cub instrument panel. The panel includes the basic cluster of engine instruments and the minimum flight instruments necessary to safely operate the aircraft.

Aircraft instruments have become significantly more complex over the years, particularly with the widespread use of modern high-speed turbine aircraft. Multiple engines, higher flight speeds and a need to operate in all weather conditions greatly increased the amount of information required by the pilot. A typical airliner cockpit from the 1960s shows this complexity (See Figure 5-1-3).

The development of the transistor and integrated circuit has also greatly changed the types of instrumentation found in modern aircraft. The latest aircraft have what is known as a glass cockpit. While the advent of modern technology has brought about major changes in the display systems with some airliner cockpits looking much like video game consoles, many of the sensors in use still operate using the same technologies as they always have. The integration of computers into the cockpit has greatly simplified the pilot's workload. Computers now monitor many of the basic functions and alert the flight crew when certain conditions are met. The crew no longer needs to intensely monitor every instrument; rather they can focus on flying the aircraft and let the computer tell them when a particular area needs special attention. Figure 5-1-4 shows the cockpit of the Cessna Citation Mustang business jet. Compare this cockpit to the DC-9 photo shown in Figure 5-1-3.

One significant difference between the early, simple aircraft systems and the later airline and digital systems is that the sensing mechanism is in many cases no longer located within the instrument. This allows sensors to be located closer to the information source and signals are transmitted electrically to indicators, thus greatly reducing the weight of the installation. System complexity has increased as a result.

Figure 5-1-3. Airline cockpits equipped with analog gauges, like this Douglas DC-9, are very complex places to work.

Figure 5-1-4. Modern computer technology has greatly simplified the cockpit. The cockpit of the Cessna Citation Mustang shows the current state-of-the-art in digital displays.

Today's AMT has the opportunity to work on all of these types of instruments during the course of his or her career. A thorough understanding of the principles of instrument operation is critical to proper maintenance of the aircraft's instrumentation systems.

Section 2
Classification of Instruments

Instruments can be grouped in two different ways. The first method is by the function that they perform. The basic classifications under this method are flight instruments, engine instruments and navigation instruments. Flight instruments are those the pilot uses to aid in controlling the flight attitude of the aircraft. Engine instruments are used to monitor the operation of the engine(s). Navigation instruments are used to steer the aircraft to its intended destination, regardless of weather conditions.

Instruments can also be categorized by the principle on which they work. These principles include changes in temperature or pressure as well as magnetic, electrical and gyroscopic indicators. Prior to the digital age, these principles where primarily mechanical in nature.

Modern digital sensors have introduced multiple methods for the workings of various types of instruments.

This chapter is organized by instrument function. The various methods by which instruments work will be explained along with their function.

Sensors and Displays

Instruments initially combined both the sensing mechanism and the display in one housing. These units are often referred to as "direct reading" instruments. Aircraft complexity and technological changes have brought about the need to separate the sensors and the displays. Digital technology has introduced further variations to the instrument world. Modern aircraft often use a combination of direct reading and remote indicating gauges. Digital systems have replaced traditional analog dials with video displays.

Types of Displays

The manner in which information has been presented to the flight crew has changed over the years. All of the types of displays fulfill one common function – to inform the flight crew as to the status of a system or operation. The earliest, simplest systems relied on direct observation by the flight crew. Later systems have introduced electrical and electronic circuits to relay or process system information.

Visual/sight gauges. The simplest type of indicator is the sight gauge. The sight gauge is most commonly used as a quantity measuring tool, typically for fuel, oil or hydraulic fluids. These come in a variety of configurations. A typical method is through the use of a clear tube or panel connected to the tank or reservoir. This lets the pilot see the fluid level within the tank. Early aircraft commonly used sight gauges to provide fuel quantity information. Most production aircraft today use a remote fuel sender, but some light aircraft still use a sight gauge. Clear panels are still commonly found on turbine aircraft hydraulic systems and some oil systems. They are used during preflight to quickly check fluid levels. (See Figure 5-2-1).

The second type of visual indicator is the mechanical type. They are also most often used for fluid level measurement. Some early aircraft provided fuel quantity information through the use of a float within the fuel tank connected to a rod sticking up out of the fuel tank. The float pushed the rod up when the tank was filled with fuel, and as the fuel was used the float dropped with the fuel level. As long as the

Figure 5-2-1. A sight gauge used to quickly check the oil level on a PW530 turbine engine

Figure 5-2-2. A mechanical fuel float used on a Beechcraft Baron

rod was visible, there was still fuel in the tank. While not in common use, this method can still be found on some early aircraft.

A variation on this method is still used on current production aircraft as a backup indicator. The Beechcraft Baron uses a float connected to a simple analog dial mounted on the wing in view of the pilot. This instrument is shown in Figure 5-2-2.

Conventional analog instruments. These types of instruments display information on a round dial (Figure 5-2-3). They typically include a face and an indicating needle. The dial is marked with a range and the needle moves radially to point at the appropriate measurement on the range. They appear much like the typical automotive speedometer.

Information can be evaluated very quickly using analog instruments. The pilot does not need to stop and read the exact number that the needle is pointing at. A quick glance shows where on the range the needle is. If it is outside the normal range, then the flight crew can pause and read it in detail. This feature allows a large number of instruments to be scanned very quickly.

Vertical scale instruments. Vertical scale instruments are a variation on the analog instrument. Initially used to save space and provide a more modern instrument panel appearance, they were only incorporated into production designs for a fairly short period of time. While they can save some space and initially looked more modern, they are difficult to read quickly. Scanning the instrument panel for critical information takes more time with vertical scale instruments. During emergency situations, this uses vital time. Vertical scales are still used in some older aircraft and for some non-critical functions in new aircraft. See Figure 5-2-4.

Figure 5-2-3. An analog fuel quantity indicator

Figure 5-2-4. A vertical scale indicator

(A)

(B)

Figure 5-2-5. (A) Direct reading gauge, (B) An altimeter counter

to make a more compact, modern instrument panel. An example is shown in Figure 5-2-6. However, they have the same shortcoming as the vertical scale: they can be difficult to read quickly. They do have one important benefit. Digital readouts give the flight crew very precise measurements. Because of this, they are no longer used as primary displays for critical information, but digital readouts are often added to analog displays to provide supplemental information.

Electronic displays. Electronic flight information systems (EFIS) use computer technology to display information. See Figure 5-2-7. Computer screens replace mechanical instruments. One advantage is simplicity. One screen displays the information from many instruments. Another advantage is flexibility. Computers manage the process and displays can be configured to best suit the crew's needs. Electronic systems can combine the best features of earlier displays. They often use analog presentations with added digital data, providing for quick scans and precise measurements when needed. EFIS systems are described in more detail later in this chapter.

Digital instruments. Digital instruments can be mechanical or electronic. They display information in numeric form, rather than by a moving needle. Mechanical digital displays are wheels of numbers that rotate. A numeric display in an altimeter in shown in Figure 5-2-5.

Digital instruments can also be electronic. The mechanical number wheel is replaced by an electronic display. Digital instruments are commonly found on older aircraft. Like the vertical scale instrument, they were an attempt

Types of Sensors

Direct reading gauges. Direct reading gauges, (Figure 5-2-8) are those instruments that contain both the sensor mechanism and the display within the same case. Early aircraft used this type of instrumentation almost exclusively.

Separate sensors and indicators. Single engine aircraft usually have the gauges mounted directly behind the firewall in the cockpit. It is a simple process to run an oil-pressure, fuel-pressure, or similar line into the cockpit and attach it to the back of the instrument.

The complexity required to route plumbing from multiple engines to the cockpit brought about the need for remote indicating instruments. Plumbing adds weight and complexity to an aircraft. Pressurized aircraft must also deal with many large diameter penetrations of the pressurized fuselage wall (Figure 5-2-9). There are also significant safety issues with multiple fuel and oil lines running into the aircraft crew and passenger compartments.

Remote sensing instruments use a sensor, located close to or at the point of measurement. This sensor sends a signal, typically electrical, to an indicator in the cockpit. A single wire bundle penetrates the fuselage and carries the signals for multiple indicators. Aircraft with

Figure 5-2-6. An electronic digital instrument

analog instruments will have a wire or wire bundle from the sensor to each individual instrument (Figure 5-2-8). The wire bundle transmits data to a central computer in glass cockpit systems.

Remote Sensing Technologies

The pilot requires a wide variety of types of information. These include information on various pressures and temperatures throughout the aircraft, the position of components, the quantities of available materials and the speeds of various items. Examples include fuel and oil pressure, landing gear position, fuel quantity and engine r.p.m. All of this information must be available to the flight crew.

There are a variety of mechanisms and technologies used to gather the needed information.

Most of these technologies can be used to signal either analog or digital instruments. Some newer digital systems have new technology solid-state sensors that transmit their data digitally and can only be used with computerized glass cockpit systems.

Simple Switches

The simplest type of remote sensor is an on-off switch. Switches can be attached to components, such as landing gear, to alert the crew when the landing gear is retracted or extended. Switches can also be used to indicate high or low pressures. A pressure switch can be used to alert the flight crew in the event of a low fuel pressure condition. Switches are not typically used to drive aircraft instruments. They are most often used to turn on or off warning lights in the cockpit. They may also be used to activate or deactivate aircraft systems or instruments.

Simple Resistance

Electrical resistance in a circuit can be made to vary. This varying resistance can be used to indicate position, temperature or pressure.

Position indicating. Variable resistors and potentiometers are used as sensors and connected to many components to indicate position. The sensor is connected to an instrument by wires. Electrical current is passed through this circuit. As the component moves, the resistance generated by the sensor changes. This causes the voltage in the circuit to vary.

The instrument is essentially a voltmeter. The needle moves as the voltage in the circuit

Figure 5-2-7. A typical EFIS display showing both analog and digital information on the same screen

Figure 5-2-8. The gauge on the left is a direct reading instrument. The gauge on the right has a remote sensor.

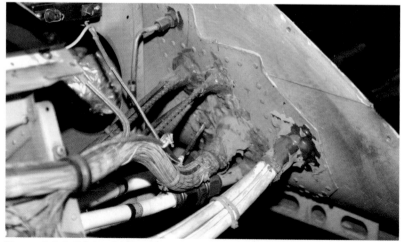

Figure 5-2-9. Wiring harness & cable bundles must be sealed where they penetrate a pressurized cabin.

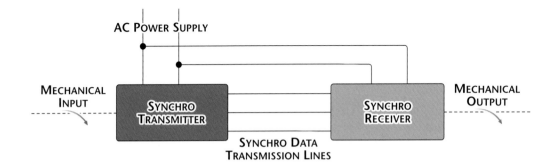

Figure 5-2-10. Synchro block diagram

changes. The instrument face is marked to correspond with the position of the component.

Temperature indicating. Many metals have the property of varying their resistance to the flow of electrons as their temperature changes. Some metals have a very predictable and consistent resistance change over a specific temperature range. Sensors made of these metals can be used to indicate temperature. An electrical current is passed through a temperature probe manufactured from a specific metal type. This temperature probe is placed in the region to be measured.

This may be an oil tank to measure oil temperature, an exhaust system to measure exhaust gas temperature or the inside of a turbine engine to measure combustion temperatures.

An indicator is connected to the temperature probe. The instrument is calibrated to display

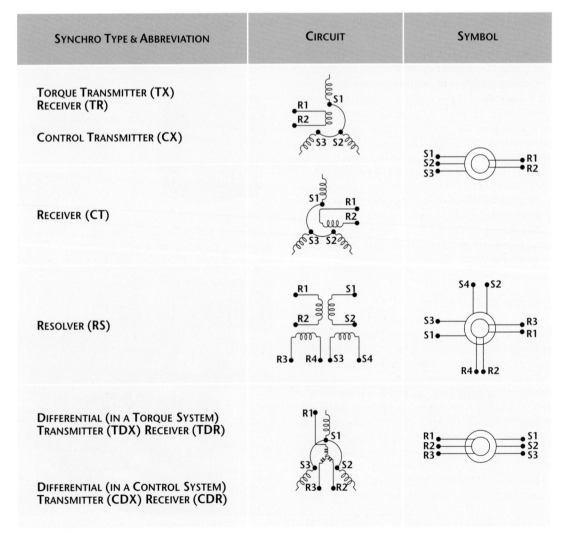

Figure 5-2-11. Types of synchros

the temperature corresponding to the resistance present at a given temperature.

Pressure indicating. The latest digital systems include silicon chips that are sensitive to pressure. The resistance generated by these chips varies with the pressure placed on them. The voltages and resistances used in these systems are too small to be used by traditional analog instruments. The extremely sensitive voltages involved require the chip to convert the electrical resistance into a digital signal that is communicated to the flight control computer.

Servo-Mechanisms (Synchro Systems)

One of the most common types of remote indicating systems is the *synchro*. The term synchro is used to refer to synchronous data-transmission systems. A synchro system consists of a transmitting and a receiving element. Their key feature is the ability to measure and indicate angular deflection. Both direct current and alternating current systems are in use.

Synchros are used extensively in modern air transport aircraft. They are commonly found in engine instruments, analog air data computers, remote-indicating compasses, landing gear and flap position indicators and in flight director systems.

AC Synchro Systems

A simple alternating current synchro system is shown in Figure 5-2-10. It consists of a synchro transmitter (generator) and a synchro receiver (repeater).

Synchro systems can be divided into four categories. These categories are:

- Torque
- Control
- Differential
- Resolver

The standard symbols for these four types are shown in Figure 5-2-11.

Torque indicating synchros. The simplest type is the torque indicating system. A basic torque indicating synchro system is shown in Figure 5-2-12. The synchro transmitter is a rotor with a single winding and a stator with three windings. The windings are arranged 120° apart.

The rotor is connected to the transmitter shaft. Its winding is excited by the AC power supply. As this winding rotates within the three stator windings, it induces voltages in the stators as a function of the rotor angle.

These voltages are transmitted to the synchro receiver via three wires. The synchro receiver is electrically identical to the transmitter. The voltages transmitted to the receiver cause its rotor to move to the same position as the rotor in the transmitter. This is caused by a magnetic force induced by the windings acting upon the rotor. One early synchro of this type was

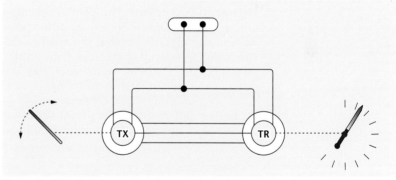

Figure 5-2-12. Synchro transmitter and receiver

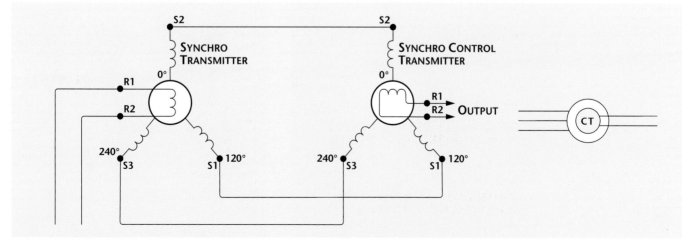

Figure 5-2-13. Synchro control schematic

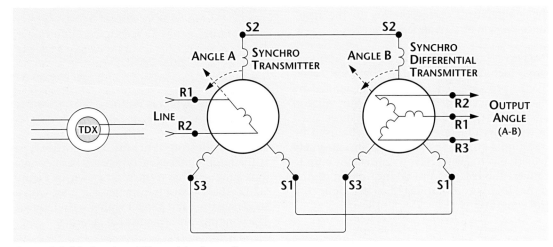

Figure 5-2-14. Synchro differential schematic

known as the *Magnesyn* system. This term is no longer in common use.

Control synchros. The control synchro system is similar to the torque indicating system above. The primary difference is in function. The control synchro generates an error signal rather than a rotational torque movement within the receiver. This error signal is amplified and output to the control phase of an AC servomotor. A typical synchro control schematic is shown in Figure 5-2-13.

Differential synchros. Differential synchros are used to measure position differences. Adding a synchro differential to a synchro circuit allows the circuit to measure two angular positions. The resulting signal indicates the difference or sum of the two positions.

A synchro differential differs from the standard synchro in the number of rotor windings.

Differential synchros have an additional rotor winding for a total of three. Both transmitters and receivers are wired the same internally.

When used as a differential transmitter, the voltage from the synchro transmitter is applied to the stators. The output voltages from the rotor windings are then applied to the synchro transmitter stator windings. When the synchro differential rotor shaft is rotated, the combined output signal will vary, indicating the difference between the two positions. This wiring schematic is shown in Figure 5-2-14.

The synchro differential can be used as a differential receiver by connecting one synchro transmitter to the rotors and another to the stators. The shaft of the differential synchro will indicate the sum of the two signals. A sample synchro differential wiring schematic is shown in Figure 5-2-15.

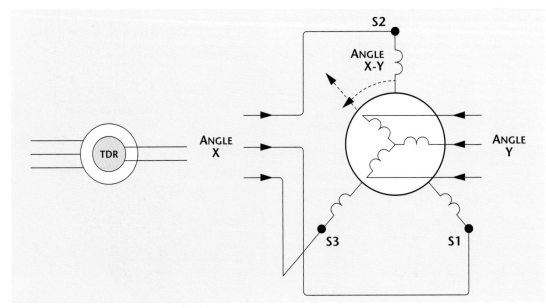

Figure 5-2-15. Synchro differential schematic

Figure 5-2-16. Synchro resolver

Synchro resolvers. Synchro resolvers generate a voltage signal indicating position rather than a rotational force. This voltage represents the angular position of the shaft. A synchro resolver has two stator windings 90° apart and two rotors also set at 90° apart. The stators act as transformer primary windings and the rotors act as transformer secondaries. A typical synchro resolver is shown in Figure 5-2-16.

The relationship between the rotors and the stators induces voltages based on their relative positions. The voltage varies as the rotor is moved. A graph of this voltage, showing one revolution of the shaft, takes on a sine-cosine waveform. Synchro resolvers can be used as differential control transformers, control transformers, or transmitters. Wiring for all three types are shown in Figure 5-2-17.

DC Synchro Systems

The DC synchro system (sometimes referred to as a DC *selsyn* system) is a widely used electrical method of indicating a remote mechanical condition. The DC synchro system remote indicating mechanism can be used to show the movement and position of retractable landing gear, wing flaps, oil cooler doors, or similar movable parts of the aircraft.

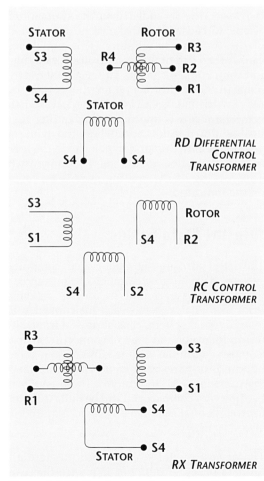

Figure 5-2-17. Synchro resolver schematic

Figure 5-2-18. DC synchro transmitter

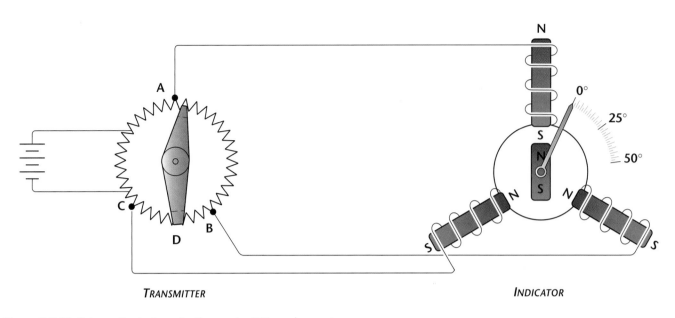

Figure 5-2-19. Schematic design of a three-wire DC synchro system

A synchro system consists of a transmitter (see Figure 5-2-18), an indicator and the connecting wires. Current supplied from the aircraft's electrical system powers the system.

Transmitter. A synchro system schematic is shown in Figure 5-2-19. The transmitter consists of a circular resistance winding and a rotating contact arm. The contact arm turns on a shaft in the center of the resistance winding. The two ends of the arm, or brushes, always touch the winding on opposite sides. The shaft to which the contact arm is fastened protrudes through the end of the transmitter housing and is attached to the unit (flaps, landing gear, etc.) whose position is to be transmitted. The transmitter is usually connected to the unit through a mechanical linkage. As the unit moves, it causes the transmitter shaft to turn. Thus, the arm can be turned so that voltage can be applied at any two points around the circumference of the winding.

As the voltage at the transmitter taps is varied, the distribution of currents in the indicator coils varies and the direction of the resultant magnetic field across the indicator is changed. The magnetic field across the indicating element corresponds in position to the moving arm in the transmitter. Whenever the magnetic field changes direction, the polarized motor turns and aligns itself with the new position of the field. The rotor thus indicates the position of the transmitter arm.

When the DC selsyn system is used to indicate the position of landing gear, an additional circuit is connected to the transmitter winding, which acts as a lock-switch circuit. The purpose of this circuit is to show when the landing gear is up and locked, or down and locked.

Lock switches are shown connected into a three-wire system in Figure 5-2-20.

Indicator. A resistor is connected between one of the taps of the transmitter at one end and to the individual lock switches at the other end. When either lock switch is closed, the resistance is added into the transmitter circuit to cause an unbalance in one section of the transmitter winding. This unbalance causes the current flowing through one of the indicator coils to change. The resultant movement in the indicator pointer shows that the lock switch has been closed. The lock switch is mechanically connected to the landing gear up- or down-locks, and when the landing gear locks either up or down, it closes the lock switch connected to the selsyn transmitter. This locking of the landing gear is repeated on the indicator.

Digital Data Buses

The latest aircraft with integrated computer systems have eliminated much of the complex wiring harnesses found in analog systems. Prior to digital systems, each instrument was an individual system, consisting of a transmitter, an indicator and a set of connecting wires. Complex, multiple engine aircraft can have hundreds of wires interconnecting all of their instruments and indicators. These systems become very complex and can be quite difficult to troubleshoot. The mass of wire required is also very heavy.

The newest digital systems incorporate digital data buses. A data bus is a wire that can transmit multiple signals at the same time.

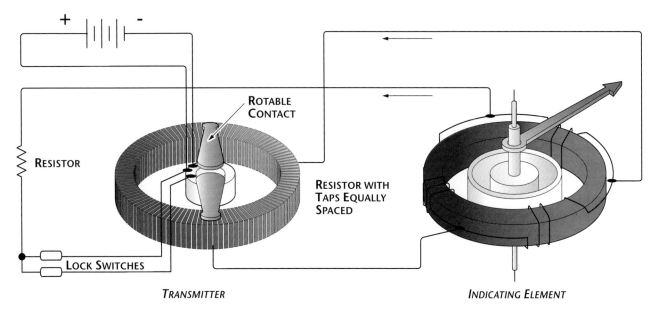

Figure 5-2-20. Double lock switch circuit

Each sensor converts its data from an analog electrical signal to a digital data packet. Each packet is encoded with an address, indicating where it came from and what type of data it contains. These data packets are then transmitted electrically on a single wire. Multiple sensors can be connected to this single wire. The data packets are transmitted to the aircraft's central computer, which then decodes the digital data, and sends the appropriate control signal to the cockpit display. A data bus is essentially a computer network.

Critical systems have multiple data buses connected to them in order to provide a backup circuit in the event of a data bus failure. A system with only a single data bus could loose communications and control with all sensors if a small portion of the bus failed. Multiple data buses provide backup communication paths.

The Boeing 777 data bus system is shown in Figure 5-2-21. The 777 has eleven data buses. Three of these data buses are dedicated to the flight control system. They are connected to

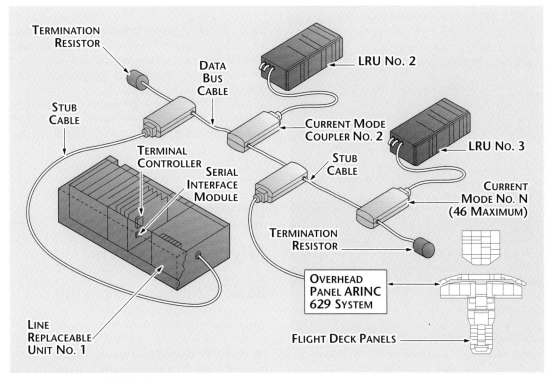

Figure 5-2-21. Digital data bus

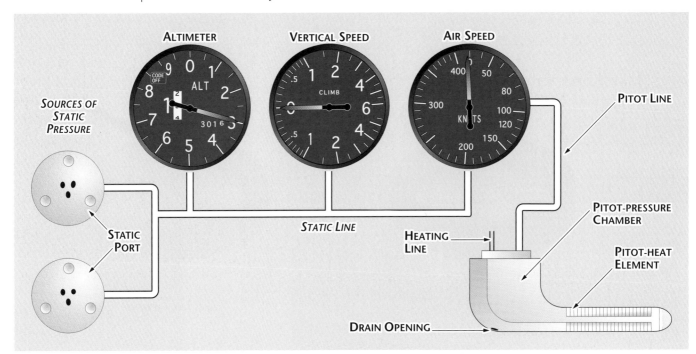

Figure 5-3-1. A basic pitot-static system includes a pitot tube, static ports and three indicators.

26 LRUs. The aircraft has a double backup (or triple redundant) system to ensure that the fly-by-wire flight control system is not disabled by a simple bus failure.

Four additional data buses provide the communication system for the avionics, electrical, electro-mechanical, environmental control and propulsion systems. All of these systems communicate over the same set of buses, with four buses provided for redundancy. These buses connect to 53 LRUs. These buses are independent of the flight control buses. This prevents the other aircraft systems from affecting the flight controls in the effect of any kind of error or glitch.

The final four buses interconnect the two airplane information management system computers and the three control display units.

Early digital systems added data converters to analog sensors. These converters read and interpreted the analog signal and converted the information to a digital signal that can be transmitted on the data bus. Newer systems use advanced technology incorporating lightweight silicon chips that sense pressures, temperatures and position and directly output digital signals.

Figure 5-3-2. Nose mounted pitot tubes can be placed under, on top, or on the sides of the nose of the aircraft. This style pitot tube can also be mounted below the wing.

Section 3

Flight Instruments

Basic Pitot-Static Systems

Successful aircraft operation is dependent on the pilot's ability to know the relationship between the aircraft and the air that it is moving through. Flying too slow, too fast, or too low can all have dangerous consequences. The pitot-static system is the key. It calculates the aircraft's speed and altitude by measuring air pressures.

A basic aircraft pitot-static system (Figure 5-3-1) includes instruments that operate on the principle of the barometer. The system consists of a pitot tube, static air ports, and three indicators, which connect with pipelines that carry air. The three indicators are airspeed, altimeter, and the vertical speed indicator (VSI). Aircraft equipped with advanced electronics instrumentation systems will use the same inputs but process and display the information using solid state circuitry. These systems will be discussed later.

The airspeed indicator shows the speed of the aircraft through the air and the altimeter shows the altitude. The VSI indicates how fast the aircraft is climbing or descending. They all operate on air that comes in from outside the aircraft during flight.

Pitot tubes & static ports. The pitot tube mounts on the outside of the aircraft at a point where the air is least likely to be turbulent. The tube points in a forward direction parallel to the aircraft's line of flight. One common type of airspeed tube mounts on a streamlined mast extending out from the fuselage or below the wing. These may extend below the nose, out from the side, or may even be mounted on the top of the nose (see Figure 5-3-2). The aircraft manufacturer determines the exact placement for each aircraft type after extensive flight testing. Another type mounts on a boom extending forward from the nose or the leading edge of the wing. This type is shown in Figure 5-3-3. Although there is a slight difference in their construction, they operate identically.

Some older aircraft use a pitot tube designed by the US military during World War II. This type, known as the shark fin type (Figure 5-3-4), has a distinctive fin sticking up about three-quarters of the way back from the pitot entrance. This type is known as a pitot-static tube because it also contains a static port. While not common, this type can still be found in service on many older aircraft.

Some high-speed aircraft use a modern version of the pitot-static tube (Figure 5-3-5). It contains a pitot pressure port and two static pressure ports. When working with pitot-static systems, care must be taken to determine if the aircraft has separate pitot and static ports or if they are integrated into a single probe. In some cases, the static ports are located some distance from the pitot tubes.

Pitot stands for impact pressure, i.e. the pressure of the outside air against the aircraft fly-

Figure 5-3-3. Another type of pitot tube extends forward from the nose or wing leading edge.

Figure 5-3-4. Some aircraft still use the older military standard shark fin style pitot-static tube.

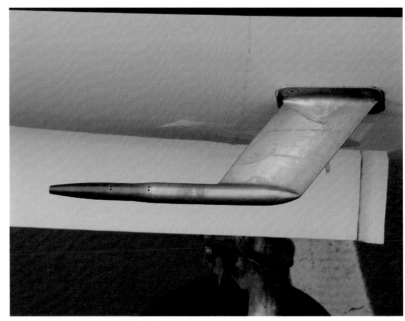

Figure 5-3-5. A modern high-speed pitot-static probe.

Figure 5-3-6. Static ports are typically mounted on both sides of the fuselage.

Figure 5-3-7. The Falcon 900 corporate jet is typical of a multiple pitot-tube configuration.

ing through it. The tube that goes from the pitot tube to the airspeed indicator applies the outside air pressure to the airspeed indicator. The airspeed indicator calibration allows various air pressures to cause different readings on the dial. The purpose of the airspeed indicator is to interpret pitot air pressure in terms of airspeed in knots.

Generally, static air ports (Figure 5-3-6) are small calibrated holes in an assembly mounted flush with the aircraft fuselage. Their position is in a place with the least amount of local airflow moving across the ports when the aircraft is flying.

Static means stationary or not changing. The static part of the pitot-static system also introduces outside air. However, it is at its normal outside atmospheric pressure as though the aircraft were standing still in the air. The static line applies this outside air to the airspeed indicators, the altimeter, and the vertical speed indicator.

Pitot-static instruments operate by (1) measuring the difference between the static pressure sensed at the static port and the impact pressure at the pitot tube, (2) by measuring one of the two pressures, or (3) by measuring the rate of change in one of the pressures.

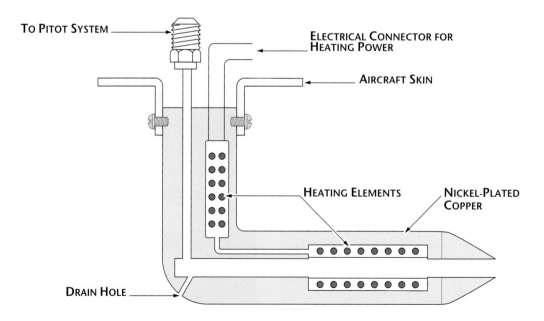

Figure 5-3-8. Heated pitot tubes contain internal electrical heating elements

Figure 5-3-9. A basic airspeed indicator

Most light aircraft have a single pitot-static system. This system typically includes one pitot tube and two static ports, one on each side of the aircraft. Providing a static port on both sides of the aircraft automatically compensates for any pressures induced by a side-slip. More advanced aircraft have multiple pitot-static systems. Some have as many as four pitot-tubes and eight static ports as well as additional sensors needed for high-speed flight. A typical multiple pitot tube configuration is shown in Figure 5-3-7.

Many pitot tubes are provided with heating elements to prevent icing during flight. See Figure 5-3-8. During icing conditions, the electrical heating elements can be turned on by means of a switch in the cockpit.

The electrical circuit for the heater element may be connected through the ignition switch. Extreme care is necessary during maintenance operations as a heated pitot tube can become extremely hot. Never touch a pitot tube when it is being heated. The safest way to check for proper pitot heat is to do measure the amperage draw with the system on. Touching the pitot tube is not necessary.

Simple Airspeed Indicators

The airspeed indicator (Figure 5-3-9) measures the impact pressure sensed at the pitot tube and compares it to the static pressure from the static port. While this information tells the pilot how fast the aircraft is moving through the air, it can also be used to estimate ground speed and to determine throttle settings for the most efficient flying speed. Airspeed also provides a basis for calculating the best climbing and gliding angles. It warns the pilot if diving speed approaches the safety limits of the aircraft's structure. Additionally, since airspeed increases in a dive and decreases in a climb, the indicator can be used as a backup to the vertical speed indicator for maintaining level flight.

An airspeed indicator has a cylindrical, airtight case that connects to the static line from

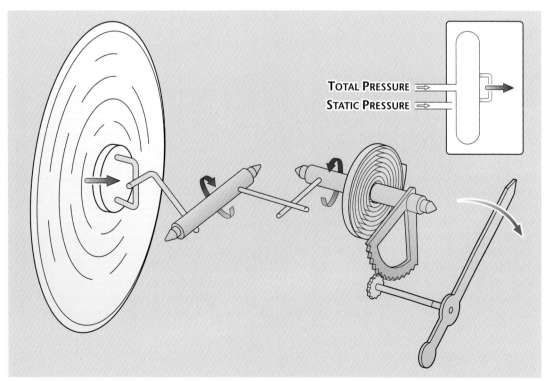

Figure 5-3-10. Pitot pressure enters the aneroid diaphragm in an airspeed indicator and causes it to expand.

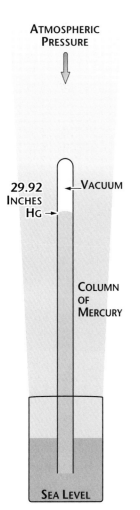

Figure 5-3-11.
Mercurial barometer

the pitot-static tube. Inside the case is a small aneroid diaphragm of phosphor bronze or beryllium copper. The diaphragm is very sensitive to changes in pressure, and it connects to the impact pressure (pitot) line. This allows air from the pitot tube to enter the diaphragm. The side of the diaphragm fastens to the case and is rigid. The needle or pointer connects through a series of levers and gears to the free side of the diaphragm. Figure 5-3-10 shows a cutaway view of a typical airspeed indicator.

The airspeed indicator is a differential pressure instrument. It measures the difference between the pressures in the impact pressure line and in the static pressure line. The two pressures are equal when the aircraft is stationary on the ground. Movement through the air causes pressure in the impact line to become greater than that in the static line. This pressure increase causes the diaphragm to expand. The expansion or contraction of the diaphragm goes through a series of levers and gears to the face of the instrument to regulate needle position. The needle shows the pressure differential in MPH or knots. In air navigation, all speeds and distances are typically in nautical miles.

Air pressure (and density) drops as altitude increases. This reduces the impact pressure to the pitot tube as the aircraft climbs. As a result, indicated airspeed falls off as altitude increases. Some altimeters, and most aircraft flight manuals, provide a means of calculating ground speed based on a combination of indicated airspeed, temperature and altitude.

Altimeters

An altimeter is an instrument that measures static pressure. Before understanding how the altimeter works, a thorough understanding of altitude is required. Remember, even though the altimeter reads in feet, it is actually measuring pressure.

The word altitude is vague, so it needs further defining. The term altitude includes altitude above mean sea level (MSL) and altitude above ground level (AGL). It also includes pressure altitude, indicated altitude, density altitude, and elevation.

Mean sea level. Since about 80 percent of the earth's surface is water, it is natural to use sea level as an altitude reference point. The pull of gravity is not the same at sea level all over the world because the earth is not perfectly round and because of tides. To adjust for this, an average (or mean) value is set; this is referred to as the mean sea level (MSL). Mean sea level is the point where gravity acting on the atmosphere produces a pressure of 14.70 pounds per square

inch. This pressure supports a column of mercury in a barometer to a height of 29.92 inches. This is the reference point from which all other altitudes are measured. See Figures 5-3-11 and 5-3-12. Altimeters indicate altitude above mean sea level.

Elevation and true altitude. *Elevation* is the height of a land mass above mean sea level. Elevation is measured with precision instruments that are far more accurate than the standard aircraft altimeter. Elevation information is found on charts or, for a particular spot, may be painted on a hangar near an aircraft ramp or taxi area.

True altitude is the actual number of feet above MSL. Precision surveying tools are used to measure the altitude. In standard day conditions, pressure altitude and true altitude are the same.

Absolute altitude. *Absolute altitude* is the distance between the aircraft and the terrain over which it is flying. It is referred to as the altitude above ground level (AGL). AGL altitudes are only significant when aircraft are flying near the ground, such as during take-off and landing. AGL can be calculated by subtracting the elevation of the terrain beneath the aircraft (referenced on the charts) from the altitude read on the altimeter (MSL).

A quicker, more precise indication can be obtained by using a *radar altimeter*. Radar altimeters send a radar signal directly down. This radar signal is used to precisely measure the distance between the aircraft and the ground. This measurement is also referred to as radar altitude.

Pressure altitude. It is not possible to have a ruler extending from an aircraft and reaching to sea level to measure altitude. To measure altitude, instruments sense air pressure and compare it to known values of standard air pressure at specific, measured altitudes. The altitude read from a properly calibrated altimeter referenced to 29.92 inches of mercury (Hg) is the *pressure altitude*.

Refer to Figure 5-3-12. If a pressure altimeter senses 6.75 pounds per square inch pressure with the altimeter set to sea level and barometric pressure 29.92 inches of mercury, the altimeter indicates 20,000 feet. This does not mean that the aircraft is exactly 20,000 feet above mean sea level. It means the aircraft is in an air mass exerting a pressure equivalent to 20,000 feet on a standard day. Pressure altitude is not true altitude.

Indicated and calibrated altitude. Unfortunately, standard atmospheric conditions very seldom exist. Atmospheric conditions and barometric pressure can vary considerably. A pressure change of one-hundredth (0.01) of an inch of mercury represents a 9-foot change in altitude at sea level. Barometric pressure changes between 29.50 and 30.50 are not uncommon (a pressure change of about 923 feet). *Indicated altitude* is the uncorrected reading of a barometric altimeter. *Calibrated altitude* is the indicated altitude corrected for inherent and installation errors of the altimeter instrument. On an altimeter without such errors, indicated altitude and calibrated altitude are identical. Assume that this is the case for the rest of this discussion.

When flying below 18,000 feet, the aircraft altimeter must be set to the altimeter setting (barometric pressure corrected to sea level) of a selected ground station within 100 miles of the aircraft. Altitude read from an altimeter set to local barometric pressure is indicated altitude. The accuracy of this method is limited because it assumes a standard lapse rate; that is, for a given number of feet of altitude, an exact change in pressure occurs. This seldom happens, which limits the accuracy of the altimeter. Above 18,000 feet, all altimeters are set to 29.92 (pressure altitude). Although the altimeter is not displaying an accurate reading of altitude, is does provide for vertical separation between aircraft, as long as all aircraft are using the same barometric pressure setting.

Density altitude. A very important factor in determining the performance of an aircraft or engine is the density of the air. The denser the air, the more horsepower the engine can

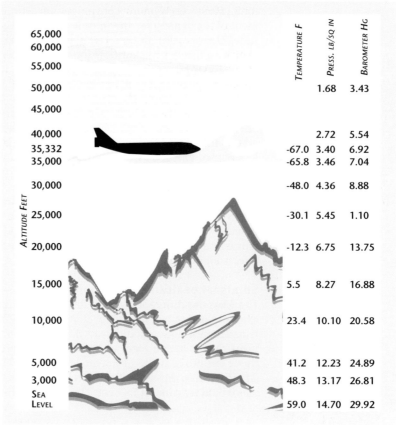

Altitude Feet	Temperature F	Press. Lb/Sq In	Barometer Hg
65,000			
60,000			
55,000			
50,000		1.68	3.43
45,000			
40,000		2.72	5.54
35,332	-67.0	3.40	6.92
35,000	-65.8	3.46	7.04
30,000	-48.0	4.36	8.88
25,000	-30.1	5.45	1.10
20,000	-12.3	6.75	13.75
15,000	5.5	8.27	16.88
10,000	23.4	10.10	20.58
5,000	41.2	12.23	24.89
3,000	48.3	13.17	26.81
Sea Level	59.0	14.70	29.92

Figure 5-3-12. The standard atmosphere

produce. Also, there is more resistance to the aircraft in flying through the air, and the airfoils produce more lift. Pressure, temperature, and moisture content all affect air density. Measurements of air density are in weight per unit volume (for example, pounds per cubic

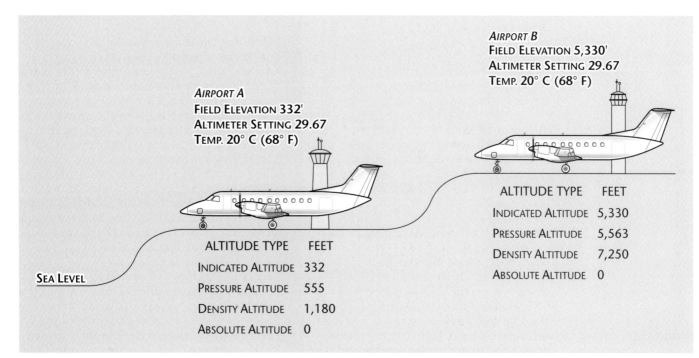

AIRPORT A
FIELD ELEVATION 332'
ALTIMETER SETTING 29.67
TEMP. 20° C (68° F)

ALTITUDE TYPE	FEET
INDICATED ALTITUDE	332
PRESSURE ALTITUDE	555
DENSITY ALTITUDE	1,180
ABSOLUTE ALTITUDE	0

AIRPORT B
FIELD ELEVATION 5,330'
ALTIMETER SETTING 29.67
TEMP. 20° C (68° F)

ALTITUDE TYPE	FEET
INDICATED ALTITUDE	5,330
PRESSURE ALTITUDE	5,563
DENSITY ALTITUDE	7,250
ABSOLUTE ALTITUDE	0

SEA LEVEL

Figure 5-3-13. Different types of altitude

foot). However, a more convenient measurement of air density for the pilot is density altitude. Density altitude is a calculated altitude. Density altitude is the altitude (or air pressure) on a standard day that is equal to the air density present on a given day at a given temperature.

On a cold day, the air is denser than on a standard day. The *density altitude* is lower than the actual altitude. On a hot, humid day the air is less dense. The density altitude is higher than the actual altitude.

Density altitude does not show on an instrument. It is usually taken from a table or computed by comparing pressure, altitude, and temperature. Although moisture content affects air density, its effect is negligible.

Summary of altitude definitions. Figure 5-3-13 shows the different types of altitude. At the example airport A, the field elevation is 322 feet. The altimeter should read 322 feet when properly set to the local altimeter setting (in this case 29.67). If the altimeter is set to 29.92 (standard day pressure at sea level), the altimeter would read about 550 feet. The density alti-

tude is such that the aircraft and engine perform the same as if they were at a standard day altitude of 1,180 feet. This means the distances required for takeoff and landing are longer for higher density altitudes (higher temperatures). The absolute altitude (AGL) for the aircraft on the runway is zero feet because the aircraft is touching the surface.

Look at the altitude for airport B (Figure 5-3-13). Notice that the difference between pressure altitude and indicated altitude remains the same as at airport A. However, at the same temperature the density altitude is much greater at the higher airport.

Altimeter Types

Several kinds of altimeters are used today. They are all constructed on the same basic principle as an aneroid barometer. They are occasionally referred to as Kollsman altimeters after their inventor, Paul Kollsman. All altimeters have pressure-responsive elements (aneroid wafers) that expand or contract with the pressure changes of different flight levels.

The heart of a pressure altimeter is its aneroid mechanism (Figure 5-3-14), which consists of one or more aneroid wafers. The expansion or contraction of the aneroid wafers with pressure changes operates the linkage. This action moves the indicating hand/counter to show altitude. Around the aneroid mechanism of most altimeters is a device called the bimetal yoke. As the name implies, this device is composed of two metals. It performs the function of compensating for the effect that temperature has on the metals of the aneroid mechanism.

The altimeter discussed in the following paragraphs is a simple one. Several complex altimeters are discussed later in this chapter, along with the automatic altitude system.

Basic pressure altimeter. The purpose of the counter pointer pressure altimeter (Figure 5-3-15) is to show aircraft height. Three hands indicate barometric pressure in feet of altitude. They operate in conjunction with the barometric scale, and indications are read on the altimeter dial. The longest hand indicates ten thousands of feet, the intermediate hand indicates thousands of feet, and the shortest hand indicates hundreds of feet.

Figure 5-3-15 shows an aircraft flying at 21,000 feet. The ten-thousands hand points past the two, indicating twenty thousand feet. The thousands hand points at one, indicating an additional one thousand feet. The hundreds hand points at zero. Combining the three indications of 20,000 feet, 1,000 feet and zero

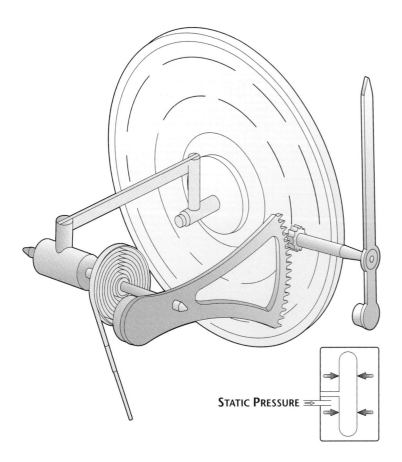

STATIC PRESSURE

Figure 5-3-14. Aneroid mechanism within an altimeter. As the pressure outside the aneroid drops, the aneroid expands. A basic Kollsman pressure altimeter.

Figure 5-3-15. A basic Kollsman pressure altimeter

Figure 5-3-16. A zero to 6,000 f.p.m. vertical speed indicator

feet yields an indicated altitude of 21,000 feet. Notice also that the barometric window is set to standard (29.92).

Atmospheric changes cause movement of the two aneroid diaphragm assemblies. These assemblies move two similar rocking shaft assemblies mutually engaged with the main pinion assembly. This movement goes to the handstaff assembly, which operates the hand assembly and drives the counter mechanism through a disk. Because of the special design of the hand assembly, the counter indication is never obscured.

Barometric corrections are made by turning the externally located knob. The knob engages the barometric dial and the main plate assembly that supports the entire mechanism. The altimeter is adjusted so that the reading on the barometric dial corresponds to the area barometric conditions in which the aircraft is flying.

Since atmospheric pressure continually changes, the barometric scale must be re-set to the local station altimeter setting before the altimeter will indicate the correct altitude of the aircraft above sea level. When the setting knob is turned, the barometric scale, the hands, and the aneroid element move to align the instrument mechanism with the new altimeter setting.

> **CAUTION:** *Do not adjust the barometer setting past the range shown in the window. To do so will make the altimeter go out of calibration and it will have to be replaced.*

All analog altimeters use an aneroid diaphragm to move the mechanism. The more complex types discussed later are built on the same basic mechanism with added features and capabilities.

Altimeter errors. Altimeters are subject to various mechanical errors. A common one is that the scale is not correctly oriented to standard pressure conditions. Altimeters should be checked periodically for scale errors in altitude chambers where standard conditions exist.

Another mechanical error is the hysteresis error. Hysteresis error is induced by the aircraft maintaining a given altitude for an extended period of time, followed by a large, quick altitude change. The resulting lag or drift in the altimeter is caused by the elastic properties of the materials that comprise the instrument. This error will eliminate itself with slow climbs and descents or after maintaining a new altitude for a reasonable period of time.

In addition to the errors in the altimeter mechanism, another error called installation error affects the accuracy of indications. The error is caused by the change of alignment of the static pressure port with the relative wind. The change of alignment is caused by changes in the speed of the aircraft and in the angle of attack, or by the location of the static port in a disturbed pressure field. Improper installation or damage to the pitot-static tube will also result in improper indications of altitude.

Small leaks can also develop in the system. These may occur when maintenance is performed and a line or fitting becomes damaged or is not attached securely. Vibration from aircraft operation can also loosen lines and fittings if not properly tightened. Small leaks can cause large errors in displayed pitot-static information. Every aircraft is required to have its pitot-static system checked, using properly calibrated test equipment, on a biannual basis.

Rate of Climb/Vertical Speed Indicators

A vertical speed indicator (VSI) shows the rate at which an aircraft is climbing or descending. They are also referred to as *rate-of-climb indicators*. This instrument is very important for night flying, flying through fog or clouds, or when the horizon is obscured. Another use is to determine the maximum rate of climb during performance tests or in actual service.

The rate of altitude change, as shown on the indicator dial, is positive in a climb and negative in a dive or glide. The dial pointer (Figure 5-3-16) moves in either direction from the zero point. This action depends on whether the aircraft is going up or down. In level flight the pointer remains at zero.

The vertical speed indicator is contained in a sealed case, and it connects to the static pressure line through a calibrated leak (see Figure 5-3-17). As the aircraft climbs, the diaphragm expands, however, the calibrated leak also allows the pressure to slowly equalize. The reverse happens when the aircraft descends. This expansion or contraction of the diaphragm moves the indicating needle through gears and levers.

Through the use of metals that have predictable temperature characteristics, the instrument automatically compensates for changes in temperature.

Although the vertical speed indicator operates from the static pressure source, it is a differential pressure instrument. The difference in pressure between the instantaneous static pressure in the diaphragm and the static pressure trapped within the case creates the differential pressure. The leak rate and the expansion or contraction rate of the diaphragm are carefully matched to give a consistent indication of the rate of change in altitude.

When the pressures equalize in level flight, the needle reads zero. As static pressure in the diaphragm changes during a climb or descent, the needle shows a change of vertical speed. However, until the differential pressure sta-

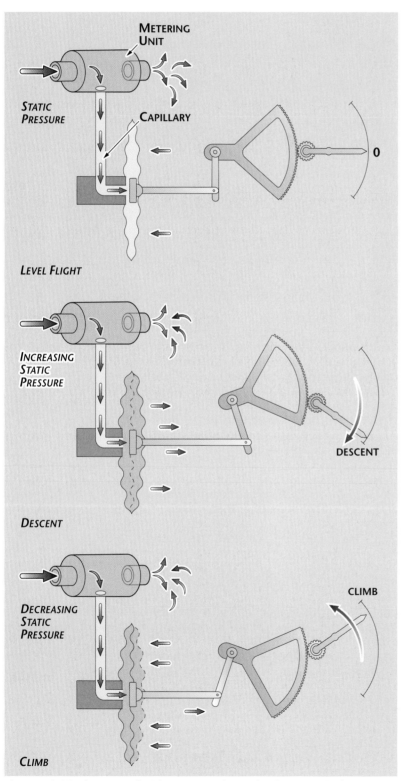

Figure 5-3-17. The aneroid within a vertical speed indicator includes a calibrated leak.

Figure 5-3-18. Instantaneous vertical speed indicators are usually marked as such.

bilizes at a definite ratio, indications are not reliable. Because of the restriction in airflow through the calibrated leak, it can take six to nine seconds before the indication is stabilized. Indications prior to that point are not reliable.

The VSI has a zero adjustment on the front of the case. This adjustment is used with the aircraft on the ground to return the pointer to zero. While adjusting the instrument, tap it lightly to remove friction effects.

Instantaneous vertical speed indicator (IVSI). The instantaneous rate-of-climb indicator is a more recent development that incorporates acceleration pumps to eliminate the limitations associated with the calibrated leak. See Figure 5-3-18 and 5-2-19. For example, during an abrupt climb, vertical acceleration causes the pumps to supply extra air into the diaphragm to stabilize the pressure differential without the usual lag time. During level flight and steady-rate climbs and descents, the instrument operates on the same principles as the conventional rate-of-climb indicator.

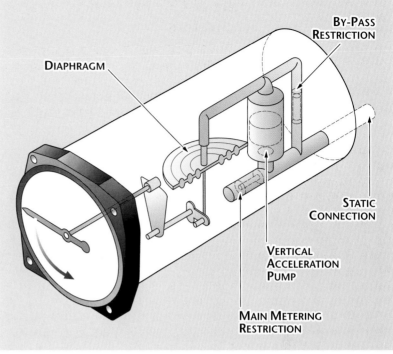

Figure 5-3-19. An IVSI contains an acceleration pump to eliminate the time lag.

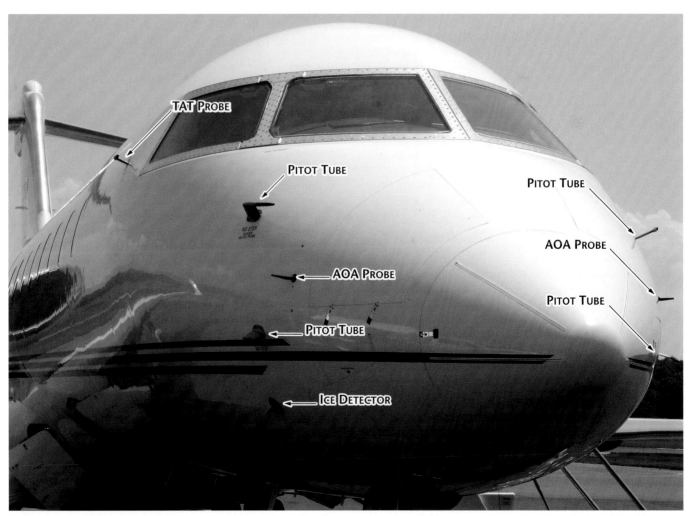

Figure 5-3-20. This Bombardier Global Express business jet shows a typical complex pitot-static system. It includes pitot tubes as well as TAT and AOA probes.

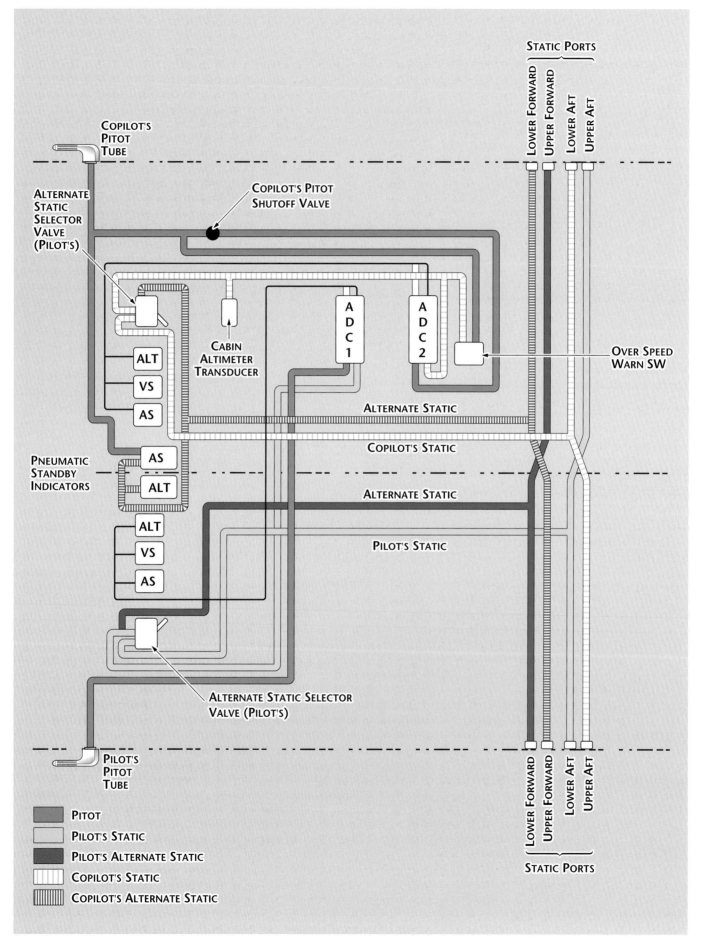

Figure 5-3-21. A typical jet aircraft pitot-static system is very complex.

A zero-setting system, controlled by a setscrew or an adjusting knob permits adjustment of the pointer to zero. The pointer of an indicator should indicate zero when the aircraft is on the ground or maintaining a constant pressure level in flight.

Complex Pitot-Static Systems

Simple pitot-static systems are found in most light aircraft. High-speed flight adds additional complexity to the measurement of airspeed and altitude. Impact and static pressure characteristics change as the speed of sound is approached. This causes significant errors in a simple pitot-static system. Air friction creates heat in the air entering the pitot-static system. This heat changes the pressures within the system. Complex pitot-static systems incorporate additional features to accurately measure airspeed and altitude during high-speed flight. The sensors shown in Figure 5-3-20 are commonly found on high-speed jet aircraft. The pitot tubes have been previously described. The other items shown, the angle of attack probe and the total air temperature sensor are described in the following sections.

Earlier high-speed jet aircraft utilizing analog gauges required the flight crew to consult tables to compensate for these errors. Air data computers (ADCs) were then developed to make these calculations and display the calculated information to the flight crew. A typical complex pitot-static system is shown in Figure 5-3-21.

These systems are required to have multiple independent pitot-static systems. This provides redundancy in the event of a system failure. The system in the illustration has two separate systems. One system is connected to the pilot's instruments, the other to the copilot's. Each system has a pitot tube as well as four static ports. Some aircraft types have multiple pitot tubes for each system as well.

Three separate altimeters and airspeed indicators are provided. Two sets, one for each pilot, are powered by each of the air data computers. A third analog standby set is provided in the event of an electrical failure. Each air data computer also drives a rate-of-climb indicator. A standby rate-of-climb indicator is not required.

Some of the instruments described, such as the complex altimeters, can also be found on aircraft with simple pitot-static systems.

Pressure Definitions

A brief summary of the key pressures, and how they vary with airspeed is helpful before proceeding.

Indicated static pressure. This pressure (P) is the atmospheric pressure as sensed at a point on the aircraft that is relatively free from airflow disturbances. At subsonic speeds, static pressure error is small and of little significance. However, at transonic and supersonic speeds, the static ports sense extreme static pressure errors. Both Mach number and angle of attack can cause significant errors in the static pressure system. Indicated static pressure (P), as detected by the aircraft static ports, deviates from true static pressure. These deviations have a definite relationship to Mach number and angle of attack. The size of the error is the ratio of true static pressure to indicated static pressure, as related to Mach number and angle of attack.

Impact pressure. As implied, impact pressure (Qc) is the force of the air against the aircraft. Qc is measured directly by use of a pitot-static probe (Figure 5-3-22) or calculated from the outputs of the static and total pressure transducers. The ADC calculates actual impact pressure (QA) as a function of Mach number squared and static pressure.

Indicated total pressure. This pressure (Pti) is the sum of static air pressure and the pressure created by aircraft motion through the air. The pitot tube senses total pressure, which is also referred to as pitot pressure.

Corrected static and corrected total pressures. These pressures, Ps and Pt, contain errors that must be corrected to get true static and true total pressures. These errors are a

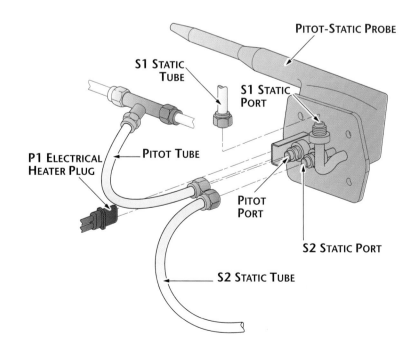

Figure 5-3-22. A modern high-speed pitot-static tube and its many connections

Figure 5-3-23. An air data computer

The air data computer (ADC) (Figure 5-3-23) was developed to compensate for the errors and display accurate information to the flight crew. Extensive aerodynamic testing is performed during the initial certification process to determine the information needed to design the air data computer. Each computer fits a specific airframe design and cannot be used on other aircraft types without reprogramming.

Most modern turbine aircraft use digital air data computers (DADC). Digital air data computers are microprocessor based digital computers that accept both digital and analog inputs, perform digital computations, and supply both digital and analog outputs. Refer back to the digital logic chapter in *Introduction to Aircraft Maintenance*, for a review of digital computer systems.

The DADC receives pitot-static pressures and total air temperature inputs for computing the standard air data functions. It also provides outputs for driving the air data displays, the transponders, flight director, autopilot, and pressurization as well as other elements of the automatic flight control and navigation systems. The typical DADC provides the following outputs.

- Altitude, corrected for position error based on Mach number signals, to the pilot and copilot altimeters

- Altitude to the transponder

- Altitude error to the flight director system

- Altitude error and airspeed to the autopilot system

- Pressure altitude, Mach number, calibrated airspeed, true airspeed, total air temperature and static and total pressure to the autothrottle system

- Mach number to the pilot and copilot Mach/airspeed indicators

- Computed true airspeed to the Mach/airspeed indicators

- Static air temperature to the SAT indicator

- Corrected pressure altitude, computed airspeed, Mach number, true airspeed, altitude change rate, and static air temperature to the flight management computers.

Depending on the type of flight data recording, some or all of these outputs are also provided to and recorded by the flight data recorder. At a minimum, the flight data recorder will have altitude and airspeed inputs from the ADC. Figure 5-3-24 illustrates these outputs.

A typical DADC has a test function. Activating the test function will cause the DADC to send

result of slope and offset errors related to Mach speeds. The computer calculates the specified slope and intercept errors as functions of the indicated pressure ratio (Pti/P) and of the indicated angle of attack (ai).

Air Data Computers

Aircraft operating below 0.8 Mach airspeed can reliably use raw pitot and static pressures to develop accurate airspeed, altitude, and vertical speed indications. Aircraft operating in this speed range use the pressures that the pitot-static ports sense.

Many modern turbine aircraft operate at speeds above Mach 0.8. At high speed, pressures build up on the external skin of the aircraft. These pressures cause a distortion of the normal flow of air, causing the pitot-static system to sense false pressures. The system then supplies erroneous information to the flight instruments. The altimeter, for instance, may show an error of more than 3,000 feet. A 3,000-foot error in altitude is intolerable and could put an aircraft in an extremely dangerous position.

Friction between the aircraft and the atmosphere during high-speed flight also creates significant air heating. Air pressures change with temperature. The friction heat causes additional pressure errors.

Flight testing has shown that these errors are predictable and consistent for any given aircraft.

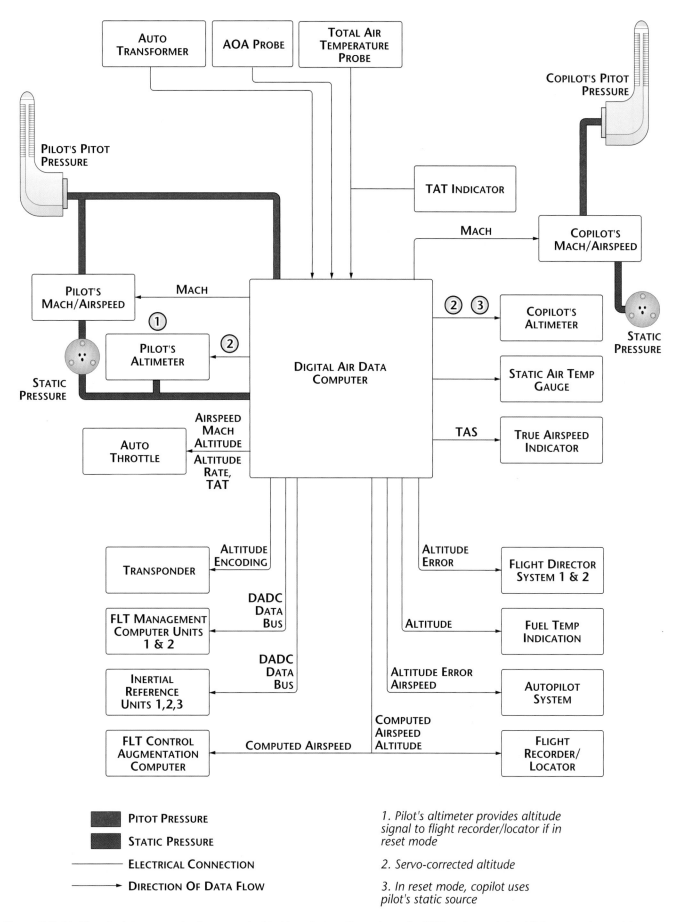

Figure 5-3-24. The air data computer is connected to the pitot-static system, the TAT probe and the AOA sensor. It provides corrected information to a number of instruments and systems.

Figure 5-3-25. High-speed aircraft use an airspeed indicator that also displays the Mach number.

test signals to each of the units controlled by the computer. In our example system, the following indications are displayed when the test switch in pressed:

- All flags are out of view
- Altimeter displays 1,000 feet
- Vertical speed indicator displays 500 feet per minute
- Indicated airspeed is 350 knots
- Vmo is indicated at 300 knots
- Mach speed is indicated at 0.79
- TAT indicates –16°C
- SAT indicates –45°C
- TAS indicates 466 knots

DADC unit number one is supplied pitot pressure from the pilot's pitot tube and static pressure from the pilot's static system or the alternate static system as selected by the pilot's static source selector switch. DADC unit number two is connected to the copilot's pitot-static inputs.

Mach/airspeed indicators. In some cases, the term Mach number is used to express aircraft speed. The Mach number is the ratio of the speed of a moving body to the speed of sound in the surrounding medium. For example, if an aircraft is flying at a speed equal to one-half the local speed of sound, it is flying at Mach 0.5. If it moves at twice the local speed of sound, its speed is Mach 2. (The term Mach number comes from the name of an Austrian physicist, Ernest

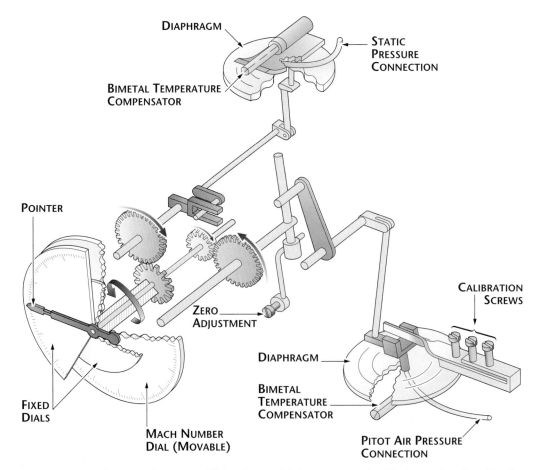

Figure 5-3-26. Mach meters have an additional aneroid that corrects the Mach number for altitude.

Mach, pioneer in the field of aerodynamics.) The speed of sound varies with air pressure. Air pressure varies with altitude, thus the speed of sound increases as an aircraft climbs, even if the barometric setting remains the same.

Figure 5-3-25 shows the front view of a typical airspeed and Mach number indicator. The instrument consists of altitude and airspeed mechanisms incorporated in a single housing. This instrument gives the pilot a simplified presentation of both indicated airspeed and Mach number. Both indications are read from the same pointer. The pointer shows airspeed at low speeds, and both indicated airspeed and Mach number at high speeds. Pitot pressure on a diaphragm moves the pointer, and an aneroid diaphragm controls the Mach number dial. The aneroid diaphragm reacts to static pressure changes because of altitude changes. Figure 5-3-26 is a mechanical schematic of an airspeed and Mach indicator.

The range of the illustrated instrument is 80 to 400 knots indicated airspeed and from 0.3 to 1.2 Mach number. Its calibrated operating limit is 50,000 feet of altitude. A stationary airspeed dial masks the upper range of the movable Mach dial at low altitudes. The stationary airspeed dial is graduated in knots. The instrument incorporates a Mach number setting index. A knob on the lower left-hand corner of the instrument adjusts the index. It can be adjusted over the entire Mach range.

Mach/airspeed indicators can also be servo controlled by an air data computer. These display computed airspeed and mach number information that has been corrected for temperature and high-mach pressure errors.

Servo controlled indicators typically have an identical face to the aneroid indicators. They differ in their internal mechanisms. The needle is moved by a servo motor that receives its position information from the air data computer. Some aircraft use indicators that only have the mach meter driven by the air data computer. The airspeed portion is operated by a conventional aneroid. Review the maintenance information for each specific aircraft to ensure a complete understanding of the system used before troubleshooting.

Complex Altimeters

Complex altimeters add a variety of additional features to the basic operation of the simple altimeter. These features include altimeters with digital readouts showing altitude, altimeters with altitude digitizers connected to transponders and altimeters with remote aneroids and servoed cockpit displays.

Figure 5-3-27. An altimeter with multiple additional readouts

The simplest addition to the basic altimeter is a digital readout. This type of altimeter, shown in Figure 5-3-27 had improved readability.

Another common addition is an altitude encoder. The altitude encoder is connected to the altimeter mechanism. It sends an electrical signal to the aircraft transponder (transponders are described in Chapter 6) providing the transponder with altitude information. Externally, the encoding altimeter is similar to the standard altimeter. It will, however, have an additional electrical connection on the back.

Servoed barometric altimeters. The servoed barometric altimeter is the most complex of the analog altimeters. A typical unit contains a pointer, an altitude counter, a failure warning flag, an altitude alert annunciator, a BARO set knob, and barometric counters in both inches of mercury and millibars. These functions will be described for the sample altimeter. Some functions and limits will vary from aircraft to aircraft.

The altitude alert annunciator illuminates to provide a visual indication when the aircraft is within 1,000 feet of the preselected altitude and extinguishes when the aircraft is within 250 feet of the preselected altitude. After capture, the light will illuminate if the aircraft departs more than 250 feet from the selected altitude and will extinguish if the aircraft altitude moves more that 1,000 feet from the preselected altitude.

The failure warning flag comes into view when the difference between the altitude displayed and the altitude signal received is too great, if the ADC goes invalid, or if the altimeter loses primary power.

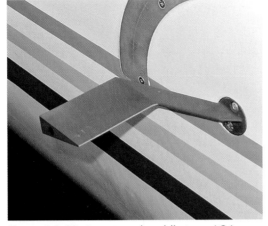

Figure 5-3-28. An external paddle type AOA transmitter

The four-drum counter displays altitude from zero to 50,000 feet. A negative (NEG) altitude shutter obscures the 10,000 and 1,000 digits of the counter at indicated altitudes below sea level. The zero position on the ten-thousands drum is black and white crosshatched to alert the pilot to altitudes below 10,000 feet.

The BARO knob sets the barometric counters. This differs from the simple altimeter, but provides for the same functionality. Rather than directly affecting the aneroid gear train, the BARO knob provides a barometric correction amount to the air data computer, which in turn adjusts the altitude signal sent back to the servoed altimeter.

Figure 5-3-29. An internal paddle type AOA transmitter

The pointer displays altitudes between 1,000 feet levels on a scale with major inches every 100 feet and minor inches every 20 feet. The pointer is read in the same manner as a simple altimeter.

Angle-of-Attack Systems

The angle of attack is the angle between the relative wind and the chord of the wing. (The chord of the wing is a straight line running from the leading edge to the trailing edge.) Increasing the angle of attack increases the pressure felt under the wing and vice versa.

The angle-of-attack indicating system consists of an airstream direction detector (transmitter), and an indicator located on the instrument panel. The airstream direction detector contains the sensing element that measures local airflow direction relative to the true angle of attack by determining the angular difference between local airflow and the fuselage reference plane. The sensing element operates in conjunction with a servo-driven balanced bridge circuit that converts probe positions into electrical signals.

Angle-of-attack transmitter (AOA). Two types of transmitters are used. One type has an external paddle, the other an internal one. The external paddle type transmitter (Figure 5-3-28) has a detector probe that senses changes in airflow. Changes in airflow cause the probe paddle to rotate. This, in turn, drives the wiper arms of the three internally mounted potentiometers.

The enclosed paddle type AOA probe contains two parallel slots that detect the differential airflow pressure (Figure 5-3-29). Air from the slots is transmitted through two separate air passages to separate compartments in a paddle chamber. Any differential pressure, caused by misalignment of the probe with respect to the direction of airflow, will cause internal paddles to rotate. The moving paddles rotate the probe through a mechanical linkage, until the pressure differential is zero. This occurs when the slots are symmetrical with the airstream direction.

Both external and enclosed paddle AOA probes have two electrically separate potentiometer wipers, rotating with the probe, provide signals for remote indications. Probe position, or rotation, is converted into an electrical signal by one of the potentiometers which is the transmitter component of a self-balancing bridge circuit. When the angle of attack of the aircraft is changed and, subsequently, the position of the transmitter potentiometer is altered, an error voltage exists between the transmitter

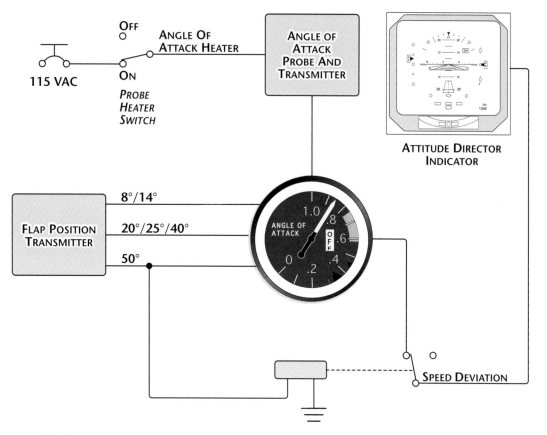

Figure 5-3-30. An angle-of-attack wiring schematic

potentiometer and the receiver potentiometer in the indicator. Current flows through a sensitive polarized relay to rotate a servomotor in the indicator. The servomotor drives a receiver/potentiometer in the direction required to reduce the voltage and restore the circuit to an electrically balanced condition. The indicating pointer is attached to, and moves with, the receiver/potentiometer wiper arm to indicate on the dial the relative angle of attack. A typical wiring schematic is shown in Figure 5-3-30.

The angle-of-attack system shows the pilot aircraft pitch attitude with respect to the surrounding air mass.

AOA Indicator. The AOA indicator presents a display of angle of attack that is indicated in percent of lift on a scale of zero to 1.0. Zero represents zero lift and one represents 100 percent. A typical AOA indicator is shown in Figure 5-3-31. The AOA indicator includes an adjustable indexer. The pilot sets the indexer to the optimum AOA for the flight mode.

Some systems include an AOA indexer. The AOA indexer is a three-light, three-color unit mounted on the aircraft glareshield. It provides a heads-up source of information with respect to deviation from the adjustable reference index on the AOA indicator. The indexer indicates when AOA is high, on-reference or low.

Newer aircraft with glass cockpit systems display the AOA information on one of the cockpit display units alongside other information.

AOA computer. The AOA computers receive information from the AOA probes and provide outputs for the AOA indicator and the stall barrier/limiter system. The AOA com-

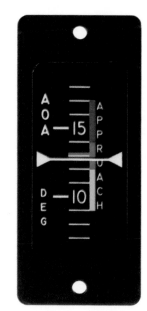

Figure 5-3-31. An angle-of-attack indicator

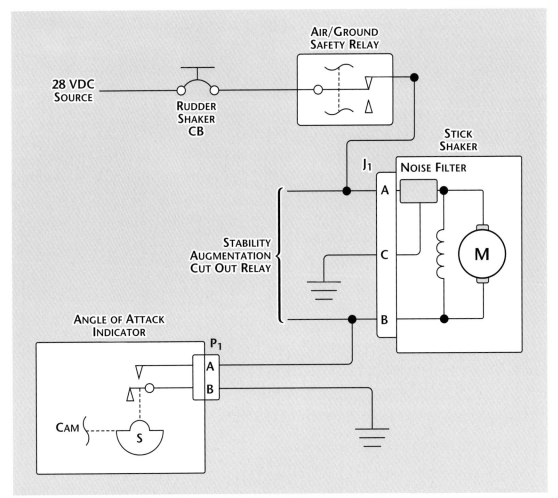

Figure 5-3-32. A wiring schematic for a stick shaker stall warning circuit.

Figure 5-3-33. A Rosemount type TAT probe

puter calculates the lift for various aircraft aerodynamic configurations, including flap and slat positions.

Because the pilot and copilot systems are independent of each other, a sideslip or other out-of-trim condition can cause the two indicators to display different information. A continual difference between the two indica-

tors usually indicates an out-of-trim condition, not an AOA computer or probe problem.

Angle-of-attack stall warning systems. Many aircraft have stall warning indicators to warn the pilot of an impending aerodynamic stall. In the past, stall warning indicators were of a pneumatic control type. They often are nothing more than a contact switch attached to a flapper in the airstream on the leading edge of the wing. When the wing stalls, the airflow across them reverses and the switch contacts are closed. These devices activated either warning horns or flashing lights.

Later, research found that a stall relates directly to the angle of attack, regardless of airspeed, power setting, or aircraft loading. The stall warning devices of most current production aircraft operate at a specified angle of attack.

At a predetermined angle of attack, the stall warning system activates. The devices operate through cams in the angle-of-attack indicator. The cam-driven switch activates a vibrator motor connected to either a rudder pedal or the control stick. Figure 5-3-32 shows a simplified

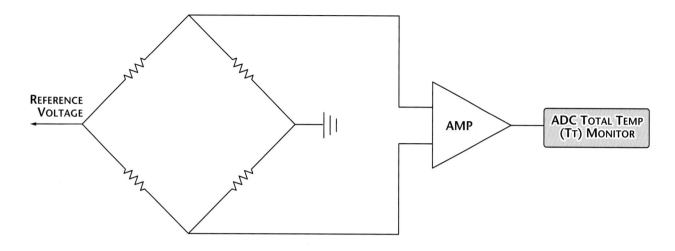

Figure 5-3-34. Total temperature circuit

schematic of the stick shaker system. When the aircraft reaches stall angle of attack, the AOA indicator cam-actuated switch completes the stick shaker motor circuit to ground. When the angle of attack returns below stall conditions, the cam deactuates the switch. The switch action removes the ground to the stick shaker motor.

Total Air Temperature (TAT) Probe

Total temperature (Tt) is the temperature of ambient air plus the temperature increase created by the motion of the aircraft. A probe senses total temperature. This probe includes a platinum resistance element inside an aerodynamic housing placed in the airstream. These probes are often referred to Rosemount probes, after the largest manufacturer of TAT probes. A Rosemount TAT probe is shown in Figure 5-3-33.

The resistive element, whose resistance varies with temperature, acts as the variable portion of a bridge circuit. The total temperature probe provides the ADC with accurate outside air temperature information. The raw information is the indicated total temperature (Tti). The computer smooths and limits computations on the Tti before using the resultant output to calculate true total temperature (Tt). Figure 5-3-34 shows a typical temperature-sensitive bridge circuit that provides temperature data to the air data computer. These temperatures are used by the air data computer to compute the corrected airspeed and altitude.

Total Air Temperature Indicator

The total air temperature (TAT) indicator provides the flight crew with the temperature

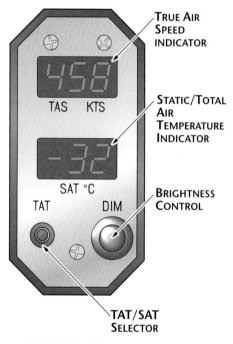

Figure 5-3-35. Total air temperature indicator

information picked up by the total air temperature probe (see Figure 5-3-35). Some indicators also display the static air temperature (SAT). The TAT/SAT indicator may be directly connected to the TAT probe or the TAT information may be relayed by the air data computer.

Most modern systems use the TAT probe to provide information to the air data computer. Prior to the advent of air data computer systems, the total air temperature probe was connected directly to the indicator. These early systems required the flight crew to make manual calculations to adjust airspeed and altitude for the induced pressure and temperature errors.

Solid State Pressure Sensors

Solid-state digital computer technology is becoming more prevalent in production aircraft. Many digital air data systems have used aneroid barometers connected to synchros to send an analog altitude signal to the air data computer. The air data computer has an analog to digital converter to change the synchro signal to a digital signal.

Some of the latest digital systems use a new type of pressure sensor that eliminates the mechanical aneroid. These sensors operate on the piezoelectric effect, which takes advantage of the fact that certain crystalline materials generate electrical signals when subject to pressure. A piezoelectric sensor consists of quartz disks with a metallic pattern placed on them. The disks are arranged in a thin stack that acts as a flexible diaphragm.

When subject to pressure, the diaphragm flexes, setting up an electrical polarization in the disks that produces an electrical charge. Polarity of the charge changes with the direction of the flexing. These voltage changes provide pressure information to the air data computer.

These sensors are the heart of the air data module (Figure 5-3-36). When connected to the static port, the air data module provides altitude information. When connected to the pitot tube, the air data computer can compare the altitude and pitot pressures to compute the airspeed. Digital air data computers also measure the rate of change in the altitude to compute vertical speed information.

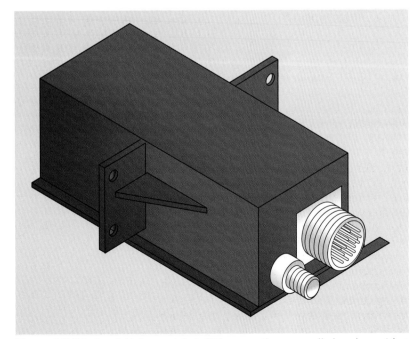

Figure 5-3-36. An air data module is little more than a small chamber with a piezoelectric sensor, electrical and pitot or static connections.

Maintenance of Pitot-Static Systems

The specific maintenance instructions for any pitot-static system are usually detailed in the applicable aircraft manufacturer's maintenance manual. However, there are certain inspections, procedures and precautions that apply to all systems.

Regulatory requirements. The requirements of CFR part 91, General Operating and Flight Rules, require aircraft operating under instrument flight rules (IFR) have their pitot-static system and altimeter tested and inspected every 24 calendar months. The altimeter requirements are listed in CFR part 43, appendix E. When a transponder with an altitude encoder is installed in the aircraft, the requirements of CFR part 43, appendix F must also be met. The test and inspection can be performed by an appropriately rated AMT, or an instrument repair station.

Inspection. Pitot tubes and their supporting structure should be inspected for security of mounting and evidence of damage. Checks should be made to ensure that electrical connections are secure. The pitot pressure entry hole, drain holes, and the static holes or ports should be inspected to ensure that they are unobstructed. The size of the drain holes and static holes is aerodynamically critical. They must never be cleared of obstruction with tools likely to cause enlargement or burring. The lines and fittings from the pitot tube and static ports to the instruments should be inspected for damage and proper routing.

Heating elements should be functionally checked to make sure the pitot tube and static ports begin to warm up when the heater is switched on. If an ammeter or load meter is installed in the circuit, a current reading should be taken.

Individual instruments should be inspected for security of mounting, visual defects, and proper functioning. The zero setting of pointers must also be checked. When inspecting the altimeter, the barometric pressure scale should be set to read field barometric pressure, the instrument should read zero within the tolerances specified. No adjustments of any kind can be made to the altimeter or any other instrument. If the reading is not within limits, the instrument must be replaced, or sent to an instrument repair shop for possible overhaul.

Pitot static tests. Aircraft pitot-static systems must be tested for leaks after the installation of any component parts, when the system malfunction is suspected, and at the periods specified in the CFR part 91.

The method of leak testing and the type of equipment to use depends on the type of aircraft and its pitot-static system. A typical system tester is shown in Figure 5-3-37. In all cases, pressure and suction must be applied and released slowly to avoid damage to the instruments. The method of testing consists of applying pressure and suction to pressure heads and static vents respectively, using a leak tester and coupling adaptors. The rate of leakage should be within the permissible tolerances for the type of aircraft. An un-pressurized aircraft system must leak no more than 100 feet in one minute with a differential pressure of 1" Hg or an altimeter setting of 1,000 feet above field elevation. A pressurized aircraft static system must have a pressure differential equivalent to the maximum cabin pressure differential and the loss must not exceed 2% of the equivalent altitude of the maximum cabin differential pressure or 100 feet, whichever is greater.

If a static pressure system leak check reveals excessive leakage, the leaks may be located by isolating portions of the line and testing each portion systematically. It is best to start at the instrument connections and working down line toward the static ports.

The tests, inspections and equipment used on aircraft with air data computers are complex and require type specific training. An air data test set can be seen in Figure 5-3-38.

Leak tests also provide a means of checking that the instruments connected to a system are functioning properly. However, a leak test does not serve as a calibration test for the altimeter or altitude reporting system.

> **CAUTION:** *Upon completion of the leak test, be sure that the system is returned to the normal flight configuration. If it was necessary to blank off various portions of the system, check to be sure that all blanking plugs, adaptors, or pieces of adhesive tape have been removed.*

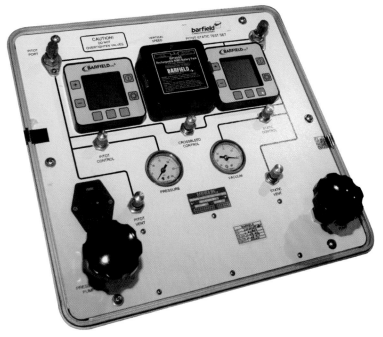

Figure 5-3-37. A typical pitot static system tester requires training before being used on an aircraft.

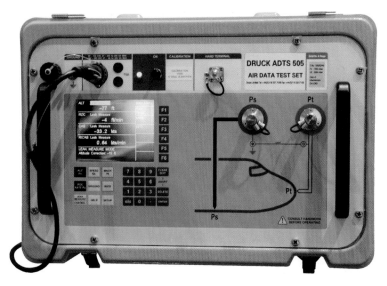

Figure 5-3-38. Aircraft that incorporate air data computers require sophisticated pitot static tests and equipment.

Section 4

Gyroscopic Instruments

Gyroscopic Principles

Early aircraft were flown by visually aligning the aircraft with the horizon. With poor visibility, it was not possible to fly the aircraft safely. The need for flight instruments to correct this condition lead to the development of gyroscopic instruments. The gyroscopic properties of a spinning wheel made precision instrument flying and precise navigation practical and reliable. Some of the instruments that use this principle are the turn-and-bank indicator, directional gyro, gyro horizon (attitude indicator), and the drift meter. Systems that use the gyroscopic principle include the autopilot, gyro stabilized flux-gate compass, flight director and the inertial navigation system. The following contains a brief description of the basic mechanical gyroscope. Solid state gyros, which sense acceleration and translate that information into attitude and heading information, are described in a later section.

• High speed rotation with low-friction

The mountings of the gyro wheels are gimbals. They can be circular rings or rectangular frames. However, some flight instruments use part of the instrument case itself as a gimbal. A simple gyroscope is shown in Figure 5-4-1.

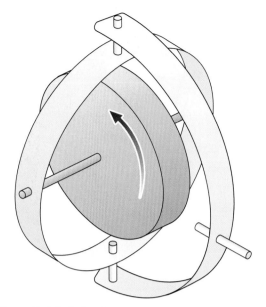

Figure 5-4-1. A simple gyroscope

The two general types of mountings for gyros are the free or universal mounting and the restricted or semirigid mounting. The type of mounting the gyro uses depends on the gyro's purpose.

A gyro can have different degrees of freedom. The degree of freedom depends on the number of gimbals supporting the gyro and the arrangement of the gimbals. Do not confuse the term degrees of freedom, as used here, with an angular value as in degrees of a circle. The term degrees of freedom, as used with gyros, shows the number of directions in which the rotor is free to move. (Some authorities consider the spin of the rotor as one degree of freedom, but most do not.)

A mechanical gyroscope is a spinning wheel or rotor with universal mounting. This mounting allows the gyroscope to assume any position in space. Any spinning object exhibits gyroscopic properties. The wheel, with specific design and mounts to use these properties, is a gyroscope. The two important design characteristics for instrument gyros are:

• High density weight for small size

A gyro enclosed in one gimbal, such as the one shown in Figure 5-4-1, has only one degree of freedom. This is a freedom of movement back and forth at a right angle to the axis of spin. When this gyro is mounted in an aircraft, with its spin axis parallel to the direction of travel and capable of swinging from left to right, it has one degree of freedom. The gyro has no other freedom of movement. Therefore, if the aircraft should nose up or down, the geometric plane containing the gyro spin axis would move exactly as the aircraft does in these directions. If the aircraft turns right or left, the gyro would not change position, since it has a degree of freedom in these directions.

A gyro mounted in two gimbals normally has two degrees of freedom. Such a gyro can assume and maintain any attitude in space. For illustrative purposes, consider a rubber ball in a bucket of water. Even though the water is supporting the ball, it does not restrict the ball's attitude. The ball can lie with its spin axis pointed in any direction. Such is the case with a two-degree-of-freedom gyro (often called a free gyro).

In a two-degree-of-freedom gyro, the base surface turns around the outer gimbal axis or around the inner gimbal axis, while the gyro spin axis remains fixed. The gimbal system isolates the rotor from the base rotation. The universally mounted gyro is an example of this type. Restricted or semirigid mounted gyros are those mounted so one plane of freedom is fixed in relation to the base.

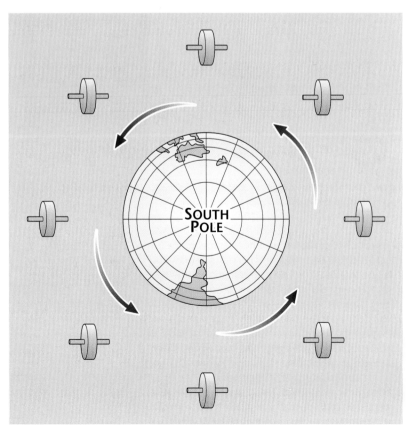

Figure 5-4-2. Action of a freely mounted gyroscope

Practical applications of the gyro are based upon two basic properties of gyroscopic action:

- Rigidity in space
- Precession

Rigidity. Newton's first law of motion states "A body at rest will remain at rest, or if in motion will continue in motion in a straight line, unless acted upon by an outside force." An example of this law is the rotor in a universally mounted gyro. When the wheel is spinning, it stays in its original plane of rotation regardless of how the base moves. Figure 5-4-2 shows this principle. The gyroscope holds its position relative to space, even though the earth turns around once every 24 hours.

The factors that determine how much rigidity a spinning wheel has are in Newton's second law of motion. This law states "The deflection of a moving body is directly proportional to the deflective force applied and is inversely proportional to its mass and speed."

Gyroscopic rigidity depends upon several design factors:

- Weight - For a given size, a heavy mass is more resistant to disturbing forces than a light mass
- Angular velocity- The higher the rotational speed, the greater the rigidity or resistance to deflection
- Radius at which the weight is concentrated- Maximum effect is obtained from a mass when its principal weight is concentrated near the rim rotating at high speed
- Bearing friction- Any friction applies a deflecting force to a gyro. Minimum bearing friction keeps deflecting forces at a minimum.

To obtain as much rigidity as possible in the rotor, it has great weight for its size and rotates at high speeds. To keep the deflective force at a minimum, the rotor shaft mounts in low friction bearings. The basic flight instruments that use the gyroscopic property of rigidity are the gyro horizon, the directional gyro, and any gyro-stabilized compass system. Therefore, their rotors must be freely or universally mounted.

Precession. Precession (Figure 5-4-3) is the resultant action or deflection of a spinning wheel caused by a deflective force applied to its rim. When a deflective force is applied to the rim of a rotating wheel, the resultant force is 90 degrees ahead of the direction of rotation and in the direction of the applied force. The rate at which the wheel precesses is inversely proportional to rotor speed and directly proportional to the deflective force. The force with which a wheel precesses is the same as the deflective force applied (minus the friction in the gimbal

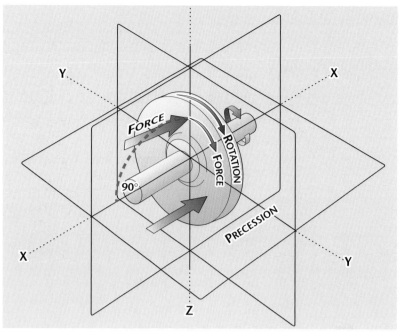

Figure 5-4-3. Precession resulting from deflective forces

ring, pivots, and bearings). If too great a deflective force is applied for the amount of rigidity in the wheel, the wheel precesses and topples over at the same time.

Any spinning mass exhibits the gyroscopic properties of rigidity in space and precession. The rigidity of a spinning rotor is directly proportional to the weight and speed of the rotor speed and inversely proportional to the deflective force.

Attitude Indicator / Gyro Horizon

Pilots determine aircraft attitude by referring to the horizon when they can see it. Often, however, the horizon is not visible. When it is dark or when there are obstructions to visibility such as overcast, smoke, or dust, the pilot cannot use the earth's horizon as a reference. When this condition exists, they refer to an instrument called the attitude indicator. This instrument is also known as a vertical gyro indicator (VGI), artificial horizon, or gyro horizon. From these instruments, pilots learn the relative position of the aircraft with reference to the earth's horizon.

The attitude indicator gyro rotor revolves with its spin axis in a vertical position to the earth's surface. This vertical position is rigidly maintained as the aircraft pitches and rolls about the space-rigid gyro (Figure 5-4-4).

The case of the gyro identically duplicates aircraft movement. The case is free to revolve around the stable gyro because of the mounting

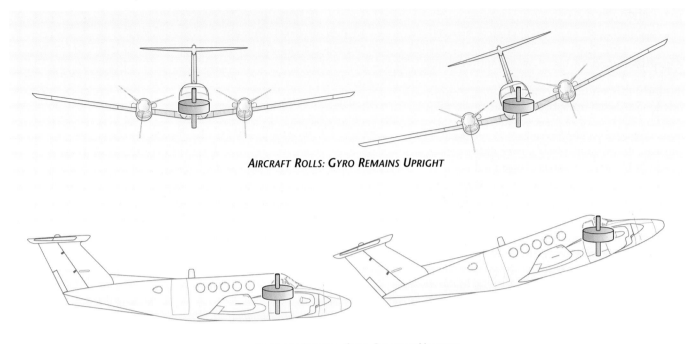

AIRCRAFT ROLLS: GYRO REMAINS UPRIGHT

AIRCRAFT PITCHES: GYRO REMAINS UPRIGHT

Figure 5-4-4. Aircraft reference and gyro stability

of the gyro rotor in gimbals. It follows, therefore, that the aircraft itself actually revolves around the rotor and is the complementing factor in establishing the indications of the instrument.

Although attitude indicators (Figure 5-4-5) differ in size and appearance, they all have the same basic components and present the same basic information. There will always be a miniature aircraft on the face of the indicator that represents the nose (pitch) and wing (bank) attitude of the aircraft. The bank pointer on the indicator face shows the degree of bank (in 10-degree increments up to 30 degrees, then in 30-degree increments to 90 degrees). The sphere is always light on the upper half and dark on the lower half to show the difference between sky and ground. Calibration marks on the sphere show degrees of pitch in 5- or 10-degree increments. Each indicator has a pitch trim adjustment for the pilot to center the horizon as necessary.

Sources of Power for Gyro Operation

Gyroscopic instruments can have an integral gyroscope or a remote gyroscope with a cockpit indicator. Most light aircraft use integral gyroscopic instruments. These instruments can be operated either by a vacuum system or an electrical power source. In some aircraft, all the gyros are either vacuum or electrically motivated; in others, vacuum (suction) systems provide the power for the attitude and heading indicators, while the electrical system drives the gyro for operation of the turn needle. Either alternating or direct current is used to power the gyroscopic instruments.

Most remote gyroscopes are either electrically powered mechanical units or solid state gyros. Solid state gyros are electrically powered, but do not contain spinning wheels. They are described separately.

Vacuum system. The purpose of the vacuum system is to draw a stream of air over the vanes of a gyro, rotating the gyro wheel (or rotor). Venturi tubes and vacuum pump systems are used to power this type of gyro. The vacuum source draws atmospheric air through an instrument air filter over the gyro, causing it to rotate at very high rate of speed.

The vacuum value required for instrument operation is usually between $3 \frac{1}{2}$ inches to $4 \frac{1}{2}$ inches Hg. A vacuum relief valve located in the supply line typically adjusts this value. The turn-and-bank indicators used in some installations require a lower vacuum setting. This is obtained using an additional regulating valve in the individual instrument supply line.

Venturi-tube systems. The advantages of the venturi as a suction source are its relatively low cost and simplicity of installation and operation. Because of their many limitations and the introduction of reliable electrical gyros, venturi tube systems are no longer in common use. A venturi vacuum system is shown in Figure 5-4-6.

15° RIGHT TURN
10° NOSE UP

Gyro remains constantly upright during maneuvers; aircraft revolves and pitches around and about the gyro.

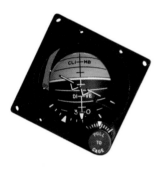

30° LEFT TURN
6° NOSE DOWN

Figure 5-4-5. Roll and pitch indications on the attitude indicator

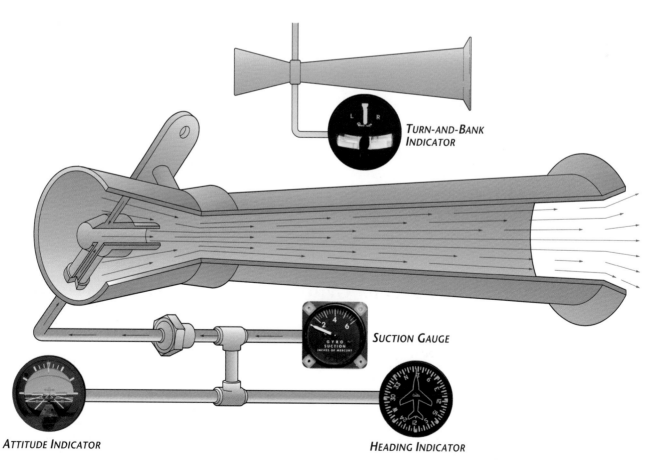

TURN-AND-BANK INDICATOR

SUCTION GAUGE

ATTITUDE INDICATOR

HEADING INDICATOR

Figure 5-4-6. Venturi vacuum system

Figure 5-4-7. Cutaway wet (oil lubricated) vacuum pump

Figure 5-4-8. Dry (carbon lubricated) vacuum pump

Figure 5-4-9. Turbine ejector valve

The line from the gyro is connected to the throat of the venturi mounted on the exterior of the aircraft fuselage. Throughout the normal operating airspeed range the velocity of the air through the venturi creates sufficient suction to spin the gyro.

The venturi system has significant limitations. It is designed to produce the desired vacuum at approximately 100 m.p.h. under standard sea-level conditions. Wide variations in airspeed or air density, or restriction to airflow by ice accretion, will affect the pressure at the venturi throat and thus the vacuum driving the gyro rotor. Additionally, preflight operational checks of venturi-powered gyro instruments cannot be made because the rotor does not reach normal operating speed until after takeoff. For this reason the system is adequate only for light-aircraft instrument training and limited flying under instrument weather conditions. Aircraft flown throughout a wider range of speed, altitude, and weather conditions require a more effective source of power independent of airspeed and less susceptible to adverse atmospheric conditions.

Engine-driven vacuum systems. Reciprocating engines can be used to turn a pump to provide a vacuum source for vacuum driven gyros. The engine driven vacuum pump is very commonly used on light aircraft. Two different types of vacuum pumps are in common use, the wet pump and the dry pump.

The wet pump uses engine oil to lubricate the pump. A cutaway wet pump is shown in Figure 5-4-7. Both wet and dry pumps have a similar vane action. The differences are in their method of lubrication, which means different materials for the pump vanes.

The dry pump, shown in Figure 5-4-8, uses vanes made of a special carbon material. This carbon, as it wears off the vanes, provides for the pump lubrication. One disadvantage of the wet type pump is that it requires an air/oil separator to remove the oil from the air before it is discharged overboard or to some other use.

Some small turbine aircraft use engine bleed air passed through an ejector valve (see Figure 5-4-9) to create a vacuum source. Air pressure moving through the ejector valve creates a suction effect much like the early venturi systems, however the pressure regulator maintains a constant airflow through the ejector. The system also becomes operational as soon as the engines are started, thus eliminating the major limitations of the early venturi system.

Apart from routine maintenance of the filters and plumbing, which are absent in the

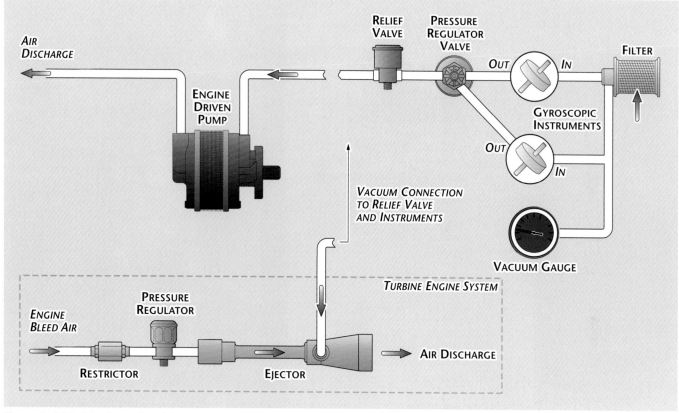

Figure 5-4-10. A basic engine driven vacuum system. A turbine ejector pump is shown at the bottom.

electric gyro, the engine-driven pump is as effective a source of power for light aircraft at low and medium altitudes as the electrical system. At higher altitudes, vacuum systems do not work well.

Pressure operated systems. Aircraft operating at higher altitudes create special problems. The air is thin enough that sufficient vacuum cannot be generated to properly spin a gyro. These aircraft require either electrically driven gyros or pressure-operated gyros.

Pressure systems push air across the gyroscope rotor rather than sucking it past them. This requires some minor changes to the rotor design and a slightly different plumbing operation. While not truly vacuum gyro systems, their operation is essentially the same as with vacuum systems.

Typical Pump-Driven Vacuum System

Figure 5-4-10 shows the components of a vacuum system. Pump capacity and pump size vary in different aircraft, depending on the number of gyros to be operated. Either an engine driven vacuum pump or a bleed air operated ejector can power this system. Both types connect to the relief valve.

Relief valve. Since the system capacity is more than is needed for operation of the instruments, the adjustable suction relief valve is set for the vacuum desired for the instruments. Excess suction in the instrument lines is reduced when the spring-loaded valve opens to atmospheric pressure.

Selector valve. In twin-engine aircraft having vacuum pumps driven by both engines, the alternate pump can be selected to provide vacuum in the event of either engine or pump failure, with a check valve incorporated to seal off the failed pump.

Air filter. The master air filter screens foreign matter from the air flowing through all the gyro instruments, which are also provided with individual filters. Clogging of the master filter will reduce airflow and cause a lower reading on the suction gauge. In aircraft having no master filter installed, each instrument has its own filter. With an individual filter system, clogging of a filter will not necessarily show on the suction gauge.

Vacuum/suction gauge. The suction gauge is a pressure gauge, indicating the difference in inches of mercury ("Hg), between the pressure inside the system and atmospheric or cockpit pressure. The desired vacuum, and the minimum and maximum limits, vary with gyro design. If the desired vacuum for the attitude and heading

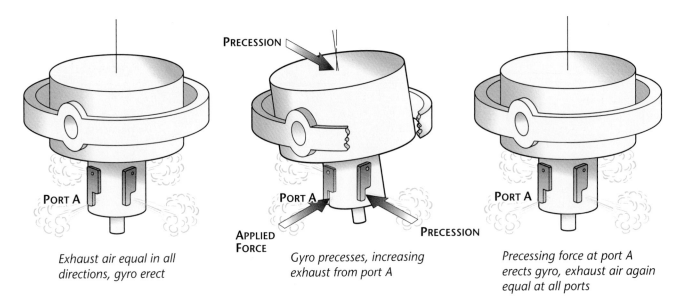

PRECESSION

PORT A   PORT A   PORT A

APPLIED
FORCE   PRECESSION

Exhaust air equal in all   *Gyro precesses, increasing*   *Precessing force at port A*
directions, gyro erect   *exhaust from port A*   *erects gyro, exhaust air again*
                                                                  equal at all ports

Figure 5-4-11. Erecting mechanism of a vacuum driven attitude indicator

indicators is 5"Hg and the minimum is 4.6" Hg, a reading below the latter value indicates that the airflow is not spinning the gyros fast enough for reliable operation. In many aircraft, the system provides a suction gauge selector valve, permitting the pilot to check the vacuum at several points in the system.

Vacuum-Driven Attitude Gyros

In a typical vacuum-driven attitude gyro system, air is sucked through the filter, then through passages in the rear pivot and inner gimbal ring, then into the housing where it is directed against the rotor vanes through two openings on opposite sides of the rotor. The air then passes through four equally spaced ports

in the lower part of the rotor housing and is sucked out into the vacuum pump or venturi (Figure 5-4-11).

The chamber containing the ports is the erecting device that returns the spin axis to its vertical alignment whenever a precessing force, such as bearing friction, displaces the rotor from its horizontal plane. Four exhaust ports are each half-covered by a pendulous vane, which allows discharge of equal volumes of air through each port when the rotor is properly erected. Any tilting of the rotor disturbs the total balance of the pendulous vanes, tending to close one vane of an opposite pair while the opposite vane opens a corresponding amount. The increase in air volume through the opening port exerts a precessing force on the rotor housing to erect the gyro, and the pendulous vanes return to a balanced condition (Figure 5-4-12).

The limits of the attitude indicator specified in the manufacturer's instructions refer to the maximum rotation of the gimbals beyond which the gyro will *tumble*. The bank limits of a typical vacuum-driven attitude indicator are from approximately 100° to 110°, and the pitch limits vary from approximately 60° to 70°, depending on the design of a specific unit. If, for example, the pitch limits are 60° with the gyro normally erected, the rotor will tumble when the aircraft climb or dive angle exceeds 60°. As the rotor gimbal hits the stops, the rotor precesses abruptly, causing excessive friction and wear on the gimbals. The rotor will normally precess back to the horizontal plane at a rate of approximately 8° per minute.

Many gyros include a *caging device*, used to erect the rotor to its normal operating position

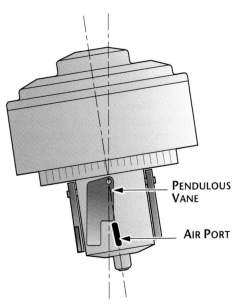

PENDULOUS
VANE

AIR PORT

Figure 5-4-12. Action of the pendulous vanes

prior to flight or after tumbling, and a flag to indicate that the gyro must be uncaged before use. Turning the caging knob prevents rotation of the gimbals and locks the rotor spin axis in its vertical position.

Vacuum system maintenance practices. Errors in the indications presented on the attitude indicator will result from any factor that prevents the vacuum system from operating within the design suction limits, or from any force that disturbs the free rotation of the gyro at design speed. These include poorly balanced components, clogged filters, improperly adjusted valves and pump malfunction. Such errors can be minimized by proper installation, inspection, and maintenance practices.

Other errors, inherent in the construction of the instrument, are caused by friction and worn parts. These errors, resulting in erratic precession and failure of the instrument to maintain accurate indications, increase with the service life of the instrument.

For the aviation technician the prevention or correction of vacuum system malfunctions usually consists of cleaning or replacing filters, checking and correcting for insufficient vacuum, or removing and replacing the instruments. A list of the most common malfunctions, together with their correction, is included in Table 5-4-1

Electric Attitude Indicator

In older aircraft, suction-driven gyros had been favored over the electric types because of their comparative simplicity and lower cost. Improved electrically driven gyros have made the vacuum gyro a thing of the past for all but the smallest general aviation aircraft. While many light aircraft still use vacuum gyros, the ease of installation and reliability of electric systems are making even that a much rarer thing.

Electrically operated gyroscopes use small electric motors to spin the rotor instead of air. These motors are mounted within the gimbals and attached directly to the rotor. The spinning rotor has exactly the same gyroscopic qualities as a vacuum driven gyro. One key difference from the flight crew's perspective is how to know when the system is powered. Vacuum systems are monitored using the vacuum/suction gauge. Electric gyros have a power warning flag to alert the crew to a failure within the gyro or its power source (see Figure 5-4-13).

Electrically driven gyros require an erection mechanism to keep the gyro axis vertical to the

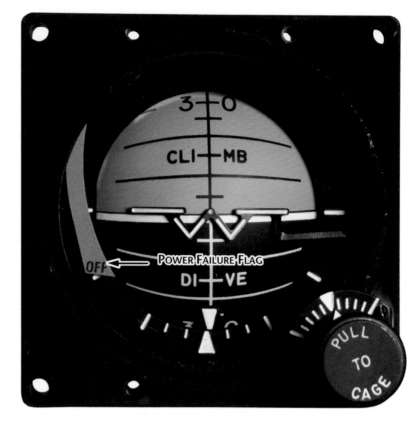

Figure 5-4-13. Power failure flag on an electric gyro

surface of the earth. To accomplish this, a high-speed rotating magnet is attached to the top of the gyro shaft. Around this magnet, but not attached, is a sleeve that is rotated by magnetic attraction. As illustrated in Figure 5-4-14, the steel balls are free to move around the sleeve. If the pull of gravity is not aligned with the axis of the gyro, the balls will fall to the low side. The resulting precession re-aligns the axis of rotation vertically.

The gyro can be caged manually by a lever and cam mechanism to provide rapid erection.

Remote Gyro Indicators

Some attitude indicators have a self-contained gyro. Other more modern indicators use pitch and roll information from the inertial system or the attitude heading reference system. These systems are accurate and reliable. They gain their reliability and accuracy from being larger since size is not limited by the space of an instrument panel. Electrical signals from the remote gyro travel via synchros. The signal is amplified in the indicator to drive servomotors and position the indicator sphere. This positioning is exactly as the vertical gyro position in the gyro case. In the newer attitude indicators, the sphere is gimbal-mounted and capable of 360-degree rotation. In contrast, the older gyros could only travel 60 degrees to 70

	POSSIBLE CAUSE	ISOLATION PROCEDURE	CORRECTION
NO VACUUM PRESSURE OR INSUFFICIENT PRESSURE	Defective valve gauge	On multi–engine aircraft check opposite engine system on the gauge	Replace faulty vacuum gauge
	Vacuum relief valve incorrectly adjusted	Change valve adjustment	Make final adjustment to proper setting valve
	Vacuum relief valve installed backwards	Visually inspect	Install properly
	Broken line	Visually inspect	Replace line
	Lines crossed	Visually inspect	Install lines properly
	Obstruction in vacuum line	Check for collapsed line	Clean and test line. Replace defective part(s)
	Vacuum pump failure	Remove and inspect	Replace faulty pump
	Vacuum regulator valve incorrectly adjusted	Make valve adjustment and note pressure	Adjust to proper pressure
	Vacuum relief valve dirty	Clean and adjust relief valve	Replace if adjustment fails
EXCESSIVE VACUUM	Relief valve improperly adjusted	---------------------------	Adjust relief valve to proper setting
	Inaccurate vacuum gauge	Check calibration of gauge	Replace faulty gauge
GYRO HORIZON BAR FAILS TO RESPOND	Instrument caged	Visually Inspect	Uncage instrument
	Instrument filter dirty	Check filter	Replace or clean as necessary
	Insufficient vacuum	Check vacuum setting	Adjust relief valve to proper setting
	Instrument assembly worn or dirty	---------------------------	Replace instrument
TURN-AND-BANK INDICATOR FAILS TO RESPOND	No vacuum supplied to instrument	Check lines and vacuum system	Clean or replace lines and replace components of vacuum system as necessary
	Instrument filter clogged	Visually inspect	Replace filter
	Defective instrument	Test with properly functioning instrument	Replace faulty instrument
TURN-AND-BANK POINTER VIBRATES	Defective instrument	Test with properly functioning instrument	Replace defective instrument

Table 5-4-1. Vacuum system troubleshooting

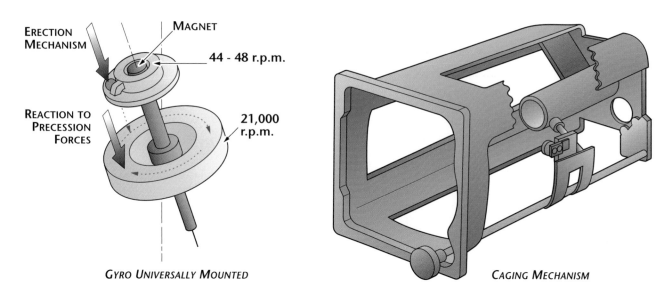

Figure 5-4-14. Erecting and caging mechanisms of an electric attitude indicator

degrees of pitch and 100 degrees to 110 degrees of roll.

The latest gyro systems have replaced the mechanical gyroscope with an electronic one. These systems, sometimes referred to as inertial navigation systems, use ring laser gyros or *micro electronic motion sensors (MEMS)*.

Most modern turbine aircraft, using either analog or digital displays, incorporate the attitude indicator into an instrument called a flight director. The flight director adds radio navigation information to the gyro display. An analog flight director is shown in Figure 5-4-15. Flight director navigation functions are covered in Chapter 6, Navigation and Communication Systems.

Solid State Gyros

Numerous efforts have been made to replace the conventional mechanical gyro. Some of these efforts have been very successful, although until recently they were all very expensive and only found in military aircraft and commercial airliners. Modern electronics advances have made solid state gyro technology available to general aviation aircraft.

Gyros are mechanically complex devices. While fairly reliable, greater reliability is always a safety issue. Directional gyros also have a tendency to precess over time, slowly drifting to the left or right. This precession is not a problem for short distance flights, but for longer flights these errors become progressively larger. These limitations are unacceptable for aircraft in long-range airline and military uses.

Some systems were initially developed to meet a military need for an accurate, long-range system that is independent of ground based references. Other systems have been made possible as a result of modern commercial electronics development.

Inertial Reference Systems

Many air forces have long desired a navigation system that is independent of any outside or ground-based references. Ground based radio

Figure 5-4-15. An analog flight director installed in a King Air corporate turboprop

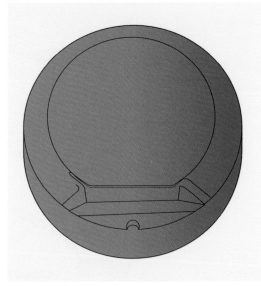

Figure 5-4-16. A ring laser gyro

systems can be jammed or distorted, preventing accurate navigation.

Precision accelerometers can be used to calculate speed and direction information. The driver of a car feels the acceleration, deceleration and turning forces. The same types of forces are present in aircraft. Sensors were developed to accurately measure these acceleration forces. When these sensors are connected to a computer, a precision navigation system can be developed that is totally independent of any outside reference.

These early systems were two-dimensional in their operation and were only used for navigation. Attitude control requires a significantly higher level of precision. Eventually, accelerometers were developed that could be used in attitude reference systems.

Accurately measuring these accelerations requires a stable, level platform, even while the aircraft maneuvers. A gyro-stabilized platform was necessary. These systems, while accurate, are very expensive to manufacture and maintain. They are typically found only on military aircraft and older long-range airliners.

Laser Ring Gyros

The ring laser gyro (RLG) made inertial systems practical for a much larger segment of aviation (see Figure 5-4-16). The ring laser gyro uses the Sagnac effect. Georges Sagnac, a French physicist, experimented in 1913 with a rotating turntable that contained rings of light shining in opposite directions. Sagnac discovered that the light travelling with the rotation took slightly longer to travel around the ring than the light travelling against the rotation.

The rotation of the turntable creates a longer path for the light travelling with the rotation. Figure 5-4-17 illustrates how this works. This longer path changes the frequency of the light. A phase detector compares the frequency of both lights. If the ring laser is not moving, both lights operate at the same frequency. When the ring laser is rotated, the lights shift frequencies. The phase detector measures this shift.

By placing a ring laser to measure yaw, pitch and roll the ring laser can act as a gyroscope. The first ring laser gyros required the same type of gimballed platform as an inertial navigation system. Advances in computer technology made it possible to rigidly mount the ring laser gyros (Figure 5-4-18) to the airframe. The computer constantly calculates the position of the laser rings and is able to determine the appropriate horizontal reference.

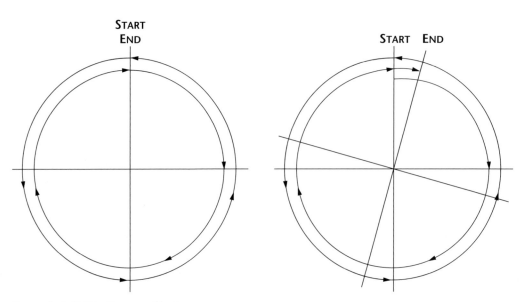

Figure 5-4-17. The Sagnac effect

These systems, known as strap-down systems, are much less expensive to manufacture than gimballed systems. The are also significantly more reliable. Strap-down ring laser gyros are the primary gyro reference system in most modern jet aircraft.

MEMS

Micro electro-mechanical sensors (MEMS) are the latest in solid-state gyro technology (Figure 5-4-19). These sensors have been developed for a number of high volume commercial applications. Adapted for aviation, MEMS gyro systems contain very small silicon chips that precisely measure acceleration. Their high volume commercial use makes them very inexpensive.

FAA certified MEMS based solid state gyros are now available. Their low cost makes them affordable for even piston powered light aircraft. Their low cost is making possible glass cockpit technology for a significant portion of the general aviation market. See Figure 5-4-20.

Directional Gyro

Gyros can also be used to provide heading information. This type of gyro is referred to as a directional gyro. It provides directional information by constraining a vertically mounted mechanical gyro to a single degree of freedom. While directional gyros are primarily used for navigation, they operate on the same gyroscopic principles as the attitude gyro.

A typical directional gyro is shown in Figure 5-4-21. It consists of a gyro rotor and a vertically mounted compass card. As the aircraft changes direction, the gyro remains oriented towards north. A mechanical linkage between the gyroscope and the vertical card indicator moves the card.

All of the gyro technologies discussed can be used to provide direction information. They use a remote indicating instrument to display the heading information.

Turn-and-Bank Gyros

Turn-and-bank indicator. The turn-and-bank indicator (Figure 5-4-22), also called the turn-and-slip indicator, shows the lateral attitude of an aircraft in straight flight. It also provides a reference for the proper executions of a coordinated bank and turn. The turn-and-bank indicator shows when the aircraft is flying on a straight course and the direction and rate of a turn. It was one of the first instruments for

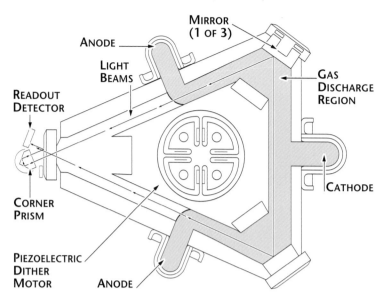

Figure 5-4-18. A ring laser gyro's internal schematic

Figure 5-4-19. A MEMS solid state inertial gyro.

Figure 5-4-20. The Cirrus single engine four place lightplane is now available with an advanced computer based flight instrument system.

Photo courtesy of Cirrus Design

controlling an aircraft without visual reference to the ground or horizon.

The indicator is a combination of two instruments, a ball and a turn pointer. The ball part of the instrument operates by natural forces (centrifugal and gravitational). The turn pointer depends on the gyroscopic property of preces-

Figure 5-4-21. Directional Gyro

Figure 5-4-22. A turn-and-bank indicator

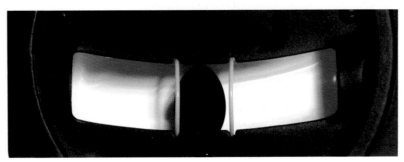

Figure 5-4-23. The ball reacts to gravity and centrifugal forces

sion for its indications. The power for the turn indicator gyro is either electrical or vacuum.

Ball. The ball portion of a turn-and-bank indicator (Figure 5-4-23) consists of a sealed, curved, glass tube. The tube contains water-white kerosene and a black or white agate or common steel ball bearing. The ball bearing is free to move inside the tube. The fluid provides a damping action and ensures smooth and easy movement of the ball. The curved tube allows the ball to seek the lowest point when in level flight. This is the tube center. A small projection on the left end of the tube contains a bubble of air. The bubble lets the fluid expand during changes in temperature. There are two markings or wires around the center of the glass tube. They serve as reference markers to show the correct position of the ball in the tube.

The only force acting on the ball during straight and level flight is gravity. The ball seeks its lowest point and stays within the reference marks. In a turn, centrifugal force also acts on the ball in a horizontal plane opposite to the direction of the turn.

When the force acting on the ball becomes unbalanced, the ball moves away from the center of the tube. In a skid, the rate of turn is too great for the angle of bank. The excessive centrifugal force moves the ball to the outside of the turn. The ball moves in the direction of the force, toward the outside of the turn. Returning the ball to center (coordinated turn) calls for increasing bank or decreasing rate of turn, or a combination of both.

In a slip, the rate of turn is too slow for the angle of bank. The ball moves toward the inside of the turn. Returning the ball to the center (coordinated turn) requires decreasing the bank or increasing the rate of turn, or a combination of both.

The ball instrument is actually a balance indicator because it shows the relationship between angle of bank and rate of turn. It lets the pilot know when the aircraft has the correct rate of turn for its angle of bank.

Turn Pointer. The turn pointer operates on a gyro. The gimbal ring encircles the gyro in a horizontal plane and pivots fore and aft in the instrument case. The major parts of the turn portion of an electrical turn-and-bank indicator, shown in Figure 5-4-24, are as follows:

- A frame assembly used for assembling the instrument.

- A motor assembly consisting basically of the stator, rotor, and motor bearings. The

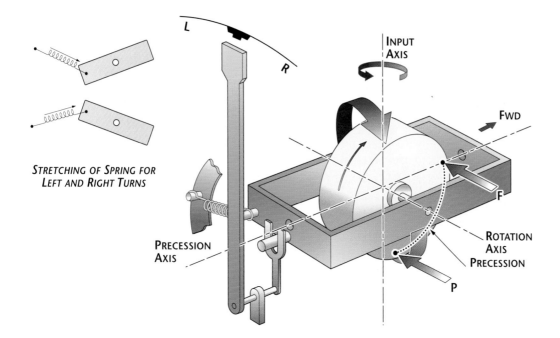

Figure 5-4-24. The gyro mechanism within the turn portion of the turn-and-bank

electrical motor serves as the gyro for the turn indicator.

- A plate assembly for mounting the electrical receptacle, pivot assembly, and choke coil and capacitors for eliminating radio interference. It also mounts the power supply of transistorized indicators.

- A damping unit that absorbs vibrations and prevents excessive oscillations of the needle. The unit consists of a piston and cylinder mechanism. The adjustment screw controls the amount of damping.

- An indicating assembly composed of a dial and pointer

- The cover assembly

The carefully balanced gyro rotates about the lateral axis of the aircraft in a frame that pivots about the longitudinal axis. When mounted in this way, the gyro responds only to motion around a vertical axis. It is unaffected by rolling or pitching.

The turn indicator takes advantage of one of the basic principles of gyroscopes – precession. Precession, as already explained, is a gyroscope's natural reaction 90 degrees in the direction of rotation from an applied force. It is visible as resistance of the spinning gyro to a change in direction when a force is applied. As a result, when the aircraft makes a turn, the gyro position remains constant. However, the frame in which the gyro hangs dips to the side opposite the direction of turn. Because of the design of the linkage between the gyro frame and the pointer, the pointer shows the correct direction of turn. The pointer displacement is proportional to the aircraft rate of turn. If the pointer remains on center, it shows the aircraft is flying straight. If it moves off center, it shows the aircraft is turning in the direction of the pointer deflection. The turn needle shows the rate (number of degrees per minute) at which the aircraft is turning.

By using the turn-and-bank indicator, the pilot checks for coordination and balance in straight and level flight and in turns. By cross-checking this instrument against the airspeed indicator, the relation between the aircraft lateral axis and the horizon can be determined. For any given airspeed, there is a definite angle of bank necessary to maintain a coordinated turn at a given rate.

Turn Coordinator

The turn coordinator (Figure 5-4-25) is a variation on the turn-and-bank indicator. The gyroscope axis is offset approximately 30° from the aircraft's longitudinal axis. This changes the precision axis, adding sensitivity to roll as well as turning.

Rolling an aircraft is the first step in a turn. The turn coordinator provides information about developing skids or slips before they occur. Turns can be flown more smoothly using this information. The turn coordinator is usually labeled as such and typically has an aircraft silhouette rather than a vertical bar display. A typical turn coordinator is shown in Figure 5-4-26.

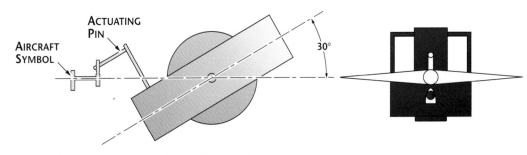

Figure 5-4-25. Turn coordinator internal gyro axis

Figure 5-4-26. A typical turn coordinator

COMMON ACRONYMS AND TERMS	
AHRS	Attitude and Heading Reference System
CRT	Cathode Ray Tube
DAIS	Digital Avionics Information System
EADI	Electronic Attitude Direction Indicator
EFIS	Electronic Flight Instrument System
EHSI	Electronic Horizontal Situation Indicator
FMC	Flight Management Computer
FMCS	Flight Management Computer System
EICAS	Engine Indicating and Crew Alerting System
LCD	Liquid Crystal Display
MFD	Multi-Function Display
PFD	Primary Flight Display

Table 5-5-1. Common acronyms and terms

Section 5

Electronic Flight Information Systems

Digital systems have dramatically changed the appearance of the modern cockpit. While the information the flight crew must monitor has not changed, the manner in which that information is managed and displayed has in many ways changed for the better.

Commonly referred to as a "glass cockpit," advanced displays come in many forms. The term "glass cockpit" refers to the expanse of flat glass panels replacing the traditional clusters of analog gauges. Different manufacturers may use different terms to describe similar units. Some of the common terms are listed in Table 5-5-1.

The first systems used *cathode ray tubes (CRTs)*. Using the same technology as a traditional television, CRT based systems require a lot of depth behind the instrument panel. While still in widespread use, CRTs have largely been replaced by *liquid crystal display (LCD)* based systems in new production aircraft. LCD's offer better reliability, lower weight and a more compact installation. Both types of displays can show similar information.

Electronic systems have a number of advantages compared to traditional analog instrumentation (Figure 5-5-1). They have fewer moving parts and a significantly smaller number of parts. The complexity of the system, its mounting and wiring is much simpler. This combination translates into lower maintenance costs and better dispatch reliability. They are also larger and can be easier to read.

Electronic flight instrument systems (EFIS) first replaced the gyro horizon and directional gyros with display screens. These screens also add airspeed and altimeter displays to the attitude display. This combination, known as an *electronic attitude direction indicator (EADI)*,

Figure 5-5-1. The DC-9/717 series airliners have been updated with modern electronic systems. View (A) shows the 1960's analog configuration, (B) shows the current electronic flight display system. *Photo (B) Courtesy of AirTran Airways*

replaces the traditional flight instrument cluster (see Figure 5-5-2). The gyro horizon, airspeed indicator, altimeter, and vertical speed indicators are all included.

The first generation of glass cockpit systems provided enlarged, combined flight information displays. Figure 5-5-3 shows a typical Honeywell system installed in a Hawker 800 business jet. The attitude gyro and directional gyros have each been replaced with a single screen. The cockpit includes a radar screen and two navigation control screens. The rest of the instrument panel uses conventional analog instrument systems. This type of system takes advantage of the benefits of digital air data systems and enhanced displays, but still has a very complex cockpit installation.

Most systems in production today use large multi-function displays. These displays can show a variety of different types of information and can include data from multiple sources on the same screen. Symbol generators receive data from several sensors and send that data to the appropriate screen. Almost all of the information displayed on conventional analog instruments is now incorporated in the electronic flight instrumentation system.

Figure 5-5-2. Electronic attitude direction indicator (EADI)

Multi-function displays also allow the flight crew to move displayed information from one screen to another. In the event of a screen failure, the displays can easily be reconfigured to compensate. Compare the lower half of the pilot's and copilot's primary displays in Figure 5-5-4. The pilot's panel is acting as horizontal situation indicator, while the copilot's is configured as a radar screen. If either screen failed, the center screen can be reconfigured to display this information.

Electronic flight information systems come in many forms. Various manufacturers have developed systems with three, four, five and six screen systems. Additional variations will surely be seen in future aircraft. The most common systems are five and six panel configurations. Each pilot has two screens. One of the screens is used for attitude and heading information, while the other is typically configured to show weather radar information. Figure 5-5-5 shows a typical five-panel system.

Figure 5-5-3. Some aircraft use a combination of electronic and analog displays.

Figure 5-5-4. Multi-function displays in the cockpit of the Raytheon Premier business jet

Photo courtesy of Raytheon Aircraft

Figure 5-5-5. A typical five-panel cockpit display

The center of the panel will have the remaining screen. They contain the engine operating information. It also monitors the other aircraft systems, including environmental, hydraulics, landing gear and electrical. Known as the *engine indicating and crew alert system (EICAS)* (Figure 5-5-6), this system is described in more detail in Section 7 (Engine Instruments) of this chapter.

From an operational standpoint, the flight crew can custom tailor the displays for the current operational needs. The newest systems offer digital databases containing airport maps, en route navigation information and many other references. The flight crew can easily search and display this data alongside the flight instrument information. Some of these systems add an additional display on the side of the cockpit for the database.

Maintenance is simplified by flight management computer systems (FMCS). System failures and intermittent faults can be recorded and replayed on the ground. Many systems offer enhanced diagnostic software systems. This combination simplifies the maintenance process.

The computer becomes the AMT's assistant in the diagnostic and troubleshooting process. In many cases, the system already knows where the failure has occurred, the technician's job is to access the computer and make the necessary system repairs.

Figure 5-5-6. The engine indicating and crew alerting system (EICAS) panel

Figure 5-6-1. Standby indicators used in the Falcon 900 business jet

Section 6

Standby and Other Flight Instruments

Standby Instruments

All aircraft that use computed airspeed and altitude information, including those that have glass cockpits, are required to have a set of standby instruments. These instruments provide a backup system in the event of a computer or electrical failure. The typical configuration includes an attitude gyro, an airspeed indicator, and an altimeter. These are shown in Figure 5-6-1.

Standby attitude gyro. The standby attitude indicator (see Figure 5-6-2) consists of a miniature aircraft symbol, a bank angle dial, and a bank index. It also includes a two-colored drum background with a horizon line dividing the two.

The indicator roll index is graduated in 10-degree increments to 30 degrees, with graduation marks at 60 degrees and 90 degrees. The example indicator is capable of displaying 360 degrees of roll, 92 degrees of climb, and 79 degrees of dive. Because of the high spin rate of the gyro, the indicator displays accurate pitch-and-roll data for 9 minutes after electrical failure. Some units are equipped with battery backups. The attitude indicator incorporates a pitch trim knob to position the miniature aircraft symbol above or below the horizon reference line. The pitch trim knob also cages the gyro. When the pitch trim knob is pulled out, the gyro will cage. Rotating the knob clockwise while extended will lock in the extended position. The attitude indicator also incorporates an OFF flag. The flag appears if electrical power fails, or if you cage the gyro.

Standby altimeter. Standby altimeters are conventional Kollsman type aneroid altimeters. The standby altimeter provides altitude information to the flight crew in the event of a system failure. It is self-contained and is not dependent on aircraft power. The standby altimeter is typically connected into the copilot pitot-static system. It is subject to error at high Mach numbers, but is accurate at slower speeds and can be used for landing in the case of an electrical or systems failure.

Standby airspeed indicator. The standby airspeed indicator is a small, conventional mechanical airspeed indicator. It requires no external power. While the standby airspeed indicator does not compensate for pressure errors at high Mach numbers, it provides a much needed backup for landing in the event of systems failure. The standby airspeed indicator is typically connected directly into the copilot pitot-static systems. In some cases the standby airspeed indicator and altimeter are combined in a single case (see Figure 5-6-3).

Other Flight Instruments

Many additional instruments are provided to the flight crew as backups, to provide additional information necessary in certain flight conditions or to monitor auxiliary systems. Those instruments that are used to provide additional flight information are described in the following section.

Compass. During the early days of aviation, direction of flight was determined chiefly by direct-reading magnetic compasses. Today, the direct-reading magnetic compass (Figure 5-6-4) is used as a standby compass. Direct-reading magnetic compasses used in most aircraft mount on the instrument panel for use by the pilot. It is read like the dial of a gauge. While technically a navigation instrument, it can also be used to provide attitude information.

A nonmagnetic metal bowl, filled with liquid, contains the compass indicating card. The card provides the means of reading compass indications. The card mounts on a float assembly and is actually a disk with numbers painted on its edge. A set of small magnetized bars or needles fastens to this card. The card-magnet assembly sits on a jeweled pivot, which lets the magnets align themselves freely with the north-south component of the earth's magnetic field. The compass card and a fixed-position reference marker (lubber's line) are visible through a glass window on the side of the bowl.

An expansion chamber in the compass provides for expansion and contraction of the liquid caused by altitude and temperature changes.

Figure 5-6-2. A standby attitude gyro is smaller than the typical primary attitude gyro.

Figure 5-6-3. A combination standby airspeed indicator and altimeter

Figure 5-6-4. A direct-reading magnetic compass

Figure 5-6-5. A typical direct reading outside air temperature indicator

The liquid dampens, or slows down, the oscillation of the card. Aircraft vibration and changes in heading cause oscillation. If suspended in air, the card would keep swinging back and forth and be difficult to read. The liquid also buoys up the float assembly, reducing the weight and friction on the pivot bearing.

Instrument-panel compasses are available with cards marked in steps of either 2 degrees or 5 degrees. Such a compass indicates continuously without electrical or information inputs. You can read the aircraft heading by looking at the card in reference to the lubber line through the bowl window.

Outside air temperature (OAT). Outside air temperature information is sometimes needed by the flight crew. An electrically powered indicator displaying uncorrected outside air temperature is sometimes located in the instrument panel. See Figure 5-6-5. A temperature-sensitive resistor (temperature bulb) is exposed to the slipstream. This resistor measures changes in temperature. The temperature of the air measurement is in the form of changing resistance. The outside air temperature indicator displays this change in resistance. The graduated indicator dial is typically marked in Celsius, from +50 degrees to -50 degrees.

Another type of OAT indicator does not require external power. It consists of a bimetallic temperature probe and a dial indicator. This type is mounted in the cockpit in a location that allows the probe to penetrate the skin to the outside air while still allowing the pilot easy access to the indicator face.

Flap position indicator. Flaps are used to change the camber of the wing during landing. They are also used on some aircraft during

Figure 5-6-6. A flap position indicator in a Beechcraft King Air

Figure 5-6-7. An accelerometer (sometimes referred to as a "G" meter)

take-off. The pilot can observe their position on many smaller aircraft. The flight crew cannot visually monitor them on larger aircraft. Flap settings are critical when operating into smaller airports. Flap position indicators are provided to enable the flight crew to accurately set the proper flap position. A typical flap position indicator is shown in Figure 5-6-6.

One or more flaps has a flap position transmitter attached to it. This may be synchro type transmitter or a variable resistor type.

Accelerometer. Accelerometers are typically found on aerobatic and military aircraft. The pilot must limit aircraft maneuvers so various combinations of acceleration, airspeed, gross weight, and altitude remain within specified values. This reduces the possibility of damaging the aircraft as a result of excessive stresses. The accelerometer shows the load on the aircraft structure in terms of gravitation (g) units. It presents information that lets the aircraft be maneuvered within its operational limits. A typical accelerometer is shown in Figure 5-6-7.

The forces sensed by the accelerometer act along the vertical axis of the aircraft. The main hand moves clockwise as the aircraft accelerates upward and counterclockwise as the aircraft accelerates downward.

The accelerometer indications are in g units. The main indicating hand turns to +1 g when the lift of the aircraft wing equals the weight of the aircraft. Such a condition prevails in level flight. The hand turns to +3 g when the lift is three times the weight. The hand turns to minus readings when the forces acting on the aircraft surfaces cause the aircraft to accelerate downward.

The accelerometer operates independently of all other aircraft instruments and installations. The activating element of the mechanism is a mass that is movable in a vertical direction on a pair of shafts (Figure 5-6-8). A spiral-wound main spring dampens the vertical movement

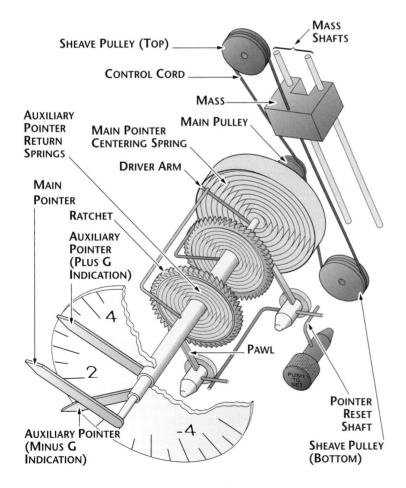

Figure 5-6-8. Accelerometer mechanical schematic

of the mass. The force of the mass travels by a string-and-pulley system to the main spring and main shaft. From here, it goes to the plus and minus assemblies. The hand assemblies mount on the plus and minus assemblies. Changes in vertical acceleration cause movement of the mass on the shafts, which translates into a turning motion of the main shaft. The turning motion pivots the indicating hands around the dial. The hand travels a distance equivalent to the value, in g units, of the upward or downward acceleration of the aircraft.

The accelerometer operates on the principle of Newton's Law of Motion. During level flight, no forces act to displace the mass from a position midway from the top and bottom of the shafts. Therefore, the accelerometer pulley system performs no work, and the indicating hands remain stationary at +1 g. When the aircraft changes from level flight, forces act on the mass. This action causes the mass to move either above or below its midway position. These movements cause the accelerometer indicating hands to change position. When the aircraft goes nose down, the hands move to the minus section of the dial. When the nose goes up, they move to the plus section.

The main hand continuously shows changes in loading. The two other hands on the accelerometer show the highest plus acceleration, and highest minus acceleration of the aircraft during any maneuver. The indicator uses a ratchet mechanism to maintain these readings. A knob in the lower left of the instrument face is used to reset the maximum- and minimum-reading hands to normal. Thus, the accelerometer keeps an indication of the highest accelerations during a particular flight phase or during a series of flights.

Clocks. Clocks come in many forms. They may have analog or digital displays. Analog clocks may be electrically or mechanically powered. Digital clocks are always electrically powered.

The typical mechanically powered analog clock is a 12-hour, elapsed time, stem-wound clock with an 8-day movement . The pull-to-set winding stem is at the lower left of the dial (see Figure 5-6-9). The dial has 60 divisions, which are read as minutes or seconds, as appropriate. The face has standard minute and hour hands, a sweep-second hand, and an elapsed-time minute hand. The clock may be started, stopped, or reset the elapsed-time minute hand by pressing a single button at the upper right of the dial.

Electrically powered analog clocks (Figure 5-6-10) are read in a similar manner, however, the are continuously powered by the aircraft elec-

Figure 5-6-9. A mechanically wound analog clock

Figure 5-6-10. An electrically powered analog clock

Figure 5-6-11. A digital clock

trical system. Mechanically wound clocks must be rewound by the flight crew on a regular basis. They must also be reset if the aircraft has not been operated recently (they only operate for eight days maximum). Electrically operated clocks do not have this limitation, but will not function in the event of an electrical failure. It

is for this reason that mechanically powered clocks are still in regular use.

Digital clocks (see Figure 5-6-11) are electrically powered, typically by the aircraft electrical system although a few battery powered types are available. They usually have a stopwatch or

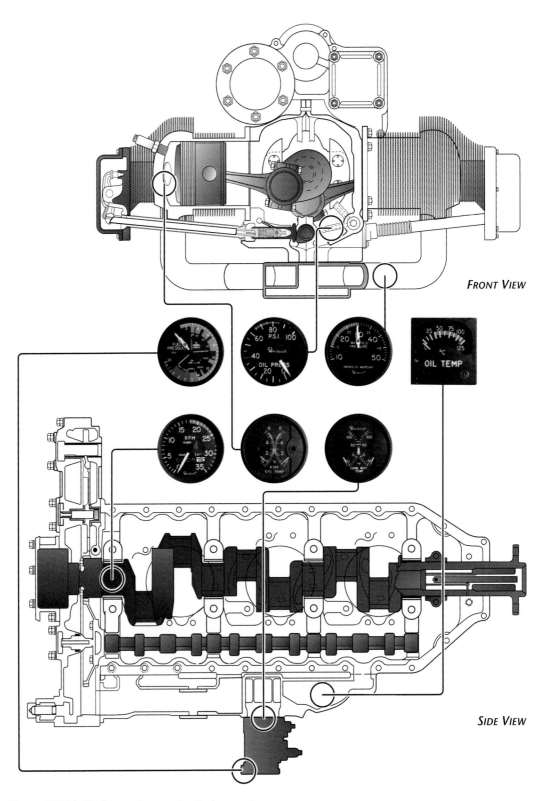

FRONT VIEW

SIDE VIEW

Figure 5-7-1A. Reciprocating engine instruments

elapsed time feature as well. While both types of clocks are very reliable, the digital style, when properly designed, is the more reliable of the two because of the lack of moving parts.

These types of clock are in the cockpit for use by the pilot or copilot. Additional clocks may be located elsewhere for use by other members of the flight crew.

Section 7

Engine Instruments

Engine instruments are used to monitor engine performance and set engine power settings. Both turbine and reciprocating engines require instrumentation. Some functions, such as r.p.m., fuel pressure and oil pressure, are common to both types of engines. Other functions are only measured on either reciprocating or turbine engines.

Safe, economical and reliable operation of aircraft engines depends on accurately measuring engine operation. In hydromechanically controlled systems, the pilot has primary responsibility for monitoring the instruments and maintaining the engine power settings within preset limits. Modern electronic control systems (primarily the FADEC and EEC) have moved the task of keeping the engine within safe operating parameters to these electronic systems. Nonetheless, the pilot must still maintain an awareness of the operating condition and settings of the engine.

Figure 5-7-1 (A) and (B) show a typical instrument grouping for a reciprocating engine as well as two types of turbine engines. Note that while many of the instruments cover similar types of systems, their ranges and styles vary somewhat.

Common Engine Instruments

Instruments commonly used on both turbine and reciprocating engines include tachometers, fuel flow indicators, fuel pressure gauges, oil temperature and pressure indicators as well as engine synchronizers. While there are some differences between turbine and reciprocating systems, in most cases the basic theory and operation of the instruments is the same.

Tachometers

All engines have maximum speed (or r.p.m.) limits. The *tachometer* shows the engine crankshaft r.p.m. Figure 5-7-2 shows a tachometer with range markings installed on the cover glass. The tachometer, often referred to as *tach*, is calibrated in hundreds with graduations at every 100-r.p.m. interval. The dial shown here starts at 0 and goes to 35 (3,500 r.p.m.).

The green arc indicates the operating r.p.m. range. The top of the green arc, 2,550 r.p.m., indicates maximum continuous power. All operation above this r.p.m. may be limited in time. The red line indicates the maximum r.p.m. permissible during takeoff; any r.p.m. beyond this value is an overspeed condition.

A turbine engine tachometer (tach) is a bit different than those used in reciprocating engine systems. To begin with, it measures in percent (% r.p.m.) of engine r.p.m. Also r.p.m. is not normally used to set engine power. Engine power is set by measuring the EPR or torquemeter. The tachometer is used mainly for engine start and to indicate an overspeed condition. Because turbine engine output can vary so much with a change in atmospheric conditions it is possible to have a 90% engine one day and a 110% engine the next day. If r.p.m. was used to set engine power you would not have full power one day and overtemp the engine the next day.

High pressure tachometer (N_2). On a single-spool engine the HP tach is the only one available. It provides the N_2 indication. On a split-spool engine the *high pressure* (HP) *tachometer* indicator (Figure 5-7-3) measures the high pressure compressor r.p.m. There is an N_2 tach for each engine. The indicators show HP compressor r.p.m. in percent. The units are self-powered by a tach generator located on the gearbox of each engine.

Low pressure tachometer (N_1). A *low pressure* (LP) *tachometer* indicator for each engine is used on all single spool engines and fan jets, either high or low bypass. The indicators show LP compressor r.p.m. in percent.

Tachometer Generators

Some light aircraft use a mechanical linkage to provide engine speed information to the tachometer. A flexible shaft is connected to the rear of the engine. The other end of the shaft connects to the back of the tachometer. As the engine rotates, so does the shaft. This drives the needle in the tachometer. This simple system is only practical in aircraft with the engine located in close proximity in front of the instrument panel.

The electrical tachometer generator system is used when a mechanical linkage cannot easily connect the engine and tachometer. Essentially, the tachometer generator system consists of an AC generator coupled to the aircraft engine and an indicator consisting of a magnetic-drag element on the instrument panel.

The generator transmits electric power to a synchronous motor, a part of the indicator. The frequency of this power is proportional to the engine speed. An accurate indication of engine speed is obtained by applying the magnetic-drag principle to the indicating element.

Generator voltage changes with engine speed. The varying voltage problem is eliminated by the generator and synchronous-motor combination. These units make a frequency-

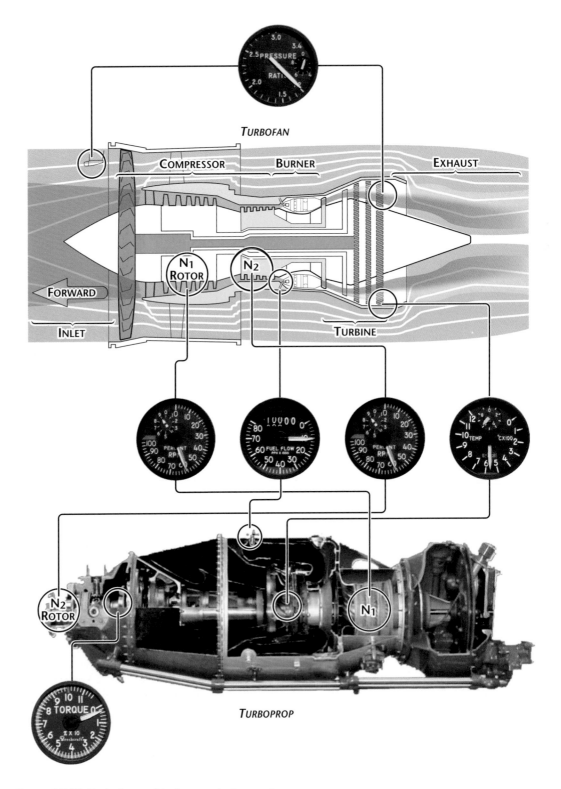

Figure 5-7-1B. Turbofan and turboprop instruments

Figure 5-7-2. A reciprocating engine tachometer

Figure 5-7-3. Turbine tachometer faces look basically the same. They are placarded as to N₁ or N₂.

sensitive system for sending an indication of engine speed to the indicator with absolute accuracy.

Tachometer generator units are small and compact (about 4 inches by 6 inches). The most common type of generator is constructed with an end shield designed so the generator can attach to a flat plate on the engine frame or reduction gearbox, with four bolts. An earlier style is attached with one large threaded nut located around the tach generator drive shaft.

Figure 5-7-4 shows a cutaway view of a tachometer generator. The generator consists of a permanent magnet rotor (1) and a stator (8) that develop three-phase power as the rotor turns.

The armature of the generator consists of a magnetized rotor. The rotor is cast directly onto the generator shaft. The generator may be of either two- or four-pole construction. The two- and four-pole rotors are identical in appearance and construction. They differ in that the two-pole rotor is magnetized north and south diametrically across the rotor, while the four-pole rotor is magnetized alternately north and south at each of the four pole faces.

The key (2) that drives the rotor is a long, slender shaft. It has enough flexibility to prevent failure under the torsional oscillations originating in the aircraft drive shaft. It will also accommodate small misalignments between the generator and its mounting surfaces.

The stator consists of a steel ring with a laminated core of ferromagnetic material. A three-phase winding goes around this core and is insulated from it. The winding is adapted for two- or four-pole construction, depending on the generator in which it is used. Figure 5-7-5

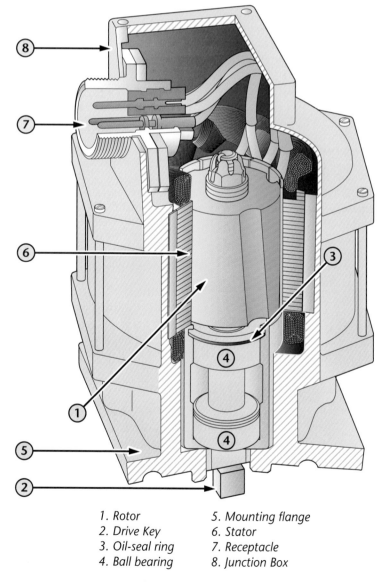

1. Rotor
2. Drive Key
3. Oil-seal ring
4. Ball bearing
5. Mounting flange
6. Stator
7. Receptacle
8. Junction Box

Figure 5-7-4. Tachometer generator cutaway

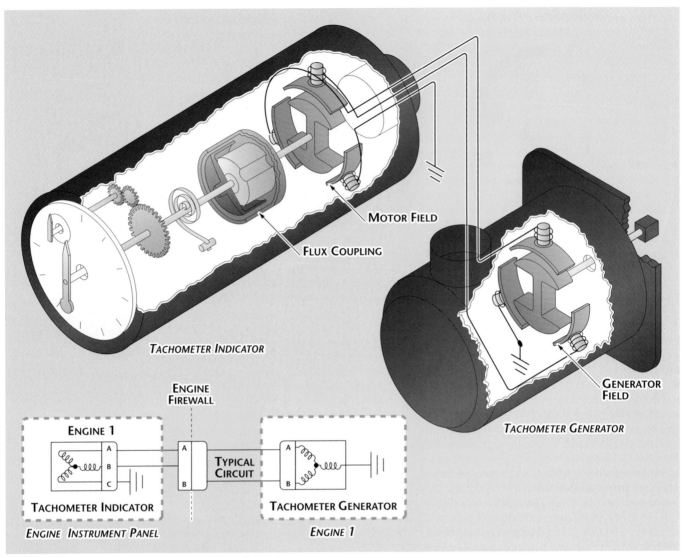

FLUX COUPLING

MOTOR FIELD

GENERATOR FIELD

TACHOMETER INDICATOR

TACHOMETER GENERATOR

ENGINE FIREWALL

ENGINE 1

A
B
C

TYPICAL CIRCUIT

A
B

A
B

TACHOMETER INDICATOR

TACHOMETER GENERATOR

ENGINE INSTRUMENT PANEL

ENGINE 1

Figure 5-7-5. Turbine tachometer generator schematic

Figure 5-7-6. Fuel flow is normally indicated in either gallons or pounds per hour. An abnormal fuel flow is always an indication that requires troubleshooting.

shows a tach generator designed for use on turbine engines.

Tachometer System Maintenance

AC 43.13-1B contains standards for accuracy and inspection criteria for tachometer accuracy. On page 8-406, the FAA establishes the need for tachometer accuracy and sets the error limit. Generally, the limit is ±2 percent. If the tachometer fails, the tach should be replaced.

Fuel Flow Indicator

Fuel flow indicators for each engine are also a necessity. They indicate the fuel flow rate for each engine. This is typically measured at the beginning of the engine fuel system (usually just before the engine driven fuel pump). Fuel flow is indicated in gallons per hour for reciprocating engines and in pounds per hour for turbine engines, although some advanced reciprocating systems use a pounds per hour indication. When troubleshooting the relationship between abnormal fuel flow and other engine parameters this gauge can be a valuable aid. A typical fuel flow indicator is shown in Figure 5-7-6 .

The fuel flow indicating system consists of a transmitter and an indicator for each engine. Some reciprocating fuel flow systems are nothing more than a fuel pressure gauge calibrated in pounds per hour. These systems do not use a transmitter, instead they have a fuel pressure line feeding them from the fuel injection distributor. This style is commonly found on Teledyne Continental Motors fuel injection systems. Troubleshooting procedures for this system are a bit different than for most others. A typical fuel flow gauge of this type is shown in Figure 5-7-7.

The fuel flow indicator is a calibrated fuel pressure gauge. When there is a blockage in a fuel injector or line, the fuel pressure increases while the fuel flow decreases. The increased pressure is displayed as an increase in fuel flow on the indicator face. Always refer to the factory maintenance manual when troubleshooting this type of system.

Turbine fuel flow systems and some reciprocating engine fuel flow systems use a system that truly measures fuel flow. These systems have an inline fuel flow transmitter. It contains a paddle or wheel that is rotated by the moving fuel. These systems provide a continuous indication of the rate of fuel delivery to the engine. Some systems include a totalizer that keeps track of the fuel used and calculates the remaining fuel in the tanks.

The measurements are transmitted electrically to the panel-mounted indicator. The use of electrical transmission ends the need for a direct fuel-filled line from the engine to the instrument panel. This minimizes the chance of fire and reduces mechanical failure rate. Many fuel flow systems are synchro based.

Vane Type Fuel Flow Transmitter

Figure 5-7-8 shows a cutaway view of a typical fuel flow transmitter. The transmitter includes both a fuel-measuring mechanism (or meter) and a synchro transmitter. You can separate these parts from one another for maintenance purposes, but they join as a single assembly for installation.

The fuel enters the inlet port of the transmitter and flows against the vane (1), causing the vane to swing. The greater the fuel flow, the greater

Figure 5-7-7. Fuel pressure gauge calibrated to act as a fuel flow indicator

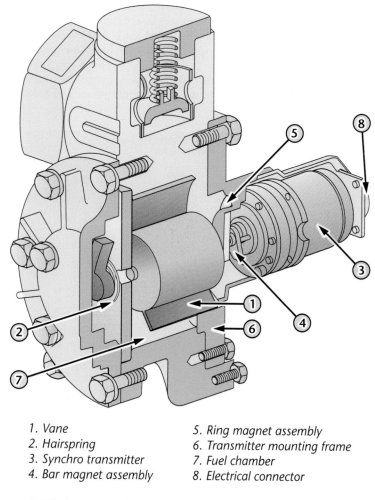

1. Vane
2. Hairspring
3. Synchro transmitter
4. Bar magnet assembly
5. Ring magnet assembly
6. Transmitter mounting frame
7. Fuel chamber
8. Electrical connector

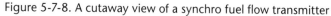

Figure 5-7-8. A cutaway view of a synchro fuel flow transmitter

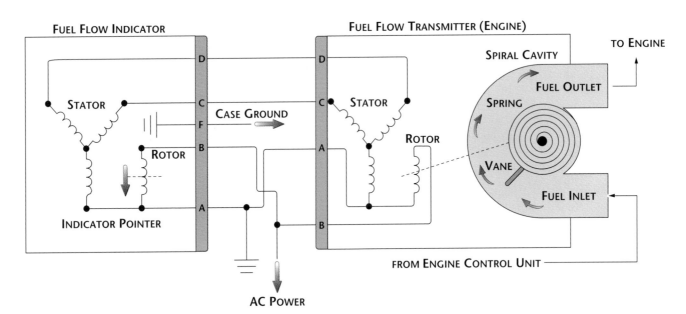

Figure 5-7-9. Synchro fuel flow indicating system

the force applied against the vane. The spiral fuel chamber design allows the distance between the vane and chamber wall to become increasingly larger as fuel flow increases so that fuel flow is not restricted. A calibrated hairspring (2) retards the motion of the vane. As the vane is rotated, the resistance in the spring increases. The vane stops moving when the forces exerted on it by the hairspring and by the fuel are equal.

The rotor shaft of the synchro-transmitter (3) connects to a bar magnet (4). Another magnet

is attached to the vane shaft (5). This magnet moves as the vane shaft moves. This design eliminates the need for a shaft entering the fuel line and the resultant problems of keeping that shaft sealed.

The two magnets keep the synchro-rotor shaft aligned with the fuel flow vane. When the vane moves, there is a corresponding movement of the synchro-rotor. The wiring schematic for this type of unit is shown in Figure 5-7-9.

The fuel flow indicator used with this type of system is a synchro-type. The rotor in the instrument synchro maintains the same orientation as the synchro-rotor in the transmitter. The face of the indicator is calibrated to display the fuel flow present for any given vane angle.

Fuel Flow Totalizing Systems

Figure 5-7-10 shows the indicator of a fuel flow totalizing system. The reading of this instrument usually shows the combined rate of fuel flow into two or more engines. Also, if only one engine is operating, the reading gives a true indication. A continuous reading of the pounds of fuel remaining in the aircraft fuel cells appears in the small window. Before starting the engines, you set the total amount of fuel in the aircraft on the pounds-fuel-remaining indicator by using the reset knob on the front of the instrument. As soon as the engines are running, the fuel flow reading shows the rate of fuel consumption. The fuel remaining indicator starts counting toward zero, giving a continuous reading of fuel remaining in the cells.

Figure 5-7-10. Fuel totalizer indicator

The entire fuel flow totalizing system consists of two or more fuel flow transmitters, an amplifier, and an indicator.

Fuel flow transmitters. The fuel flow transmitters are almost identical to those already discussed in the single system. In the fuel flow totalizing system, the transmitters connect electrically, so their combined signals go into the fuel flow amplifier as one.

Fuel flow amplifier. The fuel flow amplifier is an electronic device that supplies power of the proper size and phasing to drive the indicator. The speed at which the indicator motor runs depends on the transmitter signal going into the amplifier.

Fuel flow totalizer indicator. The fuel flow totalizer indicator contains a two-phase variable speed induction motor. This motor travels in one direction only; however, the speed varies. As the rate of fuel consumption increases, more and more power goes to the indicator motor. This causes the speed of the motor to increase proportionally to the rate of fuel consumption. The motor turns a magnetic drum-and-cup linkage (similar to the tachometer indicator hysteresis cup), which causes pointer deflection. The deflection is proportional to the motor speed, and thus proportional to the rate of fuel consumption. At the same time, a linkage having a friction clutch drives the pounds-fuel-remaining indicator. The clutch is disengaged when using the reset knob to set the reading on the pounds-fuel-remaining indicator.

Engine Oil Temperature Indicator

Many aircraft are equipped with oil temperature indicators. Most turbine engines and higher horsepower reciprocating engines transfer a moderate amount of heat to the lubricating oil during operation. Excessive oil temperatures affect the oil's viscosity. Oil cooler systems are used to maintain temperatures within the proper range.

Engine oil temperature indicators (Figure 5-7-11) are used to monitor the oil cooling system. Indicators are typically scaled in °C with a maximum temperature between 150°C and 180°C. Temperature is sensed by a probe in the oil tank or at the point that the oil is returned to the engine.

Three types of oil temperature gauges are available for use in the engine gauge unit. One unit consists of an electrical resistance type of oil thermometer, supplied with electrical current by the aircraft DC power system. A typical resistance thermometer probe is shown in Figure

Figure 5-7-11. Excessively high oil temperature is a sign of an impending problem. With high temperatures bearings and gears do not get adequate lubrication.

Figure 5-7-12. Resistance thermometer bulb

5-7-12. The third type is the mechanical bourdon tube.

Wheatstone bridge circuit. The resistance thermometer bulb is connected to a Wheatstone bridge or a ratiometer circuit. A Wheatstone bridge circuit is shown in Figure 5-7-13. The resistance bulb is one of the sides of the circuit. The other three sides of the circuit are resistors contained within the indicator.

When the temperature bulb in Figure 5-7-13 is at a temperature of 0°C, its resistance is 100 ohms. The resistance of arms X, Y, and Z are also 100 ohms each. At this temperature the Wheatstone bridge is in balance. This means the sum resistance of X and Y equals the sum resistance of the bulb and Z. Therefore, the same amount of current flows in both sides of this parallel circuit. Since all four sides are equal in resistance, the voltage drop across side X equals the drop across the bulb. Since these voltages are equal, the voltage from A to B is zero, and the indicator reads zero.

When the temperature of the bulb increases, its resistance also increases. This unbalances the bridge circuit causing the needle to deflect to the right. When the temperature of the bulb decreases, its resistance decreases. Again the bridge circuit goes out of balance. However, this time the needle swings to the left.

The galvanometer is calibrated so the amount of deflection causes the needle to point to the number of the meter scale. This number corresponds to the temperature at the location of the resistance bulb.

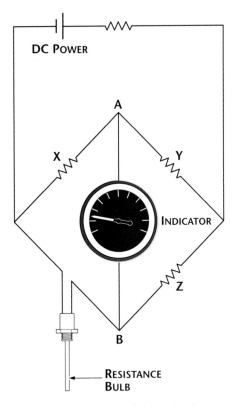

Figure 5-7-13. Wheatstone bridge circuit

This instrument requires a constant and steady supply of DC voltage. Fluctuations in the power supply affect total bridge current, which can cause an unbalanced bridge. Unless excessive heat damages the bulb, it will give accurate service indefinitely. When a thermometer does not operate properly, check carefully for loose wiring connections before replacing the bulb. An open in the voltage supply circuit will also cause the indicator to read zero.

Ratiometer system. The ratiometer is a temperature indicator that uses two coils in a balanced circuit. In some instruments, the coils turn between the poles of a permanent magnet. In other instruments, a small permanent-magnet rotor turns between stationary coils. Ratiometer circuits vary in design, but the principle of operation is very much the same for all.

Figure 5-7-14 shows a simplified circuit with a permanent-magnet rotor. The two coils are stationary in the instrument, and the indicator needle fastens to the permanent-magnet rotor. The needle position is determined by how the permanent magnet aligns itself with the resultant flux of the two coils.

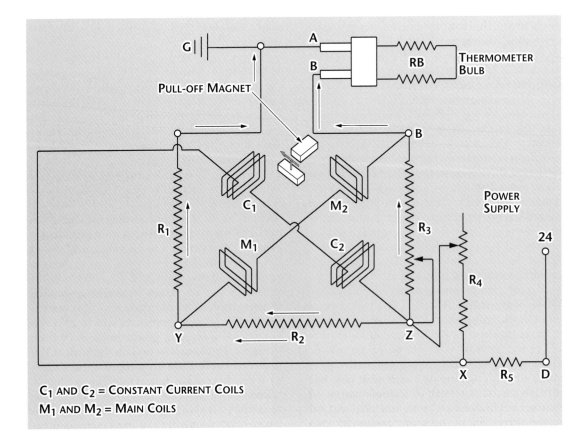

C_1 AND C_2 = CONSTANT CURRENT COILS
M_1 AND M_2 = MAIN COILS

Figure 5-7-14. Ratiometer circuit

For an understanding of how the circuit operates, trace the current through the circuit. Starting at ground, current flows up through the bulb, centering potentiometer R_5 and R_6, to point D. Current through the left leg of the bridge is from ground through R_1 to point A. Current then goes from point A through the lower part of the expansion and contraction potentiometer R_2. It also goes from pin 2 of R_2 through R_4 to point D. Here, the currents of the two legs combine and flow through R_7 to the positive 28 volts.

Note that restoring coil L_2, resistor R_3, and upper part of potentiometer R_2 forms a parallel path for current flow from point A to pin 2 of R_2. Deflection coil L_1, connects between points A and B. Therefore, any difference in potential between these two points will cause current to flow through L_1.

The ratiometer temperature indicator uses a fixed permanent magnet to pull the pointer to an off position when the indicator is not operating. Thus, current through restoring coil L_2 must compensate for the pull-off magnet when the indicator is operating. Variations in the resistance of the bulb, because of temperature changes, cause a change in voltage at point B. It also causes the resulting change in current through deflection coil L_1.

Bourdon tube system. Some systems use a direct reading instrument. This style, the capillary oil thermometer, is a vapor pressure thermometer. It consists of a bulb connected by a capillary tube to a Bourdon tube and a multiplying mechanism connected to a pointer. A Bourdon tube mechanism is shown in Figure 5-7-15 As the temperature changes, the Bourdon expands or contracts. It can also be used to measure pressure changes.

This tube expands or contracts and rotates the oil temperature pointer. The pointer shows the oil temperature on a dial. The Bourdon tube is used in many direct reading pressure and temperature indicators. Excessive oscillation of the gauge pointer indicates that there is air in the lines leading to the gauge or that some unit of the system is functioning improperly.

Oil Pressure Indicators

Oil pressure instruments show that oil is or is not circulating under proper pressure. An oil pressure drop warns of impending engine failure due to lack of oil, oil pump failure, or broken lines. Oil pressure shows on an engine gauge unit (Figure 5-7-16). This unit consists of three separate gauges in a single case – oil pressure, cylinder head temperature, and oil temperature. The oil pressure gauge can also be a single instrument.

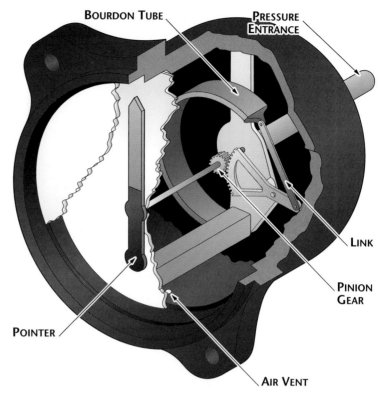

Figure 5-7-15. Bourdon tube mechanism

Figure 5-7-16. A three-in-one engine gauge unit contains multiple instruments in one case. They are quite common on reciprocating aircraft.

The gauge has a Bourdon tube mechanism for measuring fluid under pressure. The instrument's oil pressure range is from 0 to 200 p.s.i. You read the scale in graduations of 10 p.s.i. There is a single connection on the back of the case leading directly into the Bourdon tube. As with a Bourdon tube temperature gauge, excessive oscillation of the gauge pointer indicates that there is air in the lines leading to the gauge or that some unit of the system is functioning improperly.

Figure 5-7-17. A synchro driven oil pressure indicator

Figure 5-7-18. A synchronizer indicator

Figure 5-7-19. A reciprocating EGT

The synchro system is another method of measuring oil under pressure. This type of oil pressure system (see Figure 5-7-17) is used on many larger aircraft. Essentially, it is a method of directly measuring engine oil pressure. An aneroid type diaphragm is connected to a synchro-transmitter. As the oil pressure expands the aneroid, a linkage rotates the synchro.

The synchro position is electrically transmitted from the point of measurement to the synchro-indicator on the instrument panel. The synchro system ends the need for direct pressure lines from the engine to the instrument panel. It also reduces the chances of fire, loss of oil or fuel, and mechanical difficulties.

Engine Auto Synchronizer

An *engine auto synchronizer* receives signals from both engine tachometers. See Figure 5-7-18. The amplifier produces a signal that operates an actuator on the master engine throttle linkage. Thus, it varies the slave engine r.p.m. to match the master engine r.p.m. The left throttle lever does not move while the throttle linkage is being positioned by the synchronizer. The range of control of the synchronizer is between 6 and 10 percent. The system is fail-safe in that the pilot, by operating the throttles manually, will have full control of both engines.

Each different airplane system will have a specific set of instructions on how to move the throttles with the ENGINE SYNC engaged. Engine Auto Sync is not available on all aircraft.

Exhaust Temperature Gauges

Turbine and most advanced reciprocating aircraft use some type of exhaust temperature measurement. While their purpose, from a flight crew's perspective, is somewhat different, they operate using very similar technologies.

Reciprocating engines use the EGT to measure the temperature of the exhaust gas as it leaves the cylinder(s). From that reading, the fuel mixture can be leaned in-flight with no danger of overheating the engine. The primary purpose in reciprocating engine applications is fuel economy. A typical reciprocating EGT is shown in Figure 5-7-19. Some gauge systems use a multi-position switch with a single instrument.

Exhaust gas temperature indicator (EGT). A turbine engine *exhaust gas temperature (EGT)* probe can be located either between or behind the turbine stages. EGT is an engine operating limit and is used to monitor the integrity of the turbine and check engine operating conditions. It is also the first indication that an engine has started to run during a start.

Turbine inlet temperature (TIT). *Turbine inlet temperature (TIT)* is the most important consideration in engine instrumentation. TIT determines how much fuel can be burned before critical temperature is reached. Before digital instrumentation, it was very difficult to read TIT. In many applications, *Interstage Turbine Temperature (ITT)* is measured instead. Additionally, TIT would require an armload of charts to compute the proper fuel burn. Thus TOT (Total Operating Temperature) was almost always instrumented instead. In the digital age it is not difficult and is used as a modifying parameter for TOT. Basically it is another item that a FADEC system takes care of. TIT is not commonly instrumented. See Figure 5-7-20.

Figure 5-7-20. EGT and TIT are similar. EGT is the most critical temperature during engine operation.

EGT Probes

Most EGT and TIT systems use thermocouple probes to sense these temperatures. They work very well in high temperature applications.

A *thermocouple* is a junction or connection of two unlike metals; such a circuit has two junctions. When one of the junctions becomes hotter than the other, an electromotive force is produced in the circuit. By including a galvanometer in the circuit, this electromotive force can be measured. The hotter the high-temperature junction (hot junction) becomes, the greater the electromotive force. By calibrating the galvanometer dial, in degrees of temperature, it becomes a thermometer. The galvanometer contains the *cold junction*.

A typical thermocouple thermometer system consists of a galvanometer indicator, a thermocouple or thermocouples, and thermocouple leads. Some thermocouples consist of a strip of copper and a strip of constantan pressed tightly together. Constantan is an alloy of copper and nickel. Other thermocouples consist of a strip of iron and a strip of constantan. Others may consist of a strip of Chromel and a strip of Alumel.

The *hot junction* of the thermocouple varies in shape, depending on its application. Both reciprocating and turbine EGT probes have a similar appearance. Figure 5-7-21 shows both types of a EGT probe. Thermocouple leads are critical in makeup and length because the galvanometers are calibrated for a specific set of leads in the circuits.

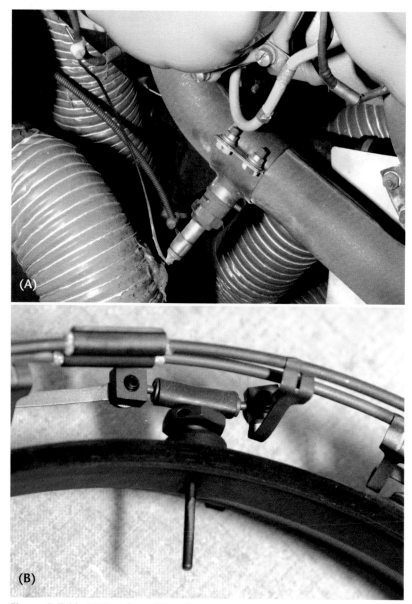

Figure 5-7-21. EGT probes: (A) reciprocating type, (B) turbine type

Figure 5-7-22. A cylinder head temperature gauge

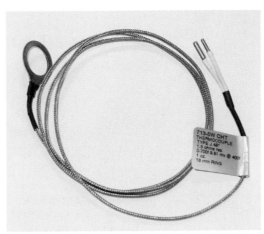

Figure 5-7-23. A gasket type cylinder head temperature probe

Reciprocating Engine Instruments

There are a few types of engine instruments that are only found on reciprocating engines. These monitor functions that are only found on reciprocating engines.

Cylinder Head Temperature Indicator. Cylinder head temperatures are indicated by a gauge connected to a thermocouple attached to the cylinder (Figure 5-7-22). Single-cylinder and multi-cylinder installations may be found. If a single thermocouple is used, it will be located on the cylinder determined by test to be the hottest.

The temperature recorded with either of these methods is merely a reference or control temperature; as long as it is kept within the prescribed limits, the temperatures of the cylinder dome, exhaust valve and piston will be within a satisfactory range. When the thermocouple is attached to only one cylinder, it can do no more than give evidence of general engine temperature. While normally it can be assumed that the remaining cylinder temperatures will be lower, conditions such as detonation will not be indicated unless they occur in the cylinder that has the thermocouple attached.

Many cylinder head temperature systems us a gasket type thermocouple (Figure 5-7-23). In the gasket thermocouple, the rings of two dissimilar metals are pressed together, forming a spark plug gasket. Each lead that connects back to the galvanometer must be of the same metal as the thermocouple part to which it connects. For example, a copper wire connects to the cop-

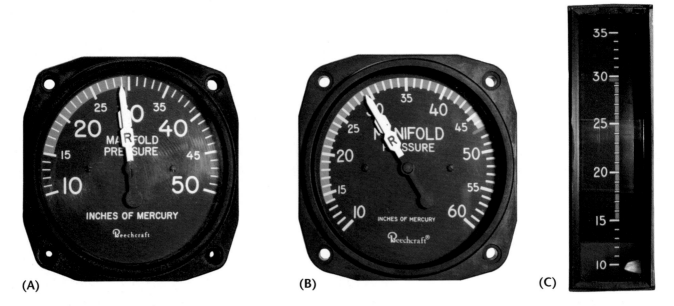

Figure 5-7-24. Manifold pressure gauges: (A) is normally used on a standard installation, (B) would most likely be found on a supercharged engine, and (C) is a vertical scale unit.

per ring, and a constantan wire connects to the constantan ring.

Manifold Pressure Indicator. The preferred type of instrument for measuring the manifold pressure is a gauge that records the pressure as an absolute pressure reading (Figure 5-7-24). This is accomplished by using a specially designed manifold pressure gauge that indicates absolute manifold pressure in inches of mercury. Manifold pressure will be below barometric pressure on normally aspirated engines and greater than barometric pressure on supercharged engines.

A blue arc represents the range within which operation with the mixture control in automatic-lean (only on large radial engines) position is permissible. A green arc indicates the range within which the engine must be operated with the mixture control in the normal or rich position. The red arc indicates the maximum manifold pressure permissible during takeoff. On installations where water injection is used, a second red line is located on the dial to indicate the maximum permissible manifold pressure for a wet takeoff. Water injection is not typically used on modern production engines.

Manifold pressure gauges are usually direct reading gauges that connect to the intake manifold. This connection is located on the output side of the supercharger or turbocharger on engines so equipped.

Turbine Engine Instruments

Turbine engines also require a few unique types of instruments. Some of these instruments are only used on specific types of turbine engines. Engine power output is measure differently on turbojet and turbofan engines than it is on turboprop and turboshaft engines.

Engine pressure ratio (EPR) indicator. An *engine pressure ratio (EPR)* indicator for each engine measures the engine pressure ratio as a measure of the thrust being developed by the engine. The indicator provides an indication of engine power in the form of the ratio of exhaust total pressure (P_7) to intake total pressure (P_1). EPR also requires the monitoring of several other parameters to be correct. Before the digital age EPR required the use of charts to actually compute the correct reading for the particular conditions. With digital fuel controls all the variations are compared and computed for you. Thus EPR is an actual presentation of engine power. See Figure 5-7-25.

Exhaust total pressure (P_7) and intake total pressure (P_1) are measured using simple pitot type probes. The basic EPR function compares

these two pressures. This differential must then be adjusted for temperature, altitude and other factors to give an accurate engine power rating. A typical intake pressure sensor probe is shown in Figure 5-7-26. EPR indicators are typically found on turbojet and turbofan engines. Torquemeters are used to measure power output on turboshaft and turboprop engines.

Torquemeter indicator. Turboprop aircraft measure engine power output through the use of a *torquemeter* (Figure 5-7-27). The electric torquemeter system in turboprop aircraft measures the horsepower produced by the engine. Two methods are in use. The first measures torque at an extension shaft. Each system con-

Figure 5-7-25. An EPR indication shows the ratio between engine inlet pressure and engine outlet pressure. It is an indication of the thrust of an engine.

Figure 5-7-26. EPR equipped turbine engines will have an EPR pressure sensor in the engine inlet. If this becomes blocked, the EPR indicator will give erroneous indications.

Figure 5-7-27. Torquemeter indicator

Figure 5-7-28. Turbine vibration indicator

Figure 5-7-29. Typical EICAS display showing N_1, TACH, ITT, N_2 TACH, fuel flow, oil temperature and pressure.

sists of a transmitter (part of the engine extension shaft), a phase detector, and an indicator. The system measures the torsional deflection (twist) of the extension shaft as it sends power from the engine to the propeller. The more power the engine produces, the greater the twist in the extension shaft. Magnetic pickups detect and measure this deflection electroni-

cally. The indicator registers the amount of deflection in shaft horsepower.

The second method in common use is a hydro-mechanical torquemeter. This system uses oil pressure to measure torque. Rotational forces within the engine gearbox are used to compress oil in a torquemeter sensing chamber. The system then measures the difference in pressure between the oil in the sensing chamber and the internal oil pressure of the gearbox. This pressure differential is transmitted to the torquemeter via an electrical signal.

Turbine vibration indicator. The *Turbine Vibration Indicators (TVI)* provide a continuous monitoring of the balance of the rotating assemblies in the engine in order to detect a possible internal failure that could result in an engine failure (see Figure 5-7-28). The indication is normally picked up from the low pressure compressor or the low pressure turbine. Vibration indications for both engines are normally presented on a single indicator with a switch to allow either compressor or turbine vibration to be displayed. Onboard vibration monitoring systems are not new. However, in the digital age they are even more useful. Not only can vibrations be monitored, but a vibration can be cross referenced to a specific set of instrument readouts, producing a better picture of what is going on at a moment in time.

Engine top temperature control. On engines with FADEC controls, this is monitored and controlled automatically. It senses the temperature from thermocouples in the turbine primary exhaust gas stream, preventing over temperature of the turbine gas. Normally one element in each thermocouple is used for cockpit *Turbine Gas Temperature (TGT)* indication. The second element passes signals to a temperature control signal amplifier. When the TGT exceeds the maximum the amplifier signals the temperature control actuator on the engine fuel flow regulator. The actuator operates, resulting in a reduced fuel flow, thus preventing the TGT from exceeding the limits.

These systems do not operate during engine ground starting. Control of the TGT during start is only by the pilot's manual operation of the throttle lever. This system can reduce the fuel flow as much as 6 to 10 percent which could slightly reduce HP compressor speed.

Engine Indicating and Crew Alert System (EICAS). Most new aircraft use digital electronic instrumentation that is presented on a CRT (Cathode Ray Tube) or LCD (Liquid Crystal Display). See Figure 5-7-29. Glass cockpit technology has all but eliminated the traditional analog gauge in new transport aircraft construction. In this system, known as the *Engine Indicating and Crew Alert System or EICAS*, all of the engine sensors transmit their information to the EICAS computer. The EICAS computer then sends digital data to the cockpit screen telling it what information to display. In most cases this information is presented on a computer monitor in a form that graphically resembles the face of an analog instrument.

All of the information presented on conventional gauges is displayed on the cockpit display. Typically there is only one cockpit display unit. This unit may replace as many as 36 conventional analog instruments. A significant benefit is greatly reduced maintenance in the cockpit and enhanced aircraft reliability. The presentation can easily be tailored to meet specific operational needs. In the event of a screen failure, the EICAS information can be displayed on a flight display screen or vice versa.

One benefit of the EICAS system is the ability to alert the flight crew of failures or abnormal conditions. The screen shows system status messages, in addition to basic engine data, during all phases of flight. It also can display color-coded alert messages communicating both the type and severity of a failure.

The crew can also set the display to only show the primary parameters full time and allow the EICAS to monitor secondary systems, presenting information on those systems only when certain parameters are exceeded. This simplifies the information the crew must absorb and reduces crew workload and fatigue.

A typical EICAS system will provide the following primary information to the flight crew:

- Actual and commanded N_1 speed
- Transient N_1 (the difference between actual and commanded N_1)
- Max potential and permissible N_1

Additional indications occur if N_1 limits are exceeded.

- Current thrust lever position
- Thrust Reverser system status
- Exhaust Gas Temperature (EGT)
- Max permissible EGT

Additional indications occur if EGT limits are exceeded.

- N_2 rotor speed
- Fuel Flow

Additional secondary data is also presented. These include fuel used, oil quantity, oil pressure, oil temperature, vibration sensor data, oil filter status, fuel filter status, ignition system

Figure 5-7-30. The right hand side of this EICAS display alerts the crew to a number of configuration and warning indications.

status, start valve position, engine bleed pressure and nacelle temperature.

The EICAS system also provides a number of advantages for the AMT. The system can automatically record trend monitoring information, eliminating hand recording (and the potential for mistakes). After an engine or system fault, all of the data listed previously can be called up and evaluated during the troubleshooting process. The AMT has much more information available than with conventional systems. Maintenance personnel can also use the EICAS on the ground to monitor many systems and access the computer's *Built-In-Test-Equipment (BITE)*. Many systems on modern digitally controlled aircraft have the ability to perform extensive self-testing. The computer can analyze circuits and system status and relay the information back to the EICAS panel.

In addition to displaying engine related information, the EICAS also presents vital information regarding pressurization, aircraft configuration and other key systems to the crew. The Crew Alerting System portion of the EICAS presents status information and alerts for a

variety of systems throughout the aircraft. Figure 5-7-30 shows some of these alerts listed in the top right-hand corner. The status of the cabin door is shown as well as a number of alerts relating to the brake system. The EICAS display shown also has a notice to the crew to set the field elevation in the flight control computer before departing.

Section 8
Maintenance of Instruments

A wide variety of skills are needed to maintain aircraft instruments and their associated equipment. It is important to check, inspect, and maintain these instruments because the aircraft will not perform properly unless the instruments present reliable information.

Instruments used in aircraft must give correct indications. The existence of errors in instrument systems directly relates to flight safety and efficient aircraft performance. Each system has an allowable tolerance for instrument error. Borderline instrument errors may not be acceptable. This is particularly important in high-speed aircraft.

Maintenance technicians regularly perform functional tests on aircraft instruments to make sure they give accurate indications. Operational and functional tests take time. When performing an inspection, it is essential that the technician know how the particular aircraft instrument operates. Having the correct, calibrated tools, equipment and technical manuals is necessary before starting and system maintenance.

General Maintenance

Instrument maintenance is primarily troubleshooting and repair of the instrumentation system, not the internal repair of individual gauges. FAA regulations restrict the repair or overhaul of most instruments to certified repair stations. Repair stations must meet stringent operating guidelines. Additionally, a given repair station is usually limited to performing maintenance only on specific types of instruments.

Cases

Figure 5-8-1. One and two piece phenolic instrument cases

Instruments typically come in one of four different kinds of cases.

Figure 5-8-2. Flanged instruments: (A) front mounted and (B) rear mounted

1. One-piece phenolic composition cases

2. Two-piece phenolic composition cases

3. Nonmagnetic all-metal cases

4. Metallic-shielded cases

See Figure 5-8-1 for an example of a one and two piece phenolic instrument cases.

The cases come in several different sizes so instruments can be easily removed and maintenance simplified. Special instruments that contain mechanisms too large for adaptation to a standard case come in specially designed cases.

Instruments easily mount on the instrument panel with locking devices molded into the instrument flange assembly, by spring locknuts, or mounting clamps. Flanged instruments may be mounted with the mounting flange either in front of or behind the panels. Mounting behind the panel gives a cleaner panel appearance, but requires access to the rear of the panel for removal. Front mounted instruments can be easily slid out for maintenance. Both styles are shown in Figure 5-8-2.

Instruments that use a mounting clamp (see Figure 5-8-3) are easily removed by unscrew-

ing the tension screw in the instrument's corner. The tension screw does not need to be fully removed to release the tension on the clamp assembly.

Markings and Graduation

Markings on the instrument face or on the glass lens help flight personnel confirm systems operations are maintained within prescribed ranges. Electronic systems provide similar markings as part of their digital displays. Instrument markings indicate ranges of operation or minimum and maximum limits, or both. Table 5-8-1 highlights the most common instrument markings.

Generally, the instrument marking system consists of five colors (red, yellow, blue, white and green) and intermediate blank spaces. Figure 5-8-4 shows an instrument with markings added to the cover glass.

The yellow arc covers a given range of operation and is an indication of caution. Generally, the yellow arc is located on the outer circumference of the instrument cover glass or dial face.

Figure 5-8-3. An instrument mounting clamp

Figure 5-8-4. An oil pressure indicator showing typical range markings

COMMON INSTRUMENT MARKINGS	
MARKING	DEFINITION
Green arc	Normal Operating Range
Yellow Arc	Caution Range
White Arc	Special Operations Range
White Radial Line	Slippage Mark
Red Arc	Prohibited Range
Red Radial Line	Do Not Exceed
Blue Arc	Special Operations Range
Blue Radial Line	Special Operating Condition
Red Triangle, Diamond, or Dot	Max Limit for High Transit Indications

Table 5-8-1. Common instrument markings

A red line or mark indicates a point beyond which a dangerous operating condition exists, and a red arc indicates a dangerous operating range. Of the two, the red mark is used more commonly and is arranged in a radial pattern on the cover glass or dial face.

The blue arc or a white arc, like the yellow, indicates a range of operation. The blue or white arc might indicate, for example, the manifold pressure gauge range in which the engine can be operated with the carburetor control set at automatic lean. The blue or white arc is used only with certain engine instruments, such as

the tachometer, manifold pressure, cylinder head temperature and torquemeter.

The green arc shows a normal range of operation. When used on certain engine instruments, however, it also means that the engine must be operated with an automatic rich carburetor setting when the pointer is in this range.

When the markings appear on the cover glass, a white line is used as an index mark, often called a slippage mark. The white radial mark indicates any movement between the cover glass and the

Figure 5-8-5. A plastic non-glare cover over an aluminum instrument panel

case, a condition that would cause dislocation of the other range and limit markings.

Panels

Instrument panels are usually made from sheet aluminum alloy, with sufficient strength to resist flexing. Aluminum panels are nonmagnetic. The panel may be painted to eliminate glare and reflection. Many instrument panels have a plastic, non-glare panel that covers the aluminum panel (Figure 5-8-5). Most installations are also provided with a glareshield to prevent direct sunlight from causing glare and reflections that would make the instruments unreadable. A basic glareshield is shown in Figure 5-8-6. Many turbine aircraft incorporate annunciator lights or additional cockpit controls into the glareshield.

Most instrument panels are provided with a lighting system for night operation. Many aircraft provide multiple lighting sources.

Instruments can be internally lit or provided with a spotlight mounted in one of the instrument case mounting holes. These spotlights are typically referred to as "post lights". Figure 5-8-7 shows a post light assembly and how it appears when installed in the panel. Additional indirect lighting may be provided from under the instrument panel glareshield or from cockpit overhead lights.

Some panels have overlay panels equipped with internal lighting. These lighting panels are constructed of clear plastic and are painted on the front and the back. The lights are placed around the edge of the panel, hence their name "edge lit" or EL panels. They are equipped with lettering and other markings made of translucent paint. When the lighting is turned on, these areas illuminate and are very easy to read at night. Figure 5-8-8 shows a typical panel. This type of panel requires special care when performing maintenance. Any scratch that penetrates the top layer of paint will result light being emitted at that spot.

Instrument panels are shock-mounted to absorb low frequency, high-amplitude shocks. The mounts consist of square-plated absorbers in sets of two, each secured to separate brackets (Figure 5-8-9). You should inspect the mounts periodically for deterioration; if the rubber is cracked, replace the pair. A shock mounted instrument panel is shown in Figure 5-8-10.

Instrument System Inspections

Instrument system maintenance will usually include the following types of inspections:

Figure 5-8-6. A single engine Cessna instrument panel equipped with a glareshield

Figure 5-8-7. The top left corner of this Omni Bearing Selector is equipped with a post light

- Check pointers for excessive errors. Some indicators should show existing atmospheric pressure, existing temperatures, etc. Others should indicate zero.

- Check instruments for loose or cracked cover glasses. Replace pitot-static instruments if damaged.

- Check instrument lights for proper operation.

- Check caging and setting knobs for freedom of movement and correct operation.

- Carefully investigate any irregularity the pilot reports.

When performing a phase/calendar inspection, make the following checks:

Figure 5-8-8. An edge lit panel overlay

- Check the mounting of all the instruments and their dependent units for security.

- Check for leaks in instrument cases, lines, and connections.

- Check the condition of operation and limitation markings against the TCD's for the aircraft. Also, check their condition. Check slippage marks.

- Check for contact and condition of bonding on instruments.

- Check shock mountings for condition of rubber and security of attachment.

- Check for freedom of motion of all lines and tubing behind the instrument panel. Also, check that they are properly clamped or taped to avoid chafing, and free from moisture, crimps, etc.

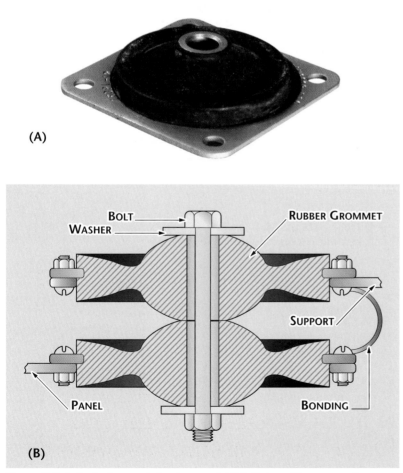

Figure 5-8-9. An instrument shock mount (A) full view (B) cross section

Compass Swing Procedures

Compass swing must be performed whenever any ferrous component of the system (i.e. flux valve compensator, or standby compass) is installed, removed, repaired, or a new compass is installed. The magnetic compass can be checked for accuracy by using a compass rose located on an airport.

The compass swing is normally effected by placing the aircraft on various magnetic headings and comparing the deviations with those on the deviation cards.

1. Have the aircraft taxied to the NORTH (0°) radial on the Compass Rose. Use a hairline sight compass (a reverse reading compass with a gun sight arrangement mounted on top of it) to place the aircraft in the general vicinity. With the aircraft facing North and the person in the cockpit running the engine(s) at 1,000 r.p.m., a mechanic, standing approximately 30 feet in front of the aircraft and facing South, "shoots" or aligns the master compass with the aircraft center line. Using hand signals, the mechanic signals the person in the cockpit to make additional adjustments to align the aircraft with the master compass. Once aligned on the heading, the person in the cockpit runs the engine(s) to approximately 1,700 r.p.m. to duplicate the aircraft's magnetic field and then the person reads the compass.

2. If the aircraft compass is not in alignment with the magnetic North of the master compass, correct the error by making small adjustments to the North-South brass adjustment screw with a nonmetallic screw driver (made out of brass stock, or stainless steel welding rod). Adjust the N-S compensator screw until the compass reads North (0°). Turn the aircraft until it is aligned with the East-West, pointing East. Adjust the E-W compensator screw until it reads 90°. Continue by turning the aircraft South 180° and adjust the N-S screw to remove one-half of the South's heading error. This will throw the North off but the total North-South should be divided equally between the two headings. Turn the aircraft until it is heading West 270°, and adjust the E-W screw on the compensator to remove one-half of the West error. This should divide equally the total E-W error. The engine(s) should be running.

3. With the aircraft heading West, start your calibration card here and record the magnetic heading of 270° and the compass reading with the avionics/electrical systems on, then off. Turn the aircraft to

align with each of the lines on the compass rose and record the compass reading every 30°. There should be not more than a plus or minus 10° difference between any of the compass' heading and the magnetic heading of the aircraft.

4. If the compass cannot be adjusted to meet the requirements, install another one.

5. When the compass is satisfactorily swung, fill out the calibration card properly and put it in the holder in full view for the pilot's reference.

Instrument System Repair

After starting the engine, check the instrument pointers for oscillation. Also, check the readings for consistency with engine requirements and speeds. On multiengine aircraft, check the instruments for the various engines against each other. Investigate any inconsistency; it may indicate a faulty engine, component, or instrument.

After a particular trouble has been diagnosed, replace the faulty unit. Remember, the following precautions when removing and installing instruments:

- Handle instruments carefully at all times. Additional damage may result from improper handling.

- Do not change the panel location of an indicator.

- Do not force the mounting screws. If the screw is cross-threaded, replace it. Do not draw the screws up too tight against the panel. This may distort the case enough to affect the operation of the instrument, crack the case, or break off the mounting lugs.

- When removing or installing tubing of a pressure-operated instrument, use a backup wrench to avoid twisting the tubing or fitting. Do not exert undue force while tightening the connection.

- Install all electrical plugs hand tight.

- Before connecting an electrical plug to an instrument, check the plug for bent or broken pins.

- Cap the open electrical receptacles, plugs, and hose connections to prevent foreign material from entering the instrument or system.

Built in test Equipment (BITE)

Most modern digital systems include Built-In-Test-Equipment (BITE). These systems can

Figure 5-8-10. A shock mounted instrument panel

monitor and test many instrument system parameters. Maintenance personnel can use the BITE function to perform troubleshooting tests while on the ground. The system can also call up operating data that was recorded in flight. This data can be used to help troubleshoot intermittent failures or to evaluate systems that can only be operated in flight.

The BITE system and related items are controlled on the maintenance page that can be accessed through one of the cockpit display units (Figure 5-8-11). Some aircraft have a dedicated maintenance access terminal instead. The maintenance page can be used to display system diagrams showing the current status of the items in the system. See Figure 5-8-12.

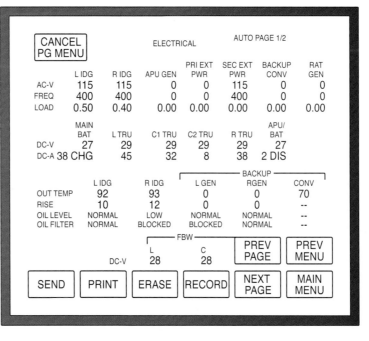
Figure 5-8-11. A maintenance page displayed on a cockpit CDU

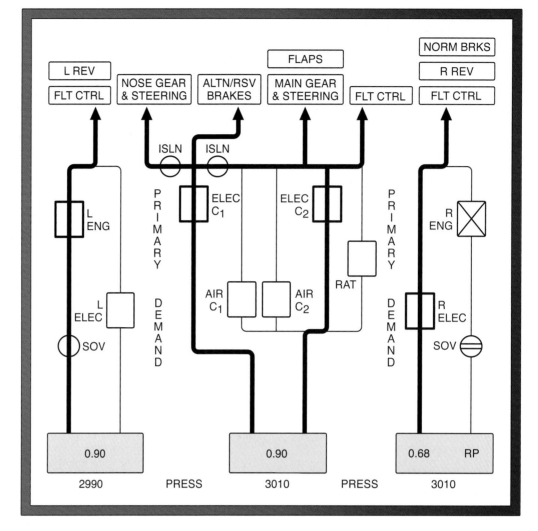

Figure 5-8-12. A maintenance CDU showing an aircraft system diagram

Section 9

Autopilot Systems

Automatic pilots are systems used to control aircraft without direct pilot control. They can range from simple wing-leveling systems to complete flight control systems that can operate the aircraft from take-off, climb, cruise, descent and landing.

The earliest and simplest systems were basic wing-levelers. These systems used a gyroscope to maintain the wings in a level attitude, thus reducing the pilot's workload on long flights. Autopilot systems soon added directional control and altitude capabilities.

Control Mechanisms

Modern autopilots use two different types of control mechanisms. The two types are posi-

tion (or horizon) based and rate based autopilots. Both types use gyros to provide the basic attitude and heading information.

Position based autopilots interpret the gyro movement and determine flight control commands that will return the aircraft to the desired pitch or roll attitude. These systems use the attitude gyro as the primary reference. A position based artificial horizon is shown in Figure 5-9-1.

Rate based systems typically use a turn coordinator for their gyro reference. These systems can detect roll rates as little as $1/16,000^{th}$ of a degree per second. The computer uses this roll information to calculate the aircraft attitude and generate a corrective action. Rate based systems also include accelerometers to measure the rate of change and whether this rate is increasing or decreasing.

Position based systems are very responsive. They can even overstress an aircraft in extreme turbulence due their rapid response rate. They are also dependent on proper attitude gyro

Figure 5-9-1. An artificial horizon used in a position based auto-pilot system

Figure 5-9-2. An analog control servo connected to the flight control cables

function. A tumbled attitude gyro will cause the autopilot to fail.

The turn coordinator used in rate based systems are more reliable than attitude gyros, thus making rate based systems slightly less prone to failure. Rate based systems are slightly less responsive, but do not risk overstressing the airframe. While both types are in common use in light aircraft, rate based systems are more common on transport category aircraft.

Basic System Components

All autopilots have a gyroscopic reference (the sensing element), a computer to calculate position and corrections (the computing element), a control console to activate and deactivate the system (the command element) and servos to move the control surfaces (the output element). Depending on the control mechanism type the gyroscopic reference will be either a directional gyro and an artificial horizon or a turn coordinator and accelerometers. The control computer may be analog or digital. Most analog systems add control servos to the basic flight controls. Figure 5-9-2 shows an analog servo connected to the control cables. Digital, fly-by-wire systems use the same flight control actuators for manual and autopilot flight. The difference is in where the control signal is generated.

Many systems include other components adding additional functions and controls. Figure 5-9-3 shows an autopilot system that has an altitude hold feature as well as a radio coupler. The radio coupler connects the autopilot to the navigation radio systems, allowing the autopilot to fly the aircraft to a specific navigation waypoint.

Today's advanced systems offer Mach-trim systems and auto throttle capabilities. They also interconnect with a variety of advanced navigation systems. Most of these advanced systems integrate the autopilot with the flight director system (described in the Navigation and Communication Systems chapter of this book).

Autopilot Axis Control

Autopilots are available to control one, two or three of axes of flight. These three axes are heading, pitch and roll. Figure 5-9-4 illustrates how the aircraft moves around these axes. For a complete review of aerodynamics, refer to the aerodynamics chapter in *Introduction to Aircraft Maintenance*.

The most basic autopilot is the single axis wing-leveler. They do little more than maintain the aircraft in a level attitude, thus relieving pilot workload on long flights. Two and three axes autopilots are much more common.

A two-axes autopilot controls roll and either pitch or heading. Three axes autopilots control roll, pitch and heading. While a three axes autopilot controls roll, pitch and heading, it may not control all three sets of flight controls. Some three axes autopilots only control ailerons and elevators. Elevators are used for pitch control and ailerons are used for both roll and heading control. A quick review of aerodynamics shows that the primary turning force is provided by bank angle, which is in turn set by the ailerons.

Yaw Dampers

This simplifies the design of the autopilot but adds some undesirable flight characteristics. Turns controlled only by ailerons are un-coordinated turns and create some side-slip in the aircraft. This can be uncomfortable for passengers. Most three axis, two-servo autopilots provide for the addition of a yaw damper. The yaw damper (Figure 5-9-5) provides two significant features, yaw damping and turn co-ordination.

Some aircraft have a tendency to oscillate about their heading axis, particularly during turbulence. These slight left and right movements can cause motion sickness for passengers. The yaw damper was developed to actively control the rudder to eliminate these oscillations. A second benefit of the yaw damper is turn-coordination.

Yaw dampers use a turn-coordinator to determine then the aircraft is slipping or skidding and move the rudder to counteract these movements. When the yaw damper is activated during autopilot controlled flight, it moves the rudder to eliminate any uncomfortable sideslip during turns. Three servo autopilots incorporate the yaw damping function into the autopilot, not as a separate system.

Autopilot Operation

The automatic pilot system flies the aircraft by using electrical signals developed in gyro-

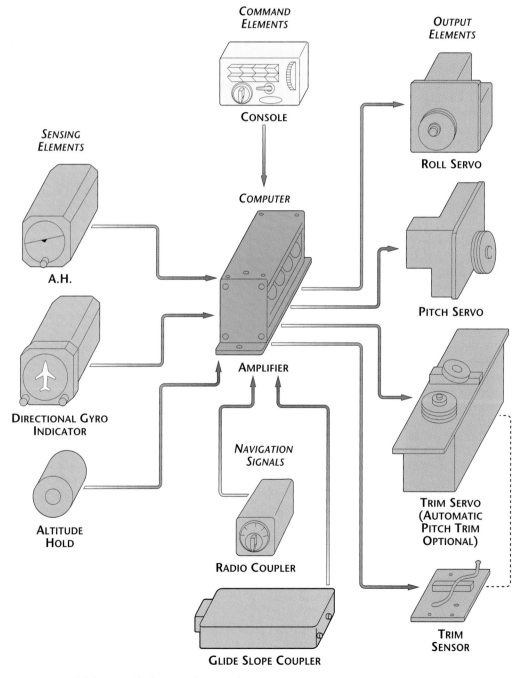

Figure 5-9-3. A full featured light aircraft autopilot system

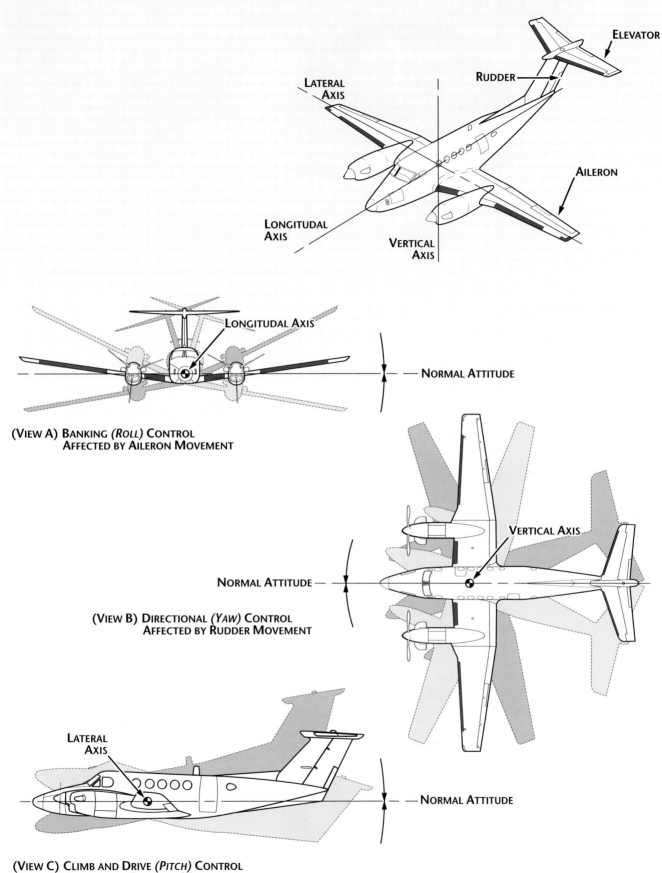

LATERAL
AXIS

ELEVATOR

RUDDER

AILERON

LONGITUDAL
AXIS

VERTICAL
AXIS

LONGITUDAL AXIS

NORMAL ATTITUDE

(VIEW A) BANKING *(ROLL)* CONTROL
AFFECTED BY AILERON MOVEMENT

VERTICAL AXIS

NORMAL ATTITUDE

(VIEW B) DIRECTIONAL *(YAW)* CONTROL
AFFECTED BY RUDDER MOVEMENT

LATERAL
AXIS

NORMAL ATTITUDE

(VIEW C) CLIMB AND DRIVE *(PITCH)* CONTROL
AFFECTED BY ELEVATOR MOVEMENT

Figure 5-9-4. Autopilots can control heading, roll and pitch

Figure 5-9-5. A yaw damper system can operate independently of the autopilot.

sensing units. These units are connected to or contained within flight instruments that indicate direction, rate-of-turn, bank, or pitch. If the flight attitude or magnetic heading is changed, electrical signals are developed in the gyro-sensing units. The signals are used to control the operation of servo units that convert electrical or hydraulic energy into mechanical motion.

The control computer reads these signals and commands the appropriate servo to move the flight controls to correct the flight attitude. Most autopilots provide for both manual and automatic operation. Both modes are set by the flight control console. The console described is a typical three axis system console. It allows for one, two or three axis control in manual or automatic modes.

The control console (Figure 5-9-6) is used to activate and deactivate the autopilot. Some systems also incorporate an additional autopilot

disconnect switch on the control wheel. This lets the pilot quickly disconnect the autopilot in the event of a system failure. Most autopilots will also automatically disengage if preset roll or pitch changes are exceeded. If the aircraft encounters turbulence or a flight attitude that the autopilot is not capable of handling, the system automatically reverts to pilot control.

Manual autopilot control. The simplest mode is roll control. Manual roll control mode will maintain the aircraft in the roll attitude selected by the pilot using the roll control knob. The pilot can select a level attitude, or use the roll control knob to roll the aircraft and turn it to a new heading. Once the new heading is reached, the pilot must then use the roll control knob to level the aircraft. The roll control knob manually controls both the roll and heading axes.

The second manual mode is roll with pitch control. In this mode, both the roll and pitch control knobs are active. The pilot uses the pitch control wheel to set the pitch attitude of the aircraft. It can be used to cause the aircraft to climb, descend or maintain a level attitude. Again, the pilot must move the pitch control wheel to change from one pitch attitude to another.

Automatic autopilot control. The autopilot can also be used in fully automatic mode. Basic automatic modes are heading and altitude. Heading mode differs a bit from manual roll control mode. Manual roll control mode keeps the wings level but does not correct for wind or turbulence, which move the aircraft heading.

Automatic heading mode causes the aircraft to maintain a heading preset on the directional gyro or flight director. If the aircraft deviates from this heading, the autopilot will roll the aircraft into a turn, move the heading back to the

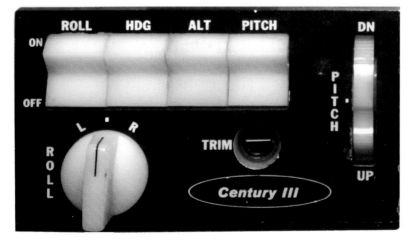

Figure 5-9-6. Control console

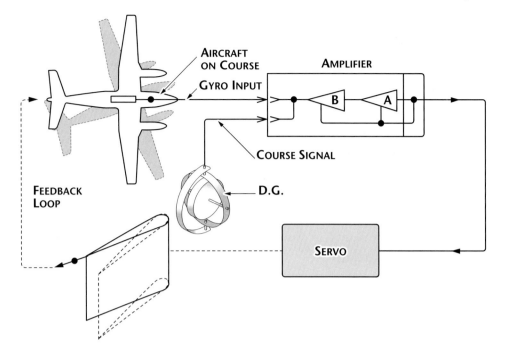

Figure 5-9-7. Autopilot feedback loop

original heading, then roll the aircraft out of the turn.

Autopilot feedback loop. The basic autopilot feedback loop, illustrated in Figure 5-9-7, is the key to automatic operation. The amplifier or control computer compares the current aircraft attitude and position with the desired attitude and position set on the autopilot. Any deviation from the desired position will cause the amplifier or computer to move a servo to return the aircraft to the desired position.

The amplifier or computer must also include circuitry to eliminate autopilot induced oscillation. During human controlled flight, the pilot must anticipate the aircraft attitude and position. As the aircraft approaches the desired heading, the pilot will reduce the roll (and rate of turn) until the aircraft stops turning as it reaches the desired heading.

If the pilot maintained the turn until the heading was reached, the aircraft would turn past the desired heading before the aircraft could be returned to a level attitude. This overshoot would then require an additional turn back to the desired heading with the same consequences. The autopilot must also be equipped with circuitry that anticipates the aircraft position and reduces the roll so that the aircraft stops turning just as it reaches the desired heading.

Altitude control. Fully integrated, three axis autopilots also include an altitude control feature. Altitude hold mode is the automatic equivalent of manual pitch control, however,

altitude hold provides for additional features. Manual pitch control maintains a level attitude but does not compensate for updrafts or downdrafts.

Altitude control capable autopilots are equipped with a barometric pressure sensor (Figure 5-9-8). The amplifier or control computer uses this barometric information to maintain the aircraft at a preset altitude in addition to maintaining a level pitch attitude. The autopilot will occa-

Figure 5-9-8. Autopilot altitude hold chamber

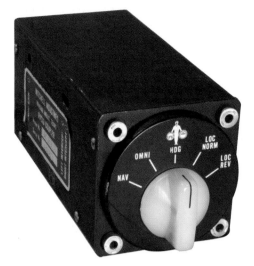

Figure 5-9-9. Radio coupler

sionally command a slight climb or descent to return the aircraft to the preset altitude.

Radio navigation control. Most advanced autopilots can be connected to the navigation radios. Radio navigation control is achieved through the use of a radio coupler (Figure 5-9-9). This radio coupler may be a stand alone unit or part of a flight director system. Both types provide a command signal to the autopilot computer.

Radio navigation control commands override heading control commands. The autopilot will fly the aircraft on a specific radio navigation course. The autopilot can fly the aircraft directly to a chosen airport, compensating for crosswinds and other weather conditions. Navigation radio system operation is covered

in detail in the *Navigation and Communication Systems* chapter of this book.

Flight computer control. The most advanced digital autopilot systems available today can control the aircraft throughout the vast majority of the flight. A flight plan, including multiple altitudes and turns can be programmed prior to flight. After takeoff, the pilot can activate the autopilot. It will then fly the aircraft to the pre-programmed altitudes, leveling the aircraft at the correct times. The autopilot will also turn the aircraft to follow a precise course. Multiple legs can be flown in succession, flying around congested or restricted airspace. The autopilot maintains the aircraft preset altitudes and courses and climbs; it descends and turns as needed.

Some of these systems also provide control to the engines and brakes as well. The engines can be throttled up or down by the autopilot for climbs and descents. Autoland equipped aircraft can even finish the flight, landing the aircraft and coming to a stop, using brakes or thrust reversers. All of this is achieved through autopilot control operation.

Autopilot System Maintenance

The information in this section does not apply to any particular autopilot system, but gives general information that relates to most autopilot systems. Maintenance of an autopilot system consists of visual inspection, replacement of components, cleaning, lubrication, and an operational checkout of the system.

With the autopilot disengaged, the flight controls should function smoothly. The resistance offered by the autopilot servos should not affect the control of the aircraft. The interconnecting mechanism between the autopilot system and the flight control system should be correctly aligned and smooth in operation. When applicable, the operating cables should be checked for tension.

An operational check is important to assure that every circuit is functioning properly. An autopilot operational check should be performed on new installations, after replacement of an autopilot component, or whenever a malfunction in the autopilot system is suspected.

After the aircraft's main power switch has been turned on, allow the gyros to come up to speed and the amplifier to warm up before engaging the autopilot. Some systems are designed with safeguards that prevent premature autopilot engagement.

AUTOPILOT SWITCH

Figure 5-9-10. Autopilot engage and disengage switches are often located on the pilot's control yoke.

While holding the control column in the normal flight position, engage the system using the engaging switch (Figure 5-9-10).

After the system is engaged, perform the operational checks specified for the particular aircraft. In general, the checks consist of:

- Rotate the turn knob to the left; the left rudder pedal should move forward, and the control column wheel should move slightly aft.

- Rotate the turn knob to the right; the right rudder pedal should move forward, and the control column wheel should move to the right. The control column should move slightly aft. Return the turn knob to the center position; the flight controls should return to the level-flight position.

- Rotate the pitch-trim knob forward; the control column should move forward.

- Rotate the pitch-trim knob aft; the control column should move aft.

If the aircraft has a pitch-trim system installed, it should function to add down-trim as the control column moves forward, and add up-trim as the column moves aft. Many pitch-trim systems have an automatic and a manual mode of operation. The above action will occur only in the automatic mode.

Check to see if it is possible to manually override or overpower the autopilot system in all control positions. Center all the controls when the operational checks have been completed.

Disengage the autopilot system and check for freedom of the control surfaces by moving the control columns and rudder pedals. Then re-engage the system and check the emergency disconnect release circuit. The autopilot should disengage each time the release button is actuated.

When performing maintenance and operational checks on specific autopilot systems always follow the procedure recommended by the aircraft or equipment manufacturer.

6

Navigation and Communication Systems

Communications and navigation systems have become so intertwined that the acronym is NAV-COM. Communication systems principally involve voice transmission and reception between aircraft, or aircraft and ground stations. Navigation equipment has reached the point that it could more properly be called electronic navigation, rather than radio navigation. However, virtually everything airborne still relies in the principles of radio transmission and reception for all information dissemination.

For many years every airplane that had a radio had to have a Aircraft Radio Station License, and every person who might operate that radio, a Radiotelephone Operators License. The process involved no fee, or a low fee, for signing up with the Federal Communications Commission (FCC). There was no testing required and no renewal needed.

The Telecommunications Act of 1996 brought about fundamental changes in the licensing of aircraft radio stations.

The FCC eliminated the individual licensing requirement for all aircraft operating domestically. This means that a pilot does not need a license to operate a two-way VHF radio, radar, or ELT aboard a domestic aircraft. Flights that operate or land outside the United States are still required to have a station license. Applications may be obtained online from the FCC.

Airborne systems range from simple radio direction finders to computerized navigational systems, and other advanced electronic techniques, to automatically solve the navigational problems for an entire flight. *Marker Beacon receivers, instrument landing systems (ILS) (involving radio signals for glideslope and direction), distance measuring equipment*

Learning Objectives:

REVIEW
- radio principles, equipment and components

DESCRIBE
- types and features of communication systems
- use and operation of airborne navigation equipment

EXPLAIN
- GPS limitations and equipment
- factors influencing the installation of communication and navigation equipment including mount, location, vibration and interference

APPLY
- calculate the profile drag of an antenna for installation

Left. This Canadair Challenger shows a typical collection of antennas. Each radio type requires an antenna designed for it's specific bandwidth.

(DME), *radar, area navigation systems* (RNAV), *Omni-directional radio receivers* (VOR) and satellite based *Global Positioning Systems* (GPS) are but a few basic applications of airborne radio navigation systems available for installation and use in aircraft.

All of these electronic installations, and more, are items that will be on an inspection checklist. Collectively, they are called avionics. Avionics is a contraction of *aviation electronics*. Originally the meaning was clear; if it was electronic and part of aviation, the name applied. With the coming of the digital age things are not so easily discernible. Nowadays, almost everything on an airplane is electronic.

In an attempt to standardize the description, many authorities have narrowed the definition to include anything that has to do with navigation, communication, or automatic flight systems. As an AMT you are not expected, in fact not allowed, to perform maintenance on these units. Repair, calibration, and in most cases installation, of electronic equipment is the realm of a certified repair station. You are expected to understand how they function, and how to inspect their basic installation and operation.

These inspections include a visual examination for security of attachment, condition of the wiring, bonding, shock mounts, radio racks and supporting structure. In addition, a functional check is usually performed to determine that the equipment is operating properly and that its operation does not interfere with the operation of other systems.

During the rest of this chapter you should think of avionics as three different groups of equipment:

- Panel mounted only. Used mostly in smaller general aviation aircraft.

- A combination of panel mounted units and remote units with the control heads panel mounted. Principally used in twin-engine and medium-sized corporate aircraft.

- Remote mounted Line Replaceable Units (LRU) (Figure 6-1-1) with panel mounted control heads. Airliners have the LRUs mounted in a separate electronic equipment bay.

A technician should possess some basic knowledge and understanding of the principles, purposes, and operation of the radio equipment used in aircraft. The information presented in this chapter is general in nature and provides a broad introduction to radio, its principles and application to aircraft from a technician's viewpoint. In no way will it make you a radio technician.

Section 1
Basic Radio Principles

Here are a few ways radio interacts with your everyday life:

- Cordless phones and cell phones
- Garage door openers and doorbells
- Television, both satellite and cable
- AM and FM radio
- Ham radio
- Military and CB radio
- Police and fire department radios
- Baby monitors and toys
- Wireless clocks and GPS
- Satellite internet access
- Microwave ovens and police radar
- Auto changing stoplights
- Everything satellite and space related
- Aviation, and many more

Frequency generation. Radio, or more correctly the propagation of radio waves, is basically very simple.

The basic concept of radio communications involves the transmission and reception of electromagnetic (radio) energy waves through space. Alternating current passing through a conductor creates electromagnetic fields around the conductor. Energy is alternately stored in these fields and returned to the conductor. As the frequency of current alternation increases, less and less of the energy stored in the field returns to the conductor. Instead of returning, the energy is radiated into space in the form of electromagnetic waves. A conductor radiating in this manner is called the *transmitting antenna*.

For an antenna to radiate efficiently a transmitter must supply it with an alternating current of the selected *frequency* (cycles). The frequency of the radio wave radiated will be equal to the frequency of the applied current.

How alternating current is produced in various frequencies is a subject for a course in basic radio. The frequency of the electrical alternation is called a sine wave. The frequency of the *sine wave* is the number of times it oscillates up and down, per second. For example, if you are listening to AM radio and are tuned to 540 on the dial, the frequency of the station is

Figure 6-1-1. The electronics bay of a Boeing 757 contains the LRUs for the avionics.

540,000 cycles per second. If you tune to 95.3 FM, the frequency is 95,300,000 hertz.

Frequency Modulation

When a transmitter is transmitting a sine wave, and that sine wave is going into space through an antenna, you have a radio station. The problem is that the wave has no information in it. To put information in the sine wave requires superimposing, or *modulating*, the information you want to transmit over the sine wave. There are three common types of modulation.

Pulse Modulation (PM). *Pulse Modulation* is simply turning the sine wave on and off. It is a process that is good for Morse code and setting radio controlled clocks; that is about all. One transmitter can cover the entire United States. See Figure 6-1-2.

Amplitude Modulation (AM). The principal use of *Amplitude Modulation* is for AM radio stations and television pictures. The peak-to-peak voltage of the sine wave changes as the amplitude of the modulation changes (Figure 6-1-3). The sine wave produced by a person's voice is the best example of AM.

Frequency Modulation (FM). Hundreds of radio applications use *frequency modulation*, including FM radio, television sound, cell phones and aircraft systems. The FM transmitters frequency (sine wave) changes very little

and is fairly immune to static and atmospheric interference (Figure 6-1-4).

The entire spectrum of possible radio frequencies is not unlimited. Specific sections of the spectrum

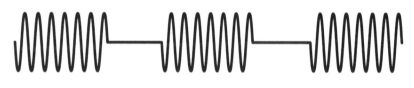

Figure 6-1-2. An example of pulse modulation

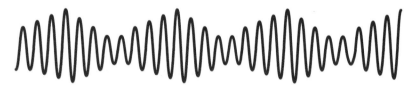

Figure 6-1-3. AM modulation, showing sine wave changes as a voice changes tone and volume.

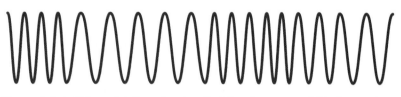

Figure 6-1-4. FM modulation showing that the sine wave width does not change, but that the frequency does.

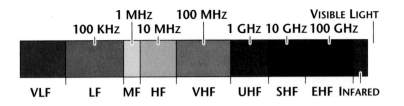
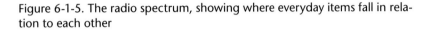

Figure 6-1-5. The radio spectrum, showing where everyday items fall in relation to each other

are reserved for certain uses (Figure 6-1-5). Those uses break down like this:

- Garage doors openers and alarm systems – around 400 megahertz

- Standard cordless phones – 40 to 60 megahertz

- Baby monitors and kids walkie talkies – 49 megahertz

- Radio control airplanes and cars – 72 to 75 megahertz

- Wildlife tracking collars – 215 to 220 megahertz

- Cell phones – 824 to 900 megahertz

- Air traffic control radar – 960 to 1,215 megahertz

- GPS systems – 1,227 to 1,575 megahertz

- Deep space communications – 2,290 to 2,300 megahertz

There are a large number of people wanting to use the airwaves, all at the same time and many on the same, or close to the same, frequency. As more products that rely on these frequencies become available, the tighter the frequency tolerances become. To keep them from interfering with each other the *Federal Communications Commission* (FCC) has established a specific *bandwidth* for every category.

Antennas

When current flows through a *transmitting antenna*, radio waves are radiated in all directions in much the same way that waves travel on the surface of a pond into which a rock has been thrown. Radio waves travel at a speed of approximately 186,000 miles per second.

When a radiated electromagnetic field passes through a conductor, some of the energy in the field will set electrons in motion in the conductor. This electron flow constitutes a current that varies with changes in the electromagnetic field. Thus, a variation of the current in a radiating antenna causes a similar varying current

in a *receiving antenna* (conductor) at a distant location. Any intelligence produced as current in a transmitting antenna will be reproduced as current in a receiving antenna.

Frequency Bands

The radio frequency portion of the electromagnetic spectrum extends from approximately 30 kilohertz (kHz) to 30,000 megahertz (MHz).

As a matter of convenience, this part of the spectrum is divided into frequency bands. Each band or frequency range produces different effects in transmission. The radio frequency bands proven most useful and presently in use are:

- Low Frequency (LF) 30 to 300 kHz

- Medium Frequency (MF) 300 to 3,000 kHz

- High Frequency (HF) 3,000 kHz to 30 MHz

- Very High Frequency (VHF) 30 to 300 MHz

- Ultra High Frequency (UHF) 300 to 3,000 MHz

- Super High Frequency (SHF) 3,000 to 30,000 MHz

In practice, radio equipment usually covers only a portion of the designated band, e.g., civil VHF equipment normally operates at frequencies between 108.0 MHz and 135.95 MHz.

Basic Equipment Components

The basic components (Figure 6-1-6) of a communication system are: transmitter, receiver, transmitting antenna, receiving antenna, microphone and a headset or speaker.

Transmitter. A transmitter may be considered a generator, which changes electrical power into radio waves and must perform these functions:

- Generate a radio frequency (RF) signal

- Amplify the RF signal

- Modulate the signal by placing intelligence on it.

The transmitter contains an oscillator circuit to generate the RF signal and amplifier circuits to increase the output of the oscillator to the power level required for operation.

Transmitters take many forms, have varying degrees of complexity and develop different levels of power. The amount of power generated by a transmitter affects the strength of the electromagnetic field radiating from the

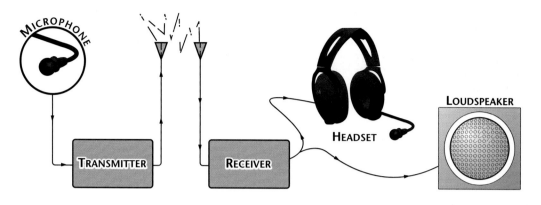

Figure 6-1-6. Primary components of a basic radio system

antenna. Thus, it follows that the higher the power output from a transmitter, the greater the distance its signal may be received.

VHF transmitters used in single engine and light twin-engine aircraft vary in power output from 1 watt to 30 watts, depending on the particular model radio. However, 3 to 5 watt ratings are used most frequently. Executive and large transport aircraft are usually equipped with VHF transmitters having a power output of 20 to 30 watts.

Aviation communication transmitters are crystal-controlled in order to meet the frequency tolerance requirements of the FCC. The frequency of the channel selected is determined by a crystal. Transmitters may have from one to 720 channels.

Receivers. The communications receiver must select radio frequency signals and convert the intelligence contained on these signals into a usable form: either audible signals for communication and audible or visual signals for navigation.

Radio waves of many frequencies are present in the air. A receiver must be able to select the desired frequency from all those present and amplify the small AC signal voltage.

The receiver contains a demodulator circuit to remove the intelligence. If the demodulator circuit is sensitive to amplitude changes, it is used in AM sets and called a detector. A demodulator circuit that is sensitive to frequency changes is used for FM reception and is known as a discriminator.

Amplifying circuits within the receiver increases the audio signal to a power level that will operate the headset or a speaker.

Transmitting antenna. An antenna is a special type of electrical circuit designed to radiate and receive electromagnetic energy. As

mentioned previously, a transmitting antenna is a conductor, which radiates electromagnetic waves when a radio frequency current is passed through it. Antennas vary in shape and design (Figure 6-1-7) depending upon the frequency to be transmitted, and specific purposes they must serve. In general, communication-transmitting stations radiate signals in all directions. However, special antennas are designed that radiate only in certain directions or certain beam patterns.

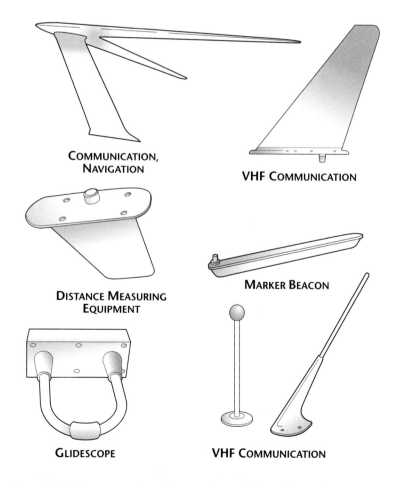

Figure 6-1-7. Because each system requires a different type of antenna, you can tell the antennas purpose by its looks.

Receiving antenna. The receiving antenna must intercept the electromagnetic waves that are present in the air. The shape and size of the receiving antenna will also vary according to the specific purpose for which it is intended. In airborne communications, the same antenna is normally used for both transmission and reception of signals.

Microphone. A microphone is essentially an energy converter that changes acoustical (sound) energy into corresponding electrical energy. When spoken into a microphone, the audio pressure waves that are generated strike the diaphragm of the microphone causing it to move in and out in accordance with the instantaneous pressure delivered to it. The diaphragm is attached to a device that causes current to flow in proportion to the pressure applied.

For good quality sound, the electrical waves from a microphone must correspond in magnitude and frequency to the sound waves that cause them. A desirable characteristic is the ability of the microphone to favor sounds coming from a nearby source over random sounds coming from a relatively greater distance. When talking into this type of microphone, the lips must be held as close as possible to the diaphragm.

Persons inexperienced in the use of the microphone are usually surprised at the quality of their own transmissions when they are taped and played back. Words quite clear when spoken to another person can be almost unintelligible over the radio. Readable radio transmissions depend on the following factors:

- Voice amplitude
- Rate of speech
- Pronunciation and phrasing.

Clarity increases with amplitude up to a level just short of shouting. When using a microphone, speak loudly, without exerting extreme effort. Talk slowly enough so that each word is spoken distinctly. Avoid using unnecessary words.

Headsets. Headsets for aircraft radios are not interchangeable with commercial audio headsets. Aircraft headsets are designed to operate in a high noise environment and are frequently included in a set of hearing protector ear cups that cover the entire ear. While not delicate, they can be damaged by rough handling. The most common damage is to the wires: mostly pinching or kinking them in a crowded cockpit.

Most headsets use a boom microphone that is positioned just ahead of the lips. The microphones are also designed for high noise environments. They should be positioned just ahead of the lips and the lips should almost touch the microphone when speaking.

Speakers. Speakers are installed in the headliners of most smaller airplanes. They are positioned so both people who are sitting at the controls can hear. Other cockpits will have speakers installed in a suitable position.

Because of their protected locations, speakers are not normally subject to rough handling. The largest cause of failure is from cracked cones brought about by aging and the environment. If a speaker is difficult to hear from clearly, check it for cracked cones and replace if necessary.

Power Supply

In many aircraft, the primary source of electric power is direct current. An *inverter* is used to supply the required alternating current. Common aircraft inverters consist of a DC motor driving an AC generator. Static, or solid-state, inverters have replaced the electromechanical inverters. Rotary inverters are today found only in much older systems. Static inverters have no moving parts, but use semiconductor devices and circuits that periodically pulse DC current through the primary of a transformer to obtain an AC output from the secondary. Many AC systems today run 120 VAC at 400 hertz.

Section 2

Communication Systems

Very High Frequency (VHF) system. The most common communication system in use today is the VHF system. In addition to VHF equipment, large aircraft are usually equipped with HF communication systems.

Airborne communications systems vary considerably in size, weight, power requirements, quality of operation and cost, depending upon the desired operation.

Almost all airborne VHF and HF communication systems use transceivers. A transceiver is a self-contained transmitter and receiver that share common circuits; i.e., power supply, antenna and tuning. The transmitter and receiver both operate on the same frequency, and the microphone button determines when there is an output from the transmitter. In the absence of transmission, the receiver is sensitive to incoming signals. Large aircraft may be equipped with transceivers or a communications system that use separate transmitters and receivers.

Figure 6-2-1. A basic VHF radio

Photo courtesy of Bendix/King

The operation of radio equipment is essentially the same whether installed on large aircraft or small aircraft. In some radio installations, the controls for frequency selection, volume and the on-off switch are integral with the radio main chassis. In other installations, the controls are mounted on a panel located in the cockpit, and the radio equipment is located in racks in another part of the aircraft.

Because of the many different types and models of radios in use, it is not possible to discuss the specific techniques for operating each in this manual. However, there are some practices that apply to all radios.

Very High Frequency (VHF) communications. VHF airborne communication sets operate in the frequency range from 108.0 MHz to 135.95 MHz. VHF receivers are manufactured that cover only the communications frequencies (not common), or both communications and navigation frequencies (common). In general, the VHF radio waves follow approximately *line-of-sight* (straight lines). Theoretically, the range of contact is the distance to the horizon and this distance is determined by the heights of the transmitting and receiving antennas. However, communication is sometimes possible many hundreds of miles beyond the assumed horizon range.

Many VHF radios have the transmitter, receiver, power supply and operating controls built into a single unit. These units are generally installed in a cutout in the instrument panel. A typical panel-mounted VHF transceiver is shown in Figure 6-2-1. Others have certain portions of the communication system mounted on the instrument panel, and the remainder remotely installed in a radio or baggage compartment.

VHF operational (ops) check. After turning the radio control switch on, allow sufficient time for the equipment to warm up before beginning the operational checks. Using the frequency selector, select the frequency of the ground station to be contacted. Adjust the volume control to the desired level.

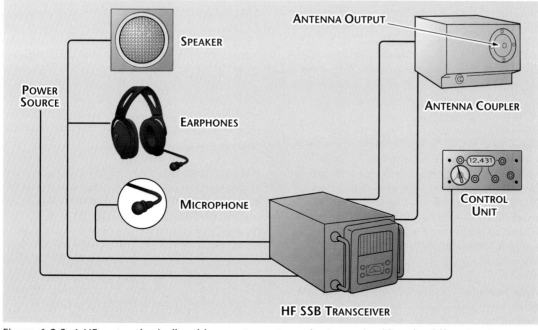

Figure 6-2-2. A HF system basically adds an antenna tuner (or tuner circuit) and a different antenna.

With the microphone held close to the mouth, press the microphone button and speak directly into the microphone to transmit; when through talking, release the button. This action will return the communication receiver to operation. When the ground station acknowledges the initial transmission, request that an operational check be made on all frequencies or channels.

High Frequency (HF) system. A high frequency communication system (Figure 6-2-2) is used for long-range communication. HF systems operate essentially the same as a VHF system, but operate in the frequency range from 3 MHz to 30 MHz. Communications over long distances are possible with HF radio because of the longer transmission range. HF transmitters have higher power outputs than VHF transmitters.

The design of antennas used with HF communication systems vary with the size and shape of the aircraft. Aircraft that cruise below 300 m.p.h. generally use a long wire antenna. Higher speed aircraft have specially designed antenna probes installed in the vertical stabilizer. Regardless of the type antenna, a tuner is used to match the impedance of the transceiver to the antenna.

An operational check of an HF radio consists of turning the control switch to on, adjusting the RF gain and volume controls, selecting the desired channel and transmitting the appropriate message to the called station. Best adjustment of the gain control can be obtained with the volume control set at half range. The gain control is used to provide the strongest signal with the least amount of noise. The volume control is used to set sound level and affects only the loudness of the signal.

Onboard Aircraft Communications

Even during the early days of aviation, communication between flight crew and ground crew personnel was extremely important. The first aircraft were started by hand as the pilot simply yelled to his ground crew. Hand signals were often used for communication during times of adverse noise conditions (See Figure 6-2-3). As aircraft began to carry passengers, they too needed to be informed of certain flight details.

Early crew to passenger communication was simply a matter of loud conversations. As aircraft grew in size and complexity, it became evident that better communications between flight and ground crews, and between crewmembers and passengers were necessary.

Today's transport category aircraft contain a variety of systems, all dedicated to communications. For example, the flight crew can communicate with passengers, ground crew, air traffic control and flight attendants. Some systems allow communication between air-

Figure 6-2-3. Communication between the tug operator and the observer, when necessary, will normally be by hand signal.

line operations or maintenance facilities and the aircraft central maintenance system. Gate changes, passenger lists or other pertinent data can be transmitted and printed using an onboard printer.

Many modern transport category aircraft also include an extensive passenger entertainment system. Entertainment systems include multi-channel audio and video programs. All of these passenger systems must be linked to the flight crew in the event the pilot (or flight attendant) needs to make an announcement.

Today's transport category aircraft employ a sophisticated audio system to coordinate communications. This section will present an overview of these information and entertainment systems by examining the B-747-400 aircraft. Most of the systems found on this aircraft are similar to those of other large aircraft.

Keep in mind, this section will present general concepts; the specific information for each aircraft must be accessed prior to performing any maintenance activities. For further information on the subject of avionics refer to *Avionics: Systems and Troubleshooting* by T.K. Eismin, available from Avotek Information Resources.

Airborne Communications Addressing and Reporting System (ACARS)

Airborne Communications Addressing and Reporting System (ACARS). *Airborne Communications Addressing and Reporting System (ACARS)* is a digital air/ground communications service designed to reduce the amount of voice communications on the increasingly crowded VHF frequencies. ACARS allows ground to aircraft communications (in a digital format) for operational flight information, such as fuel status, flight delays, gate changes, departure times and arrival times.

ACARS can also be used to monitor certain engine and system parameters and downlink relevant maintenance data to the aircraft operator. Prior to the use of ACARS, this information was transmitted using voice communications. ACARS can be thought of as e-mail for the aircraft. Since the message is transmitted in digital format via ACARS, it occupies much less time on a given frequency than conventional voice communications, and since ACARS is an automatic system, transfer of information requires virtually no flight crew efforts.

There are two major corporations that provide ACARS services worldwide: *ARINC Incorporated* and a French organization known as *SITA*.

ARINC Incorporated is a global corporation based in the United States with primary stockholders consisting of various U.S. and international airlines and aircraft operators. ARINC provides services related to a variety of aviation communication and navigation systems.

One of the services available through ARINC is called *GLOBALink*. GLOBALink is the name of the ACARS service provided by ARINC Incorporated. It should be noted there are other providers of ACARS in various parts of the world; however, ARINC and SITA provide the majority of the ACARS services. In general, airborne equipment designed to operate using GLOBALink will operate with other ACARS services throughout the world.

ACARS theory of operation. The airborne components of ACARS connect to various sensors throughout the aircraft. These sensors are used to detect parameters to be transmitted by ACARS. For example, virtually all transport category aircraft transmit 0001 (Out, Off, On and In) data. Out stands for *out-of-the-gate*. ACARS uses a parking brake (or similar) sensor to determine out-of-the-gate time. Off means aircraft *off-the-ground*; this can be determined by a landing gear sensor. On refers to the aircraft touchdown (*on-the-ground*). Once again, a landing gear sensor can detect this condition and send the information to ACARS. When it is determined when the aircraft is *in-the-gate*. ACARS could transmit In data when the parking brake is set. Remember, ACARS operates by transmitting digital data. A code of ones and zeros is transmitted to deliver all information. If one were to listen to ACARS, it might sound similar to a modem connecting to an Internet web site.

ACARS transmits all information using VHF frequencies. The airborne equipment transmits through a VHF transceiver (typically located in the aircraft's equipment bay). The future use of satellites will improve worldwide coverage and enhance ACARS performance. In the future, more data will be transmitted through some form of ACARS and it will most likely be broadcast through satellites.

The ACARS airborne equipment contains a control unit, typically located on the flight deck, a Management Unit (MU) and the necessary VHF transceiver located in an electronics equipment bay. The ground-based equipment contains antennas and VHF transceivers located at various sites, a data link via telephone lines to one or more ACARS control facilities and a data link to the various airlines. In some cases, the communications between ACARS ground facilities and the airlines are accomplished through microwave transmitters or satellite links. Of course, the airline element of ACARS is an elaborate system of command and control subsystems, which are used for maintenance, crew scheduling, gate assignments and other day-to-day operations.

The ACARS system is designed to use a variety of VHF frequencies from 129.00 MHz to 137.00 MHz. In North America, all ACARS transmissions begin on 131.55 MHz. The ACARS ground facilities may then assign a different VHF frequency (between 129.00 and 137.00 MHz) to the aircraft.

The assigned frequency will be a function of the aircraft location and the particular VHF frequencies used in that area. The ACARS ground facilities will then reassign new VHF frequencies to the aircraft ACARS as needed. Any frequency changes are totally automatic and therefore unknown to the flight crew. It is very likely that ACARS will change frequencies several times during a given flight.

Each aircraft using the ACARS system is given a specific address code. This code is used by the ground base facility whenever calling the aircraft. The airborne equipment will monitor all ACARS data transmissions on their assigned frequency and accept only those with the correct address code.

Selective Calling (SELCAL)

SELCAL is a radio system that allows an airline operator to communicate with any of his airplanes while they are in the air.

To reduce cockpit noise most flight crews turn the volume down on the radio tuned to the SELCAL frequency. When the operator wishes to call a specific airplane, they will first transmit that airplanes SELCAL code. There are 10,920 codes available.

The SELCAL code consists of four letters that serve as an identifier for each airplane. The code is used with a HF radio transmitter to basically communicate with all airline type airplanes at one time, even though the message was meant only for a specific airplane. The special code will sound a chime in the cockpit of the selected airplane only. The cockpit crew will then turn up the volume so they may transact their business.

While this may seem a bit awkward, it works fairly well. Although any cockpit crew could listen to any transmission, in practice they really do not. Everybody just keeps the volume down unless a chime goes off. There is instant communication with all transport airplanes in the air at any one time, and all on one radio frequency. Many requests for specific information concerning the flight are transmitted this way.

Section 3

Airborne Navigation Equipment

Airborne navigation equipment is a phrase encompassing many systems and instruments. These systems include VHF omni-directional range (VOR), instrument landing systems (ILS), distance-measuring equipment (DME), automatic direction finders (ADF), radar beacon transponders, Doppler systems and inertial navigation systems (INS).

When applied to navigation, the radio receivers and transmitters handle signals that are used to determine a bearing heading, and in some cases distance from geographical points or radio stations.

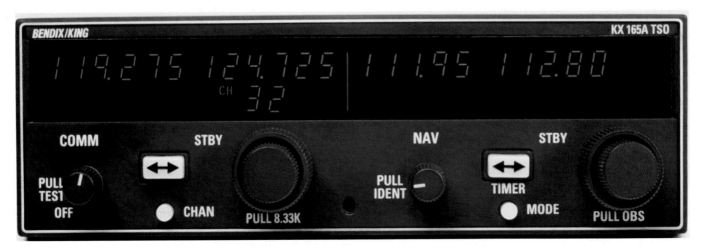

Figure 6-3-1. The VOR section of a panel mounted radio is on the right.

Photo courtesy of Bendix King

VHF Omni-directional Range (VOR) System

The *VHF omni-directional range (VOR)* is an electronic navigation system. As the name implies, the omni-directional or all-directional range station provides the pilot with courses from any point within its service range. It produces 360 usable radials or courses, any one of which is a radio path connected to the station. The radials can be considered as lines that extend from the transmitter antenna like spokes of a wheel. Operation is in the VHF portion of the radio spectrum (frequency range of 108.0 MHz/117.95 MHz) with the result that interference from atmospheric and precipitation static is negligible. The navigational information is visually displayed on an instrument in the cockpit.

The typical airborne VOR receiving system (Figure 6-3-1) consists of a receiver, visual indicator, antennas and a power supply. The antenna is most frequently mounted on the vertical stabilizer. In addition, a unit frequency selector is required and in some cases located on the receiver unit front panel. This frequency selector is used to tune the receiver to a selected VOR ground station.

The VOR receiver, in addition to course navigation, functions as a *localizer* receiver during instrument landing system (ILS) operation. Also, some VOR receivers include a *glideslope* receiver in a single case. Regardless of how individual manufacturers may design the VOR equipment, the intelligence from the VOR receiver is displayed on the *course deviation indicator (CDI)*.

The CDI, Figure 6-3-2 performs several functions. During VOR operation the vertical needle is used as the course indicator. The vertical needle also indicates when the aircraft deviates from the course and the direction the aircraft must be turned to attain the desired course.

The *TO-FROM indicator* (not shown) presents the direction to or from the station along the radial. The course deviation indicator also contains a VOR-LOC flag alarm. Normally, this is a small arm, which extends into view only in the case of a receiver malfunction or the loss of a transmitted signal.

When localizer signals are selected on the receiver, the indicator shows the position of the localizer beam relative to the aircraft and the direction the aircraft must be turned to intercept the localizer.

Ground Testing a VOR

During VOR operation the VOR radial to be used is selected by rotating the *omni-bearing*

Figure 6-3-2. The CDI is the instrument part of Omni navigation. *Photo courtesy of Bendix King*

selector (OBS). The OBS is generally located on the CDI; however, in some installations it is a part of the navigation receiver. The OBS is graduated in degrees from zero to 360. Each degree is a VOR course to be flown in reference to a ground station.

The FAA VOR *test facility (VOT)* transmits a test signal, which provides users a convenient means to determine the operational status and accuracy of a VOR receiver while on the ground where a VOT is located.

To use the VOT service, tune in the VOT frequency on your VOR receiver. With the CDI centered, the omni-bearing selector should read 0 degrees with the to/from indication showing "from" or the omni-bearing selector should read 180 degrees with the to/from indication showing "to." Should the VOR receiver operate a *radio magnetic indicator (RMI)*, it will indicate 180 degrees on any omni-bearing selector (OBS) setting. Two means of identification are used. One is a series of dots and the other is a continuous tone. Information concerning an individual test signal can be obtained from the local FSS.

If the operational check is unsatisfactory, it will be necessary to remove the VOR receiver and associated instruments from the aircraft and have them calibrated.

Instrument Landing System

The instrument landing system (ILS), one of the facilities of the Federal airways, operates in the VHF portion of the electromagnetic spectrum. The ILS can be visualized as a slide made of radio signals on which the aircraft can be brought safely to the runway.

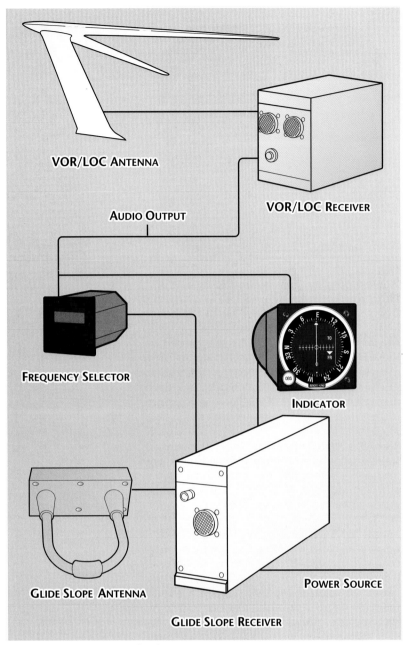

Figure 6-3-3. Instrument landing system

The entire system consists of a *runway localizer*, a *glideslope signal* and marker beacons for position location. The localizer equipment produces a radio course aligned with the center of an airport runway.

The on-course signals result from equal reception of two signals: one containing 90 Hz modulation and the other containing 150 Hz modulation. On one side of the runway center line the radio receiver develops an output in which the 150 Hz tone predominates. This area is called the blue sector. On the other side of the centerline the 90 Hz output is greater. This area is the yellow sector.

Runway localizer. The localizer facility operates in the frequency range of 108.0 MHz to 112.0 MHz on the odd tenths of the megahertz steps. The VOR receiver also operates in this frequency range on the even tenths of the megahertz steps. The airborne VOR receiver functions as the localizer receiver during ILS operation.

Glideslope. The glideslope is a radio beam, which provides vertical guidance to the pilot, assisting him in making the correct angle of descent to the runway. Glideslope signals are radiated from two antennas located adjacent to the touchdown point of the runway. Each glideslope facility operates in the UHF frequency range from 329.3 MHz to 335.0 MHz.

The glideslope and VOR/localizer receivers may be separate receivers or combined in a single case. The glideslope receiver is paired to the localizer and one frequency selector is used to tune both receivers. A component diagram of an ILS is shown in Figure 6-3-3.

The information from both localizer and glideslope receivers is presented on the CDI; the vertical needle displays localizer information and the horizontal needle displays glideslope information (Figure 6-3-4).

When both needles are centered, the aircraft is on course and descending at the proper rate. In addition, the CDI contains a red warning flag for each system that comes into view when the receiver fails or the loss of a transmitted signal occurs.

Two antennas are usually required for ILS operation. One for the localizer receiver also used for VOR navigation, and one for the glideslope. Some of the small aircraft use a single multi-element antenna for both glideslope and VOR/LOC operation. The VOR/localizer antenna is normally installed on the top of the aircraft fuselage or flush mounted in the vertical stabilizer. The glideslope antenna

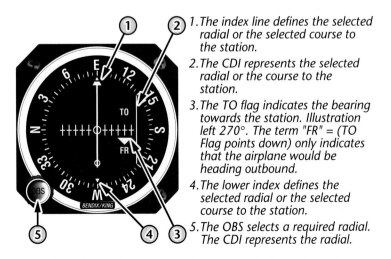

1. The index line defines the selected radial or the selected course to the station.
2. The CDI represents the selected radial or the course to the station.
3. The TO flag indicates the bearing towards the station. Illustration left 270°. The term "FR" = (TO Flag points down) only indicates that the airplane would be heading outbound.
4. The lower index defines the selected radial or the selected course to the station.
5. The OBS selects a required radial. The CDI represents the radial.

Figure 6-3-4. The CDI instrument is the principal indicator for an ILS system.

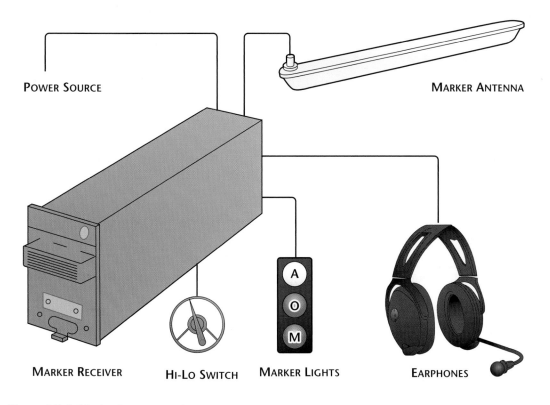

POWER SOURCE

MARKER ANTENNA

MARKER RECEIVER   HI-LO SWITCH   MARKER LIGHTS   EARPHONES

Figure 6-3-5. Marker beacon receiver

is, in most cases, installed on the nose of the aircraft. On aircraft equipped with radar, the glideslope antenna is installed under the radar.

Marker beacons. Marker beacons are used in connection with the instrument landing system. The markers are signals, which indicate the position of the aircraft along the approach to the runway. Two markers are used in each installation. The location of each marker is identified by both an aural tone and a signal lamp. The marker beacon transmitters, operating on a fixed 75 MHz frequency, are placed at specific locations along the approach pattern of an ILS facility. The antenna radiation pattern is beamed straight up.

A marker beacon receiver (Figure 6-3-5) installed in the aircraft receives the antenna signals and converts them into power to illuminate a signal lamp and produce an audible tone in a headset. The outer marker marks the beginning of the approach path. The outer marker signal is modulated by a 400-Hz signal, which produces a tone keyed in long dashes. In addition to providing aural identification, the signal lights a purple lamp in the cockpit. The middle marker is usually about 3,500 ft. from the end of the runway and is modulated at 1,300 Hz, which produces a higher-pitched tone keyed with alternate dots and dashes. An amber lamp flashes to indicate that the aircraft is passing over the middle marker.

Marker beacon receivers vary in design from simple receivers that have no operating controls and no aural output to more sophisticated receivers that produce an aural tone and have an on/off switch and a volume control to adjust the sound level of the identification code.

Where three lights are used, a white light indicates the aircraft positions at various points along the airways. In addition to the light, a rapid series of, tones (six dots per second) of 3,000 Hz are received in the headset. Distance-measuring equipment is rapidly replacing the along-route marker system. A 3,000-Hz tone and white light marker are also being used for inner markers (missed approach point) on some Category II ILS-equipped runways.

The ILS system cannot be ground tested fully without using test equipment simulating localizer and glideslope signals.

If an aircraft is located at an airport, which has an ILS-equipped runway, it may be possible to determine if the receiver is functioning by performing the following. Place the on/off switch (if so equipped) in the ON position, and adjust the frequency selector to the proper ILS channel for the airport where the aircraft is located.

Allow sufficient time for the equipment to warm up. In a strong signal area, both the local-

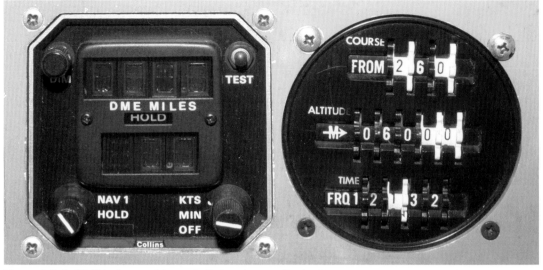

Figure 6-3-6. DME instrument in cockpit showing nautical miles

izer and glideslope warning flags will either start to move or go completely out of view. Observe that both cross pointers are deflected to their maximum displacement.

Some of the more sophisticated solid-state ILS equipment contains self-monitoring circuits. These circuits can be used for performing an operational test using the procedures in the aircraft or equipment manufacturer's service manuals.

Distance-Measuring Equipment

The purpose of Distance-Measuring Equipment (DME) is to provide a constant visual indication of the distance the aircraft is from a ground station. A DME reading is not a true indication of point-to-point distance as measured over the ground. DME indicates the slant range between the aircraft and the ground station. Slant-range error increases as the aircraft approaches the station. At a distance of 30 to 60 nautical miles, the slant range error is negligible.

DME operates in the UHF range of the radio frequency spectrum. The transmitting frequencies are in two groups, between 962 MHz to 1,024 MHz and 1,151 MHz to 1,212 MHz; the receiving frequencies are between 1,025 MHz to 1,149 MHz. Transmitting and receiving frequencies are given a channel number, which is paired with a VOR channel. In some aircraft installations, the DME channel selector is ganged with the VOR channel selector to simplify the radio operation. A typical DME control panel is shown in Figure 6-3-6.

The aircraft is equipped with a DME transceiver, which is tuned to a selected DME ground station. Usually DME ground stations are collocated with a VOR facility (called VORTAC). The airborne transceiver transmits a pair of spaced pulses to the ground station. The pulse spacing serves to identify the signal as a valid DME interrogation. After reception of the challenging pulses, the ground station responds with a pulse transmission on a separate frequency to send a reply to the aircraft.

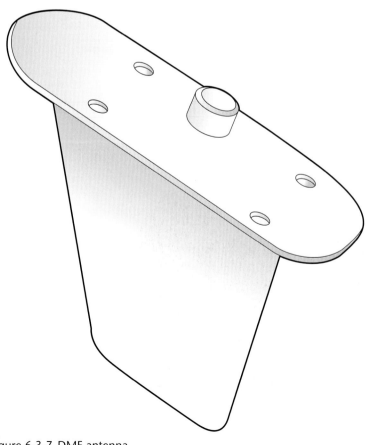

Figure 6-3-7. DME antenna

Upon reception of the signal by the airborne transceiver, the elapsed time between the challenges and the reply is measured. This time interval is a measure of the distance separating the aircraft and the ground station. This distance is indicated in nautical miles by a cockpit instrument similar to the one shown in Figure 6-3-6.

A typical DME antenna is shown in Figure 6-3-7. Most DME antennas have a cover installed to protect them from damage. The DME antenna is usually a short, stub type mounted on the lower surface of the aircraft. To prevent an interruption in DME operation, the antenna must be located in a position that will not be blanked by the wing when the aircraft is banked. The antennae are aligned with the centerline of the airplane.

To determine if the DME operates, turn the on/off switch to the on position and select the appropriate channel. Allow sufficient time for the equipment to warm up. During this period, the distance indicator, both digital and pointer, will travel from minimum to maximum readings (sweep or search). When the DME has locked on a station, the indicator will stop searching and the red warning flag (if the indicator is equipped with one) will disappear. In most installations, no functional check can be made on the ground without a DME test set. VORTAC-based DME units are rapidly being replaced by GPS navigation units.

Automatic Direction Finders

An *Automatic Direction Finder* (ADF) is a radio receiver equipped with directional antennas, which are used to determine the direction from which signals are received. Most ADF receivers provide controls for manual operation in addition to automatic direction finding. When an aircraft is within reception range of a radio station, the ADF equipment provides a means of fixing the position with reasonable accuracy. The ADF operates in the low and medium frequency spectrum from 190 kHz through 1,750 kHz. The direction to the station is displayed, on an indicator located in the cockpit, as a relative bearing to the station.

The airborne equipment (Figure 6-3-8) consists of a receiver, loop antenna, sense or nondirectional antenna, indicator and control unit.

Most ADF receivers used in general aviation aircraft are panel mounted. Their operating controls appear on the front of the radio case.

Most ADF systems use fixed, ferrite core loops in conjunction with a rotatable transformer called a resolver, or goniometer. It operates

Figure 6-3-8. Panel mounted ADF

essentially the same as the rotating loop, except that one of the windings of the goniometer rotates instead of the loop.

Most ADF installations require that the antennae be calibrated.

A general procedure for performing an operational check of the ADF system is as follows:

1. Turn on/off switch to the on position and allow the radio to warm up. On installations that use the Radio Magnetic Indicator (RMI) pointer as an ADF indicator, assure that the switch has been positioned to present ADF information (Figure 6-3-9).

2. Tune to the desired station.

3. Adjust the volume to an appropriate level.

4. Rotate loop antenna and determine that only one null is received.

5. Check that the ADF needle points towards the station. If the aircraft is situated among buildings or any other large reflecting surfaces, the ADF needle may indicate an error as a result of a reflected signal.

Figure 6-3-9. Radio magnetic indicator (RMI)

Figure 6-3-10. Transponder

Photo courtesy of Garmin

Radar Beacon Transponder

The *radar beacon transponder* system is used in conjunction with a ground based surveillance radar to provide positive aircraft identification directly on the controller's radar scope. Every transponder will show a "blip" on the radar screen. On the front of each transponder is a button labeled IDENT. To identify a specific airplane the radar operator must be in radio contact with the pilot. He will then ask the pilot to *"SQUAWK IDENT"*. The pilot will then push the IDENT button and his radar image will temporarily increase in brightness. Now the radar operator knows who he is talking to and where they are located. To give that airplane a distinct name for identification the ATC controller will direct him to reset the four-digit code on the transponder. In this way, everything on this radar screen will have a different number. There are 4096 possible number combinations.

Code changes. It is best not to make changes to a transponder code setting. If you should make an inadvertent selection of Codes 7500, 7600 or 7700 you would cause momentary false alarms at automated ground facilities. For example, when switching from Code 2700 to Code 7200, switch first to 2200 then to 7200, NOT to 7700

and then 7200. This procedure applies to non-discrete Code 7500 and all discrete codes in the 7600 and 7700 series (i.e. 7600-7677, 7700-7777), which will trigger special indicators in automated facilities. Only nondiscrete Code 7500 will be decoded as the hijack code.

Operational modes. The airborne equipment, or *transponder*, receives a ground radar interrogation for each sweep of the surveillance radar antenna, and automatically sends back a coded response. Civil transponders operate in three modes labeled *Mode A, Mode AC* and *Mode S*, which is switch controlled.

The three different modes work like this:

- Mode A is a basic general aviation transponder. It is normally set for 1200, the VFR setting for a non-instrument equipped airplane.

  With Mode A, an airplane cannot normally fly to a controlled airport, or in controlled airspace. Mode A will only give the air traffic control radar operator horizontal distance (bearing) from the radar. With each sweep of the radar, it will light up each airplane in its area of coverage so the radar operator knows where everybody is located.

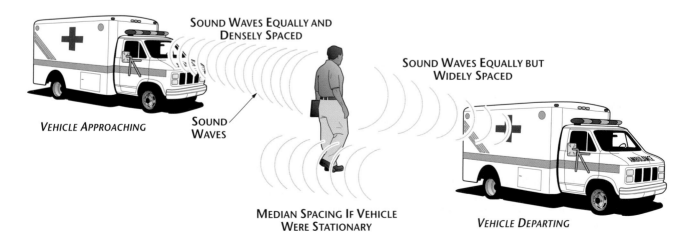

VEHICLE APPROACHING

SOUND WAVES

SOUND WAVES EQUALLY AND DENSELY SPACED

SOUND WAVES EQUALLY BUT WIDELY SPACED

MEDIAN SPACING IF VEHICLE WERE STATIONARY

VEHICLE DEPARTING

Figure 6-3-11. The frequency of sound changes as the ambulance moves past the person, therefore depicting the Doppler effect

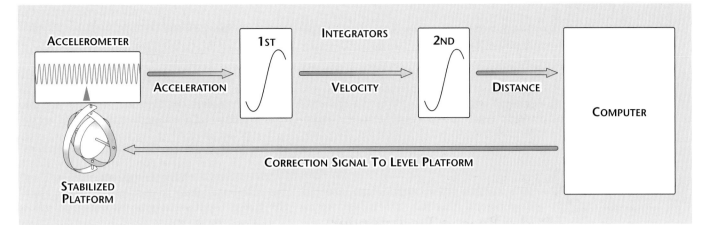

Figure 6-3-12. INS and accelerometers on platform

- Mode AC, commonly called Mode C, adds an altitude indication to the radar image. The altitude information comes from the encoding altimeter, or radio altimeter. Now the radar operator knows where you are and how high you are.

- Mode S is a refinement of both Modes A and C. Your Mode S transponder can take information from other Mode S transponders and figure out where everybody else is in relation to where you are. This is part of the Traffic Alert/Collision Avoidance (TCAS) system of air traffic control. TCAS is covered in Chapter 7 Lighting, Warning and Utility Systems.

There are several different makes of aircraft transponders in use. They all perform the same function and are basically the same electrically.

The major differences are in construction: either a single unit or a control unit for remotely operating the transponder.

A typical transponder is shown in Figure 6-3-10. The front panel of the illustrated transponder contains all the switches and dials needed for operation.

A short stub or covered stub antenna is used for transponder operation and is usually mounted on the lower surface of the aircraft fuselage.

To ground check the transponder the appropriate test equipment must be used. There is no built-in test feature.

Transponder inspection. All transponders in aircraft must be inspected for accurate operation and reporting every 24 calendar months. The inspection requires special instrumentation and can only be performed by a facility that is FAA approved for the particular equipment.

Doppler Navigation System

Doppler navigational radar automatically and continuously computes and displays ground speed and drift angle of an aircraft in flight without the aid of ground stations, wind estimates, or true airspeed data. *Doppler radar* does not sense direction as search radar does.

Instead, it is speed conscious and drift-conscious. It uses continuous carrier wave transmission energy and determines the forward and lateral velocity components of the aircraft by utilizing the principle known as the Doppler effect.

The *Doppler effect*, or frequency change of a signal, can be explained in terms of an approaching and departing sound. As shown in Figure 6-3-11, the sound emitter is a siren located on a moving ambulance and the receiver is the ear of a stationary person. Notice the spacing between the emitter when it is approaching and when it is departing from the stationary receiver. When the sound waves are closely spaced, the listener hears a sound that is higher in pitch. This happens because as the sound waves move toward you, so does the source. Thus, each sound wave is spaced closer together as the vehicle continues moving towards you. This changes the frequency and the pitch becomes higher. The reverse is also true. When the vehicle is moving away the distance it travels is added to the sound wave spacing and the frequency lowers. Doppler radar uses the frequency change phenomenon just described, except in the radar frequency range.

The doppler radar emits narrow beams of energy at one frequency, and these waves of energy strike the Earth's surface and are reflected. Energy waves returning from the earth are spaced differently than the waves striking the earth. The earth-returned energy is intercepted by the receiver and compared with the outgoing transmitter energy.

The difference, due to the Doppler effect, is used to develop ground speed and wind drift angle information.

A ground operational check of a Doppler system consists of setting a precise airspeed, and a deviation angle, which will give a distance-off-course reading. Always refer to the equipment manufacturer's instruction manual or the aircraft's operation manual for the proper test procedure.

Modern weather Doppler radar can also plot wind speed and direction. This is how storms are tracked by the weather reporting stations.

Inertial Navigation System

The *inertial navigation system* (INS) is presently being used on large aircraft as a long-range navigation aid. It is a self-contained system and does not require signal inputs from ground navigational facilities. The system derives altitude, velocity and heading information from measurement of the aircraft's accelerations. Two accelerometers are required, one referenced to North and the other to East. The accelerometers (Figure 6-3-12) are mounted on a gyro-stabilized unit, called the stable platform, to avert the introduction of errors resulting from the acceleration due to gravity.

An inertial navigation system is a complex system containing four basic components. They are:

1. A stable platform, which is oriented to maintain accelerometers horizontal to the earth's surface and provide azimuth orientation.

2. Accelerometers arranged on the platform to supply specific components of acceleration.

3. Integrators, which receive the output from the accelerometers and furnish velocity and distance.

4. A computer that receives signals from the integrators and changes distance traveled to position in selected coordinates.

The diagram in Figure 6-3-12 shows how these components are linked together to solve a navigation problem. Initial conditions are set into the system and the navigation process is begun. In inertial navigation, the term initialization is used to denote the process of bringing the system to a set of initial conditions from which it can proceed with the navigation process. These conditions include leveling the platform, aligning the azimuth reference, setting initial velocity and position and making any computations required to start the navigation.

Although all inertial navigation systems must be initialized, the procedure varies according to the equipment and the type aircraft in which it is installed. The prescribed initialization procedures are detailed in the appropriate manufacturer's manuals.

From the diagram, it can be seen that the accelerometers are maintained in a horizontal position to the Earth's surface by a gyro-stabilized platform. As the aircraft accelerates, a signal from the accelerometer is sent to the integrators. The output from the integrators, or distance, is then fed into the computer, where two operations are performed.

First, a position is determined in relation to the preset flight profile, and second, a signal is sent back to the platform to position the accelerometer horizontally to the Earth's surface. The output from high-speed gyros and accelerometers, when connected to the flight controls of the aircraft, resists any changes in the flight profile.

Loran ARNAV System

Loran is a long range navigation system presently being used by many general aviation

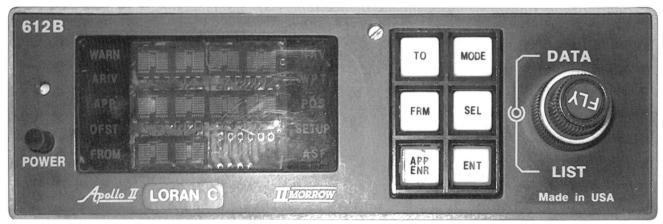

Figure 6-3-13. LORAN installation in an instrument panel

and airline aircraft as a backup to GPS systems. Although an old system, Loran has the advantage of working in areas of the world where there are no modern navigational aids. Actually, the letters in Loran, when broken down, mean long range navigation.

The Loran system was developed for the military during the second World War. Later, the Loran system first became popular with commercial fisherman. The early use of the Loran system was very complicated. However, it soon became the primary system for most commercial vessels.

As technology advanced, Loran systems became easier to use. In later Loran systems, the read-outs concerning latitude and longitude coordinates were made easier to handle because the information was being processed by computer chips. Once again, the silicon chip improved the operation of another flight system so it could handle changing coordinates, which interact with the data being fed into an integrated flight system (Figure 6-3-13).

A Loran station consists of a radio transmitting tower and a manned monitoring office. The Coast Guard has the responsibility for handling the functioning of each station as well as maintaining dual transmitters to insure that transmissions are continuous. The transmitting stations are grouped together in chains with one station set as a master and the others arranged as secondary stations. Secondary stations are identified by a letter: W – Whiskey, X – X-Ray, Y – Yankee and Z – Zulu, etc.

The Loran signals are transmitted at low frequencies of 100 kHz. Low frequency signals travel (propagate) well over the surface of the earth. Because of this, it gives them (the frequencies) a much greater range when comparing them to systems using VHF signals, such as VOR. The reason being is VHF signals are limited to line-of-sight use. LF Loran signals travel over obstructions, which would stop VHF signals. One of the excellent features of this system is soon after takeoff the pilot can make radio contact with the airport which the aircraft is flying to and the *distance away indicator* will immediately begin indicating the distance and flying time away from the airport.

The maintenance requirements of a Loran system is that the work must be done by certified repair station rated for making the repairs.

Global Positioning System Navigation

Global Positioning System (GPS). The Global Positioning System (GPS) is a satellite-based system that is capable of providing position and navigational data for ground based and airborne receivers. GPS actually determines the position of a given receiver, or receiver's antenna to be precise. The navigational capabilities of GPS are determined through multiple calculations and comparison to the *World Geodetic System* (WGD) map. The WGD map is an extremely accurate reference of the earth latitude and longitude. The global positioning system, therefore, provides location coordinates based on a latitude/longitude reference, not with reference to the known location of a ground based transmitter. Conventional forms of aircraft navigation rely on ground-based transmitters that employ limited range and relatively low frequency signals. GPS employs ultra high frequency transmitters and requires only 21 transmitters for worldwide navigation.

The global positioning system was first implemented by the U.S. military in the late 1970s. Through the early 80s, the system was tested and refined. During the 1980s and early 90s, the initial satellites were replaced by more accurate and powerful units. The FAA granted TSOs (technical standard orders) to several GPS units, which permitted installation on civilian aircraft. Today many GPS units are available and certified for en route navigation and can be used for a nonprecision IFR approach.

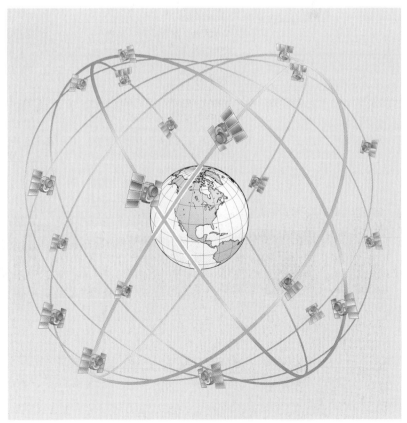

Figure 6-3-14. Six geosynchronous orbits contain 24 satellites.

In the United States, the formal name for the global positioning system is the NAVSTAR GPS. This name was first assigned by the U.S. Deputy Secretary of Defense in 1973 and the name has stuck ever since. The United States is not the only country which has developed a usable GPS. The former Soviet Union began the implementation of their satellite navigation system called Glanoss several years before the USSR's demise. Today the Commonwealth of Independent States (made up of former Soviet Union member nations) is continuing to develop their GPS. Both nations have explored the possibility of combining efforts to reduce cost and increase worldwide coverage.

GPS system elements. The NAVSTAR GPS consists of three distinct elements: the space segment, the control segment and the user segment (see Figure 6-3-14). The space segment consists of a *constellation of 24 orbiting satellites.* Twenty-one of the satellites are active and three are spares that can be moved into position in the event an active unit fails. The satellites are each placed in a near geosynchronous orbit approximately 10,900 nm

(20,200 km) above the earth. The satellites are equally spaced around six different orbits to provide worldwide coverage. Each satellite completes one orbit approximately every 12 hours. The system is designed so that a minimum of five satellites should be in view of a ground-based user at any given time, at any location on earth.

Each satellite in the system transmits position and precise time information on two frequencies known as L1 and L2. L1 operates at 1,575.42 MHz and L2 has an operating frequency of 1,227.6 MHz. The signals are digitally modulated and have a bandwidth of 20 MHz or 2 MHz depending on the type of information that is being transmitted. The individual satellites are identified by the information broadcast by the modulated carrier waves. The digital modulation of the carrier is achieved through a process known as *Phase Modulation* (PM). Phase Modulation is achieved by shifting the phase of the carrier wave to represent a change in the digital information signal. To improve receiver reception of faint GPS signals, each satellite employs 12 helical antennas arranged in a tight circular pattern.

Figure 6-3-15. GPS display as shown on a Garmin GNS 530

The control segment of the GPS consists of five ground-based monitor stations, one master control station and three ground antennas located at different sites throughout the world. The five monitor stations receive/transmit time and range data from various satellites within its region. The raw data received by these unmanned stations is sent to the *master control station* (MCS) in Colorado Springs, Colorado. The master control station computers analyze the data and provide correction signals as needed to the three ground antennas. The signals are periodically uplinked to the satellites and the necessary corrections are made by the system's software.

When GPS was first conceived, it was intended that the entire user segment would be restricted to high accuracy military applications. Today, it would be difficult to determine if there are more civilian or military GPS receivers in operation. Once released by the Department of Defense, the civilian use of GPS has skyrocketed. GPS is currently being used by farmers, surveyors, archaeologists, miners, recreational hikers and of course aircraft owners and pilots. Virtually anyone who needs to navigate, measure position, or determine velocity can become part of the GPS user segment. Anyone can buy a GPS moving map display unit for their automobile for less than a thousand dollars. As technology advances, they will be even less expensive.

The user segment is typically designed to receive time and position data from four or more satellites and process that data into the desired output. The specific equipment needed for these operations is a function of the equipment installation (if any), the desired accuracy and specific output data. Output data ranges from simple display of latitude and longitude, to moving maps and displays of local airports or ground terrain.

All GPS receivers must have at least three major elements:

- Control/display unit
- Receiver/processor circuitry
- Antenna

These elements may all be combined in one unit or consist of three individual elements. Modern hand-held receivers are available for just a few hundred dollars. These units are completely self-contained however; many are not approved for aircraft use. Many aircraft systems are designed to be permanently installed on the aircraft and often interface with other navigational equipment. On aircraft GPS equipment, the display and receiver processor are often combined in one unit; the

antenna is always mounted on the top of the aircraft fuselage to ensure proper satellite reception. See Figure 6-3-15.

Theory of operation. Two assumptions are necessary to justify the explanation that follows:

- We know exactly where each satellite is at any given time
- The distance to each satellite can be calculated by the GPS receiver/processor. In reality, the exact location of a moving satellite is easy to predict.

Once in orbit, each satellite will follow a consistent path. Exact satellite position data is transmitted to the receiver as part of the satellite message. Calculating distance from the satellite is also a simple matter. Since distance equals velocity multiplied by time, the receiver/processor need only measure the time it took for

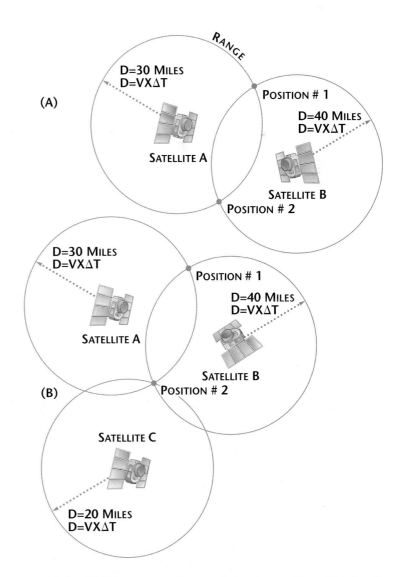

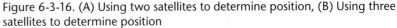

Figure 6-3-16. (A) Using two satellites to determine position, (B) Using three satellites to determine position

the GPS signal to reach the receiver. The speed at which the signal traveled to the receiver is a constant 186,000 miles/second (the speed of light). Using time and velocity to derive distance (range) is known as the time of arrival (TOA) ranging concept.

To keep things simple, a two-dimensional model will be presented first (Figure 6-3-16). Assume the GPS user segment and all satellites are located in one geometric plane. In this case, knowing the distance (range) from just two satellites would provide the location of your aircraft. In this example, the aircraft must be located somewhere on a circle with a radius of 30 miles from satellite A, and somewhere on a circle with a radius 40 miles from satellite B. In this two-dimensional model, the aircraft can be in one of two positions. To further define the location of our aircraft in the two-dimensional model, a third GPS satellite would be added. If the aircraft was 30 miles from satellite A, 40 miles from satellite B and 20 miles from satellite C, the aircraft must be in position 2. As

we all know, real aircraft can travel in three dimensions. To pinpoint position in each of these three dimensions, the aircraft must monitor at least four GPS satellites.

GPS and Aircraft Navigation

GPS provides an excellent means of general navigation; however, there are several limitations that must be addressed for civilian aircraft use. The basic GPS service fails to meet the four basic criteria:

1. Accuracy - the difference between the measured position and the aircraft's actual position. It should be noted that GPS accuracy is adequate for en route navigation, but fails to meet approach and landing requirements.

2. Availability - the ability of the system to be used for navigation whenever it is needed, and the ability to provide that service throughout the entire flight.

3. Integrity - the ability of the system to shut itself down when it is unsuited for navigation or to provide timely warnings to the pilot of the system failure.

4. Continuity - the probability that GPS service will continue to be available for a period of time necessary to complete the navigation requirements of the flight.

While aircraft are flying on an IFR flight plan, their en route separation is typically maintained at 5 miles or greater. The accuracy of GPS is sufficient for civilian en route navigation. GPS can easily provide navigational signals capable of a five mile separation ± 4%. However, while flying an approach to land, the accuracy level must be significantly higher.

The future of GPS aircraft navigation depends on improved accuracy. As for availability, the US government has stated that GPS will be consistently available for civilian use.

Aircraft GPS equipment. All aircraft GPS equipment must meet a minimum certification standard through a TSO. There are, however, several units currently available that are not TSO certified. Most of these GPS receivers were intended for non-aircraft uses such as hiking, marine, or automobile navigation. In some cases, pilots use non-TSOed equipment as a secondary reference during VFR flights.

This is perfectly legal, but only equipment approved by a TSO may be used as a source for IFR or VFR navigation. The TSO ensures a minimum quality and accuracy standard set by the FAA; hence, it provides approval for aircraft

Figure 6-3-17. (A) Radar antenna in the nose of a King Air, (B) the cockpit CRT

use. In general, the major difference between an aircraft GPS receiver and one designed for marine or other use is the database. Virtually all aircraft GPS receivers contain a database of various airports and standard navigational waypoints.

The waypoints are entered into the GPS control panel using latitude and longitude coordinates or the waypoint can be selected from the airborne GPS database. It should be noted that current civilian aircraft use of GPS navigation is limited to only two dimensions; however, the GPS is capable of providing extremely accurate position data in three dimensions. Vertical navigation (altitude) is currently provided by conventional systems.

Airborne Weather Radar System

Radar (radio, automatic detection and ranging) is a device used to see certain objects in darkness, fog, or storms, as well as in clear weather.

In addition to the appearance of these objects on the radarscope, their range and relative position are also indicated.

Radar is an electronic system using a pulse transmission of radio energy to receive a reflected signal from a target. The received signal is known as an *echo*. The time between the transmitted pulse and received echo is computed electronically and is displayed on the radarscope in terms of nautical miles.

A radar system (Figure 6-3-17) consists of a transceiver and synchronizer, an antenna installed in the nose of the aircraft, a control unit installed in the cockpit and an indicator or scope.

In the operation of a typical weather radar system, the transmitter feeds short pulses of radio-frequency energy through a waveguide to the dish antenna in the nose of the aircraft. The antenna oscillates to provide the typical radar sweep. Part of the transmitted energy is reflected from objects in the path of the beam and is received by the dish antenna. Electronic switching simultaneously connects the antenna to the transmitter and disconnects the receiver during pulse transmission. Following the completion of pulse transmission, the antenna is switched from the transmitter to the receiver. The switching cycle is performed for each transmitted pulse.

The time required for radar waves to reach the target and reflect to the aircraft antenna is directly proportional to the distance of the target from the aircraft. The receiver measures the

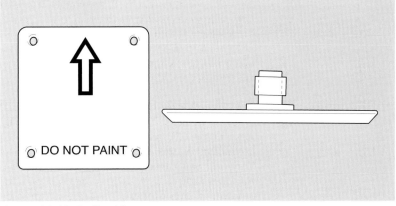

Figure 6-3-18. Radio altimeter

time interval between transmission of radar signals and reception of reflected energy and uses this interval to represent the distance, or range, of the target.

Rotation or sweep of the antenna and radar beam gives *azimuth* indications. The indicator display shows the area and the relative size of targets, whose azimuth position is shown relative to the line of flight.

The weather radar increases safety in flight by enabling the operator to detect storms in the flight path in order to chart a course around them. The terrain-mapping facilities of the radar show shorelines, islands and other topographical features along the flight path. These indications are presented on the visual indicator in range and azimuth relative to the heading of the aircraft.

An operational check consists of the following:

- Tow or taxi the aircraft clear of all buildings and parked aircraft.
- Apply power to the equipment, and allow sufficient warm-up time.
- Tilt the antenna to an upward position.
- Check the scan on the radarscope for an indication of targets.

CAUTION: *Do not operate radar in any manner that would allow the transmitted waves to strike another person. High-energy electronic wave transmission can be dangerous to humans and animals.*

Radio Altimeter

Radio altimeters are used to measure the distance from the aircraft to the ground (See Figure 6-3-18). This is accomplished by transmitting radio frequency energy to the ground and receiving the reflected energy at the aircraft.

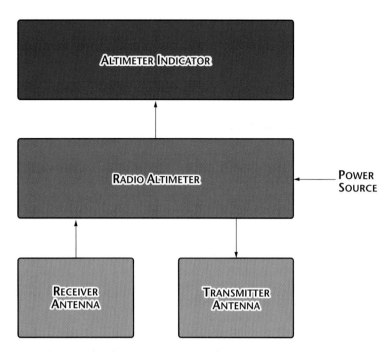

Figure 6-3-19. Radio altimeter equipment layout

Most modern day altimeters are pulse type and the altitude is determined by measuring the time required for the transmitted pulse to hit the ground and return. The indicating instrument will indicate the true altitude of the aircraft, which is its height above water, mountains, buildings, or other objects on the surface of the Earth.

The present day generation of radio altimeters are primarily used during landing and are a *Category II* requirement. The altimeter provides the pilot with the altitude of the aircraft during approach. Altimeter indications determine the decision point to either continue to land, or execute a climb-out.

A radio altimeter system (Figure 6-3-19) consists of a transceiver, normally located in an equipment rack, an indicator installed in the instrument panel and two antennas located on the belly of the aircraft.

Ground Proximity Warning System (GPWS).

GPWS is a system designed to alert pilots if their aircraft is in immediate danger of flying into the ground or an obstacle. The system monitors an aircraft's height above ground as determined by a radio altimeter as well as other systems in landing and flight configurations. In a landing configuration a computer will monitor systems such as air data computer, instrument landing system, landing gear and flap positions. The computer then keeps track of these readings, calculates trends, and will warn the captain with visual and audio messages if the aircraft is in certain defined flying configurations (modes).

The modes are:

1. Excessive descent rate (SINK RATE, PULL UP)

2. Excessive terrain closure rate (TERRAIN, PULL UP)

3. Altitude loss after takeoff or with a high power setting (DON'T SINK)

4. Unsafe terrain clearance (TOO LOW – TERRAIN, TOO LOW – GEAR, TOO LOW – FLAPS)

5. Excessive deviation below glide slope (GLIDESLOPE)

6. Excessively steep bank angle (BANK ANGLE)

7. Wind shear protection (WINDSHEAR)

More advanced systems are known as enhanced ground proximity warning systems (EGPWS), although sometimes called terrain awareness warning systems. These systems improved over the original design by looking forward as well as down to predict steep terrain ahead of the aircraft. The original GPWS had a weakness that it only looked straight down.

The EGPWS system was now combined with a worldwide digital terrain database and relies on GPS technology. On-board computers compared its current location with a database of the Earth's terrain. The Terrain Display now gave pilots a visual orientation to high and low points nearby the aircraft. EGPWS software improvements were focused on solving two common problems; no warning at all, and late or improper response. You will find these systems on all modern transport aircraft and many general aviation aircraft.

Emergency Locator Transmitter (ELT)

An *Emergency Locator Transmitter* (ELT) is a self-contained, self-powered radio transmitter designed to transmit a signal on the international distress bands of 121.5 MHz (civilian) and 243 MHz (military).

The monitoring of 121.5 MHz ELT signals by international satellites stopped in February 2009. This change results from a decision from the international organization (COSPAS/ SARSAT) to terminate satellite processing of distress signals from 121.5 MHz ELT's. Therefore, if the alerts are to be detected it is strongly recommended that aircraft operators switch to ELT's that operate at 406 MHz. As a result only ground based receiver stations such as Civil Air Patrol, ATC and airport facilities will be available to detect and respond to the 121.5 MHz signals.

The 406 MHz signal can be encoded with the owners identification, aircraft information and the last known position information, taken from the FMS or GPS, at the moment of activation. This will significantly reduce the number of false alarms that occur every year as well as improve the chances of locating a downed aircraft.

The new ELT's should conform to TSO C126 instead of the old TSO C91. Testing and inspection requirements are covered by the manufacturer's manuals and the guidelines in AC-43.13.1B. Specific aircraft requirements are covered in CFR part 91.207.

Operation is automatic on impact. The transmitter may also be activated by a remote switch in the cockpit or a switch integral with the unit.

If the G force switch in the transmitter is activated from impact in a direction parallel to the longitudinal axis of the airplane, it can be turned off only with the switch on the case. (See Figure 6-3-20).

Transmitter. The transmitter may be located anywhere within the aircraft, but the ideal location is as far aft as possible, just forward of the vertical fin. It must be accessible to permit monitoring the replacement date of the battery and for arming or disarming of the unit. A remote control arm/disarm switch may be installed in the cockpit.

The external antenna must be installed as far as practicable from other antennas to prevent interaction between avionics systems.

Batteries. Batteries are the power supply for emergency locator transmitters. When activated, the battery must be capable of furnishing power for signal transmission for at least 48 hours. The useful life of the battery is the length of time that the battery may be stored without losing its ability to continuously operate the ELT for 48 hours. This useful life is established by the battery manufacturer; batteries must be changed or recharged as required at 50% of the battery's useful life. This gives reasonable assurance that the ELT will operate if activated. The battery replacement date must be marked on the outside of the transmitter. This time is computed from the date of manufacture of the battery. If the system has accumulated more than one hour of operation in any 12 month period the batteries must also be replaced.

Batteries may be nickel-cadmium, lithium, magnesium dioxide, or dry-cell batteries. Wet cell batteries have an unlimited shelf life until liquid is added. At that time, their life in an ELT is regulated the same as dry cell batteries;

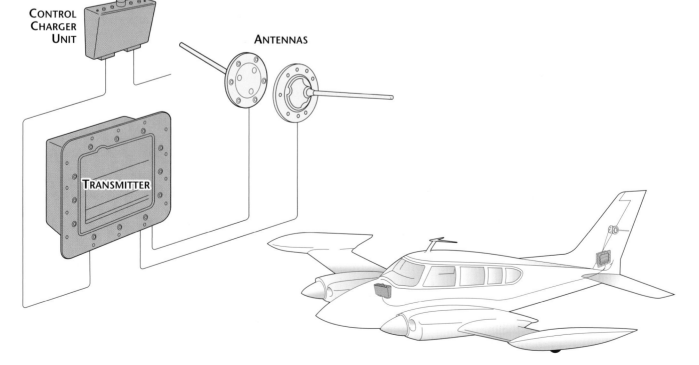

Figure 6-3-20. ELT with on/off switch

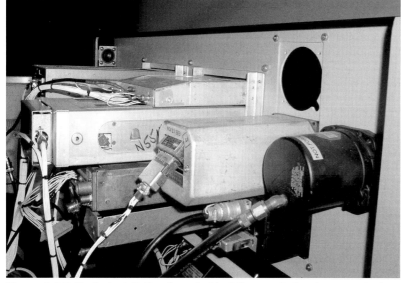

Figure 6-4-1. Radio installation from behind the panel of a classroom trainer

change them at 50% of shelf life. When replacing batteries use only those recommended by the manufacturer of the ELT.

Do not use flashlight type batteries, as their condition and useful life are unknown.

Testing. Testing of ELTs should be coordinated with the nearest FAA Tower or Flight Service Station. Tests should be conducted only during the first five minutes of any hour and should be restricted to three audio sweeps. Any time maintenance is performed in the vicinity of the ELT, the VHF communication receiver should be tuned to 121.5 MHz and listen for ELT audio sweeps. If it is determined the ELT is operating, it must be turned off immediately.

False alarms. False alarms have caused many of the problems with ELTs. Battery failures, with resulting corrosion of the unit are considered a complete failure or an unwanted transmission. Another type of unwanted transmission is the result of careless handling by the operators of the aircraft.

Test equipment. Monitors are available for identifying and/or locating unwanted ELT transmissions. A miniature scanning receiver may be mounted in the cockpit to warn the pilot if his ELT is transmitting. The other is a small portable ELT locator for use at general aviation airports to assist in finding an aircraft whose transmitter has accidentally become activated.

The operation of an ELT can be verified by tuning a communication receiver to the civil emergency frequency (121.5 MHz) and activating the ELT. Turn the ELT off immediately upon receiving a signal in the communication receiver.

In all maintenance and testing of ELTs, the manufacturers instructions must be followed.

Section 4

Installation of Communication and Navigation Equipment

There are many factors that the technician must consider prior to altering an aircraft by the addition of radio equipment. These factors include the space available, the size and weight of the equipment and previously accomplished alterations. In addition, the power consumption of the added equipment must be calculated to determine the maximum continuous electrical load. Each installation should be planned to allow easy access for inspection, maintenance and exchange of units.

The installation of radios is primarily mechanical, involving sheet metal work in mounting the radios, racks, antennas and controls. Routing of the interconnecting wires, cables, antenna leads, etc., is also an important part of the installation process. When selecting a location for the equipment, first consider the areas designated by the airframe manufacturer. If such information is not available, or if the aircraft does not contain provisions for adding equipment, select an area that will carry the loads imposed by the weight of the equipment, and which is capable of withstanding the additional inertia forces.

If the radio is to be mounted in the instrument panel and no provisions have been made for such an installation, determine if the panel is primary structure prior to making any cutouts. To minimize the load on a stationary instrument panel, install a support bracket between the rear of the radio case or rack and a nearby structural member of the aircraft.

The radio equipment must be securely mounted to the aircraft. All mounting bolts must be secured by locking devices to prevent loosening from vibration.

Adequate clearance between the radio equipment and the adjacent structure must be provided to prevent mechanical damage to electric wiring or radio equipment from vibration, chafing, or shock landing. See Figure 6-4-1. In a real aircraft panel there will not be this much room. Clearance and bracing are a necessity, but can be a problem in a live airplane.

Do not locate radio equipment and wiring near units containing combustible fluids. When separation is impractical, install baffles or shrouds to prevent contact of the combustible fluids with radio equipment in the event of a plumbing failure.

In older private airplanes radio equipment was frequently mounted under seats. FAA AC 43.13-2B lists clearances necessary. The bottom of the seat must have at least one inch of clearance when the seat is occupied.

Cooling and moisture. The performance and service life of most radio equipment is seriously limited by excessive ambient temperatures. The installation should be planned so that the radio equipment can dissipate its heat readily. In some installations, it may be necessary to produce airflow over the radio equipment, either with a blower or through the use of a venturi.

The presence of water in radio equipment promotes rapid deterioration of the exposed components. Some means must be provided to prevent the entry of water into the compartments housing the radio equipment.

Vibration isolation. Vibration is a continued motion caused by an oscillating force. The amplitude and frequency of vibration of the aircraft structure will vary considerably with the type of aircraft.

Radio equipment is sensitive to mechanical shock and vibration, and is normally shock-mounted to provide some protection against in-flight vibration and landing shock.

When special mounts (Figure 6-4-2) are used to isolate radio equipment from vibrating structure, such mounts should provide adequate isolation over the entire range of expected vibration frequencies. When installing shock mounts, assure that the equipment weight does not exceed the weight-carrying capabilities of the mounts.

Radios installed in instrument panels do not ordinarily require vibration protection, since the panel itself is usually shock-mounted. However, make certain that the added weight can be safely carried by the existing mounts. In some cases, it may be necessary to install larger capacity mounts or to increase the number of mounting points.

Radio equipment installed on shock mounts must have sufficient clearance from surrounding equipment and structure to allow for normal swaying of the equipment.

Periodic inspection of the shock mounts is required, and defective mounts should be replaced with the proper type. The factors to observe during the inspection are:

- Deterioration of the shock-absorbing material
- Stiffness and resiliency of the material
- Overall rigidity of the mount

Figure 6-4-2. Installed avionics LRUs on shock mounts

Figure 6-4-3. An avionics panel (audio panel) allows for simple control of a multitude of switches. On the left is a Marker Beacon indicator, while the center switches choose the radio that transmits.

Photo courtesy of Garmin

If the mount is too stiff, it may not provide adequate protection against the shock of landing. If the mount is not stiff enough, it may allow prolonged vibration following an initial shock.

Shock-absorbing materials commonly used in shock mounts are usually electrical insulators. Each electronic unit mounted with shock mounts must be electrically bonded to a structural member of the aircraft. This may also be accomplished by using sheets of high-conductivity metal (coppery or aluminum) where it is impossible to use a short bond strap.

Audio Panel

Most airplane installations have more than one radio installed. It would be extremely cumbersome to have to disconnect/reconnect microphones and headsets from one radio to another constantly. The answer to the problem is called an *audio panel*.

Avionics panel. An avionics panel is a small electrical panel with each item of NAVCOM equipment wired into it. This puts all on/off switches and all input/output connections in one convenient location (Figure 6-4-3). The avionics panel is normally wired to the main bus bar and allows all avionics equipment to be turned on/off at once with a master switch.

Reducing Radio Interference

Suppression of radio interference is a task of first importance. The problem has increased in proportion to the complexity of both the electrical system and the electronic equipment. Almost every component of the aircraft is a possible source of radio interference. Radio interference of any kind deteriorates the performance and reliability of the radio and electronic systems. There are three methods described are isolation, bonding and shielding.

Isolation. Isolation is the easiest and most practical method of radio noise suppression. This involves separating the source of radio noise from the input circuits of the affected equipment. In many cases, the noise in a receiver may be entirely eliminated simply by moving the antenna lead-in wire just a few

Figure 6-4-4. Various static wicks/dischargers

inches away from the noise source. Some of the sources of radio interference in aircraft are rotating electrical devices, switching devices, ignition systems, propeller control systems, AC power lines and voltage regulators.

An aircraft can become highly charged with static electricity while in flight. If the aircraft is improperly bonded, all metal parts will not have the same amount of charge. A difference of potential will exist between various metal surfaces. The neutralization of the charges flowing in paths of variable resistance, due to such causes as intermittent contact from vibration or the movement of the control surfaces, will produce electrical disturbances (noise) in the radio receiver.

Bonding. Bonding provides the necessary electrical connection between metallic parts of an aircraft. Bonding jumpers and bonding clamps are examples of bonding connectors. Bonding also provides the low-resistance return path for single-wire electrical systems.

Bonding radio equipment to the airframe will provide a low-impedance ground return and minimize radio interference from static electricity charges. Bonding jumpers should be as short as possible and installed in such a manner that the resistance does not exceed 0.003 ohm. When a jumper is used only to reduce radio noise and is not for current-carrying purposes, a resistance of 0.01 ohm is satisfactory.

The aircraft structure is also the ground for the radio. For the radio to function correctly, a proper balance must be maintained between the aircraft structure and the antenna. This means the surface area of the ground must be constant. Control surfaces, for example, may at times become partially insulated from the remaining structure. This would affect radio operation if bonding did not alleviate the condition.

Shielding. Shielding is one of the most effective methods of suppressing radio noise. The primary object of shielding is to electrically contain the radio frequency noise energy. In practical applications, the noise energy is kept flowing along the inner surface of the shield to ground instead of radiating into space. The use of shielding is particularly effective in situations where filters cannot be used. A good example of this is where noise energy radiates from a source and is picked up by the various circuits that eventually connect to the receiver input circuits. It would be impractical to filter all of the leads or units that are affected by the radiated noise energy; thus the application of effective shielding at the noise source itself is preferred, for it eliminates the radiated portion of the noise energy by confining it within the shield at its source.

Ignition wiring and spark plugs are usually shielded to minimize radio interference. If an intolerable radio noise level is present despite shielding, it may be necessary to provide a filter between the magneto and magneto switch to reduce the noise. This may consist of a single bypass capacitor or a combination of capacitors and choke coils. When this is done, the shielding between the filter and magneto switch can usually be eliminated.

The size of a filter may vary widely, depending on the voltage and current requirements as well as the degree of attenuation desired.

Filters are usually incorporated in equipment known to generate radio interference, but since these filters are often inadequate, it is frequently necessary to add external filters.

Electrostatic Discharge (ESD)

Static discharger wicks. Static dischargers are installed on aircraft to reduce radio receiver interference. This interference is caused by *corona discharge* emitted from the aircraft as a result of *precipitation static*. Corona occurs in short pulses, which produce noise at the radio frequency spectrum. Static dischargers, normally mounted on the trailing edges of the control surfaces, wing tips and vertical stabilizer, discharge the precipitation static at points a critical length away from the wing and tail extremities where there is little or no coupling of the static into the radio antenna.

Three major types of static dischargers are used:

1. Flexible vinyl-covered, silver- or carbon-impregnated braid
2. Semi-flexible metallic braid
3. Null-field

Flexible and semi-flexible dischargers are attached to the aircraft by metal screws and should be periodically checked for tightness. At least 1 inch of the inner braid of vinyl-covered dischargers should extend beyond the vinyl covering. Null-field dischargers (Figure 6-4-4) are riveted and epoxy bonded to the aircraft structure. A resistance measurement from the mount to the airframe should not exceed 0.1 ohm.

Installation of Aircraft Antenna Systems

An introductory knowledge of radio equipment is a valuable asset to the aviation technician, especially knowledge of antenna installation and maintenance, since the technician often performs these tasks.

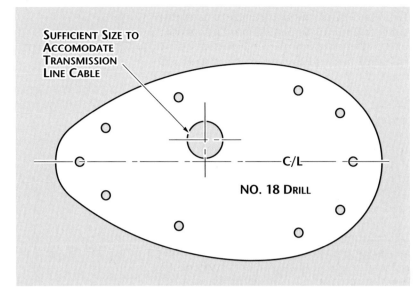

Figure 6-4-5. Pattern layout for antenna installation

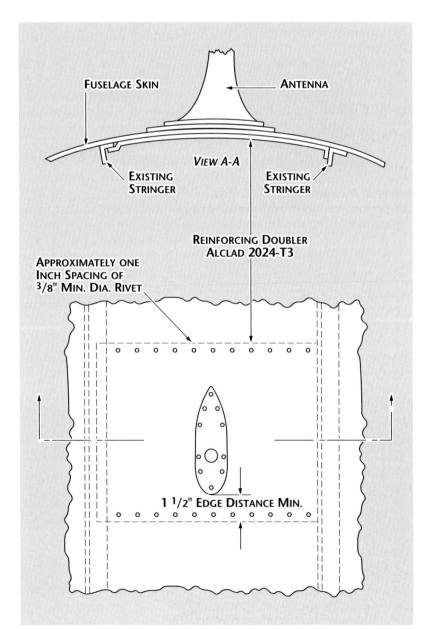

Figure 6-4-6. Doublers are sometimes necessary on skin-mounted antenna.

Antennas take many forms and sizes dependent upon the job they are to perform. Airborne antennas should be mechanically secure, mounted in interference-free locations, have the same polarization as the ground station and be electrically matched to the receiver or transmitter that they serve.

The following procedures describe the installation of a typical rigid antenna:

1. Place a template similar to that shown in Figure 6-4-5 on the fore-and-aft center line at the desired location. Drill the mounting holes and correct diameter hole for the transmission line cable in the fuselage skin.

2. Install a reinforcing doubler of sufficient thickness to reinforce the aircraft skin. The length and width of the reinforcing plate should approximate the example shown in Figure 6-4-6.

3. Install the antenna on the fuselage, making sure that the mounting bolts are tightened firmly against the reinforcing doubler and the mast is drawn tight against the gasket. If a gasket is not used, seal between the mast and the fuselage with a suitable sealer, such as zinc chromate paste, or equal.

4. Make sure the antenna is grounded to the airframe either through the mounting hardware or a separate grounding wire.

The mounting bases of antennas vary in shape and sizes; however, installation procedure is typical and may be used for most mast-type antenna installations.

The reinforcement to the mounting area is based on the profile drag of the antenna. That drag load is found by using the following formula:

$$D = .000327AV2$$

Where:

D = drag load in pounds

A = frontal area of the antenna in square feet

V = aircraft speed in m.p.h.

Antennae installed on the vertical fin could possibly change the frequency of the vibrations of the fin itself. If the unit is not a simple replacement it will at the least need an FAA 337 form. A first time installation may require an engineering approval.

Transmission lines. A transmitting or receiving antenna is connected directly to its associated transmitter or receiver by wire(s), which are shielded. The interconnecting shielded wire(s) are called a *coaxial cable*, which connects the antenna to the receiver or transmitter. The job of the transmission line (coaxial cable) is to get the energy to the place where it is to be used and to accomplish this with minimum energy loss. A transmission line connects the final power amplifier of a transmitter to the transmitting antenna.

The transmission line for a receiver connects the antenna to the first tuned circuit of the receiver. Transmission lines may vary from only a few feet to several feet in length.

Transponders, DME and other pulse type transceivers require transmission lines that are precise in length. The critical length of the transmission lines provides minimum attenuation of the transmitted or received signal. Refer to the equipment manufacture installation manual for type and allowable length of transmission lines.

When installing coaxial cable (transmission lines), secure the cables firmly along their entire length at intervals of approximately 2 ft. To assure optimum operation, coaxial cables should not be routed or tied to other wire bundles. When bending coaxial cable, be sure that the bend is at least ten times the size of the cable diameter.

Maintenance Procedures

Detailed instructions, procedures and specifications for the servicing of radio equipment

are contained in the manufacturer's operation and service manuals. In addition, instructions for removal and installation of the units are contained in the maintenance manual for the aircraft in which the equipment is installed.

Although installation appears to be a simple procedure, many radio troubles can be attributed to carelessness or oversight when replacing radio equipment. Specific instances are loose cable connections, switched cable terminations, improper bonding, lack of or improper safety wiring, or failure to perform an operational check after installation.

Two additional points concerning installation of equipment need emphasis. Prior to re-installing any unit, inspect its mounting for proper condition of shock mounts and bonding straps. After installation, safety wire as appropriate.

Section 5

Advanced Integrated Navigation Instruments

Instrument panel complexity increases pilot workload. This becomes a safety problem during critical flight phases in bad weather. Monitoring multiple flight and navigation instruments at night in a thunderstorm while trying to land an airplane is a daunting task.

Figure 6-5-1. A basic IFR instrument panel with separate flight and navigation instruments

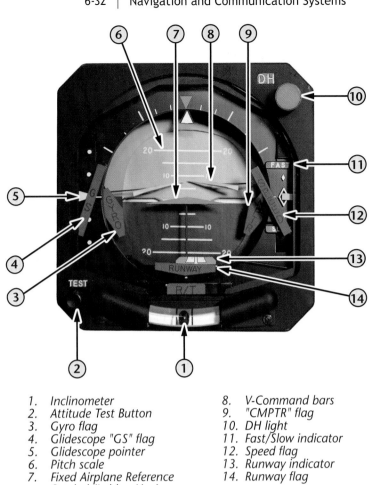

1. Inclinometer
2. Attitude Test Button
3. Gyro flag
4. Glidescope "GS" flag
5. Glidescope pointer
6. Pitch scale
7. Fixed Airplane Reference Symbol (Lubber Line)
8. V-Command bars
9. "CMPTR" flag
10. DH light
11. Fast/Slow indicator
12. Speed flag
13. Runway indicator
14. Runway flag

Figure 6-5-2. A typical Flight Director Indicator (FDI)

Light aircraft and early airliners have the navigation instruments to the side of the primary flight instruments. Figure 6-5-1 shows this type of configuration. Note that the pilot must look back and forth from the flight instruments (airspeed, attitude gyro, altimeter and vertical speed indicator) to the Course Deviation Indicator.

Two complex analog gauges were developed to combine these functions. These instruments are the flight director indicator (FDI) and the horizontal situation indicator (HSI). The flight director indicator is often called simply a flight director. Each of these instruments combines both flight instrument and navigation information on one indicator.

They are very complex mechanical devices and as such are very expensive. They were originally only used on airliners due to their high cost. Lower cost units were eventually developed. FDIs and HSIs are now used on most turbine aircraft and many high-end reciprocating aircraft.

Flight Director Indicator. A Flight Director is a multifunction instrument that combines

attitude information from the gyro horizon with navigation information from the VOR and glideslope receivers. Some units also add airspeed and altitude warnings. Figure 6-5-2 shows a typical Flight Director.

1. **Inclinometer.** The inclinometer works exactly like the ball in a turn and slip gyro. It provides slip and skid information to aid in coordinating turns.

2. **Attitude test button.** Depressing the test button moves the gyro to a preset position. The gyro should indicate 10° nose up and 20° right roll and the gyro flag should appear when it is depressed.

3. **Gyro flag.** The Flight Director receives its attitude information from a remote gyro. The gyro flag appears when gyro is unable to display accurate attitude information. This may be due to either a lack of power to the Flight Director or a signal failure from the remote gyro.

4. **Glideslope "GS" flag.** This flag appears when no glideslope signal is present or the glideslope signal is not usable.

5. **Glideslope pointer.** When the VHS Nav receiver is tuned to a frequency equipped with an ILS function the glideslope pointer will appear. When no glideslope is available, this pointer is retracted out of view.

6. **Pitch scale.** The flight director provides the primary gyro horizon attitude reference. It includes a scale showing the aircraft's pitch angle with reference to the horizon. Pitch angles of 20° up to 20° down are visible in Figure 6-5-2.

7. **Fixed airplane reference symbol.** This symbol provides the same information as the aircraft silhouette on a typical gyro horizon. It shows the position of the aircraft with reference to the horizon.

8. **V-Command bars.** The V-command bars provide steering guidance to the pilot. When the glideslope is activate, these will indicate if the aircraft is above or below the glideslope. It will also provide lateral deviation command guidance.

9. **"CMPTR" flag.** The computer or CMPTR flag monitors the computer input signals. When this flag is visible, command information is not reliable. Some Flight Directors have this labeled as the "steer" flag.

10. **DH light.** The DH or Decision Height light illuminates when aircraft reaches the altitude preset by the radio altimeter alert knob. This is known as the MDA or Minimum Decision Altitude light on some units.

11. **Fast/slow indicator.** The airspeed indicator has a knob that is used to select a desired airspeed. This pointer indicates when the aircraft is moving faster or slower than the preselected airspeed.

12. **Speed flag**. The speed flag moves into view, obscuring the fast/slow indicator, when the speed command indications are not usable.

13. **Runway indicator.** The runway indicator shows localizer deviation. This shows the lateral course deviation. As the aircraft approaches the runway, based on radar altimeter information, the runway indicator rises towards the center of the display providing the pilot with an additional visual cue. The runway indicator in Figure 6-5-2 shows that the aircraft is slightly to the left of the runway.

14. **Runway Flag.** When the VHF Nav is not tuned to a localizer, or in the event of a localizer failure, the runway indicator is obscured by a runway flag indicating that it is not providing usable data.

The flight director combines functions from many instruments and navigation sources into one easy to read display. It has greatly reduced pilot workload, however it is a complex and sometimes maintenance intensive system. When troubleshooting any flight director system, ensure that the proper maintenance and troubleshooting manuals are used.

Horizontal Situation Indicator. The horizontal situation indicator (HSI) combines the functions of a directional gyro with a VHF Nav display. Like the Flight Director, it combines many functions into one instrument. A typical HSI is shown in Figure 6-5-3.

1. **Course knob.** The course knob is used to set the desired VOR radial or localizer magnetic course. It also sets the digital display in the COURSE window.

2. **Glideslope pointer**. The glideslope pointer is visible when the VHF Nav receiver is tuned to an ILS frequency. If glideslope information is not available, the pointer is not visible.

3. **Glideslope flag**. The glideslope flap indicates the reliability of the glideslope signal. The flag is visible when the glideslope signal is not usable.

4. **Course window**. The course window displays the course set by the course knob.

5. **Course arrow.** The course arrow indicates the selected VOR radial or localizer course relative to the compass card. The course arrow is set by the course knob.

6. **Lubber line.** The lubber line aligns with the fixed airplane reference and marks the top center of the HSI.

7. **NAV flag**. The NAV flag indicates the reliability of the VOR or LOC radial signal. When the flag is visible, the signal is not reliable. This flag is labeled VOR/LOC on some units.

8. **Miles window**. DME distance is displayed in the miles window.

9. **Shutter**. The shutter obscures the DME window when DME information is not available or is not usable.

10. **HDG flag**. The heading flag is visible when the compass card information is not usable.

11. **Lateral deviation bar**. The lateral deviation bar displays the position of the selected radio course relative to the current airplane position. It provides the same information as the vertical bar on a conventional VOR CDI.

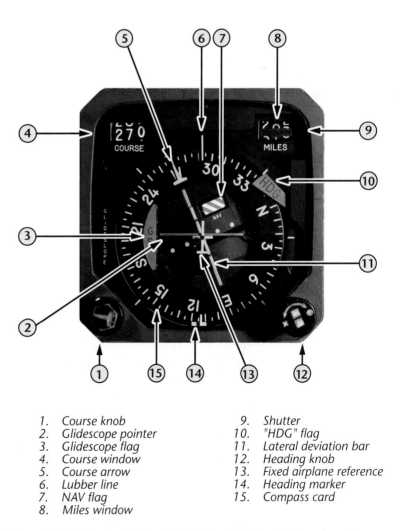

1. Course knob	9. Shutter
2. Glidescope pointer	10. "HDG" flag
3. Glidescope flag	11. Lateral deviation bar
4. Course window	12. Heading knob
5. Course arrow	13. Fixed airplane reference
6. Lubber line	14. Heading marker
7. NAV flag	15. Compass card
8. Miles window	

Figure 6-5-3. A typical horizontal situation indicator (HSI)

12. **Heading knob**. The heading knob is used to set the desired heading. Some systems couple this knob to the auto-pilot. The knob can then be used to provide steering guidance to the auto-pilot system in the auto-pilot manual steering mode.

13. **Fixed airplane reference**. The fixed airplane reference represents the current position of the airplane. Lateral deviation, course and heading information are all shown relative to this reference.

14. **Heading marker**. The heading marker indicates the selected heading. It is set with the heading knob.

15. **Compass card**. The compass card provides a continuous display of the airplane heading. Its position is determined by a remote directional gyro.

Figure 6-5-4 shows a typical corporate aircraft equipped with a flight director and HSI system. Compare the information displayed with the instrument panel shown in Figure 6-5-1. Both instrument panels provide the flight crew with the same information, however the advanced systems significantly reduce the crew's workload.

Electronic Flight Information Systems (EFIS). The next step in the development of integrated instrument and navigation displays is the electronic flight information system (EFIS). These systems provide flight director and HSI style displays combined with information from additional instruments. The basics of EFIS systems are covered in chapter 5, Aircraft Instruments.

Figure 6-5-5 shows two typical EFIS displays. Figure 6-5-5 (A) has two separate displays, one for the flight director and a second for the HSI. View (B) shows the same types of information displayed on a single screen.

All of the information displayed on old style analog gauges is incorporated into the display. The basic flight director display is in the center. Both sides of the display provides additional flight instrument information. The left hand side shows the airspeed and includes an airspeed reference marker. The right hand side provides altitude information. This includes

Figure 6-5-4. A Beechraft King Air corporate turboprop equipped with a modern Flight Director and HSI

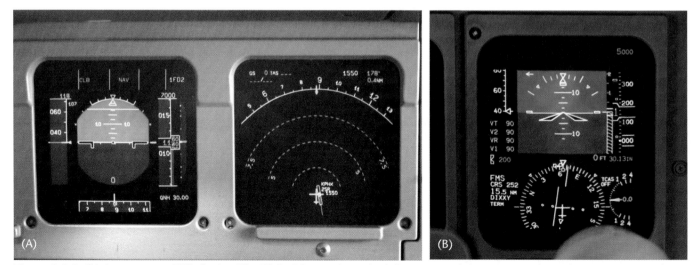

Figure 6-5-5. (A) A two screen EFIS display. The Flight Director is on the left screen and the HSI is on the right one. (B) A single screen EFIS display. The Flight Director information is displayed at the top and the HSI at the bottom.

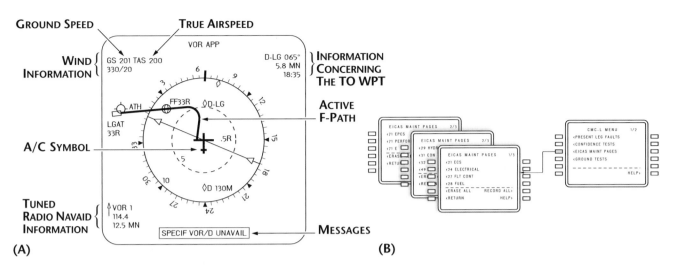

(A)   (B)

Figure 6-5-6. (A) shows a view of the actual flight path, while (B) shows a layout example of the maintenance pages available for troubleshooting and repairs. Both types of information are available on the CRT.

the current altitude, vertical speed information and an altitude reference marker. The HSI information is displayed at the bottom.

Digital systems further automate the duties of the flight crew. A flight crew can enter the cockpit and gather a pre-made-out flight plan from his/her company. The entire route of flight, including waypoints, navigation and communication frequencies, alternate airports as well as the destination airport information are all available at the touch of a button.

The EFIS also monitors a number of aircraft systems. Maintenance data can be accessed by ground personnel quickly and easily. Figure 6-5-6 shows a typical flight routing and additional maintenance pages.

Expected weather conditions, including winds aloft, can also be entered prior to the flight. A

Figure 6-5-7. A weather radar display

moving weather map can be made to roll in front of the crew to study the weather at altitude before take-off. Weather radar information can also be displayed on one of the EFIS panels during flight. Figure 6-5-7 shows a typical weather radar display.

The most advanced systems integrate with the auto-pilot and can operate the aircraft from take-off to landing with no in-flight input from the pilot. The flight management computer is accessed from a dedicated panel. A typical flight management interface is shown in Figure 6-5-8. The newest systems add a mouse-type trackball and advanced graphical interfaces similar to the latest desktop computers.

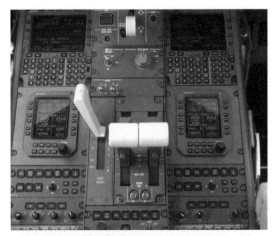

Figure 6-5-8. The flight management system is accessed through a dedicated interface. Course and waypoint information is entered into the computer using this panel.

7

Lighting, Warning and Utility Systems

Section 1

Aircraft Lighting Systems

Aircraft lighting systems is a misnomer. Like most subjects in aircraft operations and maintenance, a lighting system is actually a collection of small systems and subsystems which operate independently of each other. Each one is designed to perform a separate function. The study of lighting systems starts with a location, and follows with a function. The FAA requirements for aircraft lighting are spelled out in CFR part 23.1381/23.1401 and CFR part 25.1381/25.1403.

Exterior Lights

Exterior lights consist of position (navigation) lights, anti-collision and strobe lights, landing and taxi lights, logo lights, and wing inspection (ice) lights. Also classified as exterior lights are cargo bay, wheel well, and work and service lights found on airline equipment.

Position Lights

For many years *position lights*, also called *navigation lights*, were the minimum FAA requirement for aircraft flying at night (refer to CFR part 91.209).

Position lights consist of one red, one green, and one white light. On many aircraft each light unit contains a single lamp mounted on the surface of the aircraft. Other types of position light units contain two lamps, and are often streamlined into the surface of the aircraft structure. An example is shown in Figure 7-1-1.

Learning Objectives

DESCRIBE

- data collected by two types of data recorders and its use

- two types of advisories issued by traffic alert and collision avoidance systems

- components of the interphone system and common maintenance tasks

EXPLAIN

- location and function of types of aircraft lighting

- stall, EICAS and other aircraft warning systems

- the function of the Advanced Cabin Entertainment Service System

Left. The cockpit layout of this Falcon 900 is an example of a glass cockpit in a corporate jet. Notice the warning panel to the left of the center CRT.

Figure 7-1-1. Wing tip position lights

The green light unit is always mounted at the extreme tip of the right wing. The red unit is mounted in a similar position on the left wing. The white unit can be located on top of the vertical stabilizer, in the trailing edge of the rudder, or on the fuselage tailcone.

The purpose for the three colors of lights is quick and positive position identification. For instance, if only a white light is visible, the airplane is heading away from the viewer. A green light on the left and a red light on the right indicates the airplane is heading towards the viewer.

The wingtip and tail lights are controlled by a switch in the pilot's compartment. Some tip lights have a Plexiglas or Lucite attachment that will transmit a small amount of light as a visual in-flight indicator that they are working.

Many older navigation light systems used a fairly complicated flashing system involving an electric motor-driven switch assembly that rotated an on/off switch. Some smaller general aviation airplane systems have a system resembling an automotive emergency flasher system. These systems became outdated when wingtip strobe lights became standard equipment.

There are many variations in the position light circuits used on various aircraft. Most transport category airplanes have dual light installations for redundancy. All circuits are protected by fuses or circuit breakers.

Anti-collision Lights

Commonly called *rotating beacons*, all airplanes certified for night flight have at least one. Many airliners have two; one on top and one below

(Figure 7-1-2). These are rotating beam lights that are usually installed on top (or bottom if two are mounted) of the aft fuselage, or on top of the vertical stabilizer, in such a location that the light will not affect the vision of the crewmembers or detract from the visibility of the position lights.

An anti-collision light unit usually consists of one or two rotating lights operated by an electric motor. The light may be fixed, but mounted under rotating mirrors inside a protruding red glass housing. The mirrors rotate in an arc, and the resulting flash rate is between 40 and 100 cycles per minute. (See Figure 7-1-3.) The anti-collision light is a safety light to warn other aircraft, especially in congested areas.

Most older rotating beacons are driven by an internal electric motor and a gear assembly, while newer units are specially designed strobe lights with red lenses. It is becoming common practice to replace older motor-driven units with the new type when it is time to replace the motor.

There are no specific maintenance procedures for rotating beacons. However, that does not mean that are maintenance-free. Anti-collision lights should always be examined for any irreg-

Figure 7-1-2. Rotating beacon

ularities, such as water ingestion in the rotating assembly, leaking at the mounting flanges, loose or corroded lens retainer rings, abnormally noisy gear systems, and loose or cracked lenses. Inspect the wiring and check for loose or corroded grounds. Always make sure anti-collision lights are installed on a switch independent of the position lighting system.

Strobe Lights

Strobe lights became popular in the late 1960s, and are now required equipment on all aircraft. The systems are so dependable and draw so little power that they are normally turned on with a separate switch, any time the airplane is running.

The brilliant white light emitted is much easier to see than any combination of position lights. They are especially good in inclement weather and will show up long before any other light system (Figure 7-1-4).

Operating theory. All strobe lights work in a similar manner. A storage capacitor connected across the flashlamp is charged from a 300/400 VDC power supply.

Another small capacitor is charged from the same power supply to generate a trigger pulse.

An electrical switch, designed to produce a flash at the correct time, closes. This causes the charge on the trigger capacitor to be dumped into the primary of a pulse transformer whose secondary is connected to a wire, strip, or the

Figure 7-1-3. Anti-collision light

metal reflector in close proximity to the flash tube.

The pulse generated by this trigger is enough to ionize the xenon gas inside the flashtube. The xenon gas suddenly acquires a low resistance and the energy storage capacitor dis-

Figure 7-1-4. Strobe light

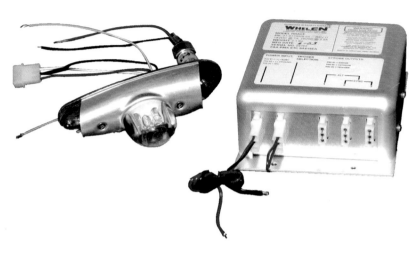

Figure 7-1-5. A strobe system, with power supply and wing tip lights

charges through the flash tube, resulting in a burst of brilliant white light for a short time (Figure 7-1-5).

The power supply recharges the capacitors and the cycle repeats.

Strobe maintenance. Simple strobe maintenance and troubleshooting involves a visual inspection only. Strobe lights do not have a maintenance program.

> **CAUTION:** *As with all high-powered light bulbs, do not touch strobe bulbs with your bare hands. Oil film left behind will help concentrate heat and lead to premature failure.*

Visually inspect for the following:

- Ground connections for corrosion and apparent movement that would indicate a loose connection.

- Look at the flash tubes to see if they are blackening internally. If so, replace them because they will fail soon.

- Check the flash tubes for cracks, as well as the light lenses and their gaskets.

Figure 7-1-6. Landing lights built into the leading edge of the wing

Figure 7-1-7. Retractable landing light mounted in the outer wing panel of a Cessna 310.
Photo courtesy of Priority Air Charter

If a single strobe doesn't work then do the following:

Place the bottom of a coffee cup or similar container against the strobe tube. At the cup opening, listen for a ping, or snap, as the trigger fires. No ping means that either the wiring has an open, a short, or the power supply isn't working. If none of the strobe lights work, one may have a grounded wire, which will cause the power supply to shut itself off rather than overheat. Disconnect all strobe leads and cycle the power supply. Check the power supply with a voltmeter to see if it is providing power; if it is then reconnect the strobe leads one at a time until the bad lead is located.

Aircraft strobe tubes have three color-coded wires:

- Pin 1 is a red wire and is the anode connection.
- Pin 2 is a black wire and is the flash tube ground.
- Pin 3 is a white wire and is the trigger wire.

If pins 1 and 3 are ever jumped for any reason, the trigger circuit in the power supply will be destroyed.

Landing Lights

Landing lights are very powerful and are designed for use during landing and, on some smaller airplanes, takeoff. They are sealed beam units that vary from lamps equal to modern automobile headlights, to more than 600 watts and 750,000 candle-power quartz halogen sealed beam units for large aircraft.

Landing lights can be located midway in the leading edge of each wing, streamlined into the aircraft surface as shown in Figure 7-1-6, or they can be retractable, as in Figure 7-1-7.

Leading edge lights. Most smaller general aviation airplanes have a landing light assembly in the wing. It contains two lights that are aimed differently; one for taxi and one for landing. An electrical relay is installed close to the light assembly to eliminate larger high amperage wiring under the instrument panel.

Retractable lights. When the retractable lamps are not in use, a motor retracts them into receptacles in the wing where the lenses are not exposed to the weather.

When the control switch is placed in the upper, or *extend*, position (Figure 7-1-8), current from the battery flows through the closed contacts of the switch and to the motor itself. When the lamp mechanism is completely lowered, the projection at the top of the gear quadrant opens the circuit to the motor, and causes the brake solenoid to release the brake. The brake is pushed against the motor shaft by the spring, stopping the motor and completing the lowering operation.

To retract the landing lamp, the control switch is placed in the *retract* position. The motor and

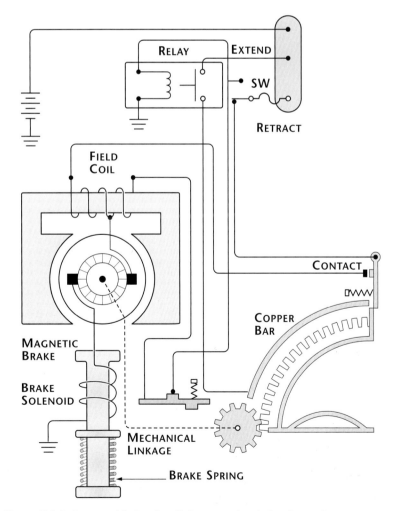

Figure 7-1-8. Retractable landing light control switch schematic

Figure 7-1-9. This Gulfstream G-III has all lights on during landing.

brake circuits are completed and the landing light mechanism is retracted. Retractable landing lights that can be extended to any position are employed on some aircraft.

Many large aircraft are equipped with four landing lights, two of which are fixed and two retractable. Fixed lights are usually located in either the wing root area or just outboard of the fuselage in the leading edge of each wing. The two retractable lights are located in the lower surface, as shown in Figure 7-1-9.

Taxi lights. Taxi lights are designed to provide illumination on the ground while taxiing or towing the aircraft to or from a runway, taxi strip, or in the hangar area.

Taxi lights are not designed to provide the degree of illumination necessary for landing lights; 150- to 250-watt taxi lights are typical on many medium and heavy aircraft.

On aircraft with tricycle landing gear, taxi lights are often mounted on the non-steerable part of the nose landing gear. In Figure 7-1-10,

these three taxi lights, mounted on a King Air 90, the outer two lights positioned at an angle to the center line of the aircraft to provide illumination directly in front of the aircraft and also some illumination to the right and left of the aircraft's path.

Taxi lights are also mounted in the recessed areas of the wing leading edge, often in the same area with a fixed landing light.

Wing inspection lights. Some aircraft are equipped with wing inspection lights to illuminate the leading edge of the wings to permit observation of icing in flight. The wing inspection light system (also called wing ice lights) are typically mounted flush on the outboard side of the fuselage (Figure 7-1-11.) These lights permit visual detection of ice formation on wing leading edges while flying at night. They are also often used as floodlights during ground servicing.

Logo lights. Logo lights are used on larger corporate and commercial airline equipment. Two lights, one on each wing tip, shine on the

Figure 7-1-10. Common method of mounting taxi lights is shown on this King Air

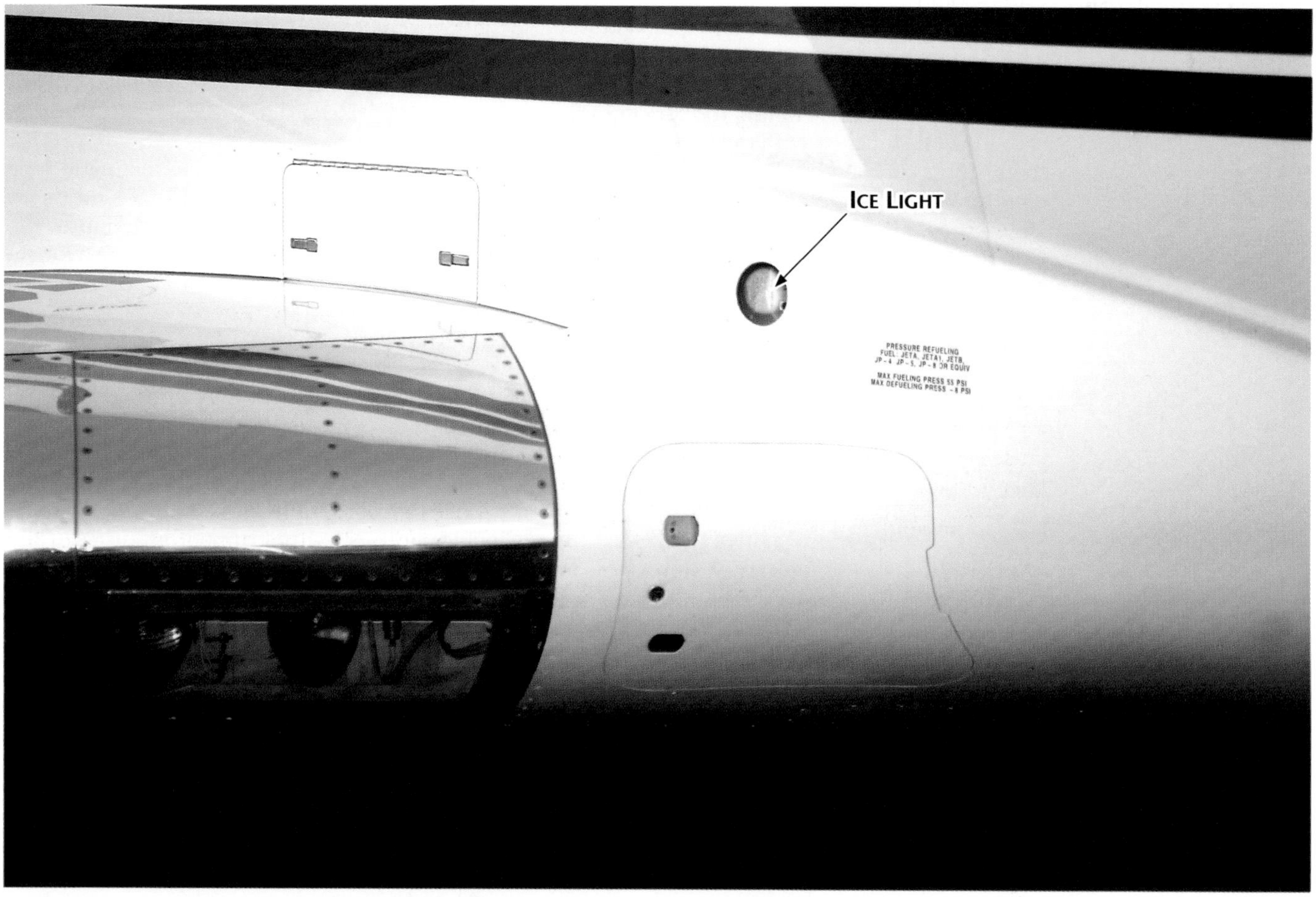

Figure 7-1-11. Ice lights are typically mounted flush on the side of the fuselage.

aft section of the fuselage. They also light up the vertical surfaces that carry the N numbers and company logo. This makes the aircraft easier to see when taxiing and during landing and take-off as well as advertising the airline. Whichever purpose is primary, it makes airport operations safer by providing clear recognition.

Maintenance and Inspection of Lighting Systems

Inspection of an aircraft's lighting systems normally includes checking the condition and security of all visible wiring, connections, terminals, fuses, and switches. A continuity light or meter is used in making these checks. Many problems can be detected by systematically testing each circuit for continuity.

- All light covers and reflectors should be kept clean and polished. Cloudy reflectors are often caused by an air leak around the lens.

- The condition of the sealing compound around position light frames should be inspected regularly. Leaks or cracks should be filled with an approved sealing compound.

Care should be exercised in installing a new bulb in a light assembly, as many bulbs fit into a socket in only one position and excessive force can cause an incomplete or open circuit in the socket area. Strobe tubes and xenon bulbs should not be touched with the hands. Leaving fingerprints on the glass becomes a heat sink, overheating the bulb.

Circuit testing, commonly known as troubleshooting is a means of systematically locating faults in an electrical system. The main faults are listed below:

- Open circuits in which leads or wires are broken.

- Shorted circuits in which grounded leads cause current to be returned by shortcuts to the source of power.

- Low power in circuits causing lights to burn dimly and relays to chatter. Electrical troubles may develop in the unit or in the wiring. If troubles such as these are carefully analyzed and systematic steps are taken to locate them, much time and energy not only can be saved, but damage to expensive testing equipment often can be avoided.

Any standard DC volt-ohmmeter (VOM) with flexible leads and test probes is satisfactory for testing circuits.

More specific information on troubleshooting electrical circuits can be found in section 19 of the chapter "Basic Electricity" in the first book of this series, *Introduction to Aircraft Maintenance.*

Section 2

Data Recorders

Under Title 14 CFR part 121 every airplane certified under CFR part 121 must have two recorders on board. One is a cockpit voice recorder and the other is a flight data recorder. These blind recorders are not accessible to the crew. The recorders are the so-called *Black Boxes* that are actually painted orange or yellow. Their purpose is to provide information in case of an accident.

Data recording media is designed to be crashworthy. In other words, it should survive any crash, in any environment, that is possible for an airplane to encounter.

Cockpit Voice Recorder (CVR). Every CFR part 121 cockpit is wired for sound. Several microphones are installed in the cockpit and can record all voice communication between the crew members. The *cockpit voice recorder* can also record any other sounds that may be present. Figure 7-2-1 shows a photo of a typical Cockpit Voice Recorder, or "Black Box".

Most CVRs are the older style magnetic tape units and record thirty minutes of sounds on a closed loop. At the end of thirty minutes the tape starts re-recording over the previously recorded tape.

Newer CVRs are digital. The digital CVRs can record up to two hours of voice and cockpit sounds before they start re-recording.

Flight Data Recorders (FDR). *Flight Data Recorders* are more complex and their purpose is to record flight and control parameters, including selected instrument readings. They can record for up to 24 hours, depending on the model of the FDR.

Older models of FDRs still in service record on Mylar tape, while current models use digital recording media. Digital media are capable of recording up to several hundred parameters, though only 88 parameters are currently recorded. Examples of items recorded are:

- Time
- Altitude

Figure 7-2-1. A cockpit voice recorder

- Airspeed
- Vertical acceleration
- Heading
- Time of each radio transmission either to or from air traffic control
- Pitch attitude
- Roll attitude
- Longitudinal acceleration
- Pitch trim position
- Control column or pitch control surface position
- Control wheel or lateral control surface position
- Rudder pedal or yaw control surface position
- Thrust of each engine
- Position of each thrust reverser
- Trailing edge flap and cockpit flap control position
- Leading edge flap and cockpit flap control position

Flight data recorders are blind recorders as well; the flight crew have no control over what is recorded. Once the airplane starts, the recording process is automatic.

Accident Investigation

The purpose behind Cockpit Voice Recorders and Flight Data Recorders is to aid *accident investigation*. While it is important to avoid accidents, no system is perfect. Accidents will happen. The National Transportation Safety Board (NTSB), as well as the FAA, have a mandate to investigate every aviation accident. Some investigations take up to a year or more.

From the findings of the NTSB and the FAA, a cause and effect for any specific accident can be identified. The findings tell us if an accident was human-induced or a mechanical failure. The information from the FDR is fed into a flight simulator and pilot investigators "re-fly" the trip. Mechanical failures can be evaluated further to determine if a re-design or modification of a part or assembly will insure that the same thing does not happen again.

Section 3

Traffic Alert and Collision Avoidance System

The Traffic Alert and Collision Avoidance System, or TCAS, is an instrument integrated into other systems in an aircraft cockpit. It is an onboard system of hardware and software that together provide a set of electronic eyes so the pilot can "see" the traffic situation in the vicinity of the aircraft (Figure 7-3-1). Information comes from the *Air Traffic Control Radar Beacon System* (ATCRBS) transponder, or from an ATC *Mode S transponder*. Part of the TCAS capability is to display the relative positions and velocities of aircraft up to 40 miles away. The instrument sounds an alarm when it determines that another aircraft will pass too closely to the subject aircraft. TCAS provides a backup to the air traffic control system's separation safeguards.

TCAS I. There are two different versions of TCAS designed for use on different classes of aircraft. The first, TCAS I, indicates the bearing and relative altitude of all aircraft within a

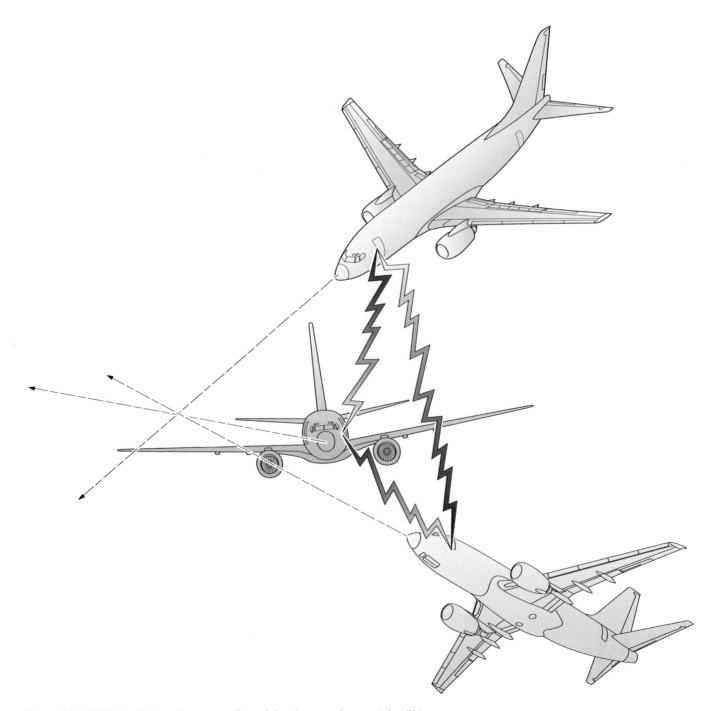

Figure 7-3-1. TCAS is designed to warn pilots of the danger of potential collision.

selected range (generally 10 to 20 miles). With color-coded symbols, the CRT display indicates which aircraft pose potential threats. This constitutes the *Traffic Advisory* (TA) portion of the system. When pilots receive a TA, they must visually identify the intruding aircraft and may alter their plane's altitude by up to 300 feet. TCAS I does not offer solutions, but does supply pilots with important data so that they can determine the best course of action. An illustration of TCAS range and altitude criteria shows the horizontal and vertical distances to monitor traffic and issue advisories to maintain safe separation of aircraft.

TCAS II. In addition to a traffic display, the more comprehensive TCAS II provides pilots with resolution advisories (RA) when needed. The system determines the course of each aircraft; climbing, descending, or flying straight and level. TCAS II then issues an RA advising the pilot to execute an evasive maneuver necessary to avoid the other aircraft, such as "Climb" or "Descend." If both planes are equipped with TCAS II, then the two computers offer compatible RA's. In other words, the pilots do not receive advisories to make maneuvers that would effectively cancel each other out, resulting in a continued threat.

The MITRE Center for Advanced Aviation Development was central to the development of TCAS due to its work on the collision avoidance logic for TCAS II. The software uses the collected data on the flight patterns of other aircraft and determines if there is a potential collision threat. The system does not just show the other planes on a display like a radar screen, but offers warnings and solutions in the form of traffic advisories (TA) and resolution advisories (RA).

Evolving technology. Although airlines were using the more advanced version of TCAS logic, some improvements were necessary as the system was too sensitive in some situations. For instance, the system issued RA's at final approach, when traffic may be closer but is safely under control. Unnecessary TA's and RA's had even been triggered by transponders on bridges and ships.

In 1992, logic version 6.04 was released to alleviate these problems. Delta Airlines, the first carrier to voluntarily use the new logic, reported an 80 percent reduction in RA's. The following year version 6.04A was developed with an additional improvement to the logic. Airlines began equipping their fleets with this version in 1994.

In 1997 a final change to the TCAS logic was made in version 7. Version 7 logic yields at least a 20 percent reduction in RA's over the previous version. It was approved by the RTCA standards committee and the FAA, and is the version installed on all new aircraft. It has also been adopted by the International Civil Aviation Organization (ICAO) as the international standard. Version 7 is required for aircraft serving European and some other countries. American carriers who fly to these countries are required to upgrade from 6.04A to 7 on their international planes, and may voluntarily upgrade the equipment in their U.S. fleets. Version 7 is required for operation in *Reduced Vertical Separation Minimum* (RVSM) airspace.

Section 4

Warning Systems

Aside from stall warning systems, discussed later in this section, small general aviation airplanes with fixed landing gear have no warning systems. Those aircraft with retractable gear systems, however, have warnings for gear down, gear in transit and gear up. Many twin and turbine airplanes with reciprocating engines have the added complexity of fire warning systems. Cabin warning systems are also in place for cabin pressurization controls. These are discussed in other chapters, but are all part of the general warning system. A typical set of warning lights is shown in Figure 7-4-1. The overhead panel of this Hawker 800XP is mostly warning lights.

Figure 7-4-1. An overhead panel is the normal location for warning lights.

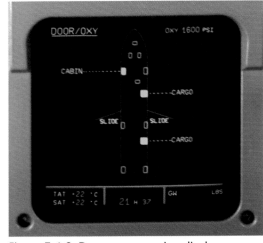

Figure 7-4-2. Door open warning display

When the size and complexity of airplanes increase, so do the various warnings. So much so, in fact, that warning systems on a part 121 airplane could fill this entire book, if studied in depth.

Door warnings. Digital cockpits make something as innocuous as *door open warning lights* a simple visual item. With CRT displays, you do not have to find the door ajar light and try to figure out which one is open or not latched. The display shown in Figure 7-4-2 shows clearly that the cabin entrance and two cargo doors are open.

Configuration Warning Systems

While most airplanes, including corporate jets, use the same types of warning systems, the differences start to show up when looking at digital airplanes.

With digital instrumentation, not only are there warning systems for the actual landing gear up and/or down, but digital electronics make possible a series of configuration warnings based

Figure 7-4-3. A micro switch operated by an airflow tab is the most common stall warning transmitter.

on gear position, flap position, throttle position, altitude, speed, and rate of descent.

Configuration warnings. If the airplane is not configured correctly for a landing or take-off, an alarm will sound. A typical warning system monitors the take off position of critical control surface systems like stabilizer trim, trailing edge flap, leading edge devices and speed brakes. When primary flight control surfaces are set for a particular phase of flight, such as landing or take off, the corresponding control-surface indicating system will show flap/slat position. Many of the takeoff warning systems on larger aircraft will self activate when thrust levers are moved to takeoff position. What the alarm is, or says, depends on the current circumstance. Table 7-4-1 shows some examples of integration seen in configuration warning systems.

Initially, the amount of warnings may seem unwieldy, but these warnings have saved many lives. A large airplane not configured properly for landing, or attempting takeoff in a landing configuration, will more than likely crash. While a misconfiguration should be caught during the performance of a checklist, the configuration warning system works acts as a backup to keep track of items that may have been missed.

Stall Warnings

The earliest stall warning was the joy stick banging back and forth between the pilots knees from the air disturbance over the ailerons just before the airplane stalled. As airplanes became more sophisticated, and control surfaces were balanced, stick-shaking became less violent, but did not disappear.

With today's hydraulically or electrically operated control systems, aircraft manufacturers still use a *stick shaker.* Never mind the fact that there is no stick, or that the complex control systems have no feel. Control feel is another system that, in the name of safety, has been built-in to today's airplanes. Part of the system of *artificial control feel* is shaking the stick, or control column, just prior to a stall.

Stall Warning Systems

Stick shakers are only one aspect of stall warnings. The mainstay is a warning system based on airflow over the wings. The three types are based on input. They are pressure input, electrical input and air-data stall.

Pressure input. The simplest stall warning is a tube that has one end in the dead

SYSTEM	WARNING SOUND	WHEN ACTIVATED
Landing gear	Klaxon	When gear is up and any throttle is retarded below 70% N_1. It also sounds when the gear is up and flaps are extended more than 20°.
Overspeed	Cricket	When airspeed has exceed V_1 by approx. 6 knots at 28,000 ft. or below. When the Mach number has exceeded M by 0.01M above 28,000 ft.
Speedbrake/takeoff	Alternating high/low	(a) Speedbrakes are deployed with flaps more than 22° or the speedbrakes are deployed with landing gear extended. (b) Takeoff is attempted with flaps out of takeoff position. (c) Takeoff is attempted with ground spoilers deployed. The ground spoiler and master caution lights will also illuminate. (d) Takeoff is attempted with the thrust reverser overcenter links having moved overcenter toward the deploy position.
Oxygen	Warbling	Cabin altitude has exceeded 28,000 ft. and passenger oxygen system has activated.
Engine/APU fire	Bell	An engine or APU fire has resulted.

Table 7-4-1. Common examples of configuration warning indications

air at the leading edge of the wing, and the other end connected to a flat rubber tube. Just before a stall the air flow over the wing shifts. This places a positive pressure on the tube and sends it to the flat rubber tube. The flat tube then makes a sound like a duck call and means "get your nose down and add some power – NOW". At the same time, the varying angle of incidence in the wing should start a shuddering in the wing roots as an additional reminder the airplane is about to stall. The input tube is easily bent during ground handling. The maintenance manual will show how to realign the tube using a template.

Electrical input. A more common stall warning works on the same airflow process, but operates a microswitch instead of blowing a "duck call" (Figure 7-4-3).

When the airflow shifts it allows the flapper to pitch up, closing the circuit. That places power on the receiver unit on the instrument panel, blinking a red light and sounding a buzzer to warn of an impending stall. Shortly thereafter the wing roots will start shuddering, and a stall would be imminent.

Maintaining an electrical input stall warning system is simple: make sure the switch unit hasn't slipped on its mounting screws, the wires are not chafed or broken, and that the system works when the switch is tripped.

Air data stall warning. The third type of stall warning system is in common use on large airplanes.

Digital electronics have made *air data systems* possible. The air data system gathers information from virtually every system on the airplane. It collects temperature, airspeed, barometric pressure, power output, flap and slat configuration, rate of climb, and most importantly, angle of attack indications (Figure 7-4-4). The flight computer compares these inputs to the aircraft's performance envelope. The computer then determines if the airplane will fly.

Figure 7-4-4. An angle-of-attack sensor

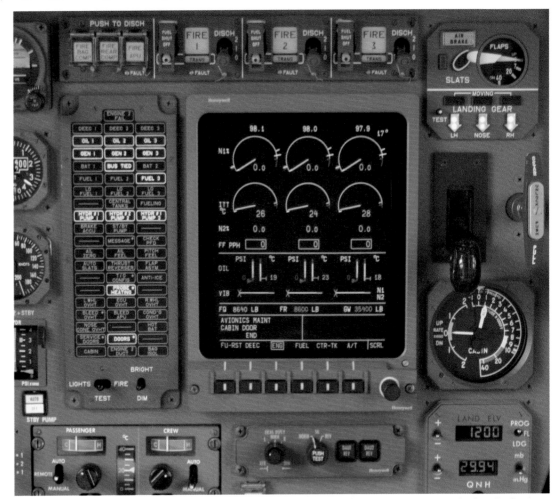

Figure 7-4-5. A typical EICAS display. Notice the warning lights on the left of this particular display.

If the computer determines that the airplane is incapable of flight, the stick starts shaking, the alarm goes off, and a loud, recorded voice warns: "STALL WARNING NOSE DOWN" repeatedly.

European Airbus airplanes were the first fly-by-wire airplanes that had air data computers programmed into the flight computers. They developed the ultimate stall warning system. All control inputs were compared to the airplane's performance envelope. The flight computer would check with the air data system and not let the pilot put the airplane in a position of stalling in the first place.

Engine Indicating and Crew Alert System Warnings (EICAS)

Any airplane with a glass cockpit has additional warning systems. EICAS is the main one. Figure 7-4-5 shows a typical EICAS.

EICAS formats. Several display formats are used by various versions of EICAS. Formats vary with aircraft model. Common formats include primary, secondary, and compact modes. The primary format is shown on the upper CRT during normal operation. Different colors are used to display information. A change in color indicates a change in system status. Six colors are used by most EICAS displays.

- White is used for display of various scales, pointers and digital readouts when the system is in the normal operating range.

- Red is used to indicate that a system has exceeded a predetermined value (Redline). If a system exceeds its limits, the entire scale, pointer, and/or digital readout will turn red. Red is also used to show certain warning messages.

- Green is used to indicate normal operating of a system.

- Amber is used for certain warning messages and some scale markings. In some cases, a system display will change to amber if that system enters the caution range.

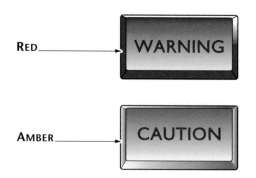

Figure 7-4-6. EICAS master warning/caution light assembly

- Magenta (pink) is used for certain messages and display parameters.
- Cyan (light blue) is used for various labels and messages.

Alert messages

During normal operation, EICAS is used to alert the flight crew as to any abnormal powerplant or airframe system operation. There are four types of alert messages used with EICAS.

- Level A warning messages are shown in red. These are the most important messages.
- Level B caution messages are displayed in amber.
- Level C advisories are displayed in amber or cyan depending on the specific model EICAS.
- Level D messages, called memos, are displayed in white.

Status and maintenance messages are also shown on some systems, and are typically displayed in white.

Warning messages. Warnings are known as level A messages. Warnings require immediate attention and immediate action by the flight crew. Level A faults include very serious failures, such as engine fire or cabin depressurization. Warnings will also activate additional aural and visual annunciators on the flight deck. For example, a fire bell sounds in the event of an engine fire. There are very few faults that require level A warnings.

Caution messages. Caution messages are known as level B alerts. Cautions appear on the EICAS CRT in amber just below any level A message. Level B alerts also create a distinct audio tone and illuminate a discrete annunciator. Cautions require immediate crew awareness and future crew action.

Advisory messages. Advisory messages (level C) require immediate crew awareness and possible future action. Memos are used for crew reminders. Typically there is no aural tone or master caution/warning associated with advisories or memos. For all levels of alerts, the most recent message appears at the top of its category. Level A messages appear at the top of the display, level B is below level A, level C is next and level D is shown on the bottom of the list.

Status and maintenance messages are used to aid the flight crew and maintenance technicians in determining the aircraft's status prior to dispatch. Maintenance messages used in conjunction with the minimum equipment list for the aircraft will determine what repairs (if any) are required prior to the next flight.

All EICAS equipped aircraft contain two master warning/caution annunciators. These annunciators consist of an illuminated switch assembly located on the instrument glare shield directly in front of the pilot and copilot.

The red warning is located in the top half of the assembly and the amber caution is located on the lower portion of the assembly (see Figure 7-4-6). The warning and caution lights illuminate whenever EICAS presents a level A and B message. If the assembly is pressed, the lamps are extinguished and any aural warning is canceled; however, the related EICAS message remains on the CRT display.

It is best not to distract the pilots with alert messages during certain phases of flight. Therefore, EICAS incorporates inhibit software to keep alert messages from being displayed during critical flight operations, such as landing and takeoff.

More information on glass cockpits and how they operate is available in the book, *Avionics: Systems and Troubleshooting*, by Tom Eismin.

Section 5

Interphone Systems

The interphone system is found on transport category aircraft to provide communications between flight crew, ground crew, flight attendants, and maintenance personnel. Due to the size of transport category aircraft, communications between different areas of the aircraft, or inside and outside of the plane are nearly impossible without the aid of the interphone system. See Figure 7-5-1. Considering that most airports are noisy, the interphone system becomes indispensable when performing maintenance and ground service activities.

The B747-400 interphone system is divided into four subsystems:

- Flight interphone
- Service interphone
- Crew call system
- Cabin interphone (see section 6)

The flight and service interphones, as well as the crew call system, operate in conjunction with the audio management unit. The cabin interphone system is not controlled directly by the audio management unit and will be discussed in the next section of this chapter.

Flight Interphone

The B747-400 flight interphone system is used to provide communications between flight crewmembers and/or other aircraft operations personnel. The flight interphone system is typically used for communications between flight crewmembers; however, the system may be interconnected to the service and cabin interphones when needed.

Microphones and headsets. There are several options regarding communication microphones (mics) and speakers/headsets. The captain and first officer choose from the following:

- Headset (mic/earphone combination)
- Hand-held mic
- Oxygen mask mic
- Headphones
- Cabin speaker

The headset contains an acoustic tube mic that carries sound waves to the transducer. The transducer converts the sound waves into an electrical signal, which is sent to the preamplifier located in the headset cord. The headset mic is keyed by one of three Press-To-Talk (PTT) switches. The headset earphone is a standard earphone assembly that allows the crewmember to monitor communications.

Controls. There are 13 audio receiver controls. Each receiver control is dedicated to a specific communication system. The receiver controls are a push on/off switch-volume control combination. When a receiver control is in the on position, the associated green indicator illuminates.

Both the captain's and first officer's control wheel contain a two-position press-to-talk switch. The switch control is a rocker assembly (trigger). Pressing the upper portion of the trigger keys the oxygen or headset mic to the selected communication system. Pressing the lower portion of the trigger will activate the interphone system regardless of the communication system selected on the Audio Control Panel.

Service Interphone

The service interphone system permits ground crew communications through various access points located throughout the aircraft. Technicians often use this portion of the interphone system during various maintenance operations. The service interphone may also be used during ground service operations. On the B747-400 there are 19 different service interphone connection points. The interphone connection points are combined into three groups: forward, mid, and aft. Each input connection of the group is paralleled and sent to the audio management unit.

The *Audio Management Unit* (AMU) receives input signals from the service interphone jacks. The AMU coordinates interphone system communications. The service interphone inputs are amplified by the AMU and sent back to the service interphone jacks or flight interphone system (if the service interphone switch is on).

Crew call system. The *crew call system* is used to alert ground crew personnel of an interphone message from the flight crew, and to alert the flight crew of a message from the ground crew. A horn, along with the interphone control panel is located in the nose wheel well. Pressing the flight deck call switch sends a signal to the Modularized Avionics and Warning Electronics Unit (MAWEA). MAWEA activates a chime through the flight deck speakers and illuminates a light on the audio control panel.

The ground crew call horn sounds any time a flight deck interphone call is made to the ground crew. The same horn is also used to alert the ground crew of an equipment cooling failure, or if the Inertial Reference (IR) unit is on and AC power is not supplied to the aircraft. If the IR is allowed to operate on battery power only, the battery will quickly discharge.

Interphone Maintenance and Troubleshooting

Maintenance and troubleshooting of the interphone system consists of wiring (connector) problems related to the various system's components, or removal and replacement of various LRUs. The AMU contains BITE circuitry that communicates with both Central Maintenance Computers (CMCs).

Mics and headsets are items that require regular maintenance. These items are constantly being moved, bounced, and dropped about the cabin. This causes wires to shake loose, to break, as well as intermittent audio transmissions. The easiest way to troubleshoot mics and headsets is to operate the system; shake the suspect unit and/or related wiring while listening for problems. Then, swap the suspect unit and try the system a second time. Service interphone connection jacks also seem to be problem components. Many of these jacks are exposed to the outside environment and are often treated roughly by ground personnel. In most cases, the jack becomes loose causing an intermittent connection. Be sure to inspect these components carefully when dealing with interphone faults.

Section 6

Advanced Cabin Entertainment Service System

The *Advanced Cabin Entertainment Service System* (ACESS) is used on the B747-400 to control five major subsystems: the passenger entertainment audio system, the cabin interphone system, cabin lighting, the passenger service system, and the passenger address system. In short, ACESS controls and distributes passenger audio and lighting. ACESS is software-driven and allows for improved cabin flexibility. That is, ACESS can easily reconfigure cabin audio and video signals to compensate for changes in cabin configuration. Through ACESS software, audio and video signals can be changed for virtually any seating arrangement. Older aircraft would have required changes in system wiring and components.

ACESS is a multiplexed digital system; therefore, the majority of ACCESS information is transmitted on a *serial data bus*. Transmission of serial digital data provides a great reduction in wiring; however, *multiplexer* and *demultiplexer* circuits are required. Multiplexers/demultiplexers are required at the beginning and end of each audio transmission route. Since the initial audio signals are analog, they must be converted to digital signals prior to multiplexing. (Multiplexing is the process of converting parallel digital data

Figure 7-5-1. Cabin communications, lighting, and rest rooms are controlled and/ or monitored from this panel.

into serial digital data.) After multiplexing, the data is then transmitted. After transmission, the data is demultiplexed. (Demultiplexing is the process of converting serial data into parallel data.) The digital to analog converter (D/A) is then used to change the digitized audio into an analog signal. The analog audio signal is then sent to the passenger headphones. Multiplexers and demultiplexers, as well as analog to digital (A/D) and D/A converters are typically circuits found within an LRU. These circuits are not stand-alone components.

The local area controllers (LACs) act as a distribution unit for each of the three main controllers and send lighting commands and digitized audio signals to the overhead electronic units and the seat electronic units for distribution to the individual passenger stations and related lighting.

The ACESS *Overhead Electronic Units* (OEU) control the overhead speakers, cabin lighting, passenger reading lights, flight attendant call lights, and passenger information lights (no smoking, and fasten seat belt signs). A maximum of 31 OEUs can be connected to one LAC. The Seat Electronic Units (SEU) are used to interface with each seat control for selecting various audio channels, volume, reading lights, and attendant call lights. The SEU also distributes the requested audio to each passenger headphone connection. One SEU is located under each group of seats, and can interface with up to four seats in a group. A maximum of 31 SEUs can connect to one OEU.

Configuration database. The configuration database is the software program which is used to inform ACESS of the current cabin layout. There are several LRUs within ACESS which contain the configuration database stored in a nonvolatile memory. The Central Management Unit (CMU) stores the configuration data base. The CMU also coordinates data base loading and storage for other ACESS LRUs.

The database can be used to identify items, such as seating configurations, cabin interphone dial codes, entertainment audio channels, passenger address areas, and the passenger address volume levels. This system provides flexibility to the airline for easy reconfiguration of the passenger compartment. The configuration database is typically modified by a shop technician or the engineering department using a personal computer. The database is then stored on a 3.5" floppy disk and transferred to the CMU using the aircraft's data loader. After the database has been downloaded to the CMU, the ACESS *Cabin Configuration Test Module* (CCTM) is used to download the database from the CMU to the various ACESS LRUs.

Cabin Interphone System

The *Cabin Interphone System* (CIS) provides communications between flight attendant stations, and from flight attendant stations to the flight deck. A handset, which resembles a telephone receiver, is located at each flight attendant's sta-

tion. From the flight deck, the pilots can interface with the CIS using either the flight deck, handset or the flight interphone system.

Cabin Lighting

The *Cabin Lighting System* (CLS) is part of ACESS which controls the following cabin lighting:

- Indirect ceiling lights
- Sidewall wash lights
- Direct ceiling lights
- Night lights

Fluorescent type lights are used for wall wash, indirect ceiling lights, and some night lighting. Incandescent lamps are used for direct lighting. The cabin lights are controlled by flight attendants using the cabin system modules.

The *Entertainment/Service Controller* (ESC) is the main ACESS control unit used to process cabin lighting commands. The Local Area Controllers (LAC) receive lighting selection commands from the Cabin Systems Module (CSM). These commands are sent to the ESC where they are distributed to the appropriate LACs. The LACs which receive the ESC commands send a signal to the appropriate Overhead Electronics Unit (OEU). The OEUs then turn on the appropriate cabin lights.

ACESS maintenance and troubleshooting. The maintenance and repair of ACESS is similar to multiplexed passenger address/entertainment/interphone systems found on other aircraft. In general, these multiplexed systems

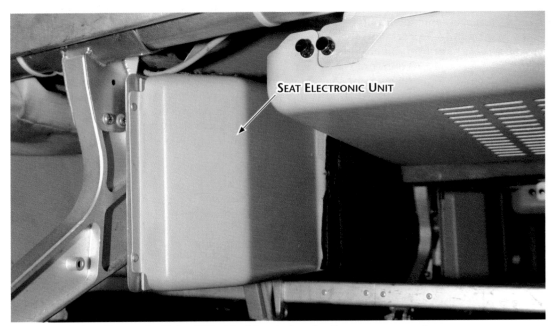

Figure 7-6-1. Seat Electronics Units (PES video) located under the passenger seats

are somewhat high maintenance simply due to their complexity. Hundreds of light bulbs and switches are linked by miles of wires controlled by dozens of electronic units. With this large number of components comes a massive number of connector plugs. Connector assemblies are often a weak link in any electronic system and the large number required for ACESS makes the system vulnerable to failures. Be sure to carefully inspect all suspect connectors whenever troubleshooting the system.

Most of the ACESS LRUs are located in the cabin behind decorative panels or mounted to seat structures. For example, *Overhead Electronic Units* (OEUs) are located behind ceiling panels directly above the passenger seating. The aircraft service manual will provide diagrams that can be used to locate various ACESS components.

On the B747-400, one electronic LRU is located under each group of passenger seats. *Seat Electronic Units* (SEUs) mount to the seat track on the aircraft floor. See Figure 7-6-1. Since passengers often rest their feet, or store baggage in this area, SEUs are often prone to wiring failures. As passengers move items (or their feet) near the SEUs, the wiring may get snagged. This causes connector problems or wiring failures due to the excess stress placed on the wiring. Whenever working on an SEU, be sure that all of the wiring is secured correctly that and all plastic covers are in place. This will help to prevent wire/connector damage.

Passenger Entertainment Video

The *Passenger Entertainment System Video* (PES video) is used to display video entertainment on various monitors and/or projectors located throughout the cabin. The system uses pre-recorded VHS format tapes that contain both the video and audio tracks. The audio data is transmitted to the passenger address controller for broadcast over cabin speakers, or to the Entertainment/Service Controller (ESC) for distribution to passenger headphones.

Architecture. The *Video System Control Unit* (VSCU) is the major control and distribution

unit for the entertainment video signals. The VSCU receives video and audio inputs from the *Video Tape Reproducer* (VTR) and a discrete signal from the decompression relay. The decompression relay monitors for sudden cabin decompression. In the event that cabin pressure is lost, all video is automatically turned off. The VSCU sends control signals to the VTR for automatic operation of the tape unit.

The VSCU is located under the stairway to the upper deck in the Video Control Center (VCC). A maximum of two video tape reproducers will also be located in the VCC. Storage for video cassettes and other materials is also located in the VCC.

Control. All video system operating controls are located on the front panel of the VSCU.

Maintenance and troubleshooting. The PES video system is a relatively simple system to maintain. Whenever troubleshooting the system, always remember which LRUs control and/or feed data to the unit in question. For example, if several monitors are inoperative, it is most likely that the area VDU which feeds those monitors is defective. Remember the VSCU controls the signals to the VDU. Therefore, if the VDU does not repair the system, the associated wiring or the VSDU should be suspected. In most cases, the LRUs are simply swapped with other units to troubleshoot the system. Connector pins are also likely fault areas. Be sure to inspect all electrical connections related to the failed system(s).

Like a household VCR, the videotape reproducer periodically requires cleaning, adjustment, or belt replacement. This is done in a service shop and the line technician simply removes and replaces the VTR. Both the video monitors and the projection units have various adjustments to improve picture quality. Brightness, color intensity, hue, and vertical hold are typically available on each monitor and projector. The projection units also have sharpness and convergence controls to adjust the picture clarity. The appropriate service information will completely describe the adjustment of these components.

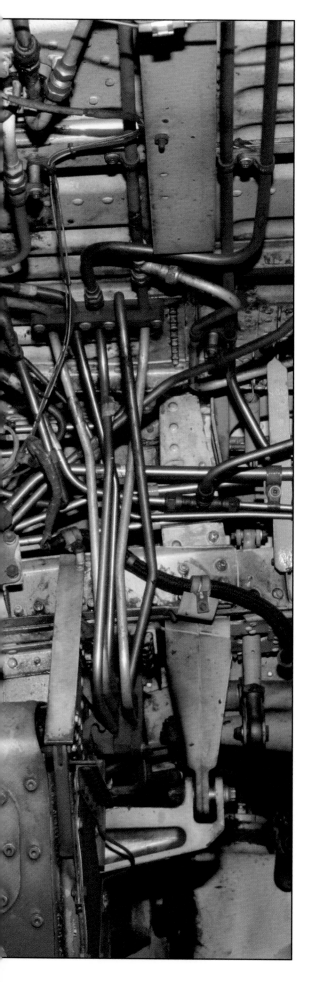

8

Fire Protection Systems

Section 1

Fire Protection Basics

Because fire is one of the most dangerous threats to an aircraft, the potential fire zones of larger aircraft are protected by a fixed fire protection system. A *fire zone* is an area or region of an aircraft designed by the manufacturer to require fire detection and/or fire extinguishing equipment and a high degree of inherent fire resistance. The term fixed describes a permanently installed system in contrast to any type of portable fire extinguishing equipment, such as a hand-held CO_2 fire extinguisher.

A complete fire protection system on modern aircraft and on many older model aircraft will include both a fire detection and a fire extinguishing system.

To identify fires or overheat conditions, detectors are placed in the various fire zones to be monitored. These units may detect temperatures above a pre-set level, rapid rise in temperature, or sense the presence of flames. These types of detectors are best suited for use in engine compartments.

Other types of detectors are also used in aircraft fire protection systems. For example, smoke detectors are better suited to monitor areas such as baggage compartments, where materials burn slowly, or smolder. Detectors in this category also include carbon monoxide detectors and chemical sampling equipment capable of detecting combustible mixtures that can lead to accumulations of explosive gases.

Additional information on fire protection, safety, and fire extinguishers may be found in

Left. With a variety of sensing devices, each tailored to a specific environment, today's fire warning and extinguishing systems are better and more dependable than ever before. The Freon agents, contained in this Boeing 737 fire bottle installation, are excellent extinguishants.

the Safety and Ground Handling chapter of *Aircraft Structural Maintenance*, .

Fire Zones

Compartments are classified into fire zones based on the airflow through them. These classifications will determine the type of detection and extinguishing equipment used in that area.

Class A zone. Zones having large quantities of air flowing past regular arrangements of similarly shaped obstructions. The power section of a reciprocating engine is usually of this type.

Class B zone. Zones having large quantities of air flowing past aerodynamically clean obstructions. Heat exchanger ducts and exhaust manifold shrouds are usually of this type (Figure 8-1-1). Also, zones where the inside of the enclosing cowling or other closure is smooth, free of pockets, and adequately drained, so leaking flammables cannot puddle, are of this type. Turbine engine compartments may be considered in this class if engine surfaces are aerodynamically clean and all airframe structural formers are covered by a fireproof liner to produce an aerodynamically clean enclosure surface.

Class C zone. Zones having relatively small air flow. An engine accessory compartment separated from the power section is an example of this type of zone.

Class D zone. Zones having very little or no air flow. These include wing compartments and wheel wells where little ventilation is provided.

Class X zone. Zones having large quantities of air flowing through them and are of unusual construction making uniform distribution of the extinguishing agent very difficult. Zones containing deeply recessed spaces and pockets between large structural formers are of this type. Tests indicate agent requirements to be double those for Class A zones.

Section 2

Fire Detection Systems

Considering the size and sophistication of modern aircraft, and the normal workload of the pilot and flight crew, it is possible that a fire could be well developed prior to being noticed. Fire detection systems are capable of monitoring many areas at the same time. This may include baggage compartments, wheel wells, and other areas not visible to the crew. Detection units are installed in locations where there are greater possibilities of a fire. These systems are designed to signal the presence of a fire early enough to provide the best opportunity for extinguishing it.

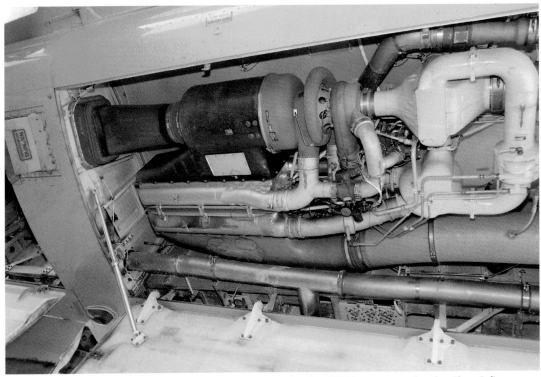

Figure 8-1-1. This air-cycle air-conditioning system in a Boeing 757 is an example of a Class B fire zone. Virtually every duct in this photograph has an overheat sensor.

Fire Detection Methods

Fires may be detected by crew or passenger observation, or using detectors employing a variety of technologies. The complete aircraft fire protection system will often incorporate several of these different detection methods.

System requirements. The ideal fire detection system will include as many as possible of the following features:

- A system which will not cause false warnings under any flight or ground operating conditions

- Rapid indication of a fire and accurate location of the fire

- Accurate indication that a fire is out

- Indication that a fire has re-ignited

- Continuous indication for duration of a fire

- Means for electrically testing fire detector system from the aircraft cockpit

- Detectors which resist exposure to oil, water, vibration, extreme temperatures, and maintenance handling

- Detectors that are light in weight and easily adaptable to any mounting position

- Detector circuitry that operates directly from the aircraft power system without inverters

- Minimum electrical current requirements when not indicating a fire

- Each detection system should actuate a cockpit light indicating the location of the fire and an audible alarm system

- A separate detection system for each engine

Common types. There are a number of detectors or sensing devices available and the aviation maintenance technician should be familiar with those most commonly seen. Many older model aircraft are equipped with some type of thermal switch system or thermocouple system.

Thermal switch system. Thermal switch, or spot detector, systems are made up of one or more lights and an audible warning device. They are powered by the aircraft electrical power system. Thermal switches trigger the warning device. These unit overheat detectors are heat sensitive devices that use contacts mounted on a bi-metallic strip to complete the circuit at a certain temperature. Non-adjustable units are available with factory-fixed settings from 100°F to 1,000°F. Adjustable units are also available for temperatures throughout this range.

The thermal switch system uses a bimetallic thermostat switch, or spot detector, similar to that shown in Figure 8-2-1 and 8-2-2. Each detector unit consists of a bimetallic thermoswitch. Most spot detectors are dual-terminal thermoswitches.

The exact number of thermal switches in the system is determined by the manufacturer and is normally covered by size of the area to be covered. In some installation, only one switch is used in the system.

Figure 8-2-3 shows that the switches are wired in parallel with each other, but in series with the warning device. When the temperature rises

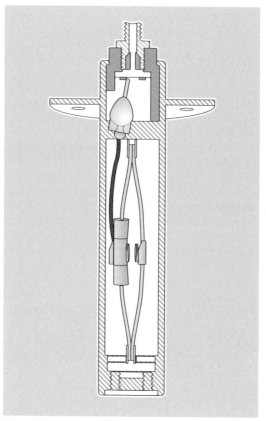

Figure 8-2-1. Fenwal spot detector

Figure 8-2-2. Adjustable type spot detector

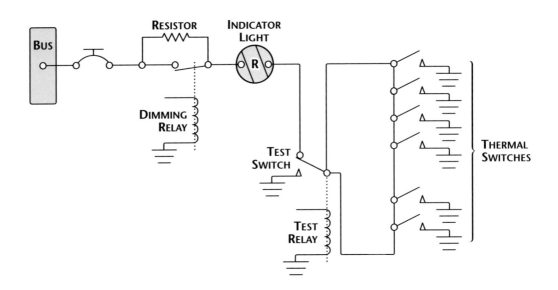

Figure 8-2-3. Thermal switch fire detection circuit

to the pre-set level, the switch will close causing the circuit to be complete, and the warning device to activate. These detectors can be connected in single-wire grounded circuits or double-wire circuits. In either system, each detector operated independently so that an open in a system conductor does not disable the detection system. Double wire systems add protection against false alarms due to wire grounding.

A double wire system is illustrated in Figure 8-2-4. As shown here the system can withstand one fault, either an electrical open circuit or a short to ground, without sounding a false fire warning. A double fault must exist before a false fire warning can occur. In case of a fire or overheat condition, the spot-detector switch closes and completes a circuit to sound an alarm.

Some warning lights are the push-to-test type. (See Figure 8-2-5). The bulb is tested by pushing it in to complete an auxiliary test circuit. The circuit in Figure 8-2-3 includes a test relay. With the relay contact in the position shown, there are two possible paths for current flow from the switches to the light. This is an additional safety feature. Energizing the test relay completes a series circuit and checks all the wiring and the light bulb.

Also included in the circuit shown in Figure 8-2-3 is a dimming relay. By energizing the dimming relay, the circuit is altered to include a resistor in series with the light. In some installations several circuits are wired through the dimming relay, and all the warning lights may be dimmed at the same time.

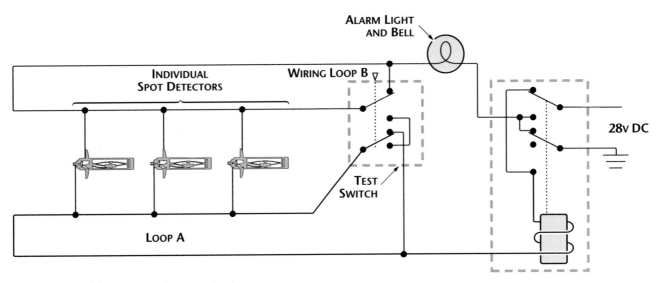

Figure 8-2-4. Double wire spot detector circuit

Thermocouple system. The *thermocouple*, or *rate-of-rise*, fire warning system operates on an entirely different principle than the thermal switch system. A thermocouple depends upon the rate of temperature rise and will not give a warning when an engine slowly overheats or a short circuit develops. The system consists of a relay box, warning lights, and thermocouples. The wiring system of these units may be divided into the following circuits (Figure 8-2-6):

- The detector circuit
- The alarm circuit
- The test circuit

The relay box contains two relays, the sensitive relay and the slave relay, and the thermal test unit. Such a box may contain from one to eight identical circuits, depending on the number of potential fire zones. The relays control the warning lights. In turn, the thermocouples control the operation of the relays. The circuit consists of several thermocouples in series with each other and with the sensitive relay.

The thermocouple is constructed of two dissimilar metals such as *chromel* and *constantan*. The point where these metals are joined and will be exposed to the heat of a fire is called a *hot junction*. There is also a *reference junction* enclosed in a dead air space between two insulation blocks. A metal cage surrounds the thermocouple to give mechanical protection without hindering the free movement of air to the hot junction.

If the temperature rises rapidly, the thermocouple produces a voltage because of the temperature difference between the reference junction and the hot junction. If both junctions are heated at the same rate, no voltage will result and no warning signal is given.

Figure 8-2-5. Warning light

If there is a fire, however, the hot junction will heat more rapidly than the reference junction. The ensuing voltage causes a current to flow within the detector circuit. Any time the current is greater than 4 milliamperes (0.004 ampere), the sensitive relay will close. This will complete a circuit from the aircraft power system to the coil of the slave relay which closes and completes the circuit to the fire-warning light.

The total number of thermocouples used in individual detector circuits depends on the size of the fire zone and the total circuit resistance. The total resistance usually does not exceed 5 ohms. As shown in Figure 8-2-6, the circuit has two resistors. The resistor connected across the terminals of the slave relay absorbs the coil's self-induced voltage. This is to prevent arcing across the points of the sensitive relay; since the contacts of the sensitive relay are so fragile they would burn or weld if arcing were permitted.

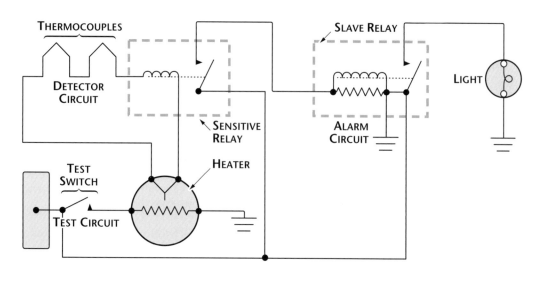

Figure 8-2-6. Thermocouple fire warning circuit

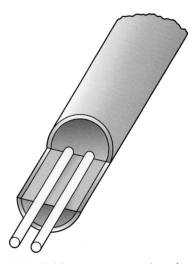

Figure 8-2-7. Kidde Aerospace sensing element

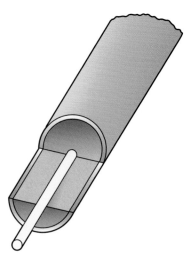

Figure 8-2-9. Fenwal sensing element

When the sensitive relay opens, the circuit to the slave relay is interrupted and the magnetic field around its coil collapses. When this happens, the coil gets a voltage through self-induction, but with the resistor across the coil terminals, there is a path for any current flow as a result of this voltage. Thus, arcing at the sensitive relay contacts is eliminated.

Continuous loop system. Greater, more complete coverage of a fire hazard can be achieved by the use of a *continuous overheat detection system*. The *continuous-loop* sensing element works much like a thermal switch, in that it completes a circuit at a certain temperature. The system is heat sensitive and provides no rate of temperature rise protection. The two commonly used types of continuous-loop systems are the *Kidde* and *Fenwal* systems.

A continuous-loop detector or sensing system permits more complete coverage of a fire hazard area than any type of spot-type temperature detectors. They are overheat detection systems where heat-sensitive units will complete an electrical circuit at a certain temperature. There is no rate-of-heat-rise sensitivity in a continuous-loop system.

Figure 8-2-8. Kidde control box

Kidde system. In the Kidde Aerospace continuous-loop system (Figure 8-2-7), two wires are imbedded in a special thermistor core within an Inconel tube. Figure 8-2-8 shows a Kidde control box.

One of the two wires in the Kidde sensing system is welded to the case at each end and acts as an internal ground. The second wire is a hot lead (above ground potential) that provides a current signal when the core material changes its resistance with a change in temperature.

When a fire or overheat condition exists and the resistance of the thermistor material decreases, a current flows between the two wires. The control box receives this signal and activates the alarm system and fire warning lights.

Fenwal system. Another continuous-loop system, manufactured by Fenwal Safety Systems (Figure 8-2-9), consists of a small, lightweight, flexible Inconel tubing with a single wire center conductor. The tubing is packed with insulation, a mineral based salt compound, and hermetically sealed.

The salt possesses the characteristic of suddenly lowering its electrical resistance as the sensing element reaches its alarm temperature. In both the Kidde and the Fenwal systems, the resistance of the ceramic or eutectic salt core material prevents electrical current from flowing at normal temperatures. In case of a fire or overheat condition, the core resistance drops and current flows between the signal wire and ground, energizing the alarm system.

Both systems continuously monitor temperatures in the affected compartments, and both will automatically reset following a fire or overheat alarm after the overheat condition is removed or the fire extinguished.

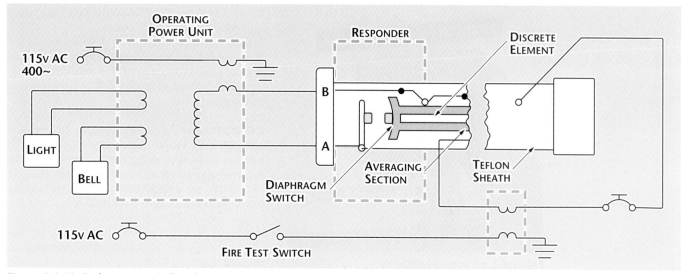

Figure 8-2-10. Early pneumatic fire detection system schematic

Pneumatic detection systems. Manufactured under several names (*Lindberg, Systron Donner, Meggitt*) over the years, this system (Figure 8-2-10) is a continuous-element type detector consisting of a stainless steel tube containing a discrete element. This element has been processed to absorb gas in proportion to the operating temperature set point. When the temperature rises (due to a fire or overheat condition) to the operating temperature set point, the heat generated causes the gas to be released from the element. Release of the gas causes the pressure in the stainless steel tube to increase. This pressure rise mechanically actuates the diaphragm switch in the responder unit, activating the warning lights and an alarm bell. The pressure diaphragm is the only moving part in the system.

A fire test switch is used to heat the sensors, expanding the trapped gas. The pressure generated closes the diaphragm switch, activating the warning system.

False alarms are a major concern with any fire detection system. The current generation of pneumatic systems are designed so that mechanical damage to the sensor tube cannot result in a false alarm. Any severe damage to the unit will provide a "no test" indication, not a false alarm.

This type of sensor is capable of two operating modes. It will respond to an overall *average high temperature* in the compartment, or to a *localized discrete temperature* caused by flame or hot gasses. Both the average and discrete temperatures are factory set and cannot be changed in the field.

The averaging function is activated by the expansion of a fixed volume of helium gas inside the detector. The pressure inside the detector will increase in proportion to the absolute temperature and will activate the alarm switch at a pre-set average temperature. This is typically between 200°F and 700°F.

Discrete sensing is accomplished using a hydrogen filled core material in the sensor tube. Large quantities of hydrogen gas are released from the detector core whenever a small section of the tube is heated to the pre-set *discrete temperature* or higher. This outgassing in the core will increase the pressure inside the detector and activate the alarm switch.

Both the averaging and discrete functions are reversible. When the sensor tube cools, the average gas pressure is reduced, and the discrete hydrogen gas returns to the core material. The reduction in pressure allows the alarm switch to return to its normal position, opening the alarm circuit.

In addition to the pressure-activated alarm switch, there is a second switch in the system.

Figure 8-2-11. Infrared detector

Figure 8-2-12. Infrared amplifier

This switch is held closed by the averaging gas pressure at all temperatures down to -65°F. If the detector should develop a leak, the loss of gas pressure would allow the integrity switch to open. The system then will not operate during a test.

Infrared detection systems. Infrared (IR) detection systems work on the principle that a fire gives off a large amount of infrared radiation. Detectors require a control amplifier connected to system voltage, and to a cockpit warning alarm (Figure 8-2-11).

Classic examples of infrared detectors are found in King Air engine nacelles. The warning system consists of three units, each with its own control amplifier as shown in Figure 8-2-12. One unit will cover the forward nacelle, one unit the upper accessory area, and one unit the lower accessory area. Radiation exposure will activate the relay circuit of the control amplifier and send power to the cockpit fire warning

alarm. Once the fire is extinguished the system will return to its standby state.

The systems are tested by closing a switch in the cockpit. When closed the warning should go off. It should stop when the test switch is opened.

False alarms. Being sensitive to IR the system can give a false indication if tested on the ground and the cowling is not closed properly. It can also produce an error if very strong light (sunshine or strong work lights) is shining directly on the nose of the airplane.

Aside from exposure to incidental IR, the systems are simple, dependable, and easy to maintain.

Overheat warning system. Overheat warning systems are used on some aircraft to indicate high area temperatures that may lead to a fire. The number of overheat warning systems varies with the aircraft. On some aircraft, they are provided for each engine turbine and each nacelle; on others they are provided for wheel well areas and for the pneumatic manifold. When an overheat condition occurs in the detector area, the system causes a light on the fire control panels to flash.

Most systems utilize some type of unit overheat detector. Each detector is operated when the heat rises to a specified temperature. This temperature depends upon the system and the type and model of the aircraft. The switch contacts of the detector are on spring struts, which close whenever the meter case is expanded by heat. One contact of each detector is grounded through the detector-mounting bracket. The other contacts of all detectors connect in parallel to the closed loop of the warning light circuit. Thus, the closed con-

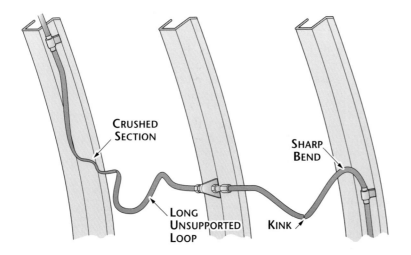

Figure 8-2-13. Sensing element defects

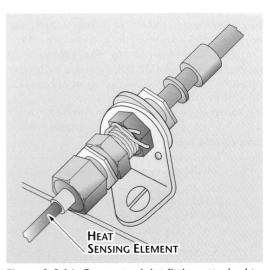

Figure 8-2-14. Connector joint fitting attached to the structure

tacts of any one detector can cause the warning lights to burn.

When the detector contacts close, a ground is provided for the warning light circuit. Current then flows from an electrical bus through the warning lights and a flasher or keyer to ground. Because of the flasher in the circuit, the lights flash on and off to indicate an overheat condition.

Control/Interface Electronics. Control units for these systems may be found in several different generations. The earliest systems utilize magnetic amplifiers and relays to control cockpit indications. Later electronic control units provide the necessary output to indicate fire or overheat conditions. The newest designs include microprocessor controls.

Digital electronic control units can now monitor the overheat and fire detection systems for overheats, fire, and fault conditions and report them over the aircraft's communication bus to the maintenance computer. Extensive Built-In-Test features with component fault/event location features may be found on some installation.

Fire Detection System Inspection and Maintenance

Fire detector sensing elements are located in many high-activity areas around aircraft engines. Their location, together with their small size, increases the chances of damage to the sensing elements during maintenance. The installation of the sensing elements inside the aircraft cowl panels provides some measure of protection not afforded elements attached directly to the engine. On the other hand, the removal and re-installation of cowl panels can easily cause abrasion or structural defects to the elements.

A well-rounded inspection and maintenance program for all types of continuous loop systems should include the following visual checks. These procedures are provided as examples and should not be used to replace approved local maintenance directives or the applicable manufacturer's instructions. Fire detection system maintenance can be found in manuals using the *ATA System* under System 26, subcode 10.

Sensing elements should be inspected for:

- Cracked or broken sections caused by crushing or squeezing between inspection plates, cowl panels, or engine components.

- Abrasion caused by rubbing of element on cowling, accessories, or structural members.

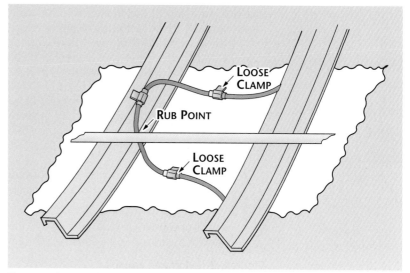

Figure 8-2-15. Rubbing interference

- Pieces of safety wire or other metal particles which may short the spot detector terminals.

- Condition of rubber grommets in mounting clamps, which may be softened from exposure to oils, or hardened from excessive heat.

- Dents and kinks in sensing element sections. Limits on the element diameter, acceptable dents or kinks, and degree of smoothness of tubing contour are specified by manufacturers. No attempt should be made to straighten any acceptable dent or kink, since stresses may be set up that could cause tubing failure. (See illustration of kinked tubing in Figure 8-2-13.)

- Loose nuts or broken safety wire at the end of the sensing elements (Figure 8-2-14). Loose nuts should be re-torqued to the value specified in the manufacturer's instructions. Some types of sensing element connections require the use of copper crush gaskets. These gaskets should be replaced any time a connection is separated.

- Broken or frayed flexible leads, if used. The flexible lead is made up of many fine metal strands woven into a protective covering surrounding the inner insulated wire. Continuous bending of the cable or rough treatment can break these fine wires, especially those near the connectors. Broken strands can also protrude into the insulated gasket and short the center electrode.

- Rubbing between a cowl brace and a sensing element (Figure 8-2-15). This interference, in combination with loose rivets holding the clamps to skin, may cause wear and short the sensing element.

- Proper sensing element routing and clamping (Figure 8-2-16). Long unsupported sections may permit excessive vibration which can cause breakage. The distance between clamps on straight runs is usually about 8 to 10 inches, and is specified by each manufacturer. At end connectors, the first support clamp is usually located about 4 to 6 inches from the end connector fittings. In most cases, a straight run of 1 inch is maintained from all connectors before a bend is started, and an optimum bend radius of 3 inches is normally adhered to.

- Correct grommet installation. The grommets are installed on the sensing element to prevent the element from chafing on the clamp. The slit end of the grommet should face the outside of the nearest bend. Clamps and grommets (Figure 8-2-16) should fit the element snugly.

Thermocouple detector mounting brackets should be repaired or replaced when cracked, corroded, or damaged. When replacing a thermocouple detector, note which wire is connected to the identified positive terminal of the defective unit and connect the replacement in the same way.

Test the fire detection system for proper operation by turning on the power supply and placing the fire detection test switch in the TEST position. The red warning light should flash on within the time period established for the system. On some aircraft, an audible alarm will also sound.

In addition, the fire detection circuits are checked for specified resistance and for an open or grounded condition. Tests required after repair or replacement of units in a fire detection system or when the system is inoperative include:

- Checking the polarity, ground, resistance and continuity of systems that use thermocouple detector units.

- Resistance and continuity tests performed on systems with sensing elements or cable detector units. In all situations, follow the recommended practices and procedures of the manufacturer of the type system with which you are working.

Fire Detection System Troubleshooting

Fire detection systems are generally reliable, and have minimal maintenance requirements. They do, occasionally, suffer damage that renders them inoperable. The following troubleshooting procedures represent the most common difficulties encountered in engine fire detection systems.

Intermittent alarms are most often caused by an intermittent short in the detector system wiring. Such shorts may be caused by a loose wire which occasionally touches a nearby terminal, a frayed wire brushing against a structure, or a sensing element rubbing long enough against a structural member to wear through the insulation. Intermittent faults can often best be located by moving wires to recreate the short.

Fire alarms and warning lights can occur when no engine fire or overheat condition exists. Such false alarms can most easily be located by disconnecting the engine sensing loop from the aircraft wiring. If the false alarm continues, a short must exist between the loop connections and the control unit.

If, however, the false alarm ceases when the engine sensing loop is disconnected, the fault is in the disconnected sensing loop, which should be examined for areas which have been bent into contact with hot parts of the engine. If no bent element can be found, the shorted section can be located by isolating and disconnecting elements consecutively around the entire loop.

Kinks and sharp bends in the sensing element can cause an internal wire to short intermittently to the outer tubing. The fault can be located by checking the sensing element with a megger while tapping the element in the suspected areas to produce the short.

Moisture in the detection system seldom causes a false fire alarm. If, however, moisture does cause an alarm, the warning will persist until the contamination is removed or boils away and the resistance of the loop returns to its normal value.

Failure to obtain an alarm signal when the test switch is actuated may be caused by a defective test switch or control unit, the lack of elec-

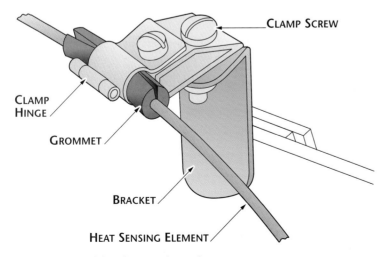

CLAMP SCREW

CLAMP HINGE

GROMMET

BRACKET

HEAT SENSING ELEMENT

Figure 8-2-16. Typical fire detector loop clamp

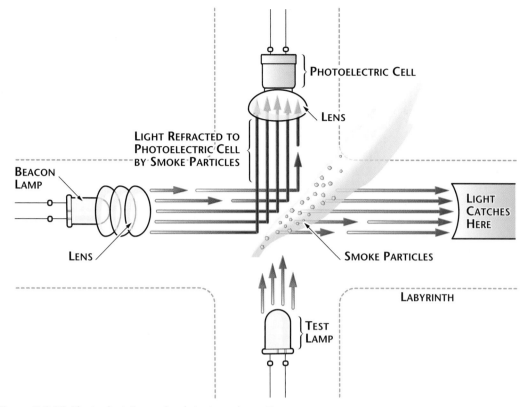

Figure 8-2-17. Photoelectric smoke detector schematic

trical power, inoperative indicator light, or an opening in the sensing element or connecting wiring. When the test switch fails to provide an alarm, the continuity of a two-wire sensing loop can be determined by opening the loop and measuring the resistance. In a single-wire, continuous-loop system, the center conductor should be grounded.

Smoke Detection Systems

Smoke detection is the primary means of fire detection used in cargo compartments. These are areas where large amounts of smoke may be present before a substantial amount of heat can be detected.

Smoke detection instruments are classified by method of detection as follows:

- Type I - Measurement of carbon monoxide gas (CO detectors)

- Type II - Measurement of light transmissibility in air (photoelectric devices)

- Type III - Visual detection of the presence of smoke by directly viewing air samples (visual devices).

While solid state electronics, new optics and new processing algorithms have been introduced, the basic mechanism used by these detectors has changed little in the past 50 years.

Visual smoke detectors. On a few older aircraft, visual smoke detectors provide the only means of smoke detection. Indication is provided by drawing smoke through a line into the indicator, using either a suitable suction device or cabin pressurization.

When smoke is present, a lamp within the indicator is illuminated automatically by the smoke detector. The light is scattered so that the smoke is rendered visible in the appropriate window of the indicator. If no smoke is present, the lamp will not be illuminated. A switch is provided to illuminate the lamp for test purposes. A device is also provided in the indicator to show that the necessary airflow is passing through the indicator.

Ionization type detector. Ionization smoke detectors were used in the early years, but are much less common today. The typical unit used is a radioactive isotope as the source to charge the combustion products (smoke). However, this source also charges everything else - including dust and fine water droplets. This resulted in frequent false alarms. The sensitivity of these detectors change with pressure (altitude) and age. As the units get older they are more prone to false alarms.

This technology has not been completely abandoned and may be found in lavatories and some cargo bays. Generally, these are more accessible areas where a false alarm can be verified.

Figure 8-2-18. Photoelectric smoke detector

to ground. A fire indication will be observed only if the beacon and test lamp, the photoelectric cell, the smoke detector amplifier, and associated circuits are operable. The amplifier contact relay takes the low voltage signal from the smoke detector and amplifies it. It then activates the fire warning system. (Figure 8-2-20).

Newer generations of photoelectric smoke detectors use a pulsed, infrared LED light source with a synchronized silicon photodiode receiver. These units feature solid state optics and circuitry and provide long service life with high reliability. Modern electronics also mean lower power requirements in this equipment.

Carbon monoxide detectors. Carbon monoxide is a by-product of incomplete combustion and is a very deadly gas. Carbon monoxide is a colorless, odorless, and tasteless gas. It is very dangerous and may cause mental dullness, physical weariness and headaches within a few hours, even at concentrations of 0.02% (2 parts per 10,000). The maximum allowable concentration, under Federal Law, for continuing exposure is 50 p.p.m. (parts per million) which is equal to 0.005% of carbon monoxide. (See Table 8-2-1.)

There are several types of portable testers (sniffers) in use. One type has a replaceable indicator tube which contains a yellow silica gel, impregnated with a complex silico-molybdate compound and is catalyzed using palladium sulfate.

In use, a sample of air is drawn through the detector tube. When the air sample contains carbon monoxide, the yellow silica gel turns to a shade of green. The intensity of the green color is proportional to the concentration of

Photoelectric type detectors. Early versions of this type of detector consisted of a photoelectric cell, a beacon lamp, a test lamp, and a light trap, all mounted on a labyrinth. An accumulation of 10% smoke in the air causes the photoelectric cell to conduct electric current. Figure 8-2-17 shows the details of the smoke detector, and indicates how the smoke particles refract the light to the photoelectric cell. Figure 8-2-18 shows a photoelectric smoke detector from a Beechcraft King Air. When activated by smoke, the detector supplies a signal to the smoke detector amplifier. The amplifier signal activates a cockpit warning light.

A test switch (Figure 8-2-19) permits checking the operation of the smoke detector. Closing the switch connects 28 VDC to the test relay. When the test relay energizes, voltage is applied through the beacon lamp and test lamp in series

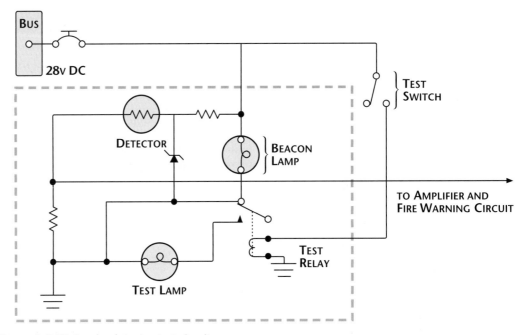

Figure 8-2-19. Smoke detector test circuit

carbon monoxide in the air sample at the time and location of the tests.

Another type indicator may be worn as a badge or installed on the instrument panel or cockpit wall. (See Figure 8-2-21) It is a button using a tablet which changes from a normal tan color to progressively darker shades of gray to black. The transition time required is relative to the concentration of CO. At a concentration of 50 p.p.m. CO (0.005%), the indication will be apparent within 15 to 30 minutes. A concentration of 100 p.p.m. CO (0.01%) will change color of the tablet from tan to gray in two to five minutes, from tan to dark gray in 15 to 20 minutes.

Optical flame detection. A relatively new development is the optical flame detection unit. These sensors use the infrared light band to sense a fire. Infrared energy generated by the excitation of CO_2 molecules in a hydrocarbon fire impinges upon the sensor, which in turn produces a small electrical signal. This signal is then amplified and digitally processed. Signal processing circuits are able to discriminate between the flicker of a fire and background sources such as a hot engine, or non-fire sources such as natural or man-made light.

Section 3

Fire Extinguishing Systems

Once a fire has been detected, it is important that it be extinguished as quickly as possible. This is as true in flight as it is on the ground. Fire extinguishing systems may be installed on the aircraft to deal with in-flight fires. Both the selection of the extinguishing agent, and the method of delivery are critical to the operation of the system.

Extinguishing Agents

Aircraft fire extinguishing agents have some common characteristics which make them compatible to aircraft fire extinguishing systems. All agents may be stored for long time periods without adversely affecting the system components or agent quality. Agents in current use will not freeze at normally expected atmospheric temperatures. The nature of the devices inside a powerplant compartment require agents that are not only useful against flammable fluid fires, but also effective on electrically caused fires. Agents are classified into two general categories based on the mechanics of extinguishing action: the inert cold gas agents and the halogenated hydrocarbon agents.

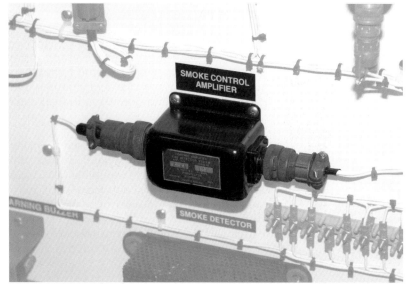

Figure 8-2-20. Photoelectric smoke control amplifier

PARTS PER MILLION	%	REACTION
50	0.005%	Maximum allowable concentration under Federal Law
100	0.01%	Tiredness, mild dizziness
200	0.02%	Headaches, tiredness, dizziness, nausea after 2 or 3 hours
800	0.08%	Unconsciousness in 1 hour and death in 2 to 3 hours
2,000	0.20%	Death after 1 hour
3,000	0.30%	Death within 30 minutes
10,000	1.00%	Instantaneous death

Note: The maximum allowable concentration under Federal Law for continuing exposure is 50 ppm (parts per million) which is equal to 0.005% of carbon monoxide

Table 8-2-1. Human reaction to carbon monoxide poisoning

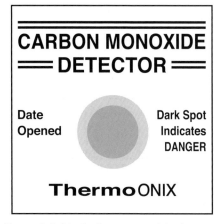

Figure 8-2-21. Carbon monoxide detector

Inert Cold Gas Agents

Both carbon dioxide (CO_2) and nitrogen (N_2) are effective extinguishing agents. Both are readily available in gaseous and liquid forms; their main difference is in the temperatures and pressures required to store them in their compact liquid phase.

Carbon dioxide. CO_2 has been used for many years to extinguish flammable fluid fires and fires involving electrical equipment. It is not combustible and does not react with most substances. It provides its own pressure for discharge from the storage vessel except in extremely cold climates where a booster charge of nitrogen may be added to winterize the system.

Normally, CO_2 is a gas, but it is easily liquefied by compression and cooling. After liquefaction, CO_2 will remain in a closed container as both liquid and gas. When CO_2 is then discharged to the atmosphere, most of the liquid expands to gas. Heat absorbed by the gas during vaporization cools the remaining liquid to -110°F and it becomes a finely divided white solid, dry ice snow. CO_2 is about $1\ ^1/_2$ times as heavy as air, which gives it the ability to replace air above burning surfaces and maintain a smothering atmosphere.

CO_2 is effective as an extinguishing agent primarily because it dilutes the air and reduces the oxygen content so that the air will no longer support combustion. Under certain conditions some cooling effect is also realized. CO_2 is considered only mildly toxic, but it can cause unconsciousness and death by suffocation if the victim is allowed to breath CO_2 in fire extinguishing concentrations for 20 to 30 minutes. CO_2 is not effective on fires involving chemicals containing their own oxygen supply, such as cellulose nitrate (some aircraft paints). Also fires involving magnesium and titanium (used in aircraft structures and assemblies) cannot be extinguished by CO_2.

Nitrogen. N_2 is an even more effective extinguishing agent. Like CO_2, N_2 is an inert gas of low toxicity. N_2 extinguishes by oxygen dilution and smothering. It is hazardous to humans in the same way as CO_2. But more cooling is provided by N_2 and pound for pound, N_2 provides almost twice the volume (of inert gas) to the fire as CO_2 resulting in greater dilution of oxygen. The main disadvantage of N_2 is that it must be stored as a cryogenic liquid which requires a Dewar bottle and associated plumbing to maintain the -320°F temperature of liquid nitrogen ($L\ N_2$).

Some large military aircraft use $L\ N_2$ in several ways. $L\ N_2$ systems are primarily utilized to inert the atmosphere in the fuel tank by replacing most of the air with dry gaseous nitrogen, thereby diluting the oxygen content. With the large quantities of $L\ N_2$ thus available, N_2 is also being used for aircraft fire control and is feasible as a practical powerplant fire-extinguishing agent. A long-duration $L\ N_2$ system discharge can provide greater safety than conventional short-duration system by cooling potential reignition sources and reducing the vaporization rate of any flammable fluids remaining after extinguishment.

Halogenated Hydrocarbon Agents

The most effective fire extinguishing agents are compounds formed by replacement of one or more of the hydrogen atoms in the simple hydrocarbons methane and ethane by halogen atoms. The probable extinguishing mechanism of Halon is a chemical interference in the combustion process between fuel and oxidizer.

Experimental evidence indicates that the most likely method of transferring energy in the combustion process is by molecule fragments resulting from the chemical reaction of the constituents. If these fragments are blocked from transferring their energy to the unburned fuel molecules the combustion process may be slowed or stopped completely (extinguished). It is believed that the Halogenated agents react with the molecular fragments, thus preventing the energy transfer. This may be termed chemical cooling or energy transfer blocking. This extinguishing mechanism is much more effective than oxygen dilution and cooling.

Halon. These agents are classified through a system of Halon numbers which describes the several chemical compounds making up this family of agents. The first digit represents the number of carbon atoms in the compound molecule; the second digit, the number of fluorine atoms; the third digit, the number of chlorine atoms; the fourth digit, the number of bromine atoms; and the fifth digit, the, number of iodine atoms, if any. Terminal zeroes are not expressed. For example, bromotrifluoromethane ($CBrF_3$) is referred to as Halon 1301.

While very popular and excellent performers production of Halon has been banned by international agreement. This is because of the ozone depleting characteristics of these chemicals. However, the use of Halons is still permitted for essential applications, such as aircraft, until a suitable replacement agent can be developed, approved, and certified for aircraft use.

Until that time, existing stocks of Halon, including that which is recovered from decommissioned industrial and consumer fire protection systems may be used. These stocks should be sufficient for many years of aircraft production and use. It is necessary, however, to restore the Halon to its original purity before re-using it in aircraft systems.

Halon recycling machines have been developed to perform this function. Used Halon is recycled, reconditioned, and then stored in a Halon bank for future use.

While this may solve the immediate problem, it does not mean that there is no need to develop a replacement for Halon. Research also needs to be pursued to develop alternative methods of extinguishing in-flight fires.

PhostrEx™. After several years of research and development, a new extinguishing agent has been produced. PhostrEx™, developed by Eclipse Aviation, is an alternative to Halon 1301. After extensive testing, Eclipse was able to meet all of the 14 CFR part 23 requirements for extinguishing agents, thereby gaining FAA acceptance of the agent. According to the testing and analysis, PhostrEx™ has no Ozone Depletion Potential (ODP) or Global Warming Potential (GWP) as defined by the Montreal Protocol and can be used without any of the associated climate and ecological consequences of the Halon agents.

Because of the chemical makeup of the agent and how it reacts during the combustion process, it is more effective than Halon 1301 in extinguishing on-board engine fires. The storage requirements are considerably less, resulting in a substantial weight and space savings. The projected replacement life for the containers is ten years, at which time they are replaced by a technician.

As is the case with all systems, the maintenance requirements for the system and the components are stipulated by the manufacturers in their maintenance manuals.

Types of Extinguishing Systems

On board fire extinguishing systems have tubing runs that carry the extinguishing agent. The identification tape marking for the tubing is brown.

Conventional systems. This term is applied to those fire-extinguishing installations first used in aircraft. Still used in some older aircraft, the systems are satisfactory for their intended use but are not as efficient as newer designs. Typically these systems utilize the perforated ring and the so-called distributor-nozzle discharge arrangement. One applica-

tion is that of a perforated ring in the accessory section of a reciprocating engine where the airflow is low and distribution requirements are not severe.

The distributor-nozzle arrangements are used in the power section of reciprocating engine installations with nozzles placed behind each cylinder and in other areas necessary to provide adequate distribution. This system usually uses carbon dioxide (CO_2) as the extinguishing agent, but may use any other adequate agent.

High-rate-of-discharge systems. This term, abbreviated HRD, is applied to the highly effective systems most currently in use. Such HRD systems provide high discharge rates through high pressurization, short feed lines large discharge valves and outlets. The extinguishing agent is usually one of the Halogenated hydrocarbons (Halons) sometimes boosted by high-pressure dry nitrogen (N_2). Because the agent and pressurizing gas of an HRD system are released into the zone in one second or less, the zone is temporarily pressurized and interrupts the ventilating airflow. The few large sized outlets are carefully located to produce high velocity swirl effects for best distribution.

Conventional System Installations

CO_2 is one of the earliest types of fire extinguisher systems for transport aircraft and is still used on many older aircraft.

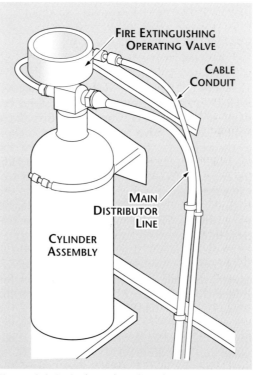

Figure 8-3-1. Carbon dioxide cylinder installation

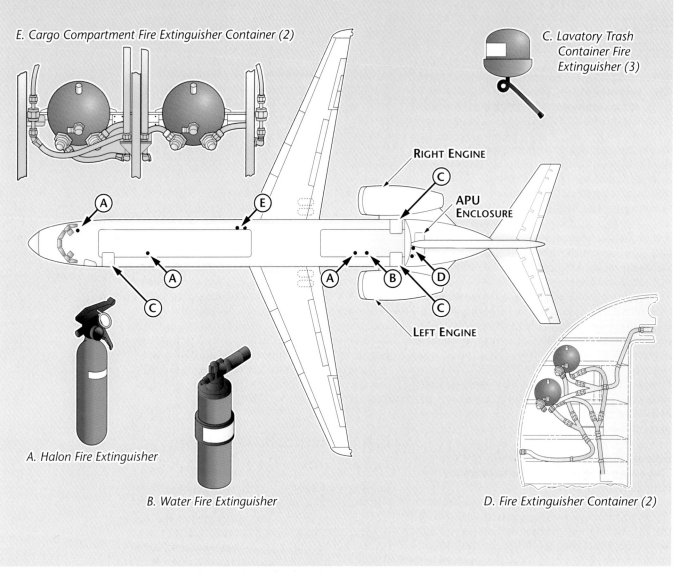

E. Cargo Compartment Fire Extinguisher Container (2)

C. Lavatory Trash Container Fire Extinguisher (3)

RIGHT ENGINE

APU ENCLOSURE

LEFT ENGINE

A. Halon Fire Extinguisher

B. Water Fire Extinguisher

D. Fire Extinguisher Container (2)

Figure 8-3-2. Transport category aircraft fire extinguisher locations

This fire extinguisher system is designed around a cylinder (Figure 8-3-1) that stores the flame-smothering CO_2 under pressure and a remote control valve assembly in the cockpit to distribute the extinguishing agent to the engines. The gas is distributed through tubing from the CO_2 cylinder valve to the control valve assembly in the cockpit, and then to the engines via tubing installed in the fuselage and wing tunnels. The tubing terminates in perforated loops which encircle the engines.

To operate this type of engine fire extinguisher system, the selector valve must be set for the engine that is on fire. An upward pull on the T-shaped control handle located adjacent to the engine selector valve actuates the release lever in the CO_2 cylinder valve. The compressed liquid in the CO_2 cylinder flows in one rapid burst to the outlets in the distribution line of the affected engine. Contact with the air converts the liquid into gas and snow that smothers the flame.

High-Rate-of-Discharge-Installations

The high-rate-of-discharge systems use Freon or Halon 1301 as the extinguishing agent. The agent is stored in cylindrical or spherical containers that are pressure tested to 1,500 p.s.i. In addition to holding the extinguishing agent, the containers are pressurized to 300 p.s.i. with nitrogen. This nitrogen charge is added to the container to ensure quick expulsion of the extinguishing agent. The bottles are activated by an explosive cartridge (squib) and discharge through tubing designed to allow rapid release of the agent. This release floods the protected compartment and extinguishes the fire. The advantage of this type of system is the speed at which the agent is delivered.

Typical Transport Category Aircraft Fire Extinguishing System

Transport category aircraft must be equipped to deal with a variety of fire hazards and carry the equipment necessary to extinguish fires, if they occur. This generally means a combination of on-board extinguishing systems and portable fire extinguishers to be used by the crew. Figure 8-3-2 provides a general idea of the fire extinguishers found on this category of aircraft.

The engine and APU fire extinguishers can be activated by the flight crew. These containers are located in the aft accessory compartment and may be discharged into the left engine, right engine, or APU enclosure.

The cargo compartment fire extinguisher system allows the flight crew to release the extinguishing agent into either the forward of aft cargo compartment. This system also uses two containers.

Each of the three lavatory trash containers are equipped with a self-contained automatic fire extinguishing system. When a trash fire occurs and the temperature exceeds the set limit, the fire extinguisher automatically discharges the extinguishing agent into the trash container. There are no flight compartment or cabin attendant alerts for this system.

There are four portable fire extinguishers on the aircraft. Three are of the Halon type and are located in the flight compartment and cabin. Additionally, in the cabin there is one water-type extinguisher for Class A fires.

Turbine Engine Ground Fire Protection

The problem of ground fires has increased in seriousness with the increased size of turbine engine aircraft. For this reason, means are usually provided for rapid access to the compressor, tailpipe, and/or burner compartments. Thus, many aircraft systems are equipped with spring-loaded access doors in the skin of the various compartments. Such doors are usually located in accessible areas, but not in a region where opening a door might spill burning liquids on the fire fighter.

Internal engine tailpipe fires that take place during engine shutdown or false starts can be blown out by motoring the engine with the starter. If the engine is running, it can be accelerated to a higher r.p.m. to achieve the same result. If such a fire persists, a fire extinguish-

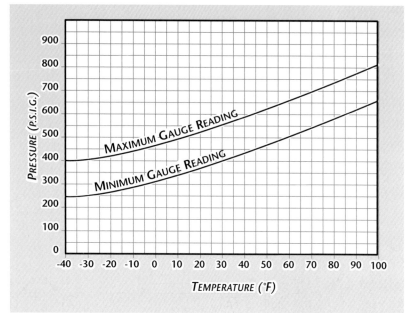

Figure 8-3-3. Fire extinguisher container pressure/temperature curve.

ing agent can be directed into the tailpipe. It should be remembered that excessive use of CO_2 or other agents which have a cooling effect can shrink the turbine housing onto the turbine and may damage the engine.

Fire Extinguishing System Inspection and Maintenance

Regular maintenance of fire extinguisher systems typically includes such items as the inspection and servicing of fire extinguisher bottles (containers), removal and re-installation of cartridge and discharge valves, testing of discharge tubing for leakage, and electrical wiring continuity tests. The following paragraphs contain details of some of the most typical maintenance procedures, and are included to provide an understanding of the operations involved.

Fire extinguisher system maintenance procedures vary widely according to the design and construction of the particular unit being serviced. The detailed procedures outlined by the airframe or system manufacturer should always be followed when performing maintenance. Manuals utilizing the ATA codes will show fire extinguishing system information under System 26, subcode 20.

Container pressure check. A pressure check of fire extinguisher containers is made periodically to determine that the pressure is between the minimum and maximum limits prescribed by the manufacturer. Changes of pressure with ambient pressure must also fall within prescribed limits. The graph shown in Figure 8-3-3 is typical of the pressure/temperature curve graphs that provide maximum and minimum

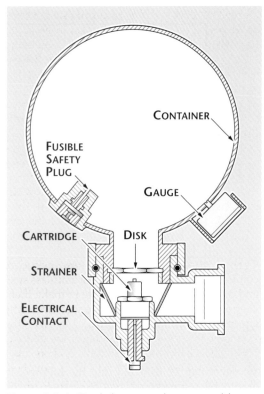

Figure 8-3-4. Single bonnet sphere assembly cross section

gauge readings. If the pressure does not fall within the graph limits, the extinguisher container should be replaced.

Freon discharge cartridges. The service life of fire extinguisher discharge cartridges is calculated from the manufacturer's date stamp, which is usually placed on the face of the car-

Figure 8-3-5. Single bonnet sphere assembly

tridge. The manufacturer's service life is usually recommended in terms of hours below a predetermined temperature limit. Many cartridges are available with a service life of approximately 5,000 hours. To determine the unexpired service life of a discharge cartridge, it is necessary to remove the electrical leads and discharge hose from the plug body, which can then be removed from the extinguisher container.

Care must be taken in the replacement of cartridge and discharge valves. Most new extinguisher containers are supplied with their cartridge and discharge valve disassembled. Before installation on the aircraft, the cartridge must be properly assembled into the discharge valve and the valve connected to the container. This is usually accomplished by means of a swivel nut that tightens against a packing ring gasket.

If a cartridge is removed from a discharge valve for any reason, it should not be used in another discharge valve assembly, since the distance the contact point protrudes may vary with each unit. Thus, continuity might not exist if a used plug which had been indented with a long contact point were installed in a discharge valve with a shorter contact point.

When actually performing maintenance, always refer to the applicable maintenance manuals and other related publications pertaining to a particular aircraft.

Freon/Halon containers. Freon/Halon extinguishing agents are stored in spherical steel containers. There are four sizes in common use today, ranging from 224 cu. in. (small) to 915 cu. in. (large). The large containers weigh about 33 lbs. The small spheres have two openings, one for the bonnet assembly (sometimes called an operating head), and the other for the fusible safety plug (Figure 8-3-4). The larger containers are usually equipped with two firing bonnets and a two-way check valve as shown in Figure 8-3-6.

The containers are charged with dry nitrogen, in addition to a specified weight of the extinguishing agent. The nitrogen charge provides sufficient pressure for complete discharge of the agent. The bonnet assembly contains an electrically ignited power cartridge which breaks the disk, allowing the extinguishing agent to be forced out of the sphere by the nitrogen charge.

A single bonnet sphere assembly is illustrated in Figure 8-3-4 and Figure 8-3-5. The function of the parts shown, other than those described in the preceding paragraph, are as follows:

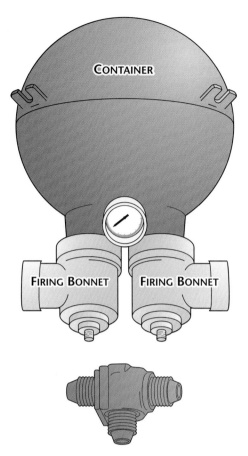

TWO WAY CHECK VALVE

Figure 8-3-6. Double bonnet sphere assembly

- The strainer prevents pieces of the broken disk from entering the system.

- The fusible safety plug melts and releases the liquid when the temperature is between 208 and 220° F.

- The gauge shows the pressure in the container. In this type of design, there is no need for siphon tubes.

On some installations the safety plug is connected to a discharge indicator mounted in the fuselage skin, while others simply discharge the fluid into the fire extinguisher container storage compartment.

The gauge on the container should be checked for an indication of the specified pressure as given in the applicable aircraft maintenance manual. In addition, make certain that the indicator glass is unbroken and that the bottle is securely mounted.

Some types of extinguishing agents rapidly corrode aluminum alloy and other metals, especially under humid conditions. When a system that uses a corrosive agent has been discharged, the system must be purged thoroughly with clean, dry, compressed air as soon as possible.

Almost all types of fire extinguisher containers require re-weighing at frequent intervals to determine the state of charge. In addition to the weight check, the containers must be hydrostatically tested, usually at 5-year intervals.

The circuit wiring of all electrically discharged containers should be inspected visually for condition. The continuity of the entire circuit should be checked following the procedures in the applicable maintenance manual. In general, this consists of checking the wiring and the cartridge, by using a resistor in the test circuit that limits the circuit current to less than 35 milliamperes to prevent detonating the cartridge.

Carbon dioxide cylinders. These cylinders come in various sizes, are made of stainless steel and are wrapped with steel wire to make them shatterproof. The normal storage pressure of the gas ranges from 700 to 1,000 p.s.i. However, the state of the cylinder charge is determined by the weight of the CO_2. In the container, about two-thirds to three-fourths of the CO_2 is liquefied. When the CO_2 is released, it expands about 500 times as it converts to gas.

The cylinder does not have to be protected against cold weather, for the freezing point of carbon dioxide is minus 110°F. However, it can discharge prematurely in hot climates. To prevent this, manufacturers put in a charge of dry nitrogen, at about 200 p.s.i. before they fill the cylinder with carbon dioxide. When treated in this manner, most CO_2 cylinders are protected against premature discharge up to 160°F. With a temperature increase, the pressure of the nitrogen does not rise as much as that of the CO_2 because of its stability with regard to temperature changes. The nitrogen also provides additional pressure during normal release of the CO_2 at low temperature during cold weather.

Carbon dioxide cylinders are equipped internally with one of three types of siphon tubes, as shown in Figure 8-3-7. Aircraft fire extinguishers have either the straight rigid or the short flexible siphon tube installed. The tube is used to make certain that the CO_2 is transmitted to the discharge nozzle in the liquid state.

CO_2 cylinders are equipped with metal safety disks designed to rupture at 2,200 to 2,800 p.s.i. This disk is attached to the cylinder release valve body by a threaded plug. A line leads from the fitting to a discharge indicator installed in the fuselage skin. Rupture of the red disk means that the container safety plug has ruptured because of an overheat condition. A yellow disk is also installed in the fuselage skin. Rupture of this disk indicates that the system has been discharged normally.

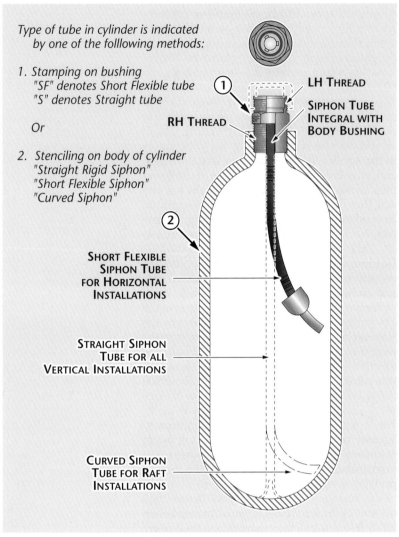

Type of tube in cylinder is indicated by one of the folllowing methods:

1. Stamping on bushing
   "SF" denotes Short Flexible tube
   "S" denotes Straight tube

   Or

2. Stenciling on body of cylinder
   "Straight Rigid Siphon"
   "Short Flexible Siphon"
   "Curved Siphon"

① RH THREAD
LH THREAD
SIPHON TUBE INTEGRAL WITH BODY BUSHING

② SHORT FLEXIBLE SIPHON TUBE FOR HORIZONTAL INSTALLATIONS

STRAIGHT SIPHON TUBE FOR ALL VERTICAL INSTALLATIONS

CURVED SIPHON TUBE FOR RAFT INSTALLATIONS

Figure 8-3-7. Typical CO_2 cylinder construction

Section 4

Fire Prevention and Protection

Leaking fuel and hydraulic, de-icing, or lubricating fluids, can be sources of fire in an aircraft. This condition should be noted, and corrective action taken, when inspecting aircraft systems. Minute pressure leaks of these fluids are particularly dangerous for they quickly produce an explosive atmospheric condition.

Carefully inspect fuel tank installations for signs of external leaks. With integral fuel tanks the external evidence may occur at some distance from where the fuel is actually escaping.

Many hydraulic fluids are flammable and should not be permitted to accumulate in the structure. Soundproofing and lagging materials may become highly flammable if soaked with oil of any kind.

Any leakage or spillage of flammable fluid in the vicinity of combustion heaters is a serious fire risk, particularly if any vapor is drawn into the heater and passes over the hot combustion chamber.

Oxygen system equipment must be kept absolutely free from traces of oil or grease, since these substances will spontaneously ignite in contact with oxygen under pressure. Oxygen servicing cylinders should be clearly marked so that they cannot be mistaken for cylinders containing air or nitrogen, as explosions have resulted from this error during maintenance operations.

Fire prevention is much more rewarding than fire extinguishing.

Cabin and Cockpit Area

All wool, cotton, and synthetic fabrics used in interior trim are treated to render them flame resistant. Tests conducted have shown foam and sponge rubber to be highly flammable. However, if they are covered with a flame-resistant fabric that will not support combustion, there is little danger from fire as a result of ignition produced by accidental contact with a lighted cigarette or burning paper.

Fire protection for the aircraft interior is usually provided by hand-held extinguishers. Four types of fire extinguishers are available for extinguishing interior fires:

- Water
- Carbon dioxide
- Dry chemical
- Halogenated hydrocarbons

Hand-Held Extinguishers

The National Fire Protection Association has classified fires into four basic types:

- Class A fires, defined as fires in ordinary combustible materials such as wood, cloth, paper, upholstery materials, etc.

- Class B fires, defined as fires in flammable petroleum products or other flammable or combustible liquids, greases, solvents, paints, etc.

- Class C fires, defined as fires involving energized electrical equipment where the electrical non-conductivity of the extinguishing media is of importance. In most cases where electrical equipment is de-energized, extinguishers suitable for use on Class A or B fires may be employed effectively.

- Class D fires, defined as any fire that involves a flammable metal, such as magnesium, or magnesium alloy. Dry chemical extinguishers should be use on class D fires.

Aircraft fires, in flight or on the ground, may encompass either or all of these type fires. Therefore, detection systems, extinguishing systems, and extinguishing agents as applied to each type fire must be considered. Each type fire has characteristics that require special handling. Agents usable on Class A fires are not acceptable on Class B or C fires. Agents effective on Class B or C fires will have some effect on Class A fires, but are not the most efficient.

Water extinguishers are for use primarily on nonelectrical fires such as smoldering fabric, cigarettes, or trash containers. Water extinguishers should not be used on electrical fires because of the danger of electrocution. Turning the handle of a water extinguisher clockwise punctures the seal of a CO_2 cartridge which pressurizes the container. The water spray from the nozzle is controlled by a trigger on top of the handle.

Carbon dioxide fire extinguishers are provided to extinguish electrical fires. A long, hinged tube with a non-metallic megaphone-shaped nozzle permits discharge of the CO_2 gas close to the fire source to smother the fire. A trigger type release is normally lock-wired and the wire can be broken by a pull on the trigger. See Figure 8-4-1.

A dry chemical fire extinguisher can be used to extinguish any type of fire. However, the dry chemical fire extinguisher should not be used in the cockpit due to possible interference with visibility and the collection of nonconductive powder on electrical contacts of surrounding equipment. The extinguisher is equipped with a fixed nozzle which is directed toward the fire source to smother the fire. The trigger is also lock-wired but can be broken by a sharp squeeze of the trigger.

The development of Halogenated hydrocarbons (Freons) as fire extinguishing agents with low toxicity for airborne fire extinguishing protection systems logically directed attention to its use in hand type fire extinguishers.

This quality allows Halon 1301 to be used in occupied personnel compartments without depriving people of the oxygen they require. Another advantage is that no residue or deposit remains after use. Halon 1301 is the ideal agent to use in airborne hand held fire extinguishers because:

- Its low concentration is very effective

- It may be used in occupied personnel compartments
- It is effective on all four types of fires
- No residue remaining after its use

Extinguishers Unsuitable as Cabin or Cockpit Equipment

The common aerosol can type extinguishers are definitely not acceptable as airborne hand type extinguishers. In one instance, an aerosol type foam extinguisher located in the pilot's seat back pocket exploded and tore the upholstery from the seat. The interior of the aircraft was damaged by the foam. This occurred when the aircraft was on the ground and the outside air temperature was 90°F. In addition to the danger from explosion, the size is inadequate to combat even the smallest fire.

A dry chemical extinguisher was mounted near a heater vent on the floor. For an unknown reason, the position of the unit was reversed. This placed the extinguisher directly in front of the heater vent. During flight, with the heater in operation, the extinguisher became overheated and exploded filling the compartment with dry chemical powder. The proximity of heater vents should be considered when selecting a location for a hand fire extinguisher.

Additional information relative to airborne hand fire extinguishers may be obtained from your local FAA District Office, and from the National Fire Protection Association.

Figure 8-4-1. A typical hand-held CO_2 fire extinguisher

9

Aircraft Fuel Systems

Aircraft fuel systems started with gravity systems that only had four parts. Now we have complex turbine engine systems with three types of pumps, several types of filters, a cross-feed system, pressure return lines, bottom pressure refueling with digital readouts, and all going into, or out of, several different tanks at the same time. Despite the apparent complexity, modern fuel systems are extremely dependable.

A Boeing 747-400 at average takeoff weight, burns 4,018 gallons of fuel per hour. That reduces to 66.97 gallons per minute, or 1.1 gallons per second. The fuel system can handle about five times that much fuel, or 5.5 gallons per second.

Section 1

Aircraft Fuel Basics

Two types of aviation fuel are in general use today for civilian aircraft. Aviation gasoline, also known as *AVGAS*, is used in virtually all reciprocating engines. A kerosene type fuel usually referred to as *jet fuel*, is used in turbine engines. With the reappearance of diesel engines for aircraft, jet fuel may also be used on these aircraft. For additional information on the types of aviation fuels, their identification, safe handling, and fueling procedures, see Chapter 3 of Aircraft Structural Maintenance in this series.

Properties of Aviation Gasoline

Aviation gasoline consists almost entirely of hydrocarbons, namely, compounds consisting of hydrogen and carbon. Some impurities, in

Left. Modern aircraft depend on clean fuel for safe operation. This Canadair regional jet is refueled using a truck-mounted pressure fueling system. Multiple filters and valves insure that clean fuel is loaded into the proper tanks.

Figure 9-1-1. Like many small airports today, this fuel farm has only two types of fuel; AVGAS LL100 and Jet A.

the form of sulphur and dissolved water, will also be present. A small amount of sulphur, always present in crude petroleum, is left after the process of turning crude oil into gasoline. The water also cannot be avoided, since the gasoline is nearly always exposed to moisture in the atmosphere. Each fuel farm, large or small, has series of quality control procedures to reduce fuel contamination. Figure 9-1-1 shows a typical small fuel farm at a general aviation airport.

Traditionally, *tetraethyl lead* (TEL) has been added to aviation gasoline to improve its performance in the engine. Organic bromides and chlorides are mixed with TEL so that during combustion volatile *lead halides* will be formed. These halides are then exhausted with the combustion by-products. TEL, if added alone, would burn to a solid lead oxide and remain in the engine cylinder. Inhibitors are added to gasoline to suppress the formation of substances that would be left as solids when the gasoline evaporates. Because of environmental concerns over the use of TEL a *low-lead* version of 100 octane AVGAS was formulated. *100LL* is currently being used in virtually all piston engine aircraft, and is often the only gasoline available at airports.

Certain properties of gasoline affect engine performance. These properties include volatility, the manner in which the fuel burns during the combustion process, and the heating value of the fuel. Also important is how corrosive the gasoline is and its tendency to form deposits in the engine during operation. These factors are important because of their effect on general cleanliness, which will have an impact on the time between engine overhauls.

Volatility is a measure of the tendency of a liquid substance to vaporize under given conditions. Gasoline is a complex blend of volatile hydrocarbon compounds that have a wide range of boiling points and *vapor pressures*. It is blended in such a way that a straight chain of boiling points is obtained. This is necessary to obtain the required starting, acceleration, power, and fuel mixture characteristics for the engine.

If the gasoline vaporizes too readily, fuel lines may become filled with vapor and cause decreased fuel flow. Taken to its extreme this will cause *vapor lock* and may result in failure of the engine. Fuel that does not vaporize readily enough can result in hard starting, slow warm-up, poor acceleration, and uneven fuel distribution to the cylinders.

Turbine Fuel

The aircraft gas turbine engine is designed to operate on a distillate fuel, commonly called jet fuel. Jet fuels, like gasoline, are also composed of hydrocarbons. The components of jet fuel contain a little more carbon and usually have a higher sulphur content than gasoline. Inhibitors may be added to reduce corrosion and oxidation. Additives are also blended with the fuel to prevent icing.

Three types of turbine fuel are used in civilian aviation. *Jet A* and *Jet A-1* are kerosene type fuels and *Jet B*, which is a blend of kerosene and gasoline type fuels. Jet fuel numbers are type numbers and have no relation to the fuel's performance in the aircraft engine.

As with gasoline, one of the most important characteristics of jet fuel is its volatility. It must, of necessity, be a compromise between several opposing factors. A highly volatile fuel is desirable to aid in starting in cold weather and to make aerial restarts easier and surer. Low volatility is desirable to reduce the possibility of vapor lock and to reduce fuel losses by evaporation.

At normal temperatures, gasoline in a closed container or tank can give off so many vapors that the fuel-air mixture may be too rich to burn. Under the same conditions, the vapor given off by Jet B fuel can be in the flammable or explosive range. Jet A fuel, however, has such a low volatility that, at normal temperatures, it gives off very little vapor and does not form flammable or explosive fuel air mixtures. Figure 9-1-2 shows the vaporization of aviation fuels at atmospheric pressure. The maximum vapor pressure allowed for aviation gasoline is 7 p.s.i. at 100°F.

Automotive Gasoline

Automotive gasoline is not normally a substitute for aviation gasoline. Neither is commercial diesel fuel. While there are a few Supplemental Type Certificates (STCs) that have been issued allowing use of automotive gasoline under certain parameters, it is not something that is in general use. Also, the vapor pressure of aviation gasoline is lower than the vapor pressure of automotive gasoline. The best advice is to not use it in certified airplanes.

Vapor Lock

Normally, the fuel remains in a liquid state until it is discharged into the air stream, and then instantly changes to a vapor. Under certain conditions, however, the fuel may vaporize in the lines, pumps, or other units. The vapor pockets formed by this premature vaporization restrict the fuel flow through components designed to handle liquids rather than gases. The resulting partial or complete interruption of the fuel flow is called *vapor lock*. The three general causes of vapor lock are the lowering of the pressure on the fuel, high fuel temperatures, and excessive fuel turbulence.

At high altitudes, the atmospheric pressure on the fuel in the tank is low. This lowers the boiling point of the fuel and causes vapor bubbles to form. This vapor trapped in the fuel may cause vapor lock in the fuel system.

Transfer of heat from the engine tends to cause boiling of the fuel in the lines and the pump. This tendency is increased if the fuel in the tank is warm. High fuel temperatures often combine with low pressure to increase vapor formation. This is most apt to occur during a rapid climb on a hot day. As the aircraft climbs, the outside temperature drops, but the fuel does not cool rapidly. If the fuel is warm enough at takeoff, it may retain enough heat to boil easily at high altitude.

The chief causes of fuel turbulence are sloshing of the fuel in the tanks, the mechanical action of the engine-driven pump, and sharp bends or rises in the fuel lines. Sloshing in the tank tends to mix air with the fuel. As this mixture passes through the lines, the trapped air separates from the fuel and forms vapor pockets at any point where there are abrupt changes in direction or steep rises. Turbulence in the fuel pump often combines with the low pressure at the pump inlet to form a vapor lock at this point.

Vapor lock can become serious enough to block the fuel flow completely and stop the engine. Even small amounts of vapor in the inlet line will restrict the flow to the engine-driven pump and reduce its output pressure.

To reduce the possibility of vapor lock, fuel lines are kept away from sources of heat; also, sharp bends and steep rises are avoided. In addition, the volatility of the fuel is controlled in manufacture so that it does not vaporize too readily. The major improvement in reducing vapor lock, however, is the incorporation of *booster pumps* in the fuel system. These pumps keep the fuel in the lines to the engine-driven pump under pressure. The slight pressure on the fuel reduces vapor formation and aids in moving a vapor pocket along. The booster pump also releases vapor from the fuel as it passes through the pump. The vapor moves upward through the fuel in the tank and out the tank vents.

To prevent the small amount of vapor that remains in the fuel from upsetting its metering action, vapor eliminators are installed in some fuel systems ahead of the metering device or are built into this unit.

Fuel System Icing

Moisture in the fuel system, in any form, poses a very real danger as the temperature

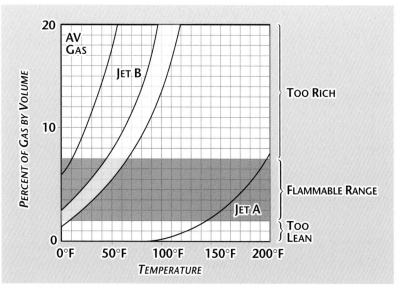

Figure 9-1-2. Vaporization of aviation fuels at atmospheric pressure

approaches the freezing point of water. Fuel system icing may be the result of either free or *entrained water.*

Free water may be present in either AVGAS or jet fuel. Careful attention during fuel handling, refueling operations, and pre-flight inspections generally result in the detection and removal of free water. If undetected, this water may accumulate in low spots, near tank outlets, in fuel filters, selector valves, and low spots of fuel lines. When frozen, this ice may prevent fuel flow and cause engine failure.

Entrained water is a danger generally associated with jet fuel. Because the specific gravity of jet fuel and water are nearly the same, water may not sink to the bottom of the tank as with gasoline. Some water may remain suspended in the fuel. When fuel temperatures approach freezing, this water will begin to change to ice. This results in ice crystals flowing along with the fuel and may result in clogged filters and lines.

To prevent engine stoppage from icing, jet aircraft may be equipped with heat exchangers to warm the fuel, and anti-icing additives may be mixed with the fuel before or during refueling. More specific information on fuel heaters will be presented later in this chapter.

Fuel System Contamination

There are several forms of contamination in aviation fuel. The higher the viscosity of the fuel, the greater is its ability to hold contaminants in suspension (Figure 9-1-3). For this reason, jet fuels having a high viscosity are more susceptible to contamination than aviation gasoline. The principal contaminants that reduce the quality of both gasoline and turbine fuels are other petroleum products, water, rust or scale, and dirt.

Water

Water can be present in the fuel in two forms:

- Dissolved in the fuel
- Entrained or suspended in the fuel

Entrained water can be detected with the naked eye. The finely divided droplets reflect light and in high concentrations give the fuel a dull, hazy, or cloudy appearance.

A cloud usually indicates a water-in-fuel suspension. Fuel, however, can be cloudy for a number of reasons. If the fuel is cloudy and the cloud disappears at the bottom, air is present. If the cloud disappears at the top, water is present. Particles of entrained water may unite to form droplets of free water.

Free water can cause icing of the aircraft fuel system, usually in the aircraft boost-pump screens and low-pressure filters. Fuel gauge readings may become erratic because the water short-circuits the aircraft's electrical fuel cell quantity probe. Large amounts of water can cause engine stoppage. If the free water is saline, it can cause corrosion of the fuel system components.

Contamination with Other Types or Grades of Fuel

The unintentional mixing of petroleum products can result in fuels that give unacceptable performance in the aircraft. An aircraft engine is designed to operate most efficiently on fuel of definite specifications. The use of fuels that differ from these specifications reduces operating efficiency and can lead to complete engine failure.

Operators of turbine-powered aircraft are sometimes forced by circumstances to mix fuels. Such mixing, however, has very definite disadvantages. When aviation gasoline is mixed with jet fuel, the TEL in the gasoline forms deposits on the turbine blades and vanes. Continuous use of mixed fuels may cause a loss in engine efficiency. However, on a limited usage basis, they have no detrimental effects on the engine. The operator must consult the aircraft operations manual before burning gasoline, or gasoline mixtures, in a turbine engine.

Aviation gasoline containing, by volume, more than 0.5 percent of jet fuel may be reduced below the allowable limits in knock rating. In reciprocating engines, this can result in detonation and failure of the engine. Gasoline contaminated with turbine fuel is unsafe for use in reciprocating engines.

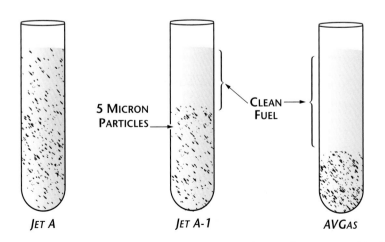

Figure 9-1-3. Comparison of particle's rate of settling in three types of fuel

Foreign Particles

Most foreign particles are found as *sediment* in the fuel. They are composed of almost any material with which the fuel comes into contact. The most common types are rust, sand, aluminum and magnesium compounds, brass shavings, and rubber.

Rust is found in two forms:

- Red rust (nonmagnetic)
- Black rust (magnetic)

They appear in the fuel as red or black powder (which may resemble a dye), rouge, or grains. Sand or dust appears in the fuel in a crystalline, granular, or glass-like form.

Aluminum or magnesium compounds appear in the fuel as a form of white or gray powder or paste. This powder, or paste, becomes very sticky or gelatinous when water is present. Brass is found in the fuel as bright gold-colored chips or dust. Rubber appears in the fuel as fairly large irregular bits. All of these forms of contamination can cause sticking or malfunctions of fuel metering devices, flow dividers, pumps, and nozzles.

Microbial growth. Microbial growth is produced by various forms of micro-organisms that live and multiply in the water interfaces of jet fuels. These organisms may form slime similar in appearance to the deposits found in stagnant water. The color of this slime growth may be red, brown, gray, or black. If not properly controlled by frequent removal of free water, the growth of these organisms can become extensive. The organisms feed on the hydrocarbons that are found in fuels, but they need free water in order to multiply.

Micro-organisms have a tendency to mat, generally appearing as a brown blanket that acts as a blotter to absorb more moisture. This mixture, or mat, accelerates the growth of micro-organisms. The buildup of micro-organisms not only can interfere with fuel flow and quantity indication, but, more importantly, it can start electrolytic corrosive action.

Today there are many antibacterial fuel additives that are available to prevent microbial formation in Jet fuel. PRIST is a common additive used in the aviation market today as a microbial growth preventer. This is added to the fuel during the fueling process of the aircraft. These fuel additives also act as an anti icing agent. The additives mix with water that condenses out of the fuel and lowers it freezing point enough that it cannot freeze on the fuel filters

Sediment. Sediment appears as dust, powder, fibrous material, grains, flakes, or stain. Specks or granules of sediment indicate particles in the visible size range, i.e., approximately 40 microns or larger in size (Figure 9-1-4). The presence of any appreciable number of such particles indicates either a malfunction of the filter/separators, a source of contamination downstream of the filter/separator, or an improperly cleaned sample container. Even with the most efficient filter/separators and careful fuel handling, an occasional visible particle will be encountered. These strays are usually due to particle migration through the filter media and may represent no particular problem to the engine or fuel control. The *sediment* ordinarily encountered is an extremely fine powder, rouge, or silt. The two principle components of this fine sediment are normally sand and rust.

Sediment includes both organic and inorganic matter. The presence of appreciable quantities of fibrous materials (close to naked eye visibility) is usually indicative of filter element breakdown, either because of a ruptured filter element or mechanical failure of a component in the system. Usually, high metal content of relatively large particles suggest a mechanical failure somewhere in the system.

In a clean sample of fuel, sediment should not be visible except upon the most meticulous inspection. Persistent presence of sediment is suspect and requires that appropriate surveillance tests and corrective measures be applied to the fuel handling system.

Sediment or solid contamination can be separated into two categories:

- Coarse sediment
- Fine sediment

Coarse sediment. Sediment that can be seen and that easily settles out of fuel or can be

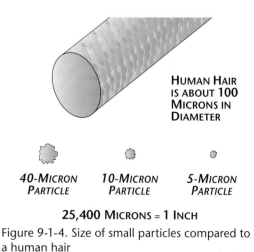

HUMAN HAIR IS ABOUT **100** MICRONS IN DIAMETER

40-MICRON PARTICLE   10-MICRON PARTICLE   5-MICRON PARTICLE

25,400 MICRONS = 1 INCH

Figure 9-1-4. Size of small particles compared to a human hair

removed by adequate filtration, is coarse sediment. Ordinarily, particles 10 microns in size and larger are regarded as coarse sediment.

Coarse particles clog orifices and wedge in sliding valve clearances and shoulders, causing malfunctions and excessive wear of fuel controls and metering equipment. They are also effective in clogging nozzle screens and other fine screens throughout the aircraft fuel system.

Fine sediment. Particles smaller than 10 microns may be defined as fine sediment. Proper settling, filtration, and centrifuging can remove 98% of the fine sediment in fuel. Particles in this range accumulate throughout fuel controls, appearing as a dark shellac-like surface on sliding valves, and may also be centrifuged out in rotating chambers as sludge-like matter, causing sluggish operation of fuel metering equipment. Fine particles are not visible to the naked eye as distinct or separate particles; they will, however, scatter light and may appear as point flashes of light or a slight haze in fuel.

Maximum possible settling time should be allowed in fuel tanks after filling to allow reasonable settling of water and sediment.

Detecting Contamination

Coarse contamination can be detected visually. The major criterion for *contamination detection* is that the fuel be clean, bright, and contain no perceptible free water. Clean means the absence of any readily visible sediment or entrained water. Bright refers to the shiny appearance of clean, dry fuels. A cloudy appearance or a water slug may indicate free water. Clouding may or may not be present when the fuel is saturated with water. Perfectly clear fuel can contain as much as three times the volume of water considered tolerable.

Several field methods for checking water content have been devised. One is the adding of a food color that is soluble in water, but not in fuel. Colorless fuel samples acquire a definite tint if water is present. Another method uses a gray chemical powder that changes color from pink to purple, if 30 or more p.p.m. (parts per million) of water are present in a fuel sample. In a third method a hypodermic needle is used to draw a fuel sample through a chemically treated filter. If the sample changes the color of the filter from yellow to blue, the fuel contains at least 30 p.p.m. of water.

Since fuel drained from tank sumps may have been cold-soaked, it should be realized that no method of water detection can be accurate while the fuel entrained water is frozen into ice crystals.

There is a good chance that water will not be drained or detected if the sumps are drained while the fuel is below 32°F, after being cooled in flight. The reason for this is that the sump drains may not be at the lowest point in the fuel tank while the airplane is in a flight attitude, and water may accumulate and freeze on other areas of the tank where it will remain undetected until it thaws.

Draining will be more effective if it is done after the fuel has been undisturbed for a period of time during which the free water can precipitate and settle to the drain point. The benefits of a settling period will be lost, however, unless the accumulated water is removed from the drains before the fuel is disturbed by internal pumps.

Section 2
Fuel System Types and Components

Fuel System Design

Design and performance improvements in aircraft and engines have increased the demands on the fuel system. Fuel systems in today's aircraft are more complicated and there has been a corresponding increase in installation, adjustment, and maintenance challenges. To ensure safe operation certificated aircraft must meet the stringent requirements established in 14 CFR Part 23.951 to 23.1001.

Airframe fuel system. The airframe fuel system begins with the fuel tank and ends at the engine fuel system. According to the regulations, the fuel system must supply fuel to the carburetor, or other metering device, under all conditions of ground and air operation. It must function properly at constantly changing altitudes and in any climate. The system should also be free of tendency to vapor lock, which can result from changes in ground and in-flight climatic conditions.

The basic components of this system include the fuel tanks, lines, selector valves, strainers, pumps and pressure gauges. We will take a look at each component and see how the type of aircraft it is installed on affects its design.

On small aircraft, a simple gravity-feed fuel system consisting of a tank to supply fuel to the engine, a shut off valve, and a fuel strainer is often installed. On multi-engine aircraft, complex systems are necessary so that fuel can be pumped from any combination of tanks to any combination of engines. Provisions for transferring fuel from one tank to another may also be included on large aircraft.

Normally an aircraft, regardless of its size, has more than one fuel tank. The type and location of these fuel tanks varies with design and construction of the aircraft, as well as the design of the fuel system. Each fuel tank is connected to a selector valve with a fuel line. The selector valve allows the flight crew to control the flow of fuel from the flight deck. The fuel system must contain at least one main strainer. This strainer will be located at the lowest point of the fuel system and serves as a collection point for the moisture in the system.

Figure 9-2-1 shows an electronic fuel system display in a glass cockpit. The display gives more information than was possible with a steam gauge system. The CRT has the following information displayed:

- Fuel used in last flight by engine
- Fuel on board in gallons (or pounds)
- Position of all fuel system valves
- Fuel in each tank by gallons (or pounds)
- Temperature of fuel in each tank

With the information available it is a simple matter to check the remote fuel valve positions.

A booster pump may be installed to ensure a positive pressure to the engine driven pump during starting and in emergencies. When starting the engine, fuel pressure supplied by the boost pump forces fuel through a bypass in the engine-driven pump to the metering device. Once the engine-driven pump is rotating at sufficient speed, it takes over and delivers fuel to the metering device at the specified pressure. Some installations allow the boost pump to be turned off during normal operations.

Engine fuel system. The engine fuel system begins where the fuel is delivered to the engine driven pump (if installed) and includes all of the fuel control and metering devices up to the point of discharge. In aircraft powered with a reciprocating engine, the fuel metering system consists of the air- and fuel-control devices from the point where the fuel enters the first control unit until the fuel is injected into the supercharger section, intake pipe, or cylinder. For example, the fuel metering system of the Teledyne-Continental IO-470 engine consists of the fuel/air control unit, the injector pump,

the fuel manifold valve, and the fuel discharge nozzles.

The fuel metering system of the gas turbine engine consists of an engine fuel control unit (FCU) and may extend to, and include, the fuel nozzles installed in the combustion section. On some turboprop engines, a temperature datum valve is a part of the engine fuel system. The rate of fuel delivery is a function of air mass flow, compressor inlet temperature, compressor discharge pressure, r.p.m., and combustion chamber pressure.

The fuel metering system must operate satisfactorily to ensure efficient engine operation as measured by power output, operating temperatures, and range of the aircraft. Because of variations in design of different fuel metering systems, the expected performance of any one piece of equipment, as well as the difficulties it can cause, will vary.

Requirements for Fuel System Design

The type of fuel system used on any given aircraft will be a function of the physical location of components, type and number of engines installed, and the anticipated operating altitudes. Fuel system designs range from the simple gravity feed system used by light aircraft to pressure feed systems used on sophisticated turbine powered aircraft.

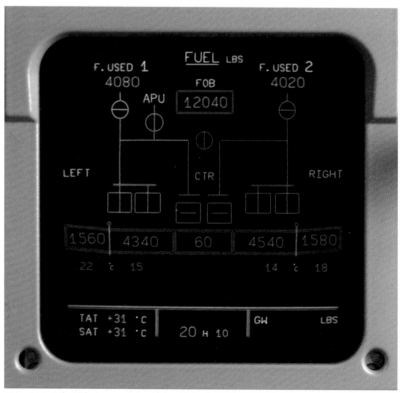

Figure 9-2-1. Electronic fuel system display

Figure 9-2-2. A Piper Super Cub with a classic gravity flow fuel system

Gravity feed. Gravity systems may be used on aircraft where the fuel tanks are located high enough above the carburetor so that gravity will provide enough fuel pressure to supply 150% of the fuel flow required for takeoff operation. This generally is limited to high wing aircraft, and certain small aircraft using a single tank mounted above the fuselage or in the nose of the aircraft. Figure 9-2-2 is a Piper Super Cub with classic gravity flow fuel system.

These systems should not be able to supply fuel to any one engine from more than one tank, unless the fuel tank airspaces are interconnected to ensure an equal feed from both

tanks. This interconnection is commonly found on Cessna single-engine aircraft that are normally operated with the fuel selector valve in the BOTH position.

Pressure feed. Pressure systems use a fuel pump to ensure that at least 125% of the actual takeoff fuel flow is available. These installations include both high wing and low wing aircraft, and may be designed to supply any number of engines.

Multi-engine aircraft fuel systems will be designed so that each engine is supplied from its own tank, pump, and fuel lines. A *fuel cross-feed system* will provide the means to transfer fuel from one tank to another, and in an emergency to operate two engines from one tank.

Fuel System Components

The basic components of an aircraft fuel system include tanks, lines, valves, pumps, filters, indicating systems, warning signals, and primer. Some systems may include central refueling provisions, a fuel jettison (dump) system, and a means for transferring fuel. These units will be discussed individually in the paragraphs that follow.

Fuel Tanks

The location, size, shape, and construction of fuel tanks vary with the type and intended use of the aircraft. Some tanks are removable, and some are integral with the construction of the wing or other structural components of the aircraft. Fuel tanks are made of materials that will not react chemically with any aviation fuel. Internal baffles prevent the fuel from sloshing around in the tank.

Metal fuel tanks. Aluminum alloy is widely used and may appear using either riveted or welded construction. Usually a *sump* and a drain are provided at the lowest point in the tank, as shown in Figure 9-2-3. When a sump or low point drain is provided in the tank, the main fuel supply is not drawn from the bottom of the tank, but from a higher point.

The space at the top of each tank is vented to the outside air to maintain atmospheric pressure within the tank. Air vents are designed to minimize the possibility of their being blocked by dirt or ice formations. In order to permit rapid changes in internal air pressure, the size of the vent is proportional to the size of the tank. This acts to prevent a collapse of the tank during a steep dive or glide. All except the very smallest tanks are fitted with internal baffles to resist fuel surging caused by changes in the

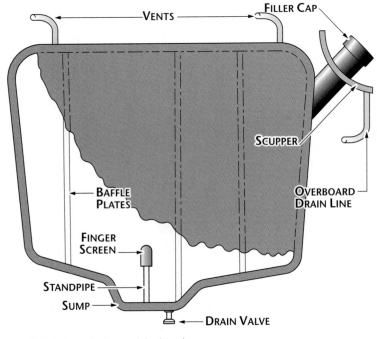

Figure 9-2-3. A typical metal fuel tank

attitude of the aircraft. Usually an expansion space is provided in fuel tanks to allow for an increase in fuel volume due to expansion.

The filler neck and cap are usually located in a recessed well, equipped with a *scupper* and drain. The scupper is designed to prevent overflowing fuel from entering the wing or fuselage structure. Fuel caps have provisions for locking devices to prevent accident loss during flight. Filler openings are clearly marked with the word AVGAS (on aircraft equipped with reciprocating engines), (see Figure 9-2-4) the tank capacity, and the grade of fuel to be used. Information concerning the capacity of each tank is usually posted near the fuel selector valve as well as the filler caps. Turbine fuel systems, even pressure filling systems, have the fuel for which the engine(s) are certified listed prominently, as in Figure 9-2-5.

Fuel Cells

Bladder fuel cells. The bladder type fuel cell is a non-self-sealing cell that is used to reduce weight. It depends entirely upon the structure of the cavity in which it is installed to support the weight of the fuel in it. These cells are removable and employ a variety of retaining systems. Removal and installation of bladder type cells must be accomplished in strict accordance with the aircraft manufacturer's instructions (Figure 9-2-6).

While designed to give many years of safe use, the bladder type fuel cell may develop leaks over time. Generally the cell can be removed and repaired by an appropriately rated repair station. While the cell is out for repair, the cavity in which the cell resides should be carefully inspected and repairs made to the materials installed to protect the bladder from chafing.

Older rubber bladder tanks were frequently covered with engine oil inside when stored for any length of time. Modern bladder tanks are made of polyethylene and require no inner protection.

Integral fuel cells. Since *integral fuel cells* are usually built into the wings of the aircraft structure, they are not removable. An integral cell is constructed inside the aircraft structure so that after the seams, structural fasteners, and access doors have been properly sealed, the cell will hold fuel without leaking. This type of construction if often referred to as a *wet wing*. Repairs to these types of tanks normally consist of patches and doublers riveted in and sealed. Repairs to sealers used in this type of fuel cell must be made in strict accordance with

Figure 9-2-4. Fuel cells on aircraft powered by reciprocating engines should be clearly marked with the word AVGAS and the minimum octane rating required.

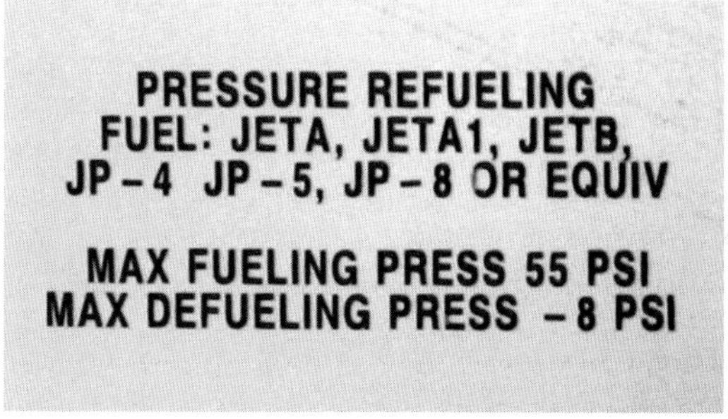

> **PRESSURE REFUELING**
> FUEL: JETA, JETA1, JETB,
> JP–4 JP–5, JP–8 OR EQUIV
>
> MAX FUELING PRESS 55 PSI
> MAX DEFUELING PRESS –8 PSI

Figure 9-2-5. This placard is an oddity. This airplane's engines are certified for seven different kinds of jet fuel.

Figure 9-2-6. Bladder type fuel cell

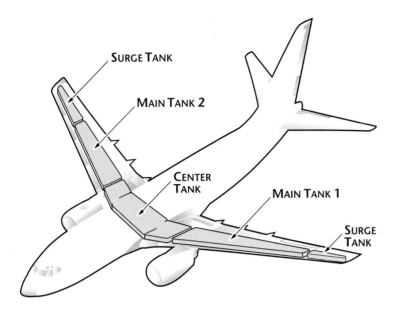

FUEL TANK CAPACITY		
	LBS	KGS
Main Tank 1	8,630	3,915
Main Tank 2	8,630	3,915
Center Tank	28,803	13,066
Total	46,063	20,896
Note: Fuel Density 6.7 lbs per US Gallon		

Figure 9-2-7. Transport airplane wings contain several integral fuel tanks

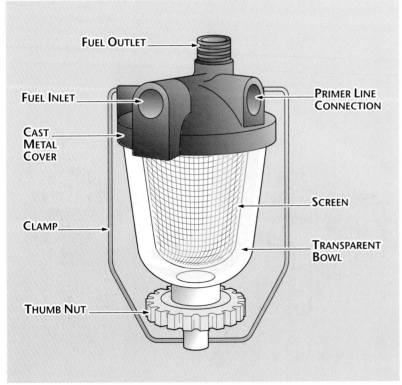

Figure 9-2-8. Main fuel strainer for a light aircraft

the manufacturer's instructions, using only the sealers specified (Figure 9-2-7).

Fuel Lines

Fuel may be supplied through a combination of rigid and flexible lines. Fuel lines must be sized so that they are capable of handling the required fuel flow under all operating conditions. The installation must also be designed so that there are no sharp bends, rapid rises or other features that might cause vapor accumulation. This could lead to vapor lock. Flexible fuel lines and rubber parts in fuel systems are made from special aromatic-resistant rubber to help with deterioration caused by aromatic aviation fuels. These fuels have a strong solvent and swelling action on some types of hose and rubber parts.

Fuel lines should be routed below electrical conduits, and should be kept as far away from hot engine areas as possible. When routing fuel lines between two rigidly mounted fittings, always incorporate at least one bend in the tubing to absorb strain caused by vibration and temperature changes. *Introduction to Aircraft Maintenance,* another book in this series contains extensive coverage of fuel lines and fittings in the chapter titled "Fluid Lines and Fittings."

Fuel Strainers and Filters

Efforts to trap or remove dirt particles from the fuel occur on a number of levels. This provides multiple opportunities to stop dirt before it causes a problem. A combination of wire mesh strainers and fuel filters are often used.

Fuel strainers may be found in the outlet of each fuel tank or integrated with the fuel boost pump assembly. These are made of relatively coarse mesh (as large as 8 mesh/inch). An additional strainer is installed at the low point between the fuel tanks the engine driven pump. This strainer incorporates a much finer mesh (usually 40 mesh/inch or higher). A third strainer is found in the inlet of the carburetor or fuel metering device.

Main fuel strainer. A typical unit used as the main fuel strainer on light aircraft is shown in Figure 9-2-8. It consists of a cast metal top, a screen, and a glass or metal bowl. The bowl is attached to the cover by a clamp and thumb nut. Fuel enters the unit through the inlet port, filters through the screen, and exits through the outlet port. At regular intervals the glass bowl must be drained, and the screen removed for inspection and cleaning. It must always be safetied after inspection.

Filters are used to provide a finer degree of separation than provided by mesh strainers. A low-pressure filter may be installed between the supply tanks and the engine fuel system to protect the engine-driven fuel pump and various control devices. An additional high-pressure fuel filter is installed between the fuel pump and the fuel control to protect the fuel control from contaminants.

The three most common types of filters in use are the *micron filter*, the *wafer screen filter*, and the plain *screen mesh filter*. The individual use of each of these filters is dictated by the filtering treatment required at a particular location.

Micron filter. The micron filter (Figure 9-2-9) has the greatest filtering action of any present-day filter type and, as the name implies, is rated in microns. (A micron is the thousandth part of 1 millimeter.) The porous cellulose material frequently used in construction of the filter cartridges is capable of removing foreign matter measuring from 10 to 25 microns. The minute openings make this type of filter susceptible to clogging; therefore, a bypass valve is a necessary safety factor.

Since the micron filter does such a thorough job of removing foreign matter, it is especially valuable between the fuel tank and engine. The cellulose material also absorbs water, preventing it from passing through the pumps. If water does seep through the filter, which happens occasionally when filter elements become saturated with water, the water can quickly damage the working elements of the fuel pump and control units. This damage occurs because these elements depend solely on the fuel for their lubrication. To reduce water damage to pumps and control units, periodic servicing and replacement of filter elements is imperative. Daily draining of fuel tank sumps and low-pressure filters will eliminate much filter trouble and prevent undue maintenance of pumps and fuel control units.

The most widely used filters are the 200-mesh and the 350-mesh micron filters. They are used in fuel pumps, fuel controls, and between the fuel pump and fuel control, where removal of micro-size particles is needed. These filters, usually made of fine-mesh steel wire, are a series of layers of wire.

Wafer screen filter. The wafer screen type of filter (Figure 9-2-10) has a replaceable element, which is made of layers of screen disks of bronze, brass, steel, or similar material. This type of filter is capable of removing micro-size particles. It also has the strength to withstand high pressure.

Screen mesh strainer filter. The screen mesh strainer-type filter is still in common use. It has long been found in internal-combustion engines of all types for fuel and oil strainers. In present-day turbojet engines it is used in units where filtering action is not so critical, such as in fuel lines before the high-pressure pump filters. The mesh size of this type of filter varies greatly according to the purpose for which it is used.

Fuel System Valves

Fuel systems must be equipped with valves or valve controls that are accessible to the pilot

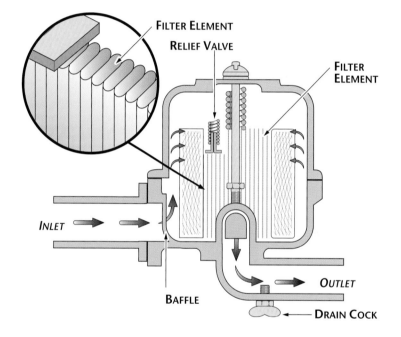

Figure 9-2-9. Micron fuel filter

Figure 9-2-10. Wafer screen fuel filter

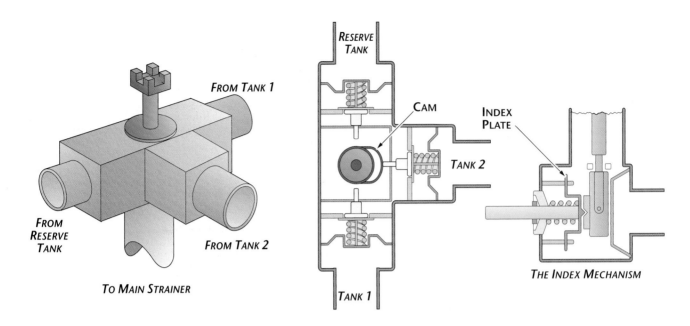

Figure 9-2-11. Poppet-type fuel selector valve

(and on large aircraft the flight engineer) that are capable of shutting off the fuel flow to any engine. The valves must accommodate the full flow capacity of the fuel line, must not leak, and must operate freely with a definite "feel" or "click" when it is in the operating position. Small aircraft may employ valves that are located in the cabin and are directly operated by the pilot. Larger aircraft generally employ electrically operated valves.

Fuel tank shutoff valves have two positions, open and closed. They are installed in the system to prevent fuel loss when a fuel system component is being removed or when a part of the system is damaged. In some installations, they are used to control the fuel flow during fuel transfer. They may be operated manually or electrically. An electrically operated fuel shutoff valve includes a reversible electric motor linked to a sliding valve assembly. The motor moves the valve gate in and out of the passage through which the fuel flows, thus shutting off or turning on the fuel flow.

Fuel selector valves will be used on systems with multiple tanks to select the tank that fuel will be drawn from during each segment of the flight. The three main types of selector valves are the poppet, cone, and disk.

Poppet selector valves. The poppet-type selector valve has an individual poppet valve at each inlet port. A cam and yoke on the same shaft act to open the selected poppet valve as the yoke is turned. Figure 9-2-11 shows how the cam lifts the upper poppet valve from its seat when the control handle

is set to the "number 2" tank. This opens the passage from the number 2 tank to the engine. At the same time, a raised portion of the index plate drops into a notch in the side of the cam. This provides the "feel" that indicates the valve is in the fully open position. The control handle should always be set by "feel" rather than by the markings on the indicator dial. The index mechanism also keeps the valve in the desired position and prevents creeping caused by vibration.

Cone selector valves. The cone-type selector valve has either an all-metal or cork-faced aluminum housing. The cone, which fits into the housing, is rotated by means of a flight deck control. To supply fuel from the desired tank, the control is turned until the passages in the cone align with the correct ports in the housing. An indexing mechanism aids in obtaining the desired setting, and also holds the cone in the selected position. Some cone-type valves have a friction release mechanism that reduces the amount of turning torque required to make a tank selection and that can be adjusted to prevent leakage.

Disk selector valves. The rotor of the disk-type selector valve fits into a cylindrical hole in the valve body. A disk-type valve is shown in Figure 9-2-12. Note that the rotor has one open port and several sealing disks – one for each port in the housing.

To select a tank, the rotor is turned until the open port aligns with the port from which fuel flow is desired. At this time all other ports are closed by the sealing disks. In this position, fuel will flow from the desired tank to the selector valve and out through the engine-feed port at the bottom of the valve. To ensure positive port alignment for full fuel flow, the indexing mechanism forces a spring-loaded ball into a ratchet ring. When the selector valve is in the closed position, the open port in the rotor is opposite a blank in the valve body, which each sealing disk covers a tank port.

Fuel Pumps

Pumps associated with aircraft fuel systems generally fall into one of two categories: boost pumps or engine driven pumps. The high rotational speed of the impeller on these pumps swirls the fuel and produces a centrifuge action that separates air and vapor from the fuel.

Boost pumps. *Boost pumps* provide a positive pressure to the inlet side of the engine drive pump and must be available for starting, takeoff, landing and high altitude operations. Centrifugal type pumps are often used as boost pumps (Figure 9-2-13). These units pressurize the fuel by drawing it into the center of a centrifugal impeller and expelling it at the outer edge. This arrangement allows fuel to flow through the pump when the pump is not in operation, and eliminates the need for any bypass mechanism.

They are generally located inside or adjacent to the fuel tank and must have enough capacity to substitute for the engine driven pump if it should fail. Pumps may be designed so that the electric motor is on the outside of the fuel tank or with the entire pump and motor located inside the tank. When the entire unit is inside the tank, it is referred to as a *submerged pump*.

Engine driven pumps. The purpose of the engine driven fuel pump is to deliver a continuous supply of fuel at the proper pressure at all times during engine operation. One type of widely used pump is the positive-displacement, rotary-vane-type pump. Small aircraft with reciprocating engines frequently use an automotive type diaphragm pump.

Vane pumps. A schematic diagram of this type of pump is shown in Figure 9-2-14. Regardless of the variations in design, the operating principles of all vane-type fuel pumps are the same.

The engine-driven pump is usually mounted on the accessory section of the engine. The rotor, with its sliding vanes, is driven by the engine through the accessory gearing. Note how the vanes carry fuel from the inlet to the outlet as the rotor turns in the direction

Figure 9-2-12. Disk-type selector valve

indicated. A seal prevents leakage at the point where the drive shaft enters the pump body, and a drain carries away any fuel that might leak past the seal. Since the fuel provides enough lubrication for the pump, no special lubrication is necessary.

Because the engine-driven fuel pump normally discharges more fuel than the engine requires, there must be some way of relieving excess fuel to prevent excessive fuel pressures at the fuel inlet of the metering device. This is accomplished through the use of a spring-loaded relief valve that can be adjusted to deliver fuel

Figure 9-2-13. Centrifugal fuel booster pump

at the recommended pressure for a particular installation. Figure 9-2-14 shows the pressure relief valve in operation bypassing excess fuel back to the inlet side of the pump. Adjustments are made by increasing or decreasing the tension of the spring.

The relief valve of the engine-driven pump is designed to open at the set pressure regardless of the pressure of the fuel entering the pump. To maintain the proper relationship between fuel pressure and air inlet pressure, the cham-

ber above the fuel pump is vented either to the atmosphere or through the balance line to the fuel-metering device. The combined pressure of spring tension and atmospheric or inlet air pressure determine the absolute pressure at which the relief valve opens.

This balanced-type relief valve has certain objectionable features that must be investigated when encountering fuel system troubles. Failure of the diaphragm allows air to enter the fuel pump on the inlet side of the pump if inlet pressure is less than atmospheric. Conversely, if the inlet pressure is higher than atmospheric pressure, fuel will be discharged from the vent. For proper altitude compensation the vent must be open. If it becomes clogged by ice or other foreign matter at altitude, the fuel pressure will decrease during descent. If the vent becomes clogged during climb-out, the fuel pressure will increase as the altitude increases.

In addition to the relief valve, the fuel pump has a bypass valve that permits fuel to flow around the pump rotor whenever the pump is inoperative. This valve, shown in Figure 9-2-15 consists of a disk that is lightly spring loaded against a series of ports in the relief valve head. When fuel is needed for starting the engine, or in the event of an engine-drive pump failure, fuel at booster-pump pressure is delivered to the fuel pump inlet. When the pressure is great enough to move the bypass disk from its seat, fuel is allowed to enter the fuel metering system. When the engine-driven pump is in operation, the pressure built up on the outlet side of the pump, together with the pressure of the bypass spring, holds the disk on its seat and prevents fuel flow through the ports.

Automotive-type diaphragm pump. Many four and six cylinder carbureted engines still use an automotive type diaphragm fuel pump, shown in Figure 9-2-16. These pumps operate by lever action developed from a special lobe on the camshaft. The most frequent problem they develop is debris in the check valves. Unfortunately, the check valves are not field reparable, nor is the pump.

Fuel Ejectors

A fuel ejector, or *ejector pump*, may be used to scavenge fuel from remote areas of fuel tanks and to provide fuel pressure to an operating engine fuel control unit (FCU). Fuel ejectors use Bernoulli's Principle to draw fuel into the unit and the motive force of excess fuel being returned from the FCU to move fuel back to the inlet side of the engine driven pump. See Figure 9-2-17.

Fuel ejectors have no moving parts, but are capable of operating only while the engine is run-

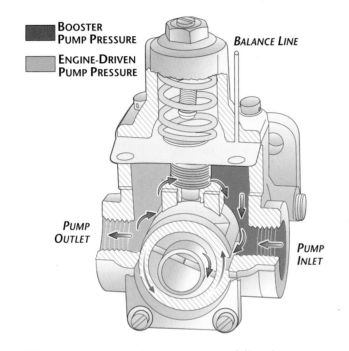

Figure 9-2-14. Engine-driven fuel pump (pressure delivery)

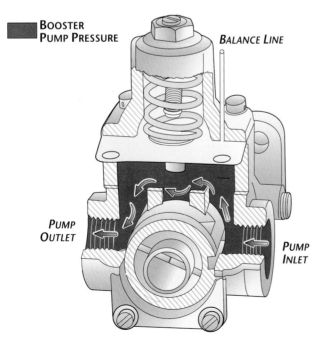

Figure 9-2-15. Engine-driven fuel pump (bypass flow)

ning. Centrifugal pumps are used in conjunction with the fuel ejectors to start the engine. With the engine running, the centrifugal pumps may be turned off and the ejectors are capable of maintaining the required fuel flow.

Heaters

As we discussed earlier, turbine engine fuel systems are very susceptible to the formation of ice in the fuel filters. When the fuel in the aircraft fuel tanks cools to 32°F, or below, entrained water in the fuel tends to freeze when it contacts the filter screen.

A fuel heater operates as a heat exchanger to warm the fuel. The heater can use engine bleed air or engine lubricating oil as a source of heat. The former type is called an air-to-liquid exchanger, and the latter type is known as a liquid-to-liquid exchanger.

The primary function of a fuel heater is to protect the engine fuel system from ice formation. However, should ice form, the heater can also be used to thaw ice on the fuel screen.

In some installations, the fuel filter is fitted with a pressure-drop warning switch, which illuminates a warning light on the cockpit instrument panel. If ice begins to collect on the filter surface, the pressure across the filter will slowly decrease. When the pressure reaches a predetermined value, the warning light flashes.

Fuel deicing systems are designed to be used intermittently. The control of the system may be manual, by a switch in the cockpit, or automatic, using a thermostatic sensing element in the fuel heater to open or close the air or oil shutoff valve. An automatic fuel heater is illustrated in Figure 9-2-18.

Fuel System Instrumentation

For a single-engine aircraft with one gravity-feed fuel tank, the only instrumentation we need is a way to tell how much fuel is in the tank. This can be accomplished using a clear tube sight gauge or a wire attached to a cork floating in the tank. Modern aircraft however, require much more sophisticated instrumentation to aid in fuel system management and ensure safe operation.

In addition to monitoring how much fuel is in the tanks, we can also monitor temperature, pressure, and flow at several locations between the tank and the fuel discharge into the engine. To aid the flight crew in monitoring fuel system parameters, warning systems are installed in conjunction with the instrumentation.

Figure 9-2-16. An automotive type diaphragm fuel pump operated by the camshaft

Fuel Quantity Indicator

The four general types of fuel quantity indicators are:

- Sight glass
- Mechanical
- Electrical
- Electronic (capacitance)

The type of system used with a particular aircraft depends on the size of the aircraft and the number and location of the fuel tanks. Since the sight glass and mechanical fuel gauges are not suitable for aircraft where fuel tanks are located an appreciable distance from the flight

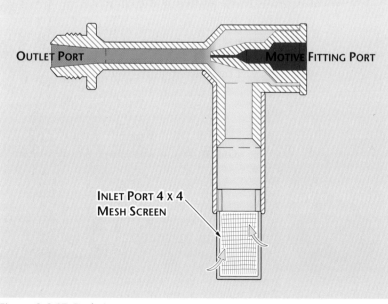

Figure 9-2-17. Fuel ejector

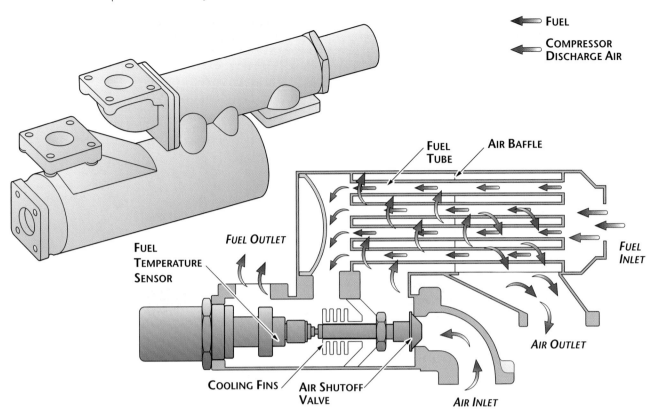

Figure 9-2-18. An automatic fuel heater using an air-to-fuel heat exchanger

deck, larger aircraft use either electrical or electronic fuel quantity indicating systems.

Sight glass. The sight glass is the simplest form of fuel quantity gauge. The indicator is a glass or plastic tube placed on the same level as the tank. It operates on the principle that liquids seek their own level. The tube is calibrated in gallons or has a metal scale near it. The sight glass may have a shutoff valve so that the fuel can be shut off to clean the gauge or to prevent loss if the tube is broken.

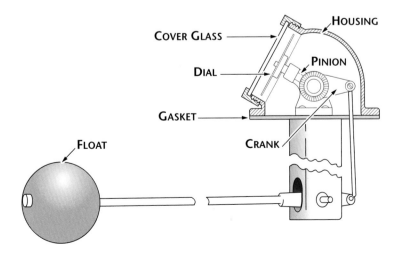

Figure 9-2-19. Float and lever type fuel quantity indicator

Mechanical-type. The mechanical-type fuel quantity system is usually located in the tank and is known as a direct reading gauge. It has an indicator connected to a float resting on the surface of the fuel. As the fuel level changes, the float mechanically operates the indicator, showing the level of fuel in the tank. One type of mechanical fuel quantity indicator is shown in Figure 9-2-19.

Electrical-type. The electrical-type fuel quantity indicating system consists of an indicator mounted on the instrument panel and a float-operated transmitter installed in the fuel tank (Figure 9-2-20). As the fuel level changes, the transmitter sends an electric signal to the indicator, which shows the changing fuel level. With this system the indicator may be located at any distance from the tank, and the fuel levels in more than one tank may be read on a single indicator.

Electronic-type. The electronic-type (capacitance) fuel quantity indicator differs from the other types in that it has no movable parts installed in the tank. Instead of floats and their attendant mechanical units, the dielectric properties of fuel and air furnish a measurement of fuel quantity. Essentially, the tank transmitter is a simple capacitor where fuel and air are the dielectric (Figure 9-2-21). The capacitance of the tank unit at any time depends on the existing proportions of fuel and air in the tank. The capacitance of the transmitter is compared to a reference capacitor in a rebalance-type bridge circuit. The related circuitry provides a signal to the panel mounted fuel indicator.

The electronic type system is more accurate in measuring fuel quantity, as it measures the fuel by weight instead of volume. Fuel volume will vary with temperature and may cause errors in systems based on float position.

Drip gauges. Drip gauges, also called measuring sticks, are provided on all 14CFR part 25 airplanes and on some corporate class airplanes as well. Their purpose is to allow the ground crew to check fuel levels when electricity is not available on the airplane.

Most installations consist of several drip tubes in each wing. As an example, a Boeing B727 has six in each wing. Drip tubes can be calibrated in inches, gallons, or kilograms. These readings can be compared with the on-board fuel indicators to determine the weight of the fuel load. The drip gauge reading must be compared to the tables of capacities of the airplane, and corrected for altitude.

To use the gauge it is unlocked and slowly lowered until fuel drips out of a hole in the lower end. The reading is then taken from the side of the gauge tube. Drip gauges can be replaced without de-fueling the tank.

Fuel Flow Indicator

Fuel flow readings will tell the flight crew the rate at which fuel is being used. Readings may be either in pounds per hour or gallons per hour. Some fuel injected reciprocating engine systems will incorporate a fuel pressure indicator that is marked in gallons per hour. While not truly a flow meter, this system provides information that is accurate enough for the safe operation of these aircraft.

A true fuel flow system consists of a *transmitter* and an *indicator*. The transmitter is installed in the fuel inlet line to the engine where it measures the rate of fuel flow. The transmitter is electrically connected to the indicator in the cockpit. Information regarding the rate at which fuel is being burned, total fuel used, fuel quantity remaining, or flight time remaining may be obtained from a fuel counter or *totalizer*. More sophisticated installations include a microprocessor to interpret the data collected and calculate the desired time or quantities.

Totalizer. The simple fuel counter, or totalizer, is similar in appearance to an automobile odometer. When the aircraft is serviced with fuel, the counter is manually set to the

total number of pounds of fuel in all tanks. As fuel passes through the measuring element of the flowmeter, it sends electrical impulses to the fuel counter. These impulses actuate the fuel counter mechanism so that the number of pounds passing to the engine is subtracted from the original reading. Thus, the fuel counter continually shows the total quantity of fuel, in pounds, remaining in the aircraft. However,

Figure 9-2-20. Resistance-type fuel quantity transmitter

Figure 9-2-21. The transmitter unit for a capacitance type fuel quantity indicator

Figure 9-2-22. A Fueltron vane-type fuel totalizer

The flow meter signal may be developed by a movable vane mounted in the fuel flow path. The impact of fuel causes the vane to swing and move against the restraining force of a calibrated spring. The final position of the vane represents a measure of the rate at which fuel is passing through the flow meter and the corresponding signal to be sent to the indicator. A vane-type flow meter is illustrated in Figure 9-2-22.

Fuel flow transmitter. The fuel flow transmitter used with turbine engines must be capable of measuring flows from 500 to 2,500 pounds per hour. One type of flow meters used on turbine engines is shown in Figure 9-2-23. This transmitter consists of two cylinders placed in the fuel stream so that the direction of fuel flow is parallel to the axes of the cylinders. The cylinders have small vanes in the outer periphery. The upstream cylinder, called the impeller, is driven at a constant angular velocity by the power supply. This velocity imparts an angular momentum to the fuel. The fuel then transmits this angular velocity to the downstream cylinder (the turbine), causing the turbine to rotate until a restraining spring force balances the force of rotation. The degree of deflection imparted on the turbine is measured and transmitted to the flight deck indicator by means of a *selsyn system*.

there are certain conditions that will cause the fuel counter indication to be inaccurate. Any jettisoned fuel is indicated on the fuel counter as fuel still available for use. Any fuel that leaks from a tank or a fuel line upstream of the flow meter is not counted and will render the time or gallons remaining reading inaccurate.

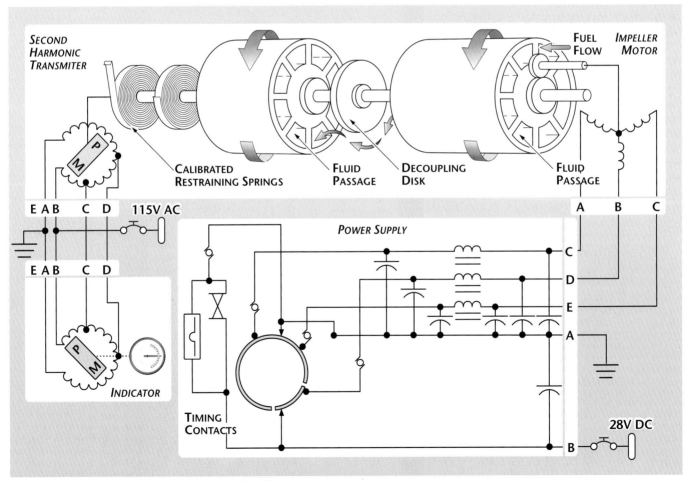

Figure 9-2-23. Schematic of a turbine engine fuel flow indicating system

Fuel Temperature

A means for checking the temperature of the fuel in the tanks and at the engine is provided on some turbine-powered aircraft. During extreme cold, especially at altitude, the gauge can be checked to determine when fuel temperatures are approaching the point at which ice crystals may begin forming in the fuel. This information can be used to determine when it is time to activate fuel heaters, if installed.

Fuel Pressure

The fuel pressure gauge usually indicates the pressure of the fuel entering the metering unit and may employ any of several measuring technologies. The simplest systems employ a means of direct measurement and a mechanical indicating system. The Bourdon tube may be used to convert changes in pressure to mechanical movement. An aneroid and bellows type instrument with a pressure line leading directly from the carburetor to the indicator may also be found. A simple type of fuel pressure gauge is seen in Figure 9-2-24.

Some installations, however, use electrical transmitters to register fuel pressure on the gauge. In this arrangement, the pressure indicating mechanism is contained in the transmitter. Fuel pressure acting on the aneroid and bellows causes motion of a *synchro transmitter*. As the transmitter unit moves, it causes a similar movement in the synchro motor that moves the indicator on the instrument panel. This arrangement makes it unnecessary for fuel to enter the flight deck, reducing fire risks.

Figure 9-2-24. Fuel pressure gauge

Fuel Pressure Warning System

In an aircraft equipped with several tanks, there is always the possible danger of allowing the fuel supply in one tank to be exhausted before the selector valve is switched to another. To prevent this, pressure-warning systems are installed in some aircraft. The installation consists of a pressure sensitive mechanism and a warning light. The warning system has both a fuel and an air connection. These are illustrated in Figure 9-2-25.

The connection marked "fuel" is connected to the fuel pressure line of the fuel metering device. The air connection is either vented to atmosphere or to the air inlet side of the fuel metering unit. This arrangement prevents the warning mechanism from acting in response to changes in the absolute pressure of the fuel.

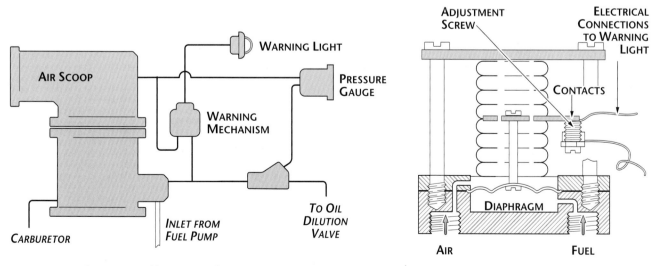

LOCATION OF UNITS IN THE SYSTEM

PRESSURE WARNING MECHANISM

Figure 9-2-25. Fuel pressure-indicating system

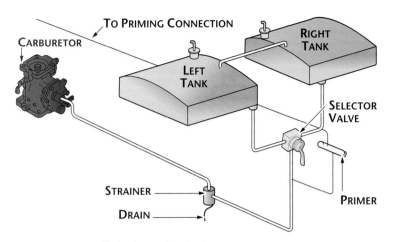

Figure 9-2-26. Small single-engine fuel system

in-transit indicator light. This light is on only during the time that the valve is in motion, and is off when movement is complete.

Typical Aircraft Fuel Systems

Aircraft fuel systems are designed to meet the particular needs of each aircraft. We can, however, take a look at a few representative systems and develop an idea of how the task of providing a reliable supply of fuel to the engine(s) is accomplished.

Single-Engine

The simplest type of fuel system is the gravity feed, which is used on many low-powered airplanes. A gravity feed system for a single-engine aircraft is shown in Figure 9-2-26.

Gravity flow. A fuel system that relies on gravity to feed fuel to the engine is common on light airplanes. Of course, the first requirement is that the fuel tanks must be mounted high enough to create the required pressure head at the carburetor. The next requirement is that the lines must be sized to allow a fuel flow of 150% of the maximum requirement for takeoff.

All lines must obviously run downhill and have no bends that could form a vapor lock. If fuel can be used from more than one tank at a time, both tanks must be vented together, as well as separately. Without proper venting the engine can starve for fuel at a critical moment.

If, for example, the absolute pressure of the fuel decreases because of a change in atmospheric or air inlet pressure, the change is also reflected at the warning mechanism. This cancels the effect of the change, and the fuel pressure warning system is not activated.

Normal fuel pressure acting against the power surface of a diaphragm holds the electrical contacts apart. When fuel pressure drops below the preset minimum, the contacts close and a warning light is illuminated. This alerts the flight crew to take whatever action is necessary to restore the fuel pressure.

Valve-in-transit indicator lights. On large multi-engine aircraft, each of the fuel crossfeed and line valves may be provided with a valve-

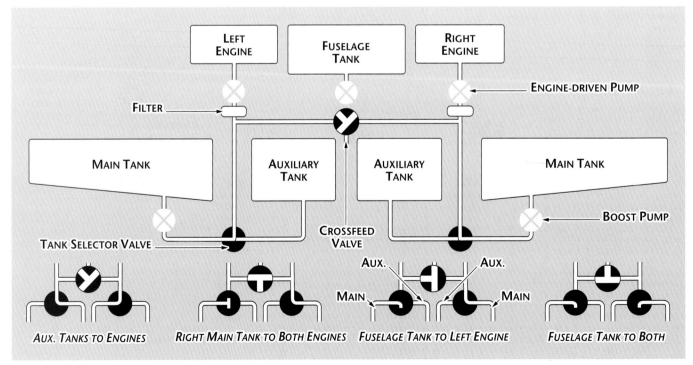

Figure 9-2-27. Multi-engine aircraft fuel system

The fuel tanks are located above the carburetor (usually in the wings), with gravity causing the fuel to flow from the tanks to the carburetor. A selector valve is provided to stop the flow or to select a particular tank in the system from which to draw the fuel. A strainer filters the fuel before it reaches the carburetor. This unit, often called a *gascolator*, provides a location for sediment to gather and a drain through which sediment and water may be removed prior to flight. A primer is installed to furnish the additional fuel required for starting. This type of primer is simply a hand-operated piston pump usually mounted on the instrument panel for pilot convenience.

Multi-Engine

The design of the fuel system for an aircraft having two or more engines presents problems not normally encountered in single-engine fuel systems. A large number of tanks are often required to carry the necessary fuel. These tanks may be located in widely separated parts of the aircraft, such as the fuselage, the inboard and outboard sections of the wings, and the engine nacelles. The individual engine fuel systems must be interconnected so that fuel can be fed from the various tanks to any engine. In case of engine failure, the fuel normally supplied to the inoperative engine must be made available to the others.

The twin-engine fuel system illustrated in Figure 9-2-27 is the simple cross-feed type. As shown, the tank selector valves are set to supply fuel from the main tanks to the engines. These valves can also be positioned to supply fuel from the auxiliary tanks. The cross-feed valve is shown in the off position. It can also be set to supply fuel from the fuselage tank to either or both engines, and to the cross-feed. A few of the many combinations than can be achieved with the three valves are also illustrated.

Integral fuel tanks will have a flapper type check valve in the lowermost fuel tank. Its purpose is to ensure fuel does not slosh outboard and starve the boost pump.

Transport Category Turbine Aircraft

Transport category aircraft generally employ several engines and many fuel tanks. They may also include single-point pressure fueling and a fuel dump system. To accommodate the many functions required by the aircraft the simple cross-feed system seen on smaller aircraft may grow into a fuel manifold system.

The main feature of the four-engine system, seen in Figure 9-2-28, is the fuel manifold. As illustrated, fuel is being supplied from the main tanks directly to the engines. The manifold valves can also be set so that all tanks feed into

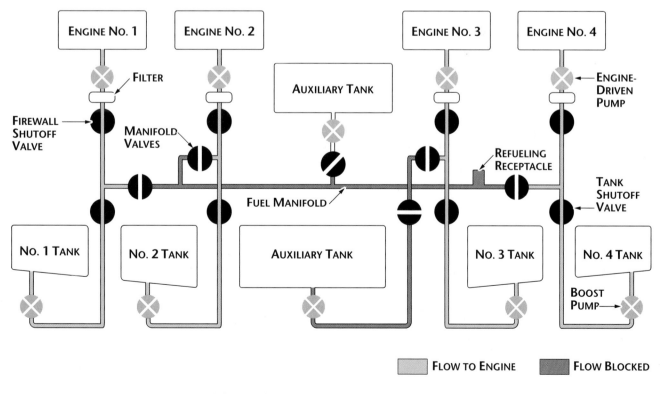

Figure 9-2-28. Transport category manifold type fuel system

the manifold and each engine receives its fuel from this line. The auxiliary fuel supply can be delivered to the engines only through the manifold. The main advantage of this system is its flexibility. Should an engine fail, its fuel is immediately available to the other engines. If a tank is damaged or the flow blocked, the corresponding engine can be supplied with fuel from the manifold.

Another advantage of this system is that all fuel tanks can be serviced at the same time through a single manifold connection (single-point fueling). This method of fueling greatly reduces the servicing time on large aircraft, and minimizes the opportunity for contamination. Some fueling stations provide manifold valve controls so that the refueler can control the amount of fuel and which tanks will be filled. These systems have detailed fueling and defueling instructions and procedures placarded on the fueling control panel access door. Always make sure the pump pressure of the fuel truck is correct for the aircraft.

Fuel jettison system. A fuel jettison (dump) system is required for transport category and general aviation aircraft if the maximum takeoff weight is greater than the maximum landing weight. The maximum takeoff and landing weights are part of the design specifications found in the Aircraft Type Certificate Data Sheets.

A fuel jettison system must be able to jettison enough fuel within 10 minutes for general aviation and 15 minutes for transport category aircraft to reach the maximum landing weight. This must be accomplished under the conditions encountered during all operations of the aircraft.

Additional specifications require that the fuel jettison must be stopped with a minimum of fuel for 45 minutes of cruise at maximum continuous power for reciprocating engines, and enough fuel for take-off and landing and 45 minutes of cruising for jet powered aircraft.

The fuel jettison system is usually divided into two separate, independent systems, one for each wing. This allows the pilot to maintain lateral stability by jettisoning fuel from the "heavy" wing, if it is necessary to do so. Normally, if an unbalanced fuel load exists, fuel will be used from the "heavy" wing by supplying fuel to engines on the opposite wing.

The system consists of lines, pumps, valves, dump chutes, and chute-operating mechanisms. Each wing contains either a fixed or extendable dump chute, depending on the system design. In either case, the fuel must discharge clear of the airplane. Fuel jettison requirements are covered in title 14 CFR part 23.1001.

Section 3

Inspection, Maintenance and Repair of Fuel Systems

Inspections performed in accordance with Part 43 of the Federal Aviation Regulations including the 100-hour and annual must include an examination of the fuel system that meets at least the criteria established in Appendix D of that regulation. The inspection of a fuel system installation consists of an examination of the system for conformity to design requirements and functional tests to prove correct operation.

Since there are considerable variations in the fuel systems used on today's aircraft, no attempt has been made here to describe any particular system. It is important to follow the manufacturer's instructions for your particular aircraft when performing any inspection or maintenance function.

Complete System Inspections

Inspect the entire system for wear, damage, or leaks. Make sure that all units are securely attached and properly safetied.

The drain plugs or valves in the fuel system should be opened to check for the presence of sediment or water. The filter and sump should also be checked for sediment, water, or slime. The filters or screens, including those provided for flow meters and auxiliary pumps, must be clean and free from corrosion.

Fuel valves may be checked for internal leakage by draining the strainer bowls, placing the valves in the off position, and turning on the fuel boost pump. If the valve leaks, fuel will flow into the strainer bowl.

All controls and control linkage should be checked for freedom of movement, security of locking, and freedom from damage due to chafing.

The fuel vents should be checked for correct positioning and freedom from obstruction; otherwise, fuel flow or pressure fueling may be affected. Fuel filler neck drains should be checked for freedom from obstruction.

If booster pumps are installed, the system should be checked for leaks by operating the pumps. During this check, the ammeter or load meter should be observed and the readings of all the pumps, where applicable, should be approximately the same. All fuel system components must be checked for proper bonding and grounding to properly drain off all static charges.

Fuel tanks. All applicable panels in the aircraft skin or structure should be removed and the tanks inspected for corrosion on the external surfaces, for security of attachment, and for correct adjustment of straps and stings. Check the fittings and connections for leaks or failures.

Some fuel tanks manufactured of light alloy materials are provided with inhibitor cartridges to reduce the corrosive effects of combined leaded fuel and water. Where applicable, the cartridge should be inspected and renewed at the specified periods.

Lines and fillings. Be sure that the fuel lines are properly supported and that the nuts and clamps are securely tightened. To tighten hose clamps to the proper torque, use a hose-clamp torque wrench. If this wrench is not available, tighten the clamp finger-tight plus the number of turns specified for the hose and clamp. If the clamps do not seal at the specified torque, replace the clamps, the hose, or both. After installing a new hose, check the clamps daily and tighten if necessary. When this daily check indicates that cold flow has ceased, inspect the clamps at less frequent intervals.

Replace the hose if the plies have separated, if there is excessive cold flow, or if the hose is hard and inflexible. Permanent impressions from the clamp and cracks in the tube or cover stock indicate excessive cold flow. Replace hose that has collapsed at the bends or as a result of misaligned fittings or lines. Some types of hose tend to flare at the ends beyond the clamps. This is not an unsatisfactory condition unless leakage is present.

Blisters may form on the outer synthetic rubber cover of hose. These blisters do not necessarily affect the serviceability of the hose. When a blister is discovered on a hose, remove the hose from the aircraft and puncture the blister with a pin. The blister should then collapse. If fluid (oil, fuel, or hydraulic) emerges from the pinhole in the blister, reject the hose. If only air emerges, pressure-test the hose at $1\frac{1}{2}$ times the working pressure. If no fluid leakage occurs, the hose is serviceable.

Puncturing the outer cover of the hose permits the entry of corrosive elements, such as water, which could attack the wire braiding and ultimately result in failure. For this reason, puncturing the outer covering of hoses exposed to the elements should be avoided.

The external surface of hose may develop fine cracks, usually short in length, which are caused by surface aging. The hose assembly is serviceable provided the cracks do not penetrate to the first braid.

Selector valves. Selector valves should be checked to determine if they rotate freely and without excessive backlash. The valve should have an obvious detent when it is in position and should be marked clearly. Check the valve for loose nuts or pins. Check worn control cables or pulleys or faulty bearings.

Pumps. During an inspection of a booster pump, check for the following conditions:

- Proper operation
- Leaks and condition of fuel and electrical connections
- Wear of motor brushes

Be sure the drain lines are free of traps, bends, or restrictions.

Check the engine-driven pump for leaks and security of mounting. Check the vent and drain lines for obstructions.

Main line strainers. All fuel strainers should be removed, checked for contamination, and cleaned at regular intervals. These intervals are listed in the aircraft maintenance manual. Check the strainers for contamination that might indicate a problem elsewhere in the fuel system. Pieces of rubber are often indications of faulty hoses.

Fuel quantity gauges. Fuel gauges and transmitters should be calibrated to indicate empty when the system has no more usable fuel in it. If a sight gauge is used, be sure that the glass is clear and that there are no leaks at the connections. Check the lines leading to it for leaks and security of attachment.

Check the mechanical gauges for free movement of the float arm and for proper synchronization of the pointer with the position of the float.

On the electrical and electronic gauges, be sure that both the indicator and the tank units are securely mounted and that the electrical connections are tight.

Fuel pressure gauge. The fuel pressure gauge should be checked for leaks around the fittings, oscillation of the needle and proper markings. It is also important to make certain that needle indicates zero when the engine and boost pumps are shut off.

Pressure warning signal. Inspect the entire installation for security of mounting and con-

Slow Seep

Seep

Heavy Seep

Area where fuel appears to flow or run following contour of skin when this area is wiped dry

Running Leak

Figure 9-3-1. Fuel leak classification

dition of the electrical, fuel, and air connections. Check the lamp by pressing the test switch to see that it lights. Check the operation by turning the battery switch on, building up pressure with the booster pump, and observing the pressure at which the light goes out. If necessary, adjust the contact mechanism.

With proper inspection and maintenance the aircraft fuel system is capable of delivering fuel to the engine reliably during normal flight conditions.

Fuel Tank Repairs

No fuel system is airworthy if it will not contain fuel. Inspection of the fuel tank bays or aircraft structure for evidence of fuel leaks is a very important part of the preflight inspection.

Defueling

Aircraft defueling must naturally occur before any repairs to a fuel tank. Depending on the airplane, defueling can be anything from draining the gas out of a J3 Cub into two five gallon cans, to draining several thousand gallons from a transport aircraft. In the process, there are some definite procedures.

- Never defuel an airplane in the hangar. The fumes are an explosion and fire hazard.

- Read the service manual for the proper procedures. Some aircraft with pressure refueling can also be defueled by the same system. Some cannot. Swept wing aircraft must have the outboard wing tanks defueled first to minimize the twisting effect on the wing caused by the fuel being located behind the wing attach points on the fuselage.

- Always ground the airplane and any containers, barrels, and trucks used for the process.

- Find out all of the requirements from the following:

  1. The aircraft manufacturer

  2. The shop where you work

  3. The airport on which you work

  4. The airport fire station (if on the airport)

Containers. All facilities have a process of handling drained fuel. Some will sell it back to fuel distributors, while others will pump it back into the fuel farm tanks to be refiltered. Still others will reuse gasoline if it is first strained through a chamois skin (gasoline will pass through, but water will not). Still others have a

hard and fast rule that says, "No fuel drained will ever be used in an airplane again".

The reason for not using drained fuel in an airplane again is that there is a considerable danger of contamination from storing fuel in barrels and cans. The possibility of water contamination from condensation is extreme. Many times, more so than when stored in underground tanks or airplane fuel tanks, plus it doesn't have the filtering and inspection processes as original fuel. The price of an airplane really isn't worth the price of the fuel saved.

Procedures. If the MMM does not give you a specific sequence of events for defueling, start with the outboard tanks first, switching as necessary to maintain balance. This could be particularly important if the airplane is on jacks. As many airplanes have jack points close to the fuselage, severe lateral imbalance could possibly cause an upset that could be catastrophic.

Welded steel tanks. Welded tanks are most common in smaller single and twin-engine aircraft. If the access plates to the fuel tank compartment are discolored, the tank should be inspected for leaks. When leaks are found, the tank must be drained and inerted.

Fuel must be drained in accordance with local policies and procedures and the manufacturer's recommendations. The draining, storage, and disposal or re-use of drained fuel may be strictly regulated in some locations. Make certain that you comply with all applicable safety laws and procedures.

Inerting the tank may be accomplished by slowly discharging a carbon dioxide fire extinguisher (5 lb. minimum size) into the tank. Dry nitrogen may be used if it is available. If the tank is to be welded, removal is necessary.

Before welding, the tank must be steamed for a minimum of 8 hours. This is to remove all traces of fuel. Air pressure not over $1/2$ p.s.i. may be used to detect the leaking area. Brush liquid soap or bubble solution over the suspected area to help identify the leak. Aluminum tanks are fabricated from weldable alloys. After riveting patches in place, the rivets may be welded to insure no leaks from that area. Pressure checks should be performed after repairs are complete to assure that all leaks have been corrected.

> **WARNING:** *Failure to properly inert a fuel tank before making a welded repair can be catastrophic. The ensuing explosion can seriously injure or kill you. Do it correctly, or not at all.*

Fuel cells. Fuel cell leaks will usually appear on the lower skin of the aircraft. A fuel stain in any area should be investigated immediately. Fuel cells suspected of leaking should be drained, removed from the aircraft and pressure checked. When performing a pressure check, $1/4$ to $1/2$ p.s.i. air pressure is adequate. All fuel cell maintenance must be accomplished in accordance with the manufacturer's specifications.

Integral fuel tanks. The integral tank is a non-removable part of the aircraft. Because of the nature of an integral tank, some leaks allow fuel to escape directly to the atmosphere. This makes it completely feasible to disregard certain minute leaks that do not represent a fire hazard or too great a loss of fuel. In order to standardize the procedures for integral tank fuel storage and maintenance, the various rates of fuel leakage are classified.

The size of the surface area that a fuel leak moistens in a 30-minute period is used as the classification standard. Wipe the leak area completely dry with clean cotton cloths. Compressed air may also be used to dry the leak area. Dust the leak area with dyed red talcum powder. The talcum powder turns red as the fuel wets it, making the wet area easier to see.

At the end of 30 minutes, each leak is classified into one of four classes of leaks: slow seep, seep, heavy seep, or running leak. The four classes of leaks are shown in Figure 9-3-1. A slow seep is a leak in which the fuel wets an area around the leak source not over $3/4$-inch in diameter. A seep is a leak that wets an area from $3/4$ inches to $1 \, 1/2$-inches in diameter. A heavy seep is a fuel leak that wets an area around the leak source from $1 \, 1/2$-inches to 3 inches in diameter. In none of these three leak classifications does the fuel run, flow, drip, or resemble any of these conditions at the end of the 30-minute time period.

The last classification, a running leak, is the most severe and the most dangerous. It may drip from the aircraft surface, it may run down vertical surfaces, or it may even run down your finger when you touch the wet area. The aircraft is unsafe for flight and must be grounded for repair. When possible, the fuel from the leaking tank should be removed after you mark the leak location. If it is impossible to *defuel* the tank immediately, the aircraft should be isolated in an approved area. Place appropriate warning signs around the aircraft until qualified personnel can defuel the leaking tank.

Grounding the aircraft for slow seeps, seeps, and heavy seeps is determined by the aircraft's handbook. This determination may depend on the location of the fuel tank. For example, can the leakage progress to a potential fire source? The number of fuel leaks in a given area is

Figure 9-3-2. Integral fuel tank access holes on a Boeing 757 wing

also a contributing factor. There is no rule of thumb for determining if the aircraft is to be grounded. Running leaks ground the aircraft regardless of location.

If the aircraft is airworthy and no repair is required, make appropriate entries on the aircraft forms and periodically observe the progress of the fuel leak. When repair is required, find the cause of the fuel leak and make an effective repair.

Integral fuel tanks on large transport category airplanes require fuel tank entry to repair some types of leaks. Entry is a complicated process of defueling, venting and drying, and testing the fuel tank to make sure it can be safely entered. Most operations require a breathing apparatus to ensure the workers safety. All require specific safety procedures, including a safety observer. Smaller integral tanks have access holes that allow for repairs after defueling (Figure 9-3-2). Though less complicated, each MMM will have a set of safety procedures for this process also.

Whether a simple gravity feed system, or a sophisticated manifold system on a Transport Category turbine jet, proper operation of the fuel system is essential to safe flight. With proper inspection and maintenance the aircraft fuel system should provide many hours of trouble free operation.

10

Ice and Rain Control Systems

All-weather flying became possible only with the advent of dependable de-icing systems. All transport category, most corporate, and many IFR private airplanes are equipped with de-ice systems. De-ice systems are as important as any other system on the airplane as they allow true all-weather operations.

Section 1

Ice Protection

Icing hazards are not limited to winter operation. Aircraft operating at high altitudes in moist air masses are just as likely to accumulate ice in August as they are in January. Therefore, year-round inspection and repair of ice projection systems is essential to the safe operation of high performance aircraft. Even though ice protection systems on smaller, low-altitude aircraft are typically considered optional equipment, proper operation is no less important. All ice protection systems are designed to maintain the stability and control of the aircraft in icing conditions. When needed they must operate properly, otherwise the aircraft and its occupants will be in severe danger.

Types of ice protection systems. There are three types of ice protection systems: *detection, anti-ice and de-ice*. Detection systems range from spotlights that shine on surfaces most likely to accumulate ice to vibration-sensitive probes that alert flight crews of icing in areas not visible from the flight deck. Anti-ice systems are designed to prevent the formation of ice while de-ice systems remove ice that has already formed. This section provides examples of these systems and discusses some basic

Learning Objectives

REVIEW
- types of ice protection systems
- visual and automatic methods of ice detection

DESCRIBE
- types and properties of de-ice and anti-ice systems
- characteristics and operation of pneumatic surface de-ice systems
- rain removal systems including properties of wiper systems and liquid repellent

Left. De-ice and anti-ice systems can make airplanes into true magic carpets. Within the limitations of take-off restrictions and airport capabilities, they can go anywhere, anytime. This Hawker business jet is equipped with a TKS wing de-ice system, and uses hot bleed air as an anti-ice agent for its engine nacelles.

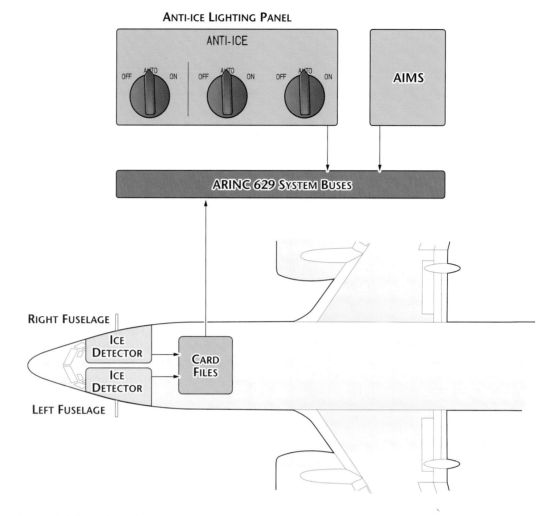

Figure 10-1-1. Ice detection system on a Boeing 777

maintenance procedures for each. Remember, however, that specific maintenance procedures are to be obtained only from current manufacturer's instructions or other FAA approved data.

Ice Detection

Visual ice detection. On smaller aircraft the wing leading edge is usually in clear view of the pilot, so ice is easily detected visually. Some small aircraft have lights installed to increase visibility of ice accumulation at night. For example, the Piper Chieftain has an ice spot light on the left engine nacelle that shines along the wing's leading edge. The pilot decides when enough ice has accumulated to activate the de-ice system.

Automatic ice detection. On larger aircraft the airframe is less visible from the flight deck. These aircraft require an automatic system to detect ice. The ice detection system on the Boeing 777 (Figure 10-1-1) is a good example from a transport jet. This system incorporates an ice detection probe on each side of the nose

Figure 10-1-2. Another example of an ice dectection device is this probe on a Global Express

of the fuselage. The probe extends into the air stream and vibrate at a particular frequency. When ice accumulates on the probe, the frequency of that vibration changes, causing the computer to activate the anti-ice system and alert the flight crew. An example on a corporate airplane is this fuselage mounted ice detector. It is on a Global Express (Figure 10-1-2).

De-ice and Anti-ice Systems

Bleed air anti-ice. Locations on the airframe that require anti-ice protection are illustrated in Figure 10-1-3. Wing and tail surfaces are heated by hot compressor bleed air ducted along the inside of the lead edges before it is exhausted overboard. In a similar fashion, bleed air is extracted from the engine anti-ice (EAI) valve and routed around the inside of each inlet (Figure 10-1-4).

Bleed air anti-ice systems are designed to prevent ice in the most extreme conditions of onrushing frigid and moist air. The heat required to do this tends to overheat the airframe and is dangerous if the system is operated while the aircraft is on the ground or the engines are not operating. Therefore, operation of the airframe surface anti-ice is prevented on the ground by a "weight on wheels"

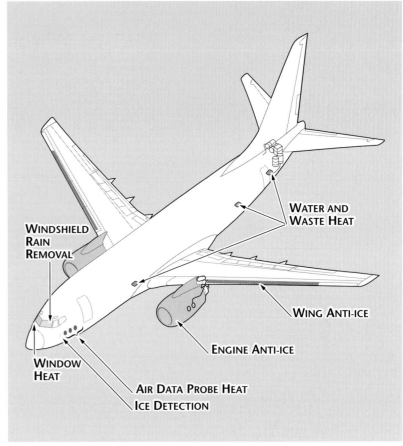

Figure 10-1-3. Areas that require anti-ice protection

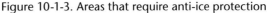

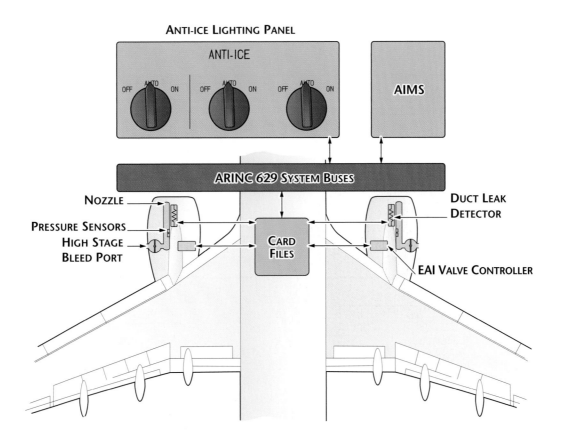

Figure 10-1-4. Bleed air extracted from the engine is routed around each inlet

Figure 10-1-5. Heated dual pitot tubes and angle of attack indicator on a Lear 45

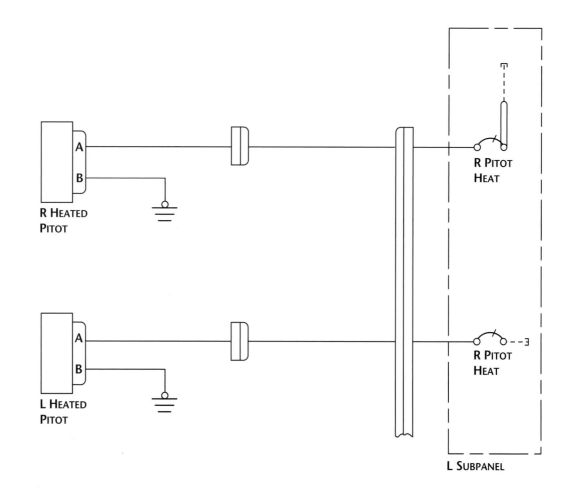

Figure 10-1-6. Typical electric anti-ice diagram for pitot systems, static ports, windshields, and other systems

or squat switch. Similarly, the engine anti-ice system incorporates an oil pressure switch auto throttle sensor or other means to prevent EAI operation when the engine is not running. It is important that such switches and control devices be taken into account during troubleshooting.

Bleed air lines and connections should be carefully inspected for evidence of failed seals and other leaks. Hot air leaking into some compartments poses a fire hazard. Exercise caution when performing maintenance on bleed air systems soon after they have been turned on because they may still be hot enough to cause burns to the skin. Care must also be taken to seal off, or otherwise isolate, bleed air ducts during such operations as engine compressor washes. Otherwise soap or other contaminants may enter the ducts causing corrosion or inhibiting operation of control valves.

Electrical anti-ice. Imbedded electrical heaters are used to anti-ice air data probes, such as pitot and static ports, fuel vents, windshields, and similar components (Figure 10-1-5). Most electrical anti-ice systems include a circuit breaker or fuse, heating element, wires and a control switch.

The wiring diagram in Figure 10-1-6 shows an example of a typical anti-ice system using dual pitot heads. The two systems are completely independent of each other. Should one system fail for any reason, the other will still be operable. All IFR certified airplanes must have more than one static source that is also anti-iced.

Observe caution when testing electrical anti-ice systems because they can become dangerously hot when operated without the cooling effect of ram air. Therefore the duration of any operational test should be minimal. For aircraft equipped with an ammeter, the preferred test method is simply to activate the system and observe a change in current draw. Otherwise the heater must be operated long enough for a change in temperature to be sensed. Do not touch the heated component directly (Figure 10-1-7). Moving your hands in close proximity to the heater is sufficient to sense a change in surrounding air temperature and to confirm operation. To reduce the risk of burns, always advise others working around the aircraft of any heater test in progress.

Pneumatic surface de-ice. As stated above, de-ice systems are designed to remove ice rather than prevent it. A pneumatic de-ice system is comprised of de-icer "boots" which are inflatable rubber tubes that are attached by glue or special rivets to the leading edges of wings and stabilizers. (Figure 10-1-8) They are periodically inflated by compressor bleed air

from turbine engines or by pneumatic pumps on reciprocating engine-powered aircraft. Ice is broken by the mechanical force of the expanding tubes and then carried away in the air stream.

Figure 10-1-7. Be careful not to touch heated probes directly.

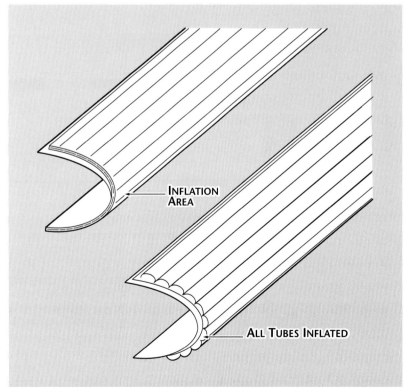

Figure 10-1-8. An example of the inflation/deflation cycle of a pneumatic de-ice boot

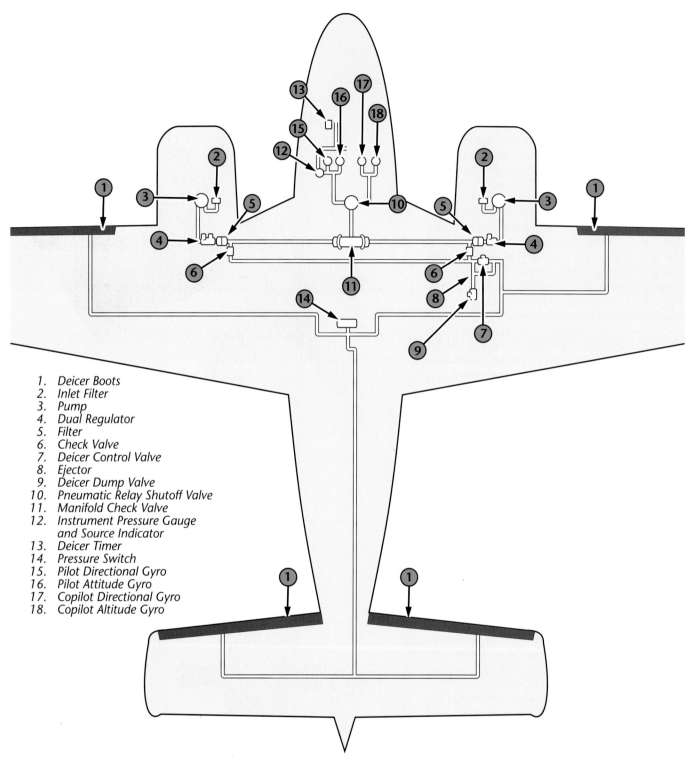

Figure 10-1-9. A twin-engine pneumatic de-ice system using dry air pumps

1. Deicer Boots
2. Inlet Filter
3. Pump
4. Dual Regulator
5. Filter
6. Check Valve
7. Deicer Control Valve
8. Ejector
9. Deicer Dump Valve
10. Pneumatic Relay Shutoff Valve
11. Manifold Check Valve
12. Instrument Pressure Gauge
    and Source Indicator
13. Deicer Timer
14. Pressure Switch
15. Pilot Directional Gyro
16. Pilot Attitude Gyro
17. Copilot Directional Gyro
18. Copilot Altitude Gyro

A negative pressure vacuum, from an ejector, is applied to the inside of the tubes between inflation cycles and when the system is not in operation. The vacuum insures complete deflation to prevent the boots from interfering with the aerodynamic qualities of the airfoil. Inflation is alternated between sets of boots rather than happening all at once. For example the wing boots may inflate alternately with the tail surface boots. Most systems allow for

an automatic mode with continuously repeating cycles controlled by a timer and a manual single-cycle mode controlled by the pilot. Fluctuations on the pressure gauge will indicate that the system is operating.

Refer to Figure 10-1-9 for an example of a pneumatic de-ice system. Pressure is supplied by two engine-driven pneumatic pumps (Item 3). These are dry air pumps which means they use carbon dust emitted from internal sliding vanes rather than engine oil for lubrication. In this particular application the pumps provide pressure to the system because they are installed near the inlet and push the air through it. In different applications, the very same pumps can be used to provide vacuum when they are installed near the exhaust end and pull the air through the system. In some applications the pumps draw vacuum on the inlet side to power certain instruments and then use the outlet exhaust to provide pressure for pneumatic de-ice.

Each pump in Figure 10-1-9 has its own *inlet filter*, (Item 2), *two-stage regulator*, (Item 4) and *in-line filter* (Item 5). The inlet filters are foam bands installed around the opening of the system and are easily replaced during any scheduled inspection. The two-stage regulators maintain a nominal pressure of 5-6 p.s.i. for instrument operation and then increase to 19-20 p.s.i. for boot inflation. The in-line filters protect the system and instruments from carbon dust and debris from the pumps or regulators when they fail. As with most filters, these should always be replaced along with pumps.

Inboard from the in-line filters the pressure lines from each engine connect at the *manifold check valve* (Item 11). In the event of a leak, pump failure, or when only one engine is operating, the manifold check valve will maintain system pressure by preventing leakage through the inoperative side.

When the system is not in operation all boots are subjected to vacuum from the *ejector valve*, (Item 8) which provides a low pressure from air exhausted through a venturi. During operation a timer controls *solenoid* valves in the distributor that alternately cycle the wing and tail boots between pump pressure and ejector vacuum (Figure 10-1-10).

Dry regulated shop air can be substituted for the engine-driven pumps or compressor bleed air during ground test operation (Figure 10-1-11). It is extremely important that only dry air be used for such tests. Excessive moisture from compressed air can freeze inside the boots and prevent them from operating.

Kits are available that include appropriate regulators, dryers and connectors suitable for several different types of aircraft. The use of test equipment enables the system to be test operated inside the hangar. Cycling the system in the hangar allows for troubleshooting to confirm operation of system components and for boot inflation to check for minor leaks. Small leaks may be patched in a similar fashion to tire inner tubes.

Figure 10-1-10. A timer and solenoid valve that controls the inflation/deflation cycles

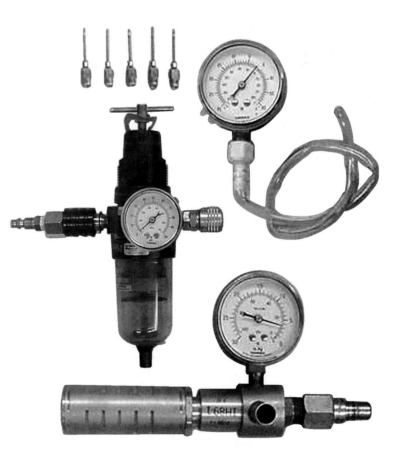

Figure 10-1-11. An Airborne pneumatic de-ice test kit. It simulates the air supply for both engine driven pump and bleed air systems.

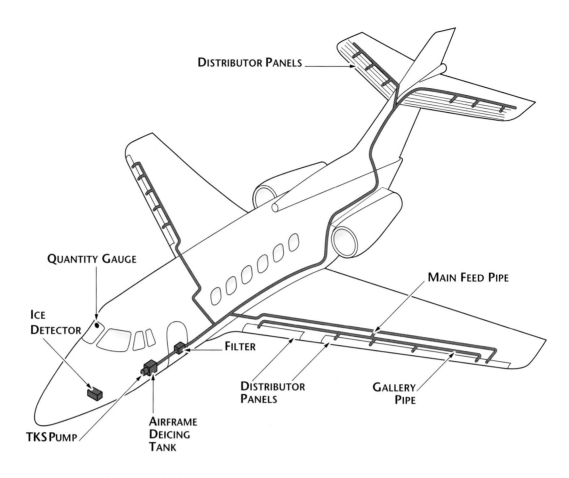

Figure 10-1-12. This pictorial schematic diagram shows the basic system layout of a Hawker TKS de-ice system.

De-ice boot maintenance. Pneumatic boots are normally glued over leading edges that have been stripped to bare metal and cleaned thoroughly. The adhesive used is a contact type and is spread on the leading edge and the boot and allowed to dry. The center of the leading edge and the center of the boot are marked with a line. The boot is then rolled up and, while one person keeps the two lines together, others can pull the boot down taut and press it into place.

Once cemented in place it entire surface needs to be rolled down for good contact. Each boot manufacturer will have their approved materials. Ice Shields™ products use either 3M EC1300L Scotch-Grip Rubber Adhesive or Bostik 1096M with 1007M primer.

All boot installations should be finished with a conductive edge sealer. The area should be taped off and the sealer applied, then the tape removed promptly. Always use the approved edge sealer so that electrostatic charges that build up can be bled off to the airframe. Otherwise, the charge may discharge between the boot and the leading edge, leaving a pinhole in the boot.

Boots will always sustain some damage over time. They can be repaired in accordance with the airframes maintenance manual. Do not use automotive silicone products to dress up a used boot. It will not dissipate static as it should and may be damaged. Keep all petroleum products away from all de-ice boots, and clean them with warm (not hot) soapy water and clear water rinse.

TKS de-ice and anti-ice. *TKS* is a glycol based *weeping membrane* system popularized by more than 26 years of service on the BAe-125 series Hawker business jets. The heart of the system is a metal membrane that covers the leading edges of the wings and empennage, where de-ice boots would be installed on a pneumatic system.

The membranes are stainless steel and have several hundred perforations per square inch. When TKS fluid is pumped into the distribution lines and fills a porous plastic sheet located under the de-ice panels, it weeps from the membrane through the almost invisible holes and covers the surface with glycol. Because ice will not form on a glycol wetted surface, and glycol can actually remove ice, TKS fluid flowing back across the wing will actually keep the wing virtually clean of ice. As glycol does not damage any finishes, there are no ill effects.

From the pilot's standpoint, TKS systems are extremely simple to operate. Simply turn on the pump switch when de- or anti-ice is needed. The fluid tank in the HS-125 will hold enough TKS for a maximum of one hour and fifty-two minutes, with an eighteen minute reserve. To service simply fill the tank.

Although later model Hawkers use 115-volt electric windshield heat, some have a methyl alcohol standby system for the left half. All models use conventional bleed air systems for engine anti-ice. Figure 10-1-12 shows the basic layout of a Hawker TKS system.

In addition to the Hawker HS-125 systems, aftermarket TKS systems are starting to appear. The system has been in service on the Beech Bonanza since 1987. It is also available for Cessna 210s and Beechcraft Barons.

Laser-drilled titanium panels are installed on the leading edges of the wings and horizontal and vertical stabilizers. A typical alcohol type slinger ring is installed on the propeller and a spray bar is mounted below on the windshield. A glycol based fluid is exuded through the panels and flows over the surfaces, just as in the Hawker system.

The glycol-based fluid is metered from a tank by a small, electrically driven pump through a micro-filter to proportioning units. The proportioning units contain calibrated capillary tubes which apportion fluid to the individual panels and the propeller slinger ring. The windshield is protected with an on-demand pump and spray bar.

A pilot needs to make only one decision while operating the ice protection system: using the anti-ice or de-ice mode. With the anti-ice mode, a protective film of glycol prevents the formation of ice. With the de-ice mode, the glycol chemically breaks the ice bond. A significant feature of both modes is the elimination of run back ice.

The outer skin of the weeping panels is manufactured from titanium, 0.9 mm thick. Titanium is lightweight, has excellent strength, durability, and corrosion resistance. The panel skin is perforated by laser-drilled holes, 0.0025 inches in diameter, 800 per square inch (see Figure 10-1-13). The backplate of a panel is manufactured from stainless steel and is formed to create a reservoir for the TKS fluid. A porous membrane between the outer skin and the reservoir assures even flow and distribution through the entire area of the panel. The porous panels are bonded as a cuff over the leading edges, utilizing a two-part, flexible adhesive. Fluid is supplied to the panels and propellers by individually selectable metering pumps. The porous area of the titanium panels is designed to cover the stagnation point travel on the appropriate leading edge over a normal operating environ-

Figure 10-1-13. A weeping de-ice panel shown on a Beechcraft Bonanza

ment, assuring fluid flow over the upper and lower surfaces of the protected area.

Windshield Icing Control Systems

The common method for controlling ice formation and fog on aircraft windows is the use of an electrical heating element built into the window. After heating, windows have a higher ability to withstand bird strikes because cold plastic and/or glass break easier than when it is warm. It is common practice to place the conductive coating towards the outside of a windshield for anti-icing, and towards the inside of side windows for defogging.

A layer of transparent conductive material is the heating element and a layer of transparent vinyl plastic adds a no shatter quality to the window. The vinyl and glass plies (Figure 10-1-14) are bonded by the application of pressure and heat. The conductive coating dissipates static electricity from the windshield in addition to providing the heating element.

On some aircraft, electric switches automatically turn the system on when the air temperature is low enough for icing or frosting to occur. The system may operate continuously, or it may operate with a pulsating pattern. Large airplanes normally supply AC electricity to the window film.

Heating to 110°F is normal for most glass/plastic-laminated materials. Normally an over-temp controller will shut the system down at 145°F. In order not to overheat a window when heat is first turned on, a ramp-up controller is used. It slows the heating process to keep the window from cracking before thermal expansion stresses can normalize.

An electrically heated windshield system includes:

- Windshield autotransformers and heat control relays
- Heat control toggle switches
- Indicating lights
- Windshield control units
- Temperature-sensing elements (thermistors) laminated in the panel

Maintenance. There are several problems associated with electrically heated windshields. These include delamination, scratches, arcing, and discoloration. Delamination (separation of the plies) although undesirable, is not harmful structurally provided it is within the limits established by the aircraft manufacturer, and is not in an area where it affects the optical qualities of the panel.

Arcing in a windshield panel usually indicates that there is a breakdown in the conductive coating. Where chips or minute surface cracks are formed in the glass plies, simultaneous release of surface compression and internal tension stresses in the highly tempered glass can result in the edges of the crack and the conductive coating parting slightly.

Arcing is produced where the current jumps this gap, particularly where these cracks are parallel to the window bus bars. Where arcing exists, there is invariably a certain amount of local overheating, which, depending upon its severity and location can cause further damage to the panel. Arcing near a temperature-sensing element is a particular problem since it can upset the heat control system.

Electrically heated windshields are transparent to directly transmitted light but they have a distinctive color when viewed by reflected light.

This color will vary from light blue, yellow tints, and light pink depending upon the manufacturer of the window panel. Normally, discoloration is not a problem unless it affects the optical qualities.

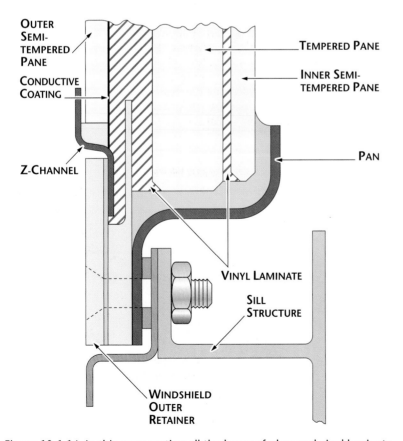

Figure 10-1-14. In this cross section all the layers of glass and vinyl laminate are visible, as is the conductive coating.

Windshield scratches are more prevalent on the outer glass ply where the windshield wipers are indirectly the cause of this problem. Any grit trapped by a wiper blade can convert it into an extremely effective grinder when the wiper is in motion. The best solution to scratches on the windshield is to prevent them; clean the windshield wiper blades as frequently as possible. Windshield wipers should never be operated on a dry panel, since this increases the chances of damaging the surface.

If visibility is not adversely affected, scratches or nicks in the glass plies are allowed within the limitations set forth in the appropriate service or maintenance manuals. Attempting to improve visibility by polishing out nicks and scratches is not recommended. This is because of the unpredictable nature of the residual stress concentrations set up during the manufacture of tempered glass.

Tempered glass is stronger than ordinary annealed glass due to the compression stresses in the glass surface, which have to be overcome before failure can occur from tension stresses in the core. Polishing away any appreciable surface layer can destroy this balance of internal stresses and can even result in immediate failure of the glass.

Window defrost system. The window defrost system directs heated air from the cabin heating system (or from an auxiliary heater, depending on the aircraft) to the pilot's and copilot's windshield and side windows by means of a series of ducts and outlets. In warm weather when heated air is not needed for defrosting, the system can be used to defog the windows. This is done by blowing ambient air on the windows using the blowers.

Windshield Alcohol De-icing Systems

An alcohol de-icing system is provided on some aircraft to remove ice from the windshield, and can be a backup to the primary de-ice system. Alcohol flows from the solenoid valve and is filtered and directed to the alcohol pumps and distributed through a system of plumbing lines to spray on the windshield.

Bleed Air Anti-Ice Systems

An effective windshield anti-ice system for turbine and turbojet airplanes is a bleed air system. By blowing hot engine bleed air across the outside of windshields it can be protected from ice, rain, and snow. The pressure air forms a protective layer over the

windshield and prevents water from contacting it, and thus ice from forming. If ice or frost has formed during ramp time, the system will eliminate it quickly.

Section 2

Rain Protection

Rain removal systems are a form of protection from adverse weather that significantly improves the safety and practicality of aircraft. The ability to take-off, fly, and land safely in rain rather than detour several miles around weather systems or cancel flights is critical to the success of almost every aircraft operator.

Types of systems. The primary types of rain removal systems are wipers and liquid repellent. Most rain removal systems contain few components and are basic in operation. Prudent routine inspection, service, and maintenance should insure their reliability.

Wiper systems. Aircraft windshield wiper systems may be powered either electrically or hydraulically. A typical electrical wiper system is shown in Figure 10-2-1. Notice there is only one motor, but two each of the flexible drive assemblies and converters. The motor provides torque that is transmitted by the flexible drive assemblies to the converters. The converters change the rotary motion of the motor to a cyclical left and right wiping action of the arm and blade. Control for the system allows for different speeds and a park mode that hides the arm assembly when the system

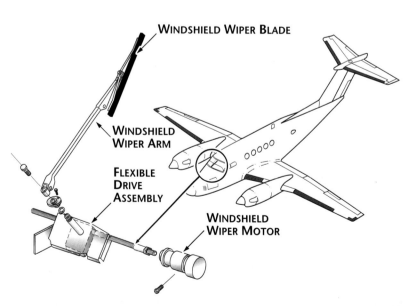

Figure 10-2-1. An electrically-operated windshield wiper system

is not in use. Some aircraft add a second motor for redundancy (Figure 10-2-2). This allows at least one window to remain clear in the event of a motor failure.

During inspections the wiper system should be test-operated in all modes. Never operate windshield wipers on a dry windshield. Using a hose or other suitable water source for lubrication will prevent damage to the blades and windows. The wiper blades should be thoroughly inspected for damage or cracking. Rather than spending the time to inspect blades, some manuals call for replacement at routine intervals. These steps should insure safe reliable operation and good visibility in the rain.

Liquid repellent. In addition to wiper systems, some aircraft have the capability of spraying a chemical on the windshield that causes the water to collect into larger beads that are more easily dispersed by the wipers and airflow. Rain repellent systems include a reservoir, pump, and spray nozzle or manifold. The electrical control for the pump includes a circuit breaker or fuse, wires, connectors, and a normally open momentary switch so that the pump only operates while the switch is depressed. (Figure 10-2-3).

The liquid repellent, if not removed immediately, can harden on the windshield and reduce visibility. Spraying repellent onto a dry windshield and without the wipers operating could require the windshield to be replaced. Handle liquid repellents carefully because they are corrosive and environmentally unfriendly. Always consult the appropriate Material Safety Data Sheet (MSDS) and read container labels carefully when using materials or chemicals you are not completely familiar with. Inspect the reservoirs regularly for leaks and clean up any spills immediately to prevent damage to surrounding areas of the aircraft. The hazards and additional equipment associated with liquid rain repellent systems are eliminated on state of the art aircraft by installing windshields that are pre-coated with a permanent repellent chemical.

Air pressure. Some airplanes use bleed air to provide a layer of compressed air that covers the outside of the windshield. The pressure prevents most rain from penetrating the layer and contacting the glass.

The air pressure simply blows the rain over the top.

Ground De-icing of Aircraft

Any deposits of ice, snow, or frost on the external surfaces of an airplane may drastically affect its performance. Reduced aerodynamic lift and increased aerodynamic drag resulting from the disturbed airflow over the airfoil surfaces, or the extra weight of the deposit over the whole aircraft is enough to make it impossible for an airplane to fly. An airplane may also be seriously affected by the freezing of moisture in controls, hinges, valves, microswitches, or by the ingestion of ice into the engine.

When aircraft are hangared to melt snow or frost, any melted snow or ice may freeze again if the aircraft is subsequently moved into sub-zero temperatures. Any measures taken to remove frozen deposits while the aircraft is on the ground must also prevent the possible refreezing of the liquid.

Frost Removal

Frost deposits can be removed by placing the aircraft in a warm hangar or by using a frost remover or de-icing fluid. These fluids normally contain ethylene glycol and isopropyl alcohol and can be applied either by spray or by hand. It should be applied within two hours of flight.

De-icing fluids may adversely affect windows or the exterior finish of the aircraft. Therefore, only the type of fluid recommended by the aircraft manufacturer should be used.

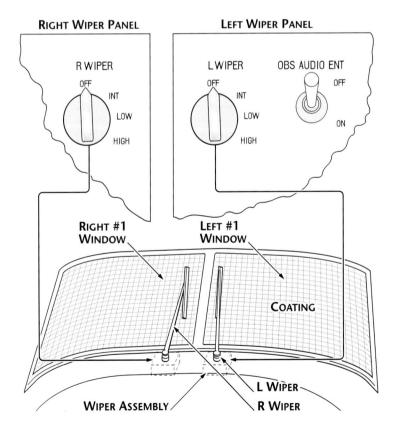

Figure 10-2-2. An example of a redundant electrical wiper system

Removing Ice and Snow Deposits

Probably the most difficult deposit to deal with is deep, wet snow when ambient temperatures are slightly above the freezing point. This type of deposit should be removed with a brush or squeegee. Use care to avoid damage to antennas, vents, stall warning devices, vortex generators, etc., which may be concealed by the snow.

Light, dry snow in subzero temperatures should be blown off whenever possible; the use of hot air is not recommended, since this would melt the snow, which would then freeze, requiring de-ice.

Moderate or heavy ice and residual snow deposits should be removed with a de-icing fluid. No attempt should be made to remove ice deposits or break an ice bond by force.

After completion of de-icing operations, inspect the aircraft to ensure that its condition is satisfactory for flight. All external surfaces should be examined for signs of residual snow or ice, particularly near the control gaps and hinges. Check the drain and pressure sensing ports for obstructions. When it becomes necessary to physically remove a layer of snow, all protrusions and vents should be examined for signs of damage.

Control surfaces should be moved to ascertain that they have full and free movement. The landing gear mechanism, doors, and bay, and wheel brakes should be inspected for snow or ice deposits and the operation of uplocks and microswitches checked.

Snow or ice can enter turbine engine intakes and freeze in the compressor. If the compressor cannot be turned by hand for this reason, hot air should be blown through the engine until the rotating parts are free.

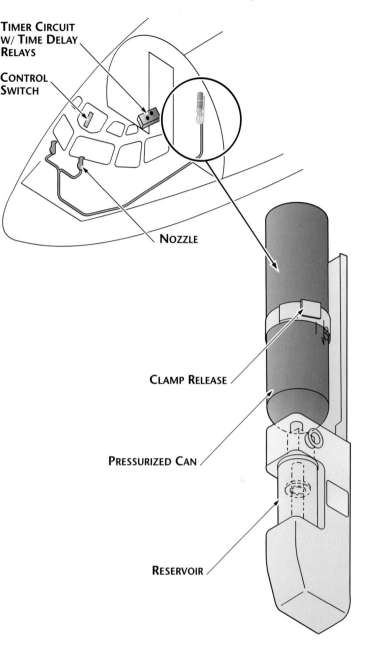

Figure 10-2-3. An example of a rain repellent system used on a Boeing 727

Pitot Tube Anti-Icing

To prevent the formation of ice over the opening in the pitot tube, a built-in electric heating element is provided. A switch, located in the cockpit, controls the power to the heater. Use caution when ground checking the pitot tube since the heater must not be operated for long periods unless the aircraft is in flight. Additional information concerning pitot tubes is found in Chapter 5.

Aircraft Instrument Systems

Heating elements should be checked for functioning by ensuring that the pitot head begins to warm up when power is applied. If an ammeter or loadmeter is installed in the circuit, the heater operation can be verified by noting the current consumption when the heater is turned on.

Water and Toilet Drain Heaters

Heaters are provided for toilet drain lines, water lines, drain masts, and waste water drains when they are located in an area that is subjected to freezing temperatures in flight. The types of heaters used are integrally heated hoses, ribbon, blanket, or patch heaters that wrap around the lines, and gasket heaters. Thermostats are provided in heater circuits where excessive heating is undesirable or to reduce power consumption. The heaters have a low voltage output and continuous operation will not cause overheating.

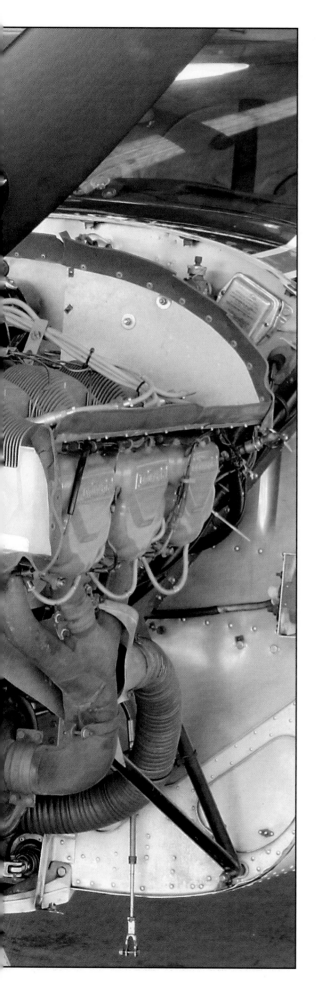

11

Aircraft Inspections

Inspections, and inspection systems, are the foundation of aircraft maintenance. They are also the most commonly performed tasks in aviation maintenance. From this firm foundation has emerged the safest method of transportation in the world.

Aircraft Inspection

How an inspection is performed. Performing an inspection on a live aircraft for the first time can be overwhelming. This is normal and it is a good thing to be cautious. Overconfidence, cockiness, or attempting to display a level of skill the technician does not possess leads to disaster. The proper combination of skill, insight and confidence can only be achieved through diligent effort over time.

Title 14 CFR part 65 (65.81) is worth highlighting:

General privileges and limitations.

(a) A certificated mechanic may perform or supervise the maintenance, preventive maintenance or alteration of an aircraft or appliance, or a part thereof, for which he is rated (but excluding major repairs to, and major alterations of, propellers, and any repair to, or alteration of, instruments), and may perform additional duties in accordance with §§65.85, 65.87, and 65.95. However, he may not supervise the maintenance, preventive maintenance, or alteration of, or approve and return to service, any aircraft or appliance, or part thereof, for which he is rated unless he has satisfactorily performed the work concerned at an earlier date. If he has not so performed that work at an earlier date, he may show his ability to do it by performing it to the satisfaction of the Administrator or under the direct supervision of a certificated and appropriately rated

Learning Objectives

DESCRIBE
- steps to perform an annual inspection, from the work order, to the inspection to signoff.
- components of and how to perform a progressive inspection
- purpose and occasion for conformity inspections

EXPLAIN
- purpose for and process of records review
- how to complete logbook sign-offs

Left. From an annual inspection on a four place Cessna in this photo, to a D check on a passenger airliner, the inspection process is the same; only the application is different. One technician can inspect a Cessna in a day, while it takes a full crew several weeks to perform a D check.

Figure 11-1-1. Searches are easier the more familiar the AMT is with the shop's record system.

Courtesy of Duncan Aviation

mechanic, or a certificated repairman, who has had previous experience in the specific operation concerned.

The purpose of this chapter is to give insight into what to expect when performing inspections in the field. Seasoned inspectors earn their reputations through diligent effort, constant application of the highest level of skill, and personal integrity.

This chapter covers an *annual/100-hour inspection* on a light aircraft, a *C check* performed on a large transport aircraft, and a *conformity inspection*.

Section 1

The Annual Inspection

Light aircraft. The following are the inspection requirements for a light aircraft, such as a Piper Cherokee: Title 14 CFR part 91.409 states that no person may operate an aircraft unless, within the preceding 12 calendar months, it has had an annual inspection in accordance with part 43 and has been approved for return to service by a person authorized by part 43.7.

If the aircraft is used for hire, such as flight instruction, 91.409(b) says the aircraft must be inspected every 100-hours of service, the scope of the inspection being the same as an annual inspection for the particular type of aircraft. The 100-hour limit may be exceeded by up to 10 hours to allow the aircraft to fly to the place where the inspection is to be performed, but the excess time must be deducted from the next 100-hour period. In other words, a privately owned light aircraft needs an annual inspection every 12 months. If the aircraft is used for hire, it needs to be inspected every 100-hours and have an annual inspection every 12 months.

How to proceed. There are several ways for an owner to have an AMT perform an annual inspection. The annual is equal to a 100-hour inspection, with some additional repairs possible, and with a bit more paperwork. Below are three basic scenarios:

- The AMT performs the 100-hour inspection/service work, and the shop's in-house IA does the inspection. Then the AMT, the IA, and the owner discuss repairs beyond the 100-hour inspection, allowing the IA or AMT to do the repairs.

- The AMT, does the 100-hour inspection, the annual inspection, and any associated repairs, with the findings approved and signed off by the in-house IA.

- The AMT does the 100-hour inspection, allowing an outside IA to do the annual inspection and associated repairs.

In general, the second method, while preferred, is not legally correct. The annual inspection can only be performed by an IA. However, the technician and the owner must decide on how to proceed before starting the work order.

The work order. An annual inspection begins and ends with paperwork. Before touching the aircraft, the technician creates a *work order* which outlines the work to be performed. The work order should be signed by the AMT and the owner. The work order is the contract that describes what will be carried out, and the fee the owner will pay. It is crucial to get the order in writing.

Maintenance research material. The following are materials that must be researched and verified to see if they are applicable to the inspection of a Piper Cherokee:

- Airworthiness Directives (AD) – 316 each

- Manufacturers Service Bulletins (SB) – 157 each

- Manufacturers Service Letters (SL) – 145 each

- Vendor Service Publications (VP) – 56 each

Researching material for the 100-hour annual inspection can mean handling 1,000 or more pieces of paper (Figure 11-1-1) not including the actual aircraft records. This is maintenance information only.

While it is true that only the AD notes are mandatory, many insurance companies insist that their clients keep up with all manufacturer's service requests. Customers that are part 135 operators may have listed all manufacturer's maintenance information as *Mandatory Compliance* items in their *Operating Specification*. It is easier for them to do it this way and takes away any choice.

Prudent customers track manufacturer's service requests as they go along. Previous technicians should have been keeping additional listings, similar to the required AD note list. If they have, there will be many entries listed as *NA by SN* (not applicable by serial number) and/or *NA by Model* (not applicable by aircraft model). For example, an AMT had seventy-nine AD notes to look up, but only ten actually applied to the Piper PA 28-180. All but two of those were previously complied with, and were listed in the *AD note compliance card* in the log book. The same thing held true for the remainder of the items. However, if items had not been tracked and logged correctly, the AMT would have to discuss it with the owner, to avoid adding considerable expense.

Records Review

The records review is sometimes the most difficult part of an inspection, but is essential. An aircraft's records tell its life story - how old it is, its accident and damage history, and any persistent maintenance problems that require attention.

Regulations. A records review is required by regulations. 14 CFR part 91.417 lists the records that must be kept, and how they must be maintained. Among the records that must be tracked are:

- Records of maintenance and inspections. This includes the last annual or 100-hour inspection, and any routine maintenance such as lubrication.

- Total times in service of the airframe, engine and propeller.

- The time since overhaul of all items on the aircraft that require overhauls on a specified time basis.

- The current status of life-limited parts, if any.

- A list of all applicable Airworthiness Directives and their current status. If they are one-time occurrences they must have been signed off as such. If the AD is repetitive, it must also be signed off as repetitive and must be complied with as stated in the AD itself.

NOTE: *An AD note usually has more than one method of compliance. When it is complied with and signed off on, the actual method of compliance must be stated, i.e. CW by Paragraph (1)(a), etc.*

Safety. Another reason for a records review is safety. Part 91.417 states that the aircraft records must show the current status of applicable Airworthiness Directives (AD notes). If a repetitive AD is not accomplished on schedule, it jeopardizes the safe operation of the aircraft. Scheduled maintenance and inspection that has not been performed or is performed late can lead to increased wear and premature failure of aircraft components.

Economy. A third reason for a review is economy. If maintenance is not documented, or documented incorrectly, it is as if the work was never done. For example, a corrosion inspection of a wing spar carry-through structure that requires removal of the wings. If accomplishment of the inspection is not recorded

Airworthiness Directive

▸ **Federal Register Information**

▾ **Header Information**
DEPARTMENT OF TRANSPORTATION

Federal Aviation Administration

14 CFR Part 39

Amendment 39-3810; AD 80-14-03

Airworthiness Directives; Piper Model PA-28-161, PA-28-181, PA-28-201T, PA-28-236, PA-28R-201, PA-28RT-201, PA-28R-201T, PA-28RT-201T, PA-32-260, PA-32-300, PA-32-301, PA-32-301T, PA-32R-300, PA-32RT-300, PA-32R-301, PA-32R-301T, PA-34-200T and PA-44-180 Airplanes
PDF Copy (If Available):

▾ **Preamble Information**
AGENCY: Federal Aviation Administration, DOT

DATES: Effective July 1, 1980.

▾ **Regulatory Information**

80-14-03 PIPER AIRCRAFT CORPORATION: Amendment 39-3810. Applies to the following airplanes certificated in all categories equipped with Bendix, King or Narco transmitters with factory installed control wheel push-to-talk switches.

(A)

Models Affected	Serial Numbers Affected
PA-28-161 Warrior II	28-7816001 through 28-8016289
PA-28-181 Archer II	28-7890001 through 28-8090266
PA-28-201T Turbo Dakota	28-7921001 through 28-7921091
PA-28-236 Dakota	28-7911001 through 28-8011096
PA-28R-201 Arrow III	28R-7837001 through 28R-7837317
PA-28RT-201 Arrow IV	28R-7918001 through 28R-8018049
PA-28R-201T Turbo Arrow III	28R-7803001 through 28R-7803373
PA-28RT-201T Turbo Arrow IV	28R-7931001 through 28R-8031074
PA-32-260 Six	32-7800001 through 32-7800008
PA-32-300 Six 300	32-7840001 through 32-7940290
PA-32-301 Saratoga	32-8006001 through 32-8006015
PA-32-301T Turbo Saratoga	32-8024001 through 32-8024007
PA-32R-300 Lance	32-7880001 through 32-7880068
PA-32RT-300 Lance II	32R-7885001 through 32R-7985105
PA-32RT-300T Turbo Lance II	32R-7787001, 32R-7887002 through 32R-7987126
PA-32R-301 Saratoga SP	32R-8013001 through 32R-8013071
PA-32R-301T Turbo Saratoga SP	32R-802900 1 through 32R-8029068
PA-34-200T Seneca II	34-7870001 through 34-8070150
PA-44-180 Seminole	44-7995001 through 44-8095020

Compliance is required as indicated unless already accomplished to prevent disruption of radio communication.

(a) Within the next 10 hours time in service after the effective date of this AD, comply with the following:

(1) Locate the audio adapter connector in the main radio harness behind the radio stack. This is the interconnect for individual radios in the audio selector panel.

(2) Locate the muting relay plug in this connector. It is a three (3) pin plug containing wires ASP-1, ASP-2, and ASP-3 and is positioned at one end of the adapter connector.

(3) Disconnect the muting relay plug and attach securely to the harness in accordance with Advisory Circular AC43.13-1A.

(4) Conduct complete operational check of all radios.

(5) Make a maintenance record entry.

(b) Within the next 100 hours time in service after the effective date of this AD, comply with the following:

(1) On PA-32-260, PA-32-3O0, PA-32-301, PA-32-301T, PA-32R-300, PA- 32RT-300, PA-32RT-300T, PA-32R-301, PA-32R-301T and PA-34-200T model aircraft gain access by removing four (4) screws from the speaker grille ring. On all other affected models, gain access by lowering the overhead dome panel.

(B)

Figure 11-1-2. Microfiche, CD-ROM, or computer-based systems, make the AD research process faster but must be used efficiently.

---

properly in the aircraft log book (or if the log book is lost or destroyed), the inspection must be done again at considerable expense. As a result, missing, incomplete or incorrect aircraft logs dramatically reduce an aircraft's resale value.

Records review procedure. First, assemble all available log books and records and make sure the records are complete in accordance with part 91.417(a). Then, group the log books and records for the airframe, engine and propeller. Note the part and serial numbers for these items, and compare them to the numbers on the data plates on the aircraft.

Next, verify the total times in service for each component, paying particular attention to proper arithmetic and watching for any gaps in the records. Many older aircraft in service will have several log books. Place the logs for each airframe, engine and propeller in chronological order. Then ensure that the last entry in one log matches the first entry in the following log, with no errors or gaps in time. An aircraft can sit for long periods of time, and the logs should reflect this. However, gaps followed by many hours of utilization without corresponding maintenance entries are suspicious.

Determine accurate times for components. Airframe total time is determined by verifying log hours. The last log book entry should show the airframe total time in service. Now, find the engine time in service by subtracting the airframe time at engine installation from the engine time at installation. Add this to the current airframe time to get the current engine total time in service. Use the same method to determine the propeller's time in service.

Accuracy is critical when noting times in service. Though not mandatory on a private reciprocating engine airplane, engines and propellers are generally overhauled on a fixed time schedules and the times should be correct. When an overhaul is budgeted at $30 or $40 per flight hour, small errors can cost big money.

The next step is to determine the inspection program used for the aircraft being inspected. The inspection program tells what scheduled inspection and maintenance must be performed to maintain airworthiness. These tasks must be listed to refer to later. Then a search must be performed for ADs and *Service Bulletins* (SBs) which apply to the aircraft (Figure 11-1-2 and 11-1-3). In the past paper listings have been used, but now information is compiled on Microfiche, compact discs, or computer programs and the search process is much faster.

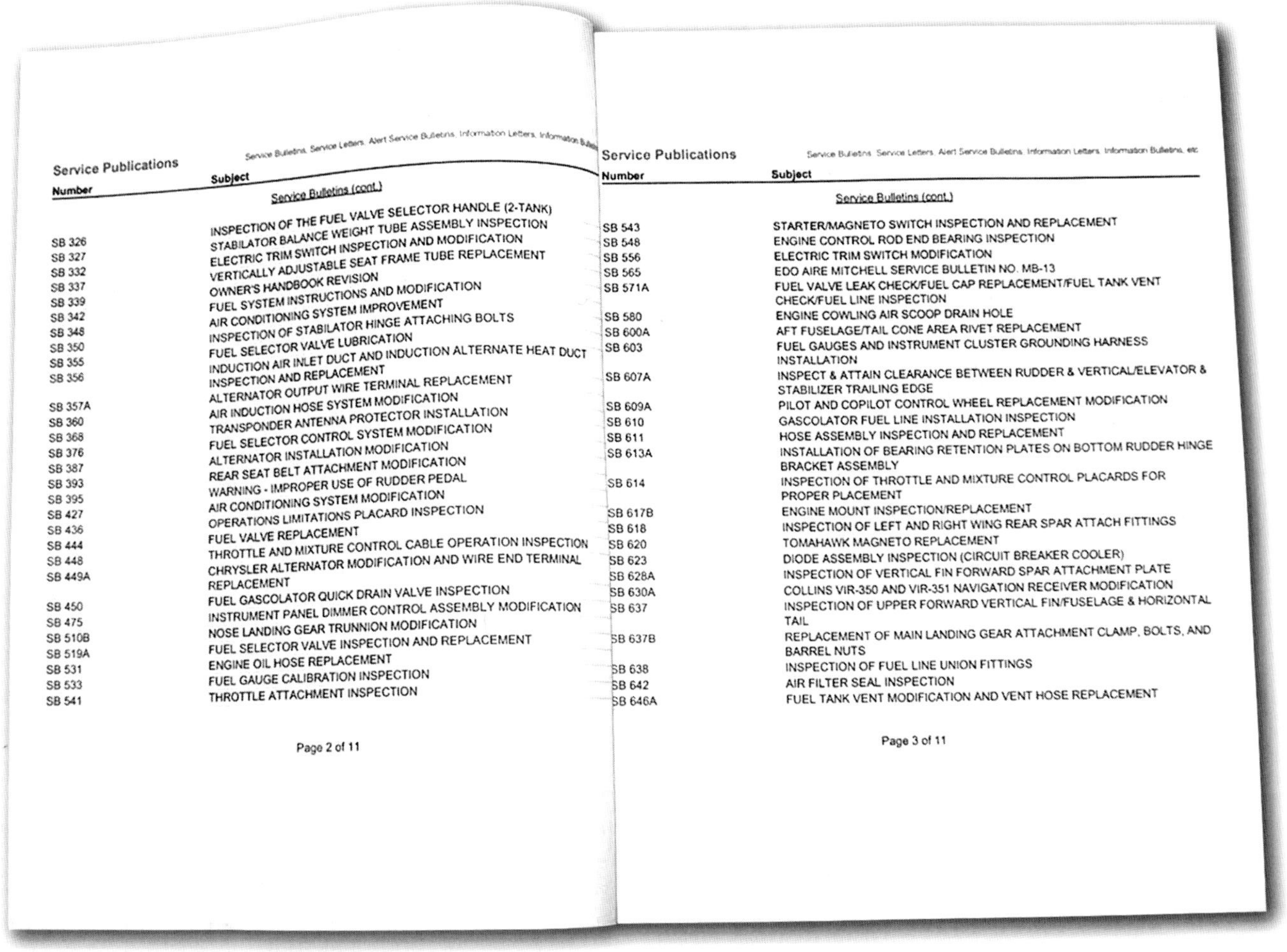

Service Publications — Service Bulletins, Service Letters, Alert Service Bulletins, Information Letters, Information Bulletins, etc.

Number	Subject
	Service Bulletins (cont.)
SB 326	INSPECTION OF THE FUEL VALVE SELECTOR HANDLE (2-TANK)
SB 327	STABILATOR BALANCE WEIGHT TUBE ASSEMBLY INSPECTION
SB 332	ELECTRIC TRIM SWITCH INSPECTION AND MODIFICATION
SB 337	VERTICALLY ADJUSTABLE SEAT FRAME TUBE REPLACEMENT
SB 339	OWNER'S HANDBOOK REVISION
SB 342	FUEL SYSTEM INSTRUCTIONS AND MODIFICATION
SB 348	AIR CONDITIONING SYSTEM IMPROVEMENT
SB 350	INSPECTION OF STABILATOR HINGE ATTACHING BOLTS
SB 355	FUEL SELECTOR VALVE LUBRICATION
SB 356	INDUCTION AIR INLET DUCT AND INDUCTION ALTERNATE HEAT DUCT INSPECTION AND REPLACEMENT
SB 357A	ALTERNATOR OUTPUT WIRE TERMINAL REPLACEMENT
SB 360	AIR INDUCTION HOSE SYSTEM MODIFICATION
SB 368	TRANSPONDER ANTENNA PROTECTOR INSTALLATION
SB 376	FUEL SELECTOR CONTROL SYSTEM MODIFICATION
SB 387	ALTERNATOR INSTALLATION MODIFICATION
SB 393	REAR SEAT BELT ATTACHMENT MODIFICATION
SB 395	WARNING - IMPROPER USE OF RUDDER PEDAL
SB 427	AIR CONDITIONING SYSTEM MODIFICATION
SB 436	OPERATIONS LIMITATIONS PLACARD INSPECTION
SB 444	FUEL VALVE REPLACEMENT
SB 448	THROTTLE AND MIXTURE CONTROL CABLE OPERATION INSPECTION
SB 449A	CHRYSLER ALTERNATOR MODIFICATION AND WIRE END TERMINAL REPLACEMENT
SB 450	FUEL GASCOLATOR QUICK DRAIN VALVE INSPECTION
SB 475	INSTRUMENT PANEL DIMMER CONTROL ASSEMBLY MODIFICATION
SB 510B	NOSE LANDING GEAR TRUNNION MODIFICATION
SB 519A	FUEL SELECTOR VALVE INSPECTION AND REPLACEMENT
SB 531	ENGINE OIL HOSE REPLACEMENT
SB 533	FUEL GAUGE CALIBRATION INSPECTION
SB 541	THROTTLE ATTACHMENT INSPECTION

Page 2 of 11

Service Publications — Service Bulletins, Service Letters, Alert Service Bulletins, Information Letters, Information Bulletins, etc.

Number	Subject
	Service Bulletins (cont.)
SB 543	STARTER/MAGNETO SWITCH INSPECTION AND REPLACEMENT
SB 548	ENGINE CONTROL ROD END BEARING INSPECTION
SB 556	ELECTRIC TRIM SWITCH MODIFICATION
SB 565	EDO AIRE MITCHELL SERVICE BULLETIN NO. MB-13
SB 571A	FUEL VALVE LEAK CHECK/FUEL CAP REPLACEMENT/FUEL TANK VENT CHECK/FUEL LINE INSPECTION
SB 580	ENGINE COWLING AIR SCOOP DRAIN HOLE
SB 600A	AFT FUSELAGE/TAIL CONE AREA RIVET REPLACEMENT
SB 603	FUEL GAUGES AND INSTRUMENT CLUSTER GROUNDING HARNESS INSTALLATION
SB 607A	INSPECT & ATTAIN CLEARANCE BETWEEN RUDDER & VERTICAL/ELEVATOR & STABILIZER TRAILING EDGE
SB 609A	PILOT AND COPILOT CONTROL WHEEL REPLACEMENT MODIFICATION
SB 610	GASCOLATOR FUEL LINE INSTALLATION INSPECTION
SB 611	HOSE ASSEMBLY INSPECTION AND REPLACEMENT
SB 613A	INSTALLATION OF BEARING RETENTION PLATES ON BOTTOM RUDDER HINGE BRACKET ASSEMBLY
SB 614	INSPECTION OF THROTTLE AND MIXTURE CONTROL PLACARDS FOR PROPER PLACEMENT
SB 617B	ENGINE MOUNT INSPECTION/REPLACEMENT
SB 618	INSPECTION OF LEFT AND RIGHT WING REAR SPAR ATTACH FITTINGS
SB 620	TOMAHAWK MAGNETO REPLACEMENT
SB 623	DIODE ASSEMBLY INSPECTION (CIRCUIT BREAKER COOLER)
SB 628A	INSPECTION OF VERTICAL FIN FORWARD SPAR ATTACHMENT PLATE
SB 630A	COLLINS VIR-350 AND VIR-351 NAVIGATION RECEIVER MODIFICATION
SB 637	INSPECTION OF UPPER FORWARD VERTICAL FIN/FUSELAGE & HORIZONTAL TAIL
SB 637B	REPLACEMENT OF MAIN LANDING GEAR ATTACHMENT CLAMP, BOLTS, AND BARREL NUTS
SB 638	INSPECTION OF FUEL LINE UNION FITTINGS
SB 642	AIR FILTER SEAL INSPECTION
SB 646A	FUEL TANK VENT MODIFICATION AND VENT HOSE REPLACEMENT

Page 3 of 11

Figure 11-1-3. A service bulletin list is normally available on a current microfiche or as a computer printout on the airplane.

Compare the records in the logbooks against the examples spelled out in the inspection program. Find each inspection and maintenance task signed off as accomplished in the last year's log book entries. The logs must show any ADs and SBs complied with (Figure 11-1-4). Continue this process until all tasks are accounted for, and all hours tally properly.

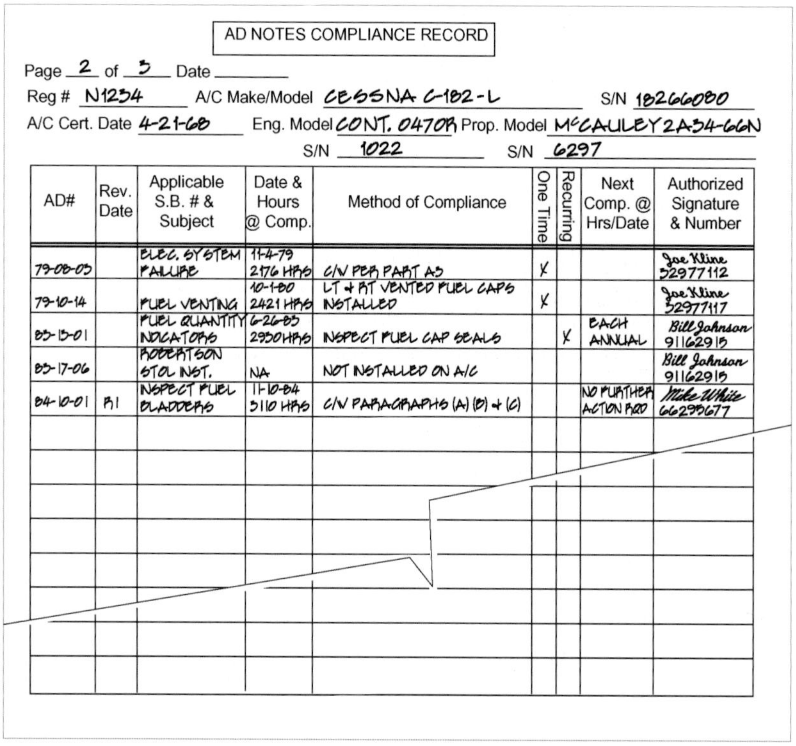

AD NOTES COMPLIANCE RECORD

Page 2 of 3 Date ______
Reg # N1234 A/C Make/Model CESSNA C-182-L S/N 18266080
A/C Cert. Date 4-21-68 Eng. Model CONT. 0470R Prop. Model McCAULEY 2A34-66N
S/N 1022 S/N 6297

AD#	Rev. Date	Applicable S.B. # & Subject	Date & Hours @ Comp.	Method of Compliance	One Time	Recurring	Next Comp. @ Hrs/Date	Authorized Signature & Number
79-08-03		ELEC. SYSTEM FAILURE	11-4-79 2176 HRS	C/W PER PART A3	X			Joe Kline 52977112
79-10-14		FUEL VENTING	10-1-80 2421 HRS	LT + RT VENTED FUEL CAPS INSTALLED	X			Joe Kline 52977117
85-13-01		FUEL QUANTITY INDICATORS	6-26-85 2930 HRS	INSPECT FUEL CAP SEALS		X	EACH ANNUAL	Bill Johnson 91162915
85-17-06		ROBERTSON STOL INST.	NA	NOT INSTALLED ON A/C				Bill Johnson 91162915
84-10-01	R1	INSPECT FUEL BLADDERS	11-10-84 5110 HRS	C/W PARAGRAPHS (A) (B) + (C)			NO FURTHER ACTION REQD	Mike White 66299677

Figure 11-1-4. A list of AD notes and their status is a required item. The list is to be kept with, and preferably attached to, the log book(s). This example above is from the FAA AI study guide.

Cleaning

After reviewing the aircraft records, work on the aircraft can begin. Part 43 appendix D requires the aircraft structure to be clean in order to see defects. First, all fairings, cowlings, and inspection panels must be removed. Accumulations of grease, dirt and debris trap moisture and lead to corrosion, or cause wear between moving parts. Before cleaning, note any accumulations that could indicate a leaking component or loose fastener.

Inspecting the Aircraft

Checklists. Part 43.15(c) requires a *checklist* to be used when performing an annual or 100-hour inspection. The checklist can be one designed by the technician, one used by another inspector, or one provided by the equipment manufacturer, which is the most common checklist. In any case, a checklist must include, at the very least, the scope and detail of the items in *Part 43 Appendix D*. Keep in mind that Appendix D gives a very basic and generic list of items to be checked.

Factory checklists. The checklist for a light aircraft is specific to a certain make and model and will go into much more detail. The factory checklist (Figure 11-1-5) is the preferred one to use. They are often cross-referenced to Service Bulletins and Service Letters, both for the airplane and for all appliances. Most factory checklists also have a notes section as the last page. This page may contain information not listed on the checklist.

The checklist breaks the inspection into separate areas such as landing gear, wings and empennage, cabin, cockpit and engine. This inspection example looks at a typical reciprocating engine, following the guidelines of Appendix D.

Spectrometric oil analysis program (SOAP). If the engine in the airplane is on an oil analysis program, there will be a specific time and method to take the oil sample. Now is the time to make that determination. If a sample is required, it must be taken and sent to the analysis laboratory quickly. Most labs will do the analysis and fax back the results. If the samples are not taken at the proper time, the inspection may be completed before the results are returned, thus delaying completion. Figures 11-1-6 shows examples of two different types of SOAP sample kits. (A) is the complete collection kit and (B) is the sample container and paperwork submitted.

While many labs will provide results as shown in Figure 11-1-7, it can be difficult to track results from them. A better way is to enter the results on a computer spreadsheet as this makes it simpler to find trends and variations and does not require much additional time.

Starting the inspection. Before cleaning the engine, it should be checked visually for signs of leakage. Thoroughly inspect it for excessive oil, fuel, or hydraulic leaks.

- Check studs and nuts for obvious damage, looseness, and improper torque. Checking the torque of every nut is not required, but look, feel, push and pull on the various components in the engine compart-

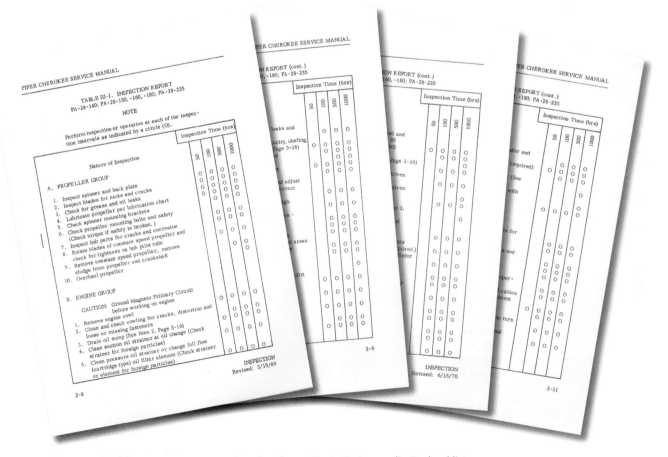

Figure 11-1-5. Factory checklists are more comprehensive than a Part 43, Appendix D checklist.

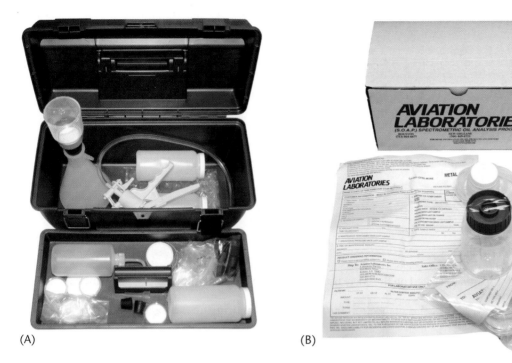

(A)

(B)

Figure 11-1-6. SOAP sample kits: (A) Complete collection kit, (B) Sample container and paperwork

ment. This helps to find loose studs and backed-off nuts.

- Inspect the engine internally. Perform a cylinder compression check. If weak cylinder compression is encountered, check internal condition and tolerances to determine the cause. The reciprocating engine chapter of *Aircraft Powerplant Maintenance* explains the process. Check for evidence of metal particles and foreign matter on screens and drain plugs. Remove the oil filter, cut it open and inspect the filter element for particles and debris. Excessive amounts of metal particles warrant further investigation.

- Check the engine mount for cracks, looseness of mounting to the airframe, and looseness of the engine in the mount. Check flexible vibration dampers for poor condition and deterioration.

- Inspect the engine controls for defects, correct travel, and improper safetying. Make sure that control cables are properly routed and secured, and that they are not chafing on adjacent structures.

- Check lines, hoses, and clamps for leakage, condition and looseness. Gently pull and flex the lines and hose to help find looseness, deterioration and kinks.

- Check exhaust stacks for cracks, defects and improper attachment. The exhaust system is exposed to high temperatures and corrosive gasses. Check for corrosion and cracks, particularly at welds.

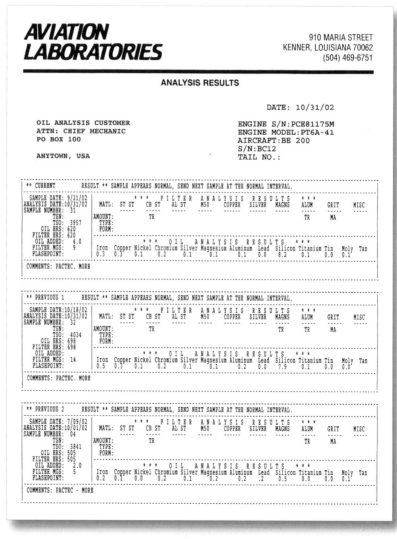

Figure 11-1-7. A spectrometric oil analysis program report

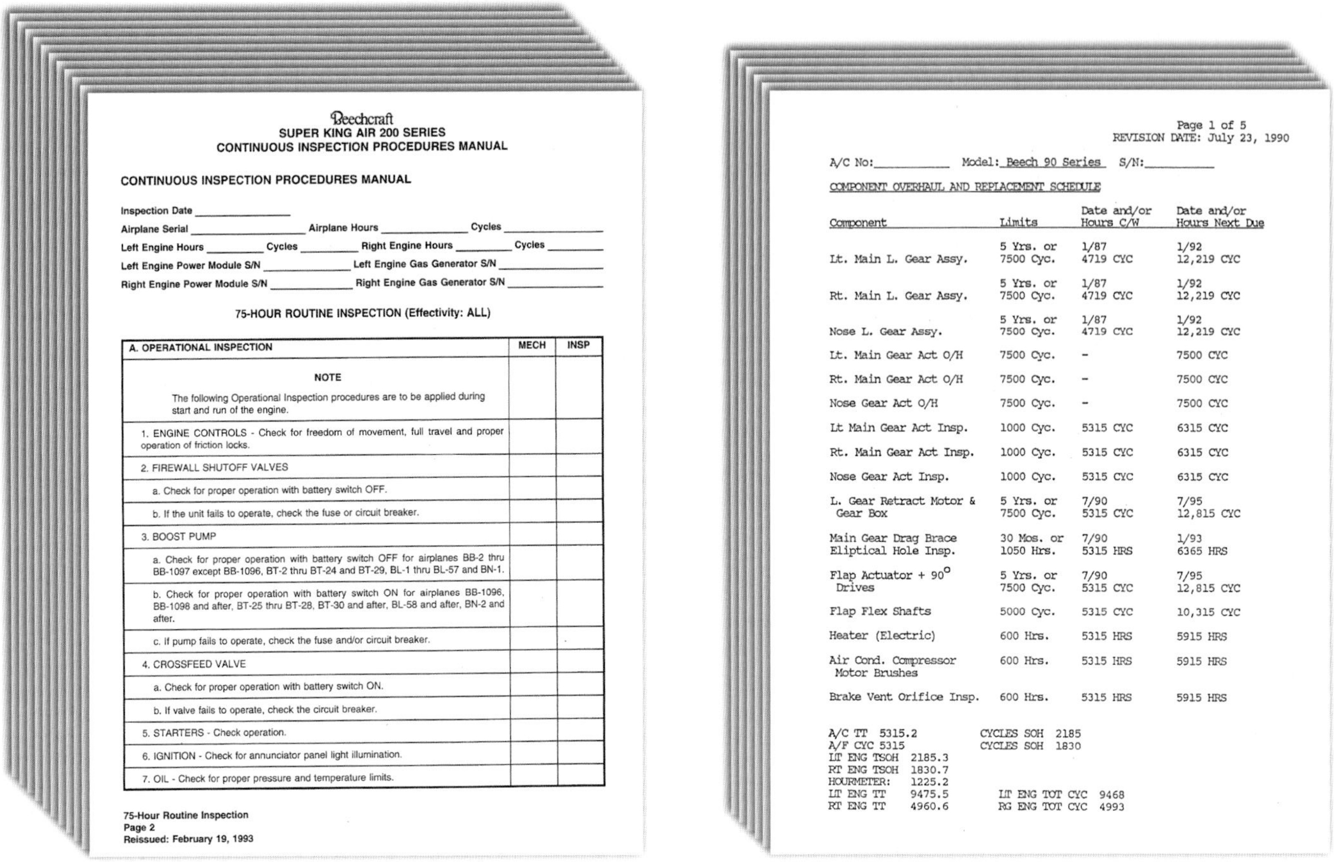

Beechcraft
SUPER KING AIR 200 SERIES
CONTINUOUS INSPECTION PROCEDURES MANUAL

CONTINUOUS INSPECTION PROCEDURES MANUAL

Inspection Date ______________
Airplane Serial ______________ Airplane Hours ______________ Cycles ______________
Left Engine Hours __________ Cycles __________ Right Engine Hours __________ Cycles __________
Left Engine Power Module S/N ______________ Left Engine Gas Generator S/N ______________
Right Engine Power Module S/N ______________ Right Engine Gas Generator S/N ______________

75-HOUR ROUTINE INSPECTION (Effectivity: ALL)

A. OPERATIONAL INSPECTION	MECH	INSP
NOTE The following Operational Inspection procedures are to be applied during start and run of the engine.		
1. ENGINE CONTROLS - Check for freedom of movement, full travel and proper operation of friction locks.		
2. FIREWALL SHUTOFF VALVES		
a. Check for proper operation with battery switch OFF.		
b. If the unit fails to operate, check the fuse or circuit breaker.		
3. BOOST PUMP		
a. Check for proper operation with battery switch OFF for airplanes BB-2 thru BB-1097 except BB-1096, BT-2 thru BT-24 and BT-29, BL-1 thru BL-57 and BN-1.		
b. Check for proper operation with battery switch ON for airplanes BB-1096, BB-1098 and after, BT-25 thru BT-28, BT-30 and after, BL-58 and after, BN-2 and after.		
c. If pump fails to operate, check the fuse and/or circuit breaker.		
4. CROSSFEED VALVE		
a. Check for proper operation with battery switch ON.		
b. If valve fails to operate, check the circuit breaker.		
5. STARTERS - Check operation.		
6. IGNITION - Check for annunciator panel light illumination.		
7. OIL - Check for proper pressure and temperature limits.		

75-Hour Routine Inspection
Page 2
Reissued: February 19, 1993

Page 1 of 5
REVISION DATE: July 23, 1990

A/C No:__________ Model: Beech 90 Series S/N:__________

COMPONENT OVERHAUL AND REPLACEMENT SCHEDULE

Component	Limits	Date and/or Hours C/W	Date and/or Hours Next Due
Lt. Main L. Gear Assy.	5 Yrs. or 7500 Cyc.	1/87 4719 CYC	1/92 12,219 CYC
Rt. Main L. Gear Assy.	5 Yrs. or 7500 Cyc.	1/87 4719 CYC	1/92 12,219 CYC
Nose L. Gear Assy.	5 Yrs. or 7500 Cyc.	1/87 4719 CYC	1/92 12,219 CYC
Lt. Main Gear Act O/H	7500 Cyc.	-	7500 CYC
Rt. Main Gear Act O/H	7500 Cyc.	-	7500 CYC
Nose Gear Act O/H	7500 Cyc.	-	7500 CYC
Lt Main Gear Act Insp.	1000 Cyc.	5315 CYC	6315 CYC
Rt. Main Gear Act Insp.	1000 Cyc.	5315 CYC	6315 CYC
Nose Gear Act Insp.	1000 Cyc.	5315 CYC	6315 CYC
L. Gear Retract Motor & Gear Box	5 Yrs. or 7500 Cyc.	7/90 5315 CYC	7/95 12,815 CYC
Main Gear Drag Brace Eliptical Hole Insp.	30 Mos. or 1050 Hrs.	7/90 5315 HRS	1/93 6365 HRS
Flap Actuator + 90° Drives	5 Yrs. or 7500 Cyc.	7/90 5315 CYC	7/95 12,815 CYC
Flap Flex Shafts	5000 Cyc.	5315 CYC	10,315 CYC
Heater (Electric)	600 Hrs.	5315 HRS	5915 HRS
Air Cond. Compressor Motor Brushes	600 Hrs.	5315 HRS	5915 HRS
Brake Vent Orifice Insp.	600 Hrs.	5315 HRS	5915 HRS

A/C TT 5315.2 CYCLES SOH 2185
A/F CYC 5315 CYCLES SOH 1830
LT ENG TSOH 2185.3
RT ENG TSOH 1830.7
HOURMETER: 1225.2
LT ENG TT 9475.5 LT ENG TOT CYC 9468
RT ENG TT 4960.6 RG ENG TOT CYC 4993

Figure 11-2-1 Progressive inspection forms

- Check the engine accessories for apparent defects and security of mounting. Check all systems for improper installation, poor condition, defects and insecure attachment.
- Check the engine cowling for cracks and defects.

When the engine inspection is completed, start and run the engine. Part 43.15 (c)2 states that you must determine satisfactory performance in accordance with the manufacturer's recommendations for static and idle r.p.m., power output, magneto operation, fuel and oil pressure, and cylinder and oil temperatures. After the engine run, again check the engine for fuel, oil, or hydraulic leaks.

Logbook Signoffs

When the inspection is completed, make an entry and sign off in each airframe, engine, propeller, and appliance logbook. First, note the date the inspection was performed and the aircraft total time in service. (Also enter the time in service for the engine, propeller, and appliances in the respective logbooks.) Next, you will enter one of two different signoffs into the logbooks.

Airworthy. If the aircraft is found to be airworthy and approved for return to service, enter this statement: *"I certify that this aircraft has been inspected in accordance with an annual inspection and was determined to be in an airworthy condition."* Then sign the log, including printed name, signature, certificate type and number. Remember, a technician must hold an Inspection Authorization (IA) to perform an annual inspection and return the aircraft to service. AMTs without an IA require the services of a qualified IA to return an airplane to service after an annual inspection.

Unairworthy. Sometimes the aircraft will not be approved for return to service. The owner may want a different technician to perform the required repairs, or for some other reason may not be able to correct defects discovered during the annual inspection. In that case, you will enter a statement similar to the following: *"I certify that this aircraft has been inspected in accordance with an annual inspection and a list of discrepancies and unairworthy items dated dd/mm/yy has been provided to the aircraft owner or operator".* Enter name, signature, certificate type and number to complete the inspection.

Section 2

Progressive Inspections

General aviation. A *progressive inspection* program is an alternative to an annual or 100-hour inspection for light aircraft. Some operators may not want to ground their aircraft for several days to perform a complete annual. Part 91.409 allows them the option of using a progressive inspection program.

The progressive inspection has the same scope and detail as an annual inspection, but is broken down into segments or phases. First, a complete annual inspection must be performed. Then, depending on the particular requirements of the progressive program, the first phase inspection will be performed at a certain number of flight hours (usually 50 hours) or calendar time, say two or three months. Figure 11-2-1 shows a set of progressive inspection forms for a Beechcraft King Air. The form is 25 pages long. In addition is a five page listing of component times in service.

The next phase is inspected in its order of rotation. At the end of one year the entire aircraft will have been inspected, and the cycle starts again. Unlike an annual inspection, a progressive inspection may be performed by a technician who does not hold an Inspection Authorization, as long as he is being supervised by an IA.

Commercial Aircraft. The size and complexity of large commercial aircraft make it impractical to conduct a single, all-encompassing inspection once a year. Because of this, airlines operating under CFR part 121 use a system of progressive airframe inspections as the backbone of their maintenance program. A progressive inspection can be thought of as a series of smaller, less intensive inspections which get more in depth in terms of which structural area or system is looked at, and how detailed the inspection of that area or system is. This reduces the amount of time the aircraft is out of service and still meets the inspection requirements specified by the manufacturer and the Federal Aviation Administration. These checks are called A, B, C, and D checks. The D check is the most time consuming of all the checks. See Figure 11-2-2.

A progressive inspection program affords the operator greater flexibility in scheduling maintenance tasks. Some of the less detailed inspections can be performed when the aircraft overnights away from a major maintenance base.

Figure 11-2-2. A Boeing 757 undergoing a D check

Figure 11-2-3. Inspections on transport category airplanes start at the airport-loading gate. Here run and taxi qualified AMTs start the trip to the run-pad.

To accomplish the heavier inspections, the operator must have access to a facility with enough space to accommodate the aircraft and support functions, such as the parts inventory and component repair shops, necessary to carry out the inspection. Typically, an air carrier maintains its own facilities for this purpose, but the use of contract maintenance companies that specialize in aircraft inspections is also common.

Whether the carrier performs the heavy checks at its own facility (in-house), or sends the aircraft to a contract maintenance vendor (outsourced), the requirements for a progressive airframe inspection are the same. The following paragraphs describe the way a typical C check inspection is conducted.

Engine run checks. The inspection process starts at a maintenance base when a group of technicians (maintenance crew) goes to pick up the aircraft at the gate (Figure 11-2-3). Technicians rated for engine run and taxi will start the engines and taxi the aircraft to a designated run area on the airport called a *run pad*.

Once at the run pad, the engines are run at different output levels and engine performance information is recorded on the inspection paperwork. Systems are also checked during the pre-run. Some of the systems that might be checked are the hydraulic, environment control, pressurization, and avionics.

When all pre-inspection engine run checks have been completed the aircraft is taxied to the check hangar and the engines are shut down. Pre-inspection engine runs are important because they establish a *base line* of engine and systems performance and expose any problems in these areas so that they can be addressed during the check.

Aircraft docking. The second phase of the inspection process is *aircraft docking*. This may seem like a simple operation, but it is extremely important that it be performed with care since it is easy to damage an aircraft when it is moved

Figure 11-2-4. The process of docking a large aircraft requires close coordination between many people.

in close quarters (Figure 11-2-4). The aircraft is attached to a *tug* via a tow bar and moved into the hangar. This requires observers, or spotters, who watch for obstructions and help the tug driver guide the aircraft into position.

Each technician involved in the process communicates with the tug driver with either hand signals or radios. Two technicians remain in the cockpit to use the aircraft brakes to stop the aircraft if it comes loose from the tug. Most damage to aircraft moving into or out of a hangar can be directly tied to a lack of spotters and/or lack of personnel riding the brakes on the flight deck (Figure 11-2-5).

Included in the docking phase are tasks that prepare the aircraft for inspection. Work platforms and stands necessary to allow personnel access to the inspection areas are carefully moved into place. The aircraft structure is grounded using wire cables clamped to the *grounding lugs* on the aircraft's landing gear and special *grounding points* located in the floor of the hangar. A ground power source is then hooked up to provide power to the aircraft for lighting and system operational checks. The aircraft is then placed on jack stands and raised and leveled. The aircraft remains on the jacks for the remainder of the inspection process. Finally, the wing tanks are drained of fuel and vented to remove harmful vapors.

After the aircraft is docked it receives a thorough washing. Pitot and static ports must be covered to prevent contamination. Landing gear doors and fairings and engine cowling will be opened for cleaning access.

Inspection Zones. The next phase of the inspection process is the opening of inspection zones. Each crew works to open the inspection panels, floorboards, fairings, and cowls necessary to perform the inspection in their area or zone. Whenever something is removed or opened to gain access to a portion of the aircraft, panel open-up paperwork is generated stating exactly which part was removed and its specific location on the aircraft. Access panels are labeled by the manufacturer with a unique number for each panel to identify them for removal and installation.

The inspection paperwork will call out only the panels that need to be opened to do the tasks needed in that area. This prevents unnecessary panel removal which saves time

Figure 11-2-5. This airplane is waiting its turn to be towed.

and confines inspection to only the areas which require it. The opening of a large aircraft for inspection purposes is no small task. There are hundreds of panels that need to be opened during a heavy airframe inspection, and this phase can take from one to four days depending on the size of the aircraft and the level of the inspection. Figure 11-2-6 is a training only manual produced by Northwest Airlines for familiarization of access panel locations. The manual contains 102 pages, giving an example of the complexity of access panel locations.

Now the inspection itself begins. In this phase *general inspectors* with inspection paperwork perform visual checks of all open areas of the aircraft, paying special attention to areas/items called out in the paperwork. The inspectors are looking for evidence of in-service wear, corrosion, leaks, and structural defects such as cracks or unusual wear.

Card System

Job cards. Both the inspection and service directions are placed on 3" x 6" manila cards. They consist of the entire inspection process broken down by the individual job. The jobs represented by the individual cards contain any additional information necessary for each separate function. Figure 11-2-7 illustrates a set of job cards.

Non-routine work cards. As examples of these discrepancies are found, they are documented on *non-routine work cards* which will be used by technicians to record the repair. The time required for a crew of inspectors to complete this phase depends on the extent to which the aircraft is opened up and the inspection level the airframe requires. Typically a heavy airframe initial inspection could last two days, or as long as a week, before it is completed.

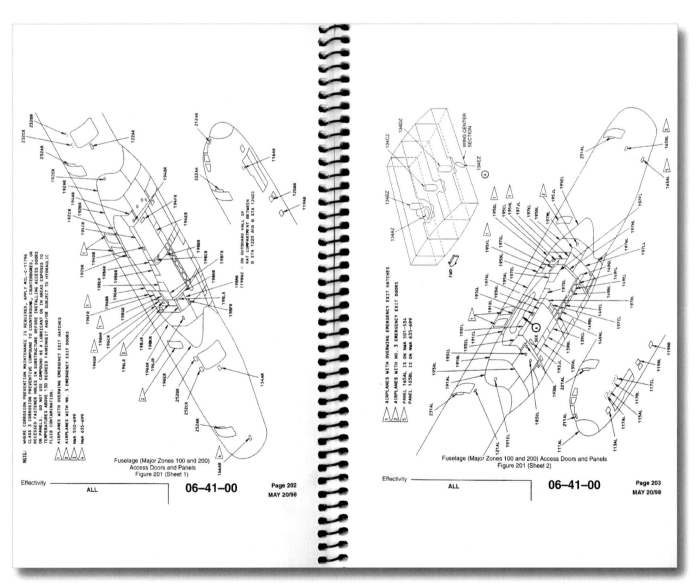

Figure 11-2-6. This illustration shows only two pages of an access panel identification book. The sheer numbers of panels that must be opened for inspection and maintenance can be astounding.

CARD NO.		WORK AREA(S)	TYPE MECHANIC REQUIRED		MECH NO.	CARD TIME	SCHEDULED STARTING TIME
1		0	Aircraft		1	80	

MAN MIN.	WORK AREA	SYS	PERIODIC INSPECTION REQUIREMENTS PRE-DOCK ELECTRICAL POWER OFF ☒ ON ☐
	6	7	Assisted by mechanics Nos. 2 and 3, accomplish the following: 1. Engine cover removed; cowling removed or opened.
	5	14	2. All switches "OFF" (except generator switch).
	5	4	3. Landing gear handle in "DOWN" position; position indicators for "DOWN" indication.
	5	3	4. DD Form 781 for discrepancies.
	5-6	3	5. Wing flaps "DOWN".
	6	14	6. Disconnect and remove battery; check for leakage or overflow of electrolyte.
	0	3	7. Aircraft (fuselage exterior, wings, wheel wells, engine, and empennage) for cleanliness; drain holes for obstructions.
	2-5	3	8. Relief tubes and horns cleaned with disinfectant-deodorant solution; adjacent areas neautralized (if they have been used).
	6	8	9. Fuel strainer for damaged screen and cleanliness.
	6	7	10. Carburetor screen for cleanliness and damage; float chamber for water and foreign matter.
	6	7	11. Engine oil screen for metal particles, damage, carbon, and cleanliness.
	6	11	12. Oil sump plug for metal particles and cleanliness.
	6	7	13. Carburetor air filter cleaned and lubricated (spray with a light film of oil, spec. MIL–C–6529, Type II).
	6	7	14. Temporarily reinstall, or close, engine cowling.

CARD SET PUBLICATION NO.	CARD SET DATE	CARD CHANGE DATE
WC 1T–6C–6–PE		REV. 20

CARD NO.		WORK AREA(S)	TYPE MECHANIC REQUIRED		MECH NO.	CARD TIME	SCHEDULED STARTING TIME
1		0	Aircraft		1	80	

Figure 11-2-7. A sample job card

As the inspection phase is being performed, technicians begin routine service tasks on the aircraft, using job cards. Specialized maintenance crews perform scheduled maintenance tasks and preventive maintenance such as fluid and filter changes, lubrication and removal and replacement of *life-limited components*. These include items such as landing gear, engines, or any other item that must be replaced with a new or overhauled unit due to time in service.

When the inspectors have completed their initial inspection, technicians can begin to evaluate and repair the resulting discrepancies. To evaluate a write-up the technician must estimate the repair time and check availability of parts. If parts are not in stock they must be ordered. Since some parts require long lead times to obtain, it is important that job evaluation and parts ordering be accomplished as soon as possible. Technicians continue with routine tasks and correct the non-routine items as parts become available.

Okay to close. When all routine and non-routine items have been completed, the heavy check enters the closing phase of the inspection. In this phase, the panels that were opened to gain access for inspection are reinstalled and the aircraft is restored to its original configuration. Before an access panel, floorboard, or cowl, can be closed it must receive an *ok to close* from an inspector. This is a critical step because in giving the okay to close the inspector is saying that all work cards for that area, whether routine or non-routine, have been completed and that there are no tools or other debris in the area. The okay to close also means that all corrosion prevention measures have been done in the area and the panel can be installed.

Zone final. Once all the panels in an area or zone have been reinstalled, the area receives a last look called a *zone final*. Zone finals are inspections to ensure that every panel has been closed in a zone and that all required equipment is installed and functioning properly. When all zones have completed their zone finals the aircraft is basically ready for revenue flight.

Undocking. With all the inspection and repairs complete and the aircraft restored to its original configuration, the crews prepare to *undock* the aircraft. In this procedure the plane is taken off the jack stands and refueled. Also, the grounding cables are removed and the ground electrical power is disconnected from the aircraft. It is also at this time that all plugs

Figure 11-2-8. When the post-runs are completed successfully, the airplane can return to revenue service.

and covers are removed that were used to protect areas which are sensitive to foreign object damage (FOD).

An example of this is the covers used to protect the pitot and static ports during the washing operation. There have been incidents involving aircraft that have taken off with these covers still in place resulting in crashes with fatalities.

Now a tow bar is connected to the aircraft and to a tug to tow the aircraft out of the hangar. As in the docking phase of the check, all crew members are involved in the movement of the aircraft out of the hangar. There are *wing walkers* at each wing tip, a *tail spotter* walking directly under the tail of the aircraft and tech-

nicians on the flight deck ready to apply the aircraft's brakes in case an emergency stop is needed.

Post-inspection Engine Run

The final phase of the heavy check is the *post inspection engine run*. With the aircraft parked on the ramp in front of the hangar, the engine crew jumps into action. Qualified run and taxi technicians take their place on the flight deck and the aircraft's engines are started for the first time in weeks. After carefully monitoring the warning systems for any sign of trouble during the engine start sequence, the technicians taxi the aircraft to the run pad for the *post-runs*.

The checks involved in the post inspection runs are essentially the same as those for the pre-runs. Engine performance is recorded at various output levels and the hydraulic system and environmental control system are checked for proper operation. The aircraft is also pressurized at this time and checked for leaks, particular attention is paid to the cabin doors since this is a common area for pressurization leakage.

If discrepancies are noted in any of the operational checks the aircraft is taxied back to the hangar for troubleshooting and repair of the problem. If all systems are found to be operating normally, the aircraft's logbook is signed-off by the technicians as *okay for return to service* and the aircraft is taxied to the gate at the passenger terminal (Figure 11-2-8). This ends the heavy check portion of the progressive inspection program.

Section 3

Conformity Inspections

As its name implies, a *conformity inspection* is performed to determine if an aircraft or its related components and associated equipment conform to a specified standard. For example, suppose an Airworthiness Directive has been issued on oxygen generators that requires all generators manufactured before a certain date to be removed from service. An air carrier would conduct a fleet campaign to determine the manufacture date of each oxygen generator in its fleet. In other words, the carrier is inspecting its generators to ensure they conform to the requirements of the AD.

New aircraft. When an aircraft operator brings a new aircraft onto its operating certificate the FAA requires proof of airworthiness. This is accomplished by an inspection to show the aircraft is in its *original type design or properly altered condition,* meeting all additional operational regulations applicable for its intended use, and is in a safe condition for flight. The operator uses a checklist to determine that the airworthiness certificate, radio station license, and aircraft registration are present, that all passenger emergency equipment is present and in serviceable condition, and that logbooks and the *minimum equipment list* (MEL) are present.

Leased aircraft. Conformity inspections are performed when an aircraft is leased and when it is returned from its lease. In this case, the conformity inspection is done twice. The first inspection is implemented upon receipt of the aircraft from the manufacturer or leasing company. It consists of a packing list of critical components installed on the airframe. Items considered critical to the safety of flight of the aircraft include the engines, landing gear, control surface actuators, and navigational equipment. Also included are cabin safety items such as crew and passenger oxygen systems, fire extinguishers, first aid supplies, and evacuation systems.

During this inspection the packing list part numbers and their corresponding serial numbers are checked against what is actually installed on the aircraft. All of the items on the list should be verified through aircraft records to be in new or newly overhauled condition. If all the items match the equipment list and are in operable condition the aircraft is accepted by the operator and put into revenue service.

The second conformity inspection is performed when the aircraft is returned following the expiration of the lease. In this case the lease agreement often states that all critical components shall be in the same or newer condition upon return of the aircraft.

Some lease agreements may also require the exact component (same serial number) be returned. The operator returning the aircraft conducts a thorough check, both visually and through the aircraft's records, to ensure that all components installed on the aircraft meet the stipulations in the lease language. This may involve the changing of some or many components and is much more involved than the original conformity inspection.

Landing gear and engines typically are replaced with freshly overhauled assemblies at this time, as are hydraulic system components critical to flight safety. Essentially, if the aircraft arrived at the operator with zero time on its timed components, then it is returned to the leasing company in the same condition.

Conformity inspections are conducted to protect both parties in an aircraft sale or lease. Such an inspection ensures that the equipment installed on the aircraft matches its packing list and that all critical components are in a condition acceptable for safe flight. A commercial operator would not accept an aircraft which needed an immediate engine change nor would a leasing company take back an aircraft that requires millions of dollars in maintenance to put the plane back into service.

In essence, conformity inspections are a method of verification that the equipment and the condition of the components listed in the aircraft's records are actually what are installed on the airframe.

Index

A

above ground level (AGL) 5-18
absolute altitude 5-18
absolute pressure 2-6. *See also cabin pressurization systems, terms and definitions*
absolute temperature 2-6. *See also cabin pressurization systems, terms and definitions*
accelerometer 5-56
accident investigation 7-9
accumulator (hydraulic) 1-12
  bladder 1-13
  diaphragm 1-13
  piston 1-14
AC generator 4-4
AC synchro system 5-9
  categories 5-9
    control 5-10
    differential 5-10
    resolvers 5-11
    torque indicating 5-9
actuating cylinder (hydraulic) 1-15
  double-action 1-15
  single-action 1-15
actuators (hydraulic) 1-14
adiabatic 2-6. *See also cabin pressurization systems, terms and definitions*
advanced cabin entertainment service system (ACESS) 7-17
  maintenance and troubleshooting 7-18
advisory messages 7-15
air bleeding. *See hydraulic contamination*
Airborne Communications Addressing and Reporting System (ACARS) 6-9. *See also communication systems*
  theory of operation 6-9
airborne weather radar system 6-23. *See also navigation systems*
aircraft altitude 2-6. *See also cabin pressurization systems, terms and definitions*
aircraft hydraulic systems 1-7
aircraft instrument systems **10-13**
aircraft pneumatic systems 1-56
air-cycle 2-12
  air conditioning systems 2-12
  machine 2-14
  operation 2-17
  pack 2-14
  system 2-14
air data computer (ADC) 5-26

air data stall warning 7-13
airframe fuel system 9-6
air pressure **10-12**
airspeed indicator 5-17
air valves 3-8
  Schrader valve 3-8
airworthiness directives (ADs) 11-3
alert messages 7-15
  advisory messages 7-15
  caution messages 7-15
  warning messages 7-15
alternating current 4-4
alternating current (AC) 4-2
alternator 4-4, 4-22
  bench testing 4-23
    field current draw 4-23
    inspection 4-24
    testing internal field circuit for a ground 4-24
    testing rectifiers 4-24
  checking generator or alternator belt tension 4-26
  description of 4-22
  kinds of 4-22
  over voltage relay 4-26
  solid state voltage regulator 4-25
  test procedure 4-23
  voltage regulator fusible wire replacement 4-25
  voltage regulator servicing 4-24
    air gap adjustment 4-24
    contact clearance adjustment 4-25
altimeter 5-18
  errors 5-21
  types of 5-20
    basic pressure 5-20
    complex 5-29
    standby 5-54
altitude definitions 5-20
altitude sickness 2-2
ambient pressure 2-6. *See also cabin pressurization systems, terms and definitions*
ambient temperature 2-6. *See also cabin pressurization systems, terms and definitions*
amphibian landing gear 3-4
amplifier circuits 6-4
angle-of-attack stall warning system 5-32
angle-of-attack system 5-30
  computer 5-31
  indicator 5-31
  transmitter 5-30
angular piston pumps. *See axial piston pump*
annual inspection 11-2
  checklists 11-5
    factory checklists 11-6

  cleaning 11-5
  logbook signoffs 11-8
  maintenance research material 11-3
    airworthiness directives (ADs) 11-3
    manufacturers service bulletins (SBs) 11-3
    manufacturers service letters (SLs) 11-3
    vendor service publications (VPs) 11-3
  records review 11-3
  spectrometric oil analysis program (SOAP) 11-6
  work order 11-3
antenna 6-4
  receiving 6-4
  transmitting 6-4
antenna systems
  installation of 6-29
  transmission lines 6-30
anti-collision lights 7-2
antioxidants 1-38
anti-skid system 3-70
  components of
    fail-safe protection 3-72
    locked wheel skid control 3-71
    normal skid control 3-71
    pilot control 3-71
    skid control box 3-71
    skid control generators 3-71
    skid control valves 3-71
    touchdown protection 3-71
  control valve 3-72
  operation 3-72
  subsystem features 3-72
apparent power 4-8
ARINC Incorporated 6-9
atmosphere, composition of 2-2
atmosphere, layers of 2-3
  exosphere 2-4
  mesosphere 2-4
  stratosphere 2-4
  thermosphere 2-4
  troposphere 2-3
atmospheric pressure 2-3
attitude indicator 5-37
attitude test button 6-32
audio panel 6-28
Automatic Direction Finder (ADF) system 6-15. *See also navigation system*
automatic temperature control 2-23
automotive gasoline 9-3
automotive-type diaphragm pump 9-14
auto-pilot axis control 5-81
auto-pilot operation 5-82
  altitude control 5-85
  automatic autopilot control 5-84
  feedback loop 5-85

flight computer control 5-86
manual autopilot control 5-84
radio navigation control 5-86
auto-pilot system 5-80
maintenance of 5-86
auxiliary air pressure 2-9. *See also positive-displacement cabin compressors (superchargers)*
auxiliary hydraulic system 1-55
aviation gasoline (AVGAS) 9-1
properties of 9-1
lead halides 9-2
tetraethyl lead 9-2
turbine fuel 9-2
vapor lock 9-3
volatility of fuel 9-2
Avionics 6-2
axial piston pump 1-10. *See also constant-delivery piston pump*
operation 1-10
azimuth 6-23

B

backup rings 1-29
identification of 1-29
installation of 1-29
storage of 1-29
types of 1-29
leather 1-29
Teflon 1-29
balance 3-35. *See also uneven tread wear*
ball check valve 1-19–1-20. *See also servo (hydraulic)*
bandwidth 6-4, 6-20
basic pitot static system 5-14
basic radio principles 6-2
basic system components 5-81
batteries 4-9
lead-acid 4-9
maintenance of 4-10
nickel-cadmium (NiCad) 4-10
battery 3-46
bead 3-27. *See also tires, aircraft*
breaking the bead 3-32
Beech King Air electrical power supply and distribution system 4-79
electrical system controls 4-82
bladder fuel cells 9-9
bleed air anti-ice 10-3
systems 10-11
bleed air systems 2-9. *See also positive-displacement cabin compressors (superchargers)*
bleeder bomb 3-62
bleeder valve 3-50

bleeding procedures 3-61. *See also brake systems, aircraft*
bottom-up method 3-62
top-down method 3-61
bleeding the system 1-6. *See also hydraulics*
Boeing 737 power supply 4-84
AC electrical power 4-84
DC electrical power 4-84
wiring diagrams 4-85
bogie 3-6, 3-57. *See also multiple wheeled landing gear*
bonding 6-29. *See also radio interference*
boost pumps 9-13
bootstrap air-cycle system 2-15
Bourdon tube system 5-67
brake accumulator 1-56
brake assemblies 3-47
disc 3-49
dual 3-47, 3-50
multiple 3-47, 3-50
trimetallic 3-51
single 3-47, 3-49
main parts 3-48
brake blocks 3-48
brake frame 3-48
clearance adjuster 3-48
expander tube 3-48
return springs 3-48
maintenance 3-63
single and dual disc brakes 3-63
manufacturers of 3-47
Cleveland 3-47
Goodrich 3-47
Goodyear 3-47
pucks 3-47
types of 3-47
disc 3-49
expander tube 3-48
shoe 3-49
brake control valves 3-53
types of 3-53
anti-skid 3-72
pressure ball check 3-53
sliding spool 3-54
brake debooster cylinder 3-54
brake system component maintenance 3-63
brake shuttle valves 3-63
independent system reservoirs 3-63
power brake valves 3-63
brake systems, aircraft 3-44
Airbus A300-600 3-44
air powered 3-44
bleeding procedures 3-61
brake wear check 3-60
component maintenance 3-63

general maintenance 3-59
hydraulic 3-44
independent-type 3-45
master cylinders 3-45
mechanical 3-44
brake wear check 3-60
Cleveland brakes 3-60
Goodrich and Goodyear 3-60
wear check method (No. 1) 3-60
wear check method (No. 2) 3-60
breaker 3-28
Built-In-Test-Equipment (BITE) 5-74, 5-79
bungee cord landing gear 3-7
bus tie breakers (BTB) 4-33
butterfly-type valve 1-61
bypass valve 1-19. *See also servo (hydraulic)*

C

cabin altitude 2-6. *See also cabin pressurization systems, terms and definitions*
cabin and cockpit area fire prevention and protection. *See fire prevention and protection*
cabin heaters 2-24
exhaust gas heaters 2-24
cabin interphone system 7-18
cabin lighting 7-18
cabin pressure dump valves 2-6. *See also cabin pressurization problems*
cabin pressurization problems 2-5
cabin pressurization pump 1-60
bleed air system 1-60
centrifugal supercharger 1-61
roots-type supercharger 1-61
cabin pressurization systems, terms and definitions 2-6
absolute pressure 2-6
absolute temperature 2-6
adiabatic 2-6
aircraft altitude 2-6
ambient pressure 2-6
ambient temperature 2-6
cabin altitude 2-6
differential pressure 2-6
gauge pressure 2-6
ram-air temperature rise 2-6
standard barometric pressure 2-6
cabin pressurization systems, testing of 2-11
cabin pressurization systems, troubleshooting 2-12
cable 4-35
cable clamp installation 4-52
caging device 5-42
calibrated altitude 5-19
cam-piston pumps 1-11

rotating-cam 1-11
stationary-cam 1-11
capacitive reactance 4-8
carbon brakes 3-51
carbon dioxide 2-2
carbon monoxide detectors 8-12
carbon pile voltage regulator 4-30
card system 11-12
carry through , 3-3
CAT 2-24. *See also exhaust gas heaters*
cathode ray tube (CRT) 5-50
caution messages 7-15
cavitation 1-40, 1-43, 1-46
CEET. *See exhaust gas heaters*
centrifugal cabin compressors 2-8. *See also positive-displacement cabin compressors (superchargers)*
centrifugal supercharger 1-61. *See also cabin pressurization pump*
chafing strip 3-27
check valve 1-20, 1-61
  in-line
    orifice 1-20
    simple 1-20
    integral 1-20
    pneumatic 1-61
chine 3-28
Chlorofluorocarbons (CFCs) 2-36. *See also Freon*
  pollution prevention 2-37
  regulations 2-36
circuit breakers 4-77
circuit limiting devices 4-77
Class A fire zone. *See fire zones*
Class B fire zone. *See fire zones*
Class C fire zone. *See fire zones*
Class D fire zone. *See fire zones*
Class X fire zone. *See fire zones*
Cleveland brakes 3-67
  automatic adjusting 3-69
  initial adjustment 3-68
  reassembly 3-68
  removing the caliper 3-68
  troubleshooting 3-69
clocks 5-57
  forms 5-57
    digital 5-58
    electrically powered analog 5-57
    mechanically powered analog 5-57
closed air-cycle system 2-15
closed-center selector valve 1-16. *See also selector valve (hydraulic)*
  rotor-type 1-16
  spool-type 1-16
closed-center system 1-18

CMPTR flag 6-32
coaxial cable 6-31
cockpit voice recorder (CVR) 7-8. *See also data recorders*
cold junction 5-69
combined hydraulic system 1-49
combustion chamber and radiator assembly 2-29
  inspection 2-29
  leakage test 2-29
combustion heaters 2-25. *See also cabin heaters*
  inspection and service 2-26
communications
  onboard aircraft 6-8
communication systems 6-1
  basic components 6-4
    headsets 6-6
    microphone 6-6
    receivers 6-5
    receiving antenna 6-6
    speakers 6-6
    transmitting antenna 6-5
  installation of 6-26
    cooling and moisture 6-27
    vibration isolation 6-27
  maintenance procedures 6-31
  reducing radio interference 6-28
  types of 6-6
    Airborne Communications Address and Reporting 6-9
    High Frequency (HF) 6-8
    Selective Calling 6-10
compass 5-54
compass card 6-34
compensator 1-11–1-12, 1-22
compensator valve 3-46
complex altimeter 5-29
  servoed barometric 5-29
complex pitot-static system 5-25
complex shock strut 1-34
  metering pin 1-34
  plunger 1-35
compressed air 1-13, 1-34, 1-60, 1-62, 1-63
conductor termination 4-58
  emergency splicing repairs 4-62
  junction boxes 4-64
  solderless splices 4-60
  stripping wire insulation 4-58
  terminal block and barrier strip connections 4-63
  terminal connectors 4-59
  wire splices 4-59
cone selector valves 9-12
configuration warning systems 7-12

conformity inspections 11-15
  leased aircraft 11-15
  minimum equipment list (MEL) 11-15
  new aircraft 11-15
connectors 4-64
  coaxial 4-67
  fabrication 4-67
  finding part numbers 4-65
    MS3102A14S-5S 4-66
  installation of 4-67
  type identification 4-65
    AN 4-65
      Class A 4-65
      Class B 4-65
      Class C 4-65
      Class D 4-65
      Class K 4-65
    MIL C-5015 4-65
    MIL-C-26482 4-65
constant-delivery piston pump 1-10. *See also hydraulic pump*
  axial piston pump 1-10
  gear pump 1-10
constellation of 24 orbiting stellites 6-20
contacts
  burned 4-12
contamination, fuel system 9-4
  detecting 9-6
  foreign particles 9-5
  microbial growth 9-5
  other types or grades of fuel 9-4
  sediment 9-5
  water 9-4
continuous loop system. *See fire detection systems*
control/interface electronics. *See fire detection systems*
control mechanisms 5-80
control temperature sensor. *See temperature control*
  high-limit temperature sensor 2-23
control valve. *See selector valve (hydraulic)*
conventional analog instruments 5-5
Conventional extinguishing system. *See fire extinguishing systems*
conventional landing gear. 3-2
cord body 3-27
corona discharge 6-29
corporate aircraft hydraulic system 1-46
course arrow 6-33
Course Deviation Indicator (CDI) 6-11, 6-32
course knob 6-33
course window 6-33
cracking pressure 1-22
crew call system 7-16

CSD disconnect mechanism 4-35
Cuno filter element. 1-8. *See also filter (hydraulic)*
current-carrying ability 4-36
current carrying capacity 4-46
  computing 4-46
current regulator 4-16. *See also regulator*
cutout relay 4-15, 4-20. *See also regulator*
cylinder head temperature indicator 5-70

D

damper 1-30
  displacement 1-30
    piston-type 1-30
    vane-type 1-30, 1-31
  nose landing gear 1-30
  rotor blade 1-30
  shear 1-30, 1-32
    linear-type 1-33
    rotary-type 1-32
  stabilizer bar 1-32
data recorders 7-8
  cockpit voice recorder (CVR) 7-8
DC generators 4-8
DC starter-generator 4-27
  generator operation 4-27
    generator control 4-27
    overexcitation protection 4-29
    reverse current protection 4-28
    undervoltage protection 4-29
    voltage regulation 4-27
  troubleshooting 4-29
    generator control panel test unit 4-29
DC synchro system 5-11
  indicator 5-12
  transmitter 5-12
DC to AC static inverters 4-5
decontamination 1-41. *See also hydraulic contamination*
  methods
    flushing 1-41, 1-42
    purging 1-41, 1-43
    purifying 1-41, 1-43
    recirculation cleaning 1-41
  selection of method 1-43
de-icer boots **10-5**. *See also pneumatic surface de-ice*
  maintenance of **10-8**
demand principle 1-11
density altitude 5-19
derating 4-75
design, fuel system 9-6
DH light 6-32

differential pressure 2-6. *See also cabin pressurization systems, terms and definitions*
differential pressure ratio 2-5. *See also cabin pressurization systems*
digital air data computers (DADC) 5-26
digital data buses 5-12
digital instruments 5-6
direct current (DC) 4-2
directional control valve. *See selector valve (hydraulic)*
direct reading gauges 5-6
disc brakes 3-49
  dual 3-47
  multiple 3-47
    trimetallic 3-50
  single 3-47
disk selector valves 9-12
dismounting 3-32
  divided (split) wheels 3-32
  remountable flange wheels 3-32
displays 5-4
  types of 5-4
    electronic 5-6
    mechanical type 5-4
    visual/sight gauges 5-4
dissolved air 1-40
distance away indicator 6-19
Distance Meauring Equipment (DME) 6-1, 6-14
docking 11-10
Doppler 6-17
  effect 6-17
  radar 6-17
Doppler navigation system 6-17. *See also navigation systems*
double-pole double-throw (DPDT) switch 4-77. *See also switch*
double-pole single-throw (DPST) switch 4-76. *See also switch*
downstream pressure 1-60
drip gauges 9-17
drive coupling 1-9–1-10. *See also hydraulic pump*

E

echo 6-23
EGT probe 5-69
ejector pump 9-14
ejector valve **10-7**. *See also pneumatic surface de-ice*
electrical anti-ice **10-5**
electrical components 4-74
electrical input stall warning system 7-13
electrical load 4-78
  controlling 4-79

  limits 4-79
  monitoring 4-79
electrically operated retraction system 3-9
electrical power sources and monitoring 4-78
electrical solenoids. *See pneumatic surface de-ice*
electrical theory overview 4-2
electrical wire chart 4-44
  instructions for use of 4-44
  selecting the correct size 4-44
electric attitude indicator 5-43
electromagnetism 4-2
electromotive force 4-3
electronic attitude direction indicator (EADI) 5-50
electronic flight information system (EFIS) 5-50
Electronic Flight Information Systems (EFIS) 6-34
Electrostatic Discharge (ESD). *See also radio interference*
  static discharger 6-29
elevation 5-18
emergency brake system 3-55
  contamination check 3-61
  parts of 3-55
    air release valve 3-55
    compressed air bottle 3-55
    pressure gauge 3-55
    T-handle 3-55
Emergency Locator Trasmitter 6-24
  batteries 6-25
  false alarms 6-26
  test equipment 6-26
  testing of 6-26
  transmitter 6-25
engine anti-ice (EAI) **10-3**
engine auto synchronizer 5-68
engine driven pumps 9-13
engine fuel system 9-7
engine indicating and crew alert system (EICAS) 5-53, 5-72, 7-14
engine instruments 5-59
engine oil temperature indicator 5-65
engine pressure ratio (EPR) 5-71
engine run checks 11-10
entrained air 1-40, 1-45–1-46
entrained water 9-4
exhaust gas heaters 2-24. *See also cabin heaters*
  CAT 2-24
  CEET 2-24
  flexible ducting 2-24
  inspection 2-25
  SCAT 2-24

SCEET 2-24

exhaust gas temperature (EGT) 5-68

   indicator 5-68

exhaust temperature gauge 5-68

exosphere 2-4

expander tube brakes 3-48

expansion turbine 2-19

exterior lights 7-1

extinguishing agents 8-13. *See also* **fire extinguishing systems**

   Halogenated hydrocarbon agents 8-14

   inert cold gas agents 8-14

F

fast/slow indicator 6-33

Federal Communications Commission (FCC) 6-4

feedback 1-6, 1-19

field relays (FR) 4-33

filter (hydraulic) 1-8

   construction 1-8

   types of

      Cuno 1-8

      micronic 1-8

fire detection methods 8-3

fire detection systems 8-2

   continuous-loop 8-6

   control/interface electronics 8-9

   infrared detection 8-8

   inspection and maintenance 8-9

   overheat warning 8-8

   pneumatic detection 8-7

   smoke detection systems 8-11

   thermal switch 8-3

   thermocouple 8-5

   troubleshooting 8-10

      failure to obtain alarm signal 8-10

      false alarms 8-10

      intermittent alarms 8-10

fire extinguishing systems 8-13

   conventional system installation 8-15

   extinguishing agents 8-13

   high-rate-of-discharge installations 8-16

   inspection and maintenance 8-17

      container pressure check 8-17

      Freon/Halon containers 8-18

      transport category aircraft 8-17

   types of 8-15

      conventional 8-15

      high-rate-of-discharge 8-15

fire prevention and protection 8-20

   cabin and cockpit area 8-20

   hand-held extinguishers 8-20

turbine engine ground 8-17

   unsuitable extinguishers for cabin/cockpit equipment 8-21

fire zones 8-1, 8-2, 8-5

   Class A 8-2

   Class B 8-2

   Class C 8-2

   class D 8-2

   Class X 8-2

fixed airplane reference 6-34

fixed airplane reference symbol 6-32

fixed landing gear 3-1. *See also* **landing gear, aircraft**

   air-oil shock strut 3-2

   bungee cord type 3-2

   retractable 3-3

fixed resistance method 4-18

flap position indicator 5-55

flexible ducting 2-24. *See also* **exhaust gas heaters**

flight data recorders (FDR) 7-8. *See also* **data recorders**

Flight Director Indicator (FDI) 6-32

   components of

      attitude test button 6-32

      "CMPTR" flag 6-32

      DH Light 6-32

      Fast/slow indicator 6-33

      fixed airplane reference symbol 6-32

      glideslope "GS" flag 6-32

      glideslope pointer 6-32

      gyro flag 6-32

      inclinometer 6-32

      pitch scale 6-32

      runway flag 6-33

      runway indicator 6-33

      speed flag 6-33

      V-command bars 6-32

flight hydraulic system 1-50

   shutoff valves 1-51

flight interphone 7-16

flow limiting 2-18

fluid contamination. *See* **hydraulic contamination**

fluids, characteristics of 1-4

   effects of temperature on liquids 1-5

   movement of fluid under pressure 1-5

force 1-2

   direction of 1-2

   magnitude of 1-2

   transmission of 1-4

      through confined liquids 1-4

      through solids 1-4

force-area-pressure formulas 1-3–1-4

foreign particles 9-5

four-way valve. *See* **selector valve (hydraulic)**

free air 1-40, 1-46

free water 9-4

freon. *See* **vapor-cycle air conditioning system**

   system components 2-33

      compressor 2-33

      condenser 2-34

      evaporator 2-35

      expansion valve 2-34

      receiver 2-34

      sight glass 2-34

frequency 4-7, 6-2

   bands 6-4

   generation 6-2

Frequency Modulation 6-3

   Amplitude Modulation (AM) 6-3

   Frequency Modulation (FM) 6-3

   Pulse Modulation (PM) 6-3

frost removal **10-12**

fuel cells 9-9

fuel dump system 9-22

fuel ejectors 9-14

fuel filters 9-10

   micron filter 9-11

   screen mesh strainer filter 9-11

fuel flow amplifier 5-65

fuel flow indicator 5-62, 9-17

   fuel flow transmitter 9-18

   totalizer 9-17

fuel flow totalizer indicator 5-65

fuel flow totalizing system 5-64

fuel flow transmitter 5-65, 9-18

fuel heaters 9-15

fuel lines 9-10

fuel pressure 9-19

fuel pumps 9-13

fuel quantity indicator 9-15

   electrical-type 9-16

   electronic-type 9-16

   mechanical-type 9-16

   sight glass 9-16

fuel strainers 9-10

   main 9-10

   screen mesh strainer filter 9-11

fuel systems; inspection, maintenance and repair 9-22

   fuel pressure gauge 9-23

   fuel quantity gauges 9-23

   fuel tank repairs 9-24

   fuel tanks 9-23

   lines and fillings 9-23

   main line strainers 9-23

   pressure warning signal 9-23

   pumps 9-23

   selector valves 9-23

fuel tank repairs 9-24
  fuel cells 9-24
  Integral fuel tanks 9-25
  welded steel tanks 9-24
fuel temperature 9-19
fuses 4-77
fusible plug 3-42

G

galvanometer 5-66
gaseous oxygen systems 2-40. *See also oxygen systems*
  cylinders 2-41
  leak-testing 2-41
gaskets (hydraulic) 1-27
  crush washer 1-27
  fabricating 1-27
  installing 1-27
  o-ring 1-27
gasper 2-19
gauge pressure 2-6. *See also cabin pressurization systems, terms and definitions*
gear compartment doors 1-21
gear indicators 3-23. *See also main landing gear*
  down position 3-23
  up position 3-23
gear pump. *See constant-delivery piston pump*
generator 4-11, 4-32
  adjustments of 4-12
  checking 4-11
  checking defective 4-13
    armature service 4-15
    excessive output 4-14
    noisy generator 4-15
    no output 4-13
    unsteady or low output 4-14
  control panels 4-33
  controls and indicators 4-33
  description of 4-11
  exciter section 4-33
  frequency control 4-34
  inspection of 4-12
  location 4-32
  main generator section 4-33
  maintenance of 4-12
    specified speeds 4-17
  paralleling 4-35
  polarizing 4-15
  speed switch 4-34
generator breakers (CB) 4-33
generator paralleling 4-35. *See also generator*
  automatic 4-35

manual 4-35
  synchronizing lights 4-35
gimbals 5-36
glass cockpit 5-50
glideslope 6-12
  receiver 6-11
  signal 6-12
glideslope flag 6-32, 6-33
glideslope pointer 6-32, 6-33
GLOBALink 6-9
Global Positioning Systems (GPS) 6-2, 6-19. *See also navigation systems*
  aircraft navigation 6-22
    equipment 6-22
  system elements 6-20
  theory of operation 6-21
gravity feed 9-8
gravity flow 9-20
ground deicing of aircraft **10-12**
ground handling wheels 3-6
ground locks 3-23
ground loop 3-3
ground power unit (GPU) 4-83
gyro flag 6-32
gyro horizon 5-37
gyroscope 5-36
  degree of freedom 5-36
  mountings of 5-36
    free 5-36
    restricted or semirigid 5-36
    universal 5-36
  properties of gyroscopic action 5-36
    precession 5-37
    rigidity 5-37
      design factors 5-37
  sources of power 5-38
    engine-driven vacuum system 5-40
    pressure operated system 5-41
    vacuum system 5-38
    venturi-tube system 5-38
  types of
    directional 5-47
    ring laser 5-46
    standby attitude 5-54
    suction-driven 5-43
    turn & bank 5-47
    vacuum-driven attitude 5-42
gyroscopic instrument 5-35
  principles 5-35

H

halogenated (chlorinated) solvent 1-40
Halogenated hydrocarbon agents 8-14. *See*

also *extinguishing agents*
  Halon 8-14
  PhostrExTM 8-15
hand-held extinguishers. *See fire prevention and protection*
Handling oxygen 2-40. *See also oxygen systems*
HDG flag 6-33
heading knob 6-34
heading marker 6-34
headsets 6-6. *See also communication systems*
helicopter landing gear 3-5. *See also landing gear, aircraft*
  fixed-wheel gear 3-5
  retractable-wheel gear 3-5
  skid-gear 3-5
  skid-type gear 3-5
hertz (Hz) 4-7
High Frequency (HF) system 6-8. *See also communication systems*
high-rate-of-discharge installations. *See fire extinguishing systems*
Horizontal Situation Indicator (HSI) 6-32, 6-33
  components of
    compass card 6-34
    course arrow 6-33
    course knob 6-33
    course window 6-33
    fixed airplane reference 6-34
    glideslope flag 6-33
    "HDG" flag 6-33
    heading knob 6-34
    lateral deviation bar 6-33
    lubber line 6-33
    miles window 6-33
    NAV flag 6-33
    shutter 6-33
hot junction 5-69
hydraulic contamination 1-36
  air bleeding 1-45
  applying hydraulic power 1-45
    operational checks 1-45
    shutdown procedure 1-45
  contamination control sequence 1-43
  control 1-46
    cleanliness 1-46
  control program 1-37
    fluid sampling 1-37
    maintenance practices 1-37
    maintenance procedures 1-37
  decontamination 1-41
  fluid contamination 1-40
    air 1-40

dissolved 1-40
entrained 1-40
free 1-40
foreign fluids 1-40
sampling point 1-41
solvent 1-40
water 1-40
sampling 1-41
filter bowl contents analysis 1-41
types of 1-38
inorganic solid 1-40
metallic solid 1-38
organic 1-38
particulate and fluid 1-38
hydraulic fluids 1-43
checking aircraft fluid levels 1-44
fluid analysis 1-44
hydraulic freight elevator 1-1
hydraulic landing gear 3-11
main gear 3-11
nose gear 3-11
system operation 3-11
hydraulic pressure indicator 1-49
hydraulic pump 1-9
cam-piston 1-11
constant-delivery 1-10
hand-operated 1-9
double-action 1-9
handle to the left 1-9
handle to the right 1-9
power-driven 1-9
drive coupling 1-9
shear section 1-9
types of
diaphragm 1-9
gear 1-9
piston 1-9
vane 1-9
variable-delivery 1-11
hydraulics
mechanical advantage of 1-5
application 1-5
rate 1-5
principles of 1-1
the role of air in 1-6
air 1-6
air and nitrogen 1-6
application 1-6
bleeding the system 1-6
malfunctions caused by 1-6
hydraulic systems 1-46
characteristics of 1-2
control sensitivity 1-2
dependability 1-2

efficiency 1-2
flexibility of installation 1-2
low maintenance requirements 1-2
low space requirements 1-2
low weight 1-2
self-lubricating 1-2
reservoirs and fluid quantity 1-48
subsystems 1-46
emergency pneumatic system 1-48
hydraulic terms, definition of 1-2
direction of force 1-2
force 1-2
magnitude of force 1-2
non-technical 1-3
pressure 1-3
technical 1-3
hydrochloric acid 1-40
hydrochlorofluorocarbons (HCFCs) 2-36
regulations 2-36
hypoxia 2-4

I

ice and snow deposits **10-13**
removing **10-13**
ice protection systems, types of **10-1**
anti-ice **10-1**
bleed air **10-3**
electrical anti-ice **10-5**
TKS **10-8**
automatic **10-2**
de-ice **10-1**
pneumatic surface **10-5**
TKS **10-8**
ice detection **10-1**
visual **10-2**
icing, fuel system 9-3
impedance (Z) 4-8
inclinometer 6-32
indicated altitude 5-19
indicator 5-12
types of
AOA 5-31
electric attitude 5-43
electronic attitude direction 5-50
engine oil temperature 5-65
exhaust gas temperature 5-68
flap position 5-55
fuel flow 5-62
fuel flow totalizer 5-65
instantaneous vertical speed (IVSI) 5-23
mach/airspeed 5-28
oil pressure 5-67
rate of climb 5-22

Remote gyro 5-43
simple airspeed 5-17
standby airspeed 5-54
turn-and-bank 5-47
vertical gyro 5-37
vertical speed (VSI) 5-22
inductive reactance 4-8
inert cold gas agents 8-14. *See also* ***extinguishing agents***
carbon dioxide 8-14
nitrogen 8-14
Inertial Navigation System (INS) 6-18
basic components of 6-18
infrared detection system. *See* ***fire detection systems***
inlet filter **10-7**. *See also* ***pneumatic surface de-ice***
in-line check valve. *See* ***check valve***
in-line filter **10-7**. *See also* ***pneumatic surface de-ice***
inspection work order 11-3
inspection zones 11-11
instantaneous vertical speed indicator (IVSI) 5-23
Instrument Landing Systems (ILS) 6-1, 6-11. *See also* ***navigation systems***
glide slope 6-12
marker beacons 6-13
runway localizer 6-12
instruments
classification of 5-4
conventional analog 5-5
digital 5-6
engine 5-59
flight 5-14
gyroscopic 5-35
standby 5-54
vertical scale 5-5
maintenance of 5-74
cases 5-74
general maintenance 5-74
graduation 5-75
instrument system inspection 5-77
instrument system repair 5-78
markings 5-75
panels 5-77
integral fuel cells 9-9
integrated drive generator (IDG) 4-3
interphone systems 7-15
cabin interphone system 7-18
crew call 7-16
flight interphone 7-16
maintenance and troubleshooting 7-16
service 7-16
in-the-gate 6-9
inverter 6-6

ionization type detector. *See smoke detection systems*

irreversible valve 1-19. *See also servo (hydraulic)*

isolation 6-28. *See also radio interference*

J

jet A-1 fuel 9-2

jet A fuel 9-2

jet B fuel 9-2

jet pump 2-17

job cards 11-12

Johnson bar 3-45

jumper lead method 4-21

K

King Air AC power 4-83

King Air cockpit system 3-10

King Air DC power 4-81

King Air electrical system controls 4-82

  battery monitor 4-82

  DC external power source 4-83

  generator out warning light 4-82

  volt-loadmeters 4-82

KVAR meter 4-32. *See also large aircraft generators*

KW meter 4-31. *See also large aircraft generators*

L

landing gear 1-7, 1-21, 1-30, 1-31, 1-32, 1-34, 1-48, 1-49, 1-54, 1-55, 1-56, 1-57, 1-62, 3-11

  actuator 3-11

  adjustment of 3-14

    gear latch 3-15

  door clearances 3-16

  drag 3-17

  motor 3-11

  retraction check 3-18

  rigging 3-14

  side brace adjustment 3-17

  system maintenance 3-26

    lubrication 3-26

landing gear, aircraft 3-1

  types of 3-1

    bungee cord 3-7

    helicopter 3-5

    leaf type 3-7

    multiple wheeled 3-6

    non-retractable shock struts 3-8

    retractable 3-3

    seaplanes and amphibians 3-4

    skiplanes 3-5

landing lights 7-5

large aircraft brake systems 3-55

  backup operation 3-58

  maintenance practices 3-57

  major subsystem sequence 3-58

  normal sequence 3-58

  operation control sequence 3-58

  physical description 3-57

  subsystem features 3-56

  system operation 3-56

large aircraft generators 4-31

  KW/KVAR explanation 4-31

    KVAR meter 4-32

    KW meter 4-31

lateral deviation bar 6-33

lead halides 9-2

leading shoe 3-49

leaf type gear legs 3-7

leakage test 2-31

life-limited components 11-13

lighting systems, maintenance and inspection 7-8

lights, types of 7-6

  anti-collision 7-2

  cabin 7-18

  exterior 7-1

  landing 7-5

  logo 7-6

  position 7-1

  strobe 7-3

  taxi 7-6

  wing inspection 7-6

linear-type shear damper 1-33

  servicing of 1-33

line-of-sight 6-7

liquid crystal display (LCD) 5-50

liquid rain repellent 10-11. *See also rain protection, types of systems*

localizer 6-11

lockout deboosters 3-55

logbook signoffs 11-8

logo lights 7-6

Loran ARNAV system 6-18. *See also navigation systems*

lubber line 6-33

M

main gear 3-10

main landing gear 3-22

  alignment 3-22

emergency extension systems 3-23

gear indicators 3-23

ground locks 3-23

nosewheel centering 3-23

safety devices 3-23

support 3-23

maintenance research material 11-3

  airworthiness directives (ADs) 11-3

  manufacturers service bulletins (SBs) 11-3

  manufacturers service letters (SLs) 11-3

  vendor service publications (VPs) 11-3

manifold check valve **10-7**. *See also pneumatic surface de-ice*

manifold pressure indicator 5-71

marker beacon 6-13

  receivers 6-1, 6-13

master control station (MCS) 6-21

master cylinders 3-45. *See also brake systems, aircraft*

  types of 3-46

    Goodyear 3-47

    vertical 3-46

master warning lights panel 1-49

maximum differential pressure 2-5. *See also cabin pressurization systems*

mean sea level (MSL) 5-18

mechanical linkage 1-2, 1-12

mechanical type display 5-4

mesosphere 2-4

metal fuel tanks 9-8

microbial growth 9-5

micro electronic motion sensors (MEMS) 5-45, 5-47

micron filter 9-11

micronic-type filter 1-8

microns 1-8, 1-38, 1-42

Microorganisms 1-40

microphone 6-6. *See also communication systems*

microswitches 1-21

miles window 6-33

minimum equipment list (MEL) 11-15

misalignment 3-35. *See also uneven tread wear*

mixer valve 2-15

motors (hydraulic) 1-14

mounting 3-32

  divided (split) wheels 3-33

  remountable flange wheels 3-33

multi-engine fuel system 9-21

multiple wheeled landing gear 3-6

  bogie 3-6

  truck 3-6

N

national transportation safety board (NTSB)
7-9. *See also accident investigation*
NAV-COM 6-1
NAV flag 6-33
navigation systems 6-10
  Airborne navigation equipment 6-10
  types of
    Automatic Direction Finder (ADF) 6-15
    Doppler 6-17
    Global Positioning System (GPS) 6-19
    Inertial 6-18
    Instrument Landing 6-11
    Loran ARNAV 6-18
    Omni-directional Range 6-11
    Radar Beacon Transponder 6-16
    World Geodetic System (WGD) 6-19
new generation aircraft 4-35
nitrogen 1-6, 1-13, 1-56, 1-63
non-retractable shock strut landing gear 3-8
non-routine work cards 11-12
nose gear 3-9
nosewheel steering system 3-24
  heavy aircraft 3-25
  light aircraft 3-24
nylon flat spotting 3-36
nylon sickness 3-36

O

off-the-ground 6-9
Ohm's Law 4-2
Oil pressure indicator 5-67
okay for return to service 11-15
Omni-Bearing Selector (OBS) 6-11
Omni-directional
  ground testing 6-11
  Range (VHS VOR) system. *See also naviga-
    tion systems*
on-the-ground 6-9
open air-cycle system 2-15
open-center selector valve 1-17. *See also se-
    lector valve (hydraulic)*
  rotor-type 1-17
  spool-type 1-18
open center system (hydraulic) 1-18
operational check 3-50, 3-59
Operational checks. *See hydraulic contamina-
    tion*
optical flame detection. *See smoke detection
    systems*
organic solids 1-38
o-rings 1-28
  identification 1-28

removal of 1-28
shelf life 1-28
storage of 1-28
oscillator circuit 6-4
outflow valve 2-5. *See also cabin pressuriza-
    tion systems*
out-of-the-gate 6-9
outside air temperature (OAT) 5-55
overheated wheel brakes 3-62
overheat warning system. *See fire detection
    systems*
overinflation 3-30, 3-35
oxidation 1-38
oxygen systems 2-39
  characteristics 2-40
  cleaning 2-41
  cylinder testing 2-41
  draining 2-41
  FARs 2-51
  gaseous oxygen systems 2-40
  generators 2-51
  handling oxygen 2-40
  inspections 2-48
  maintenance 2-46
  masks 2-44
  portable 2-46
  regulators 2-43
  safety 2-50
  solid 2-49
  tubing 2-44
  types 2-39
  valves 2-45

P

packings 1-26, 1-27, 1-29
parking brake lever 3-46
Pascal's law 1-4
passenger entertainment video 7-19
  maintenance and troubleshooting 7-19
phase 4-7
Phase Modulation (PM) 6-20
photoelectric type detectors. *See smoke detec-
    tion systems*
pilot 1-12, 1-16, 1-18, 1-19, 1-25, 1-32
pintle 1-12
piston pumps (hydraulic) 1-9
  constant delivery 1-9
    angular 1-9
    cam 1-9
    variable delivery 1-9
pitch scale 6-32
pitot tube 5-15
pitot tube anti-icing **10-13**

ply rating 3-28. *See also tires, aircraft*
pneumatic detection systems. *See fire detec-
    tion systems*
pneumatic surface de-ice **10-5**
pneumatic system components 1-61
  air bottle 1-62
  filters 1-62
  lines and tubing 1-62
  maintenance of 1-62
polymerization 1-38
poppet 1-23, 1-40
  valve 1-61
poppet selector valves 9-12
position lights 7-1
positive-displacement cabin compressors
  (superchargers) 2-7
post-inspection engine run 11-14
potentiometers 5-7
pound 1-2
power 4-8
power brake control valve 3-52
power-driven hydraulic pumps 1-9
  types of 1-9
power factor 4-8
power loss (IR loss) 4-36
power supply 6-6
precipitation static 6-29
preset reference resistor 2-21. *See also tem-
    perature control*
pressure 1-3, 5-25
  definitions 5-25
    corrected static 5-25
    corrected total 5-25
    impact 5-25
    indicated static 5-25
    indicated total 5-25
pressure adjustment screw 1-22
pressure altitude 5-18
pressure-control piston 1-12
pressure-control valve 1-12
pressure feed 9-8
pressure input stall warning system 7-12
pressure measurement 1-3
  amount of force 1-3
  examples of 1-3
  force-area-pressure formulas 1-3
  unit area 1-3
pressure reducer 1-23
  relieving pressure 1-23
  withholding pressure 1-23
pressure switch (hydraulic) 1-25
  diaphragm 1-25
  operation of 1-25
  piston 1-25
pressure system 1-57

high 1-57

low 1-57

medium 1-57

pressurization safety valve 2-6. *See cabin pressurization problems*

primary heat exchanger 2-18

heat-exchanger bypass valve. 2-18

progressive inspections 11-9

card system 11-12

job cards 11-12

non-routine work cards 11-12

commercial aircraft 11-9

docking 11-10

engine run checks 11-10

general aviation 11-9

inspection zones 11-11

post-inspection engine run 11-14

undocking 11-13

zone final 11-13

pump control circuit operation 1-54

pressure reset button 1-55

Q

quick-disconnect couplings 1-29

R

radar 6-16, 6-23

radar, area navigation systems (RNAV) 6-2

radar beaon transponder system 6-16. *See also navigation systems*

code changes 6-16

operational modes 6-16

Mode A 6-16

Mode AC 6-16

Mode S 6-16

radio altimeters 6-23

radio by-pass condenser 4-21

radio interference 6-28

reducing 6-28

bonding 6-29

isolation 6-28

shielding 6-29

radio magnetic indicator (RMI) 6-11

rain protection, types of systems **10-11**

liquid repellent **10-12**

wipers **10-11**

electric **10-11**

hydraulic **10-11**

ram-air temperature rise 2-6. *See also cabin pressurization systems, terms and definitions*

ratchet valve 1-19. *See also servo (hydraulic)*

ratiometer system 5-66

receivers 6-5

reciprocating engine instruments 5-70

cylinder head temperature indicator 5-70

manifold pressure indicator 5-71

recirculating air 2-17. *See also ventilation air*

recirculation cleaning. *See decontamination*

records review 11-3

procedure 11-4

regulations 11-3

refrigeration cycle 2-32. *See also vapor-cycle air conditioning system*

regulator 4-15

checks and adjustments 4-18

air gap 4-18, 4-21

closing voltage 4-20

current regulator 4-21

current setting 4-21

cutout relay 4-20

high points 4-21

voltage regulator 4-18

voltage setting 4-18

current regulator 4-16

cutout relay 4-15

description of 4-15

maintenance of 4-17

cleaning contact points 4-17

polarizing the generator 4-17

specified generator speeds 4-17

operation of 4-15

polarity 4-17

resistances 4-17

temperature compensation 4-17

relay contact points

burned 4-12

relays 4-75

relief valve 1-18, 1-19, 1-22

four-port 1-22

high pressure system 1-61

low pressure system 1-61

medium pressure system 1-61

two-port 1-22

remote gyro indicator 5-43

remote sensing technologies 5-7

remote sensor 5-7

reservoir (hydraulic) 1-7

additional components 1-8

pressurized 1-8

vented 1-7

resistance (R) 4-8, 4-39

calculation methods 4-39

resistances

burned 4-12

restrictors 1-61

orifice 1-61

variable 1-62

retractable landing gear 3-3

actuator 3-11

basic retractable systems 3-9

carry through 3-3

cockpit controls 3-10

electrically operated 3-9

emergency extension 3-11

main gear 3-10

motor 3-11

nose gear 3-9

position indicators 3-10

wheel well 3-4

return stream 1-16

ring laser gyro (RLG) 5-46

roots-type blower 2-7. *See also positive-displacement cabin compressors (superchargers)*

roots-type supercharger 1-61. *See also cabin pressurization pump*

rotating-cam pump. *See cam-piston pump*

runway flag 6-33

runway indicator 6-33

runway localizer 6-12. *See also Instrument Landing Systems (ILS)*

S

sampling 1-41. *See hydraulic contamination*

SCAT. *See exhaust gas heaters*

SCEET. *See exhaust gas heaters*

scissors 3-8

screen mesh strainer filter 9-11

seals (hydraulic) 1-26

composition 1-27

cup 1-26

dynamic type 1-27

identification of 1-26

oil 1-26

shaft 1-29

static type 1-27

storage 1-27

u-ring 1-26

v-ring 1-26

wiper 1-27

seaplane landing gear 3-4

secondary heat exchanger 2-18

sediment 9-5

selective calling (SELCAL) 6-10. *See also communication systems*

selector valve (hydraulic) 1-15

closed-center 1-16

control valve 1-16

directional control valve 1-16

four-way 1-16

open-center 1-17

self seal 1-57

semisolids 1-38

sensor 5-4

  types of 5-6

    direct reading gages 5-6

    remote 5-7

    separate 5-6

    solid state pressure 5-34

sequence valve 1-21

  solenoid-operated 1-21

service interphone 7-16

servo (hydraulic) 1-18

  ball check valve 1-19

  bypass valve 1-19

  irreversible valve 1-19

  ratchet valve 1-19

  sloppy link 1-19

  valve operation with no pressure 1-20

  valve operation with pressure applied 1-20

servo-shoe principal 3-49

shear section 1-9. *See also hydraulic pump*

shielding 6-29. *See also radio interference*

shimmy damper 3-25

  piston-type 3-25

  steering 3-25

  vane-type 3-25

shock cord 3-7

shock strut 3-19

  bleeding 3-21

  non-retractable 3-8

  servicing 3-20

shock strut (hydraulic) 1-33

  functions of

    cushioning jolts 1-35

    reducing shock 1-35

    supporting static loads 1-35

  maintenance of 1-36

  mechanical type 1-34

  pneudraulic type 1-34

  servicing instructions 1-36

shoe brakes 3-49

shunt generator output 4-13

shutdown procedure. *See hydraulic contamination*

shutter 6-33

sidewall 3-27

sight glass 9-16

simple resistance 5-7

  position indicating 5-7

  pressure indicating 5-9

  temperature indicating 5-8

simple shock strut (hydraulic) 1-34

sine wave 6-2

single-engine fuel system 9-20

single-pole double-throw (SPDT) switch 4-76. *See also switch*

single-pole single-throw (SPST) switch 4-76. *See also switch*

SITA 6-9

skid control box 3-71

skid control generator 3-71

skid-gear 3-5. *See also helicopter landing gear*

skiplane landing gear 3-5

Skydrol 1-44

sludge 1-38

smoke detection systems 8-11. *See also fire detection systems*

  ionization type detector 8-11

  photoelectric type detectors 8-12

  visual smoke detectors 8-11

solid state gyro

  inertial reference system 5-45

sonic venturi 2-18

speakers 6-6. *See also communication systems*

spectrometric oil analysis program (SOAP) 11-6

speed flag 6-33

spool 1-16–1-18, 1-38

square inch 1-2–1-5

squat switch 3-10

stall warning systems 7-12

standard barometric pressure. *See cabin pressurization systems, terms and definitions*

standby airspeed indicator 5-54

standby attitude gyro 5-54

static air vents 5-16

static discharger 6-29. *See also Electrostatic Discharge (ESD)*

  types of 6-29

    flexible 6-29

    null-field 6-29

    semi-flexible 6-29

static port 5-15

stationary-cam pump. *See cam-piston pumps*

stratosphere 2-4

strobe lights 7-3

  maintenance 7-4

stroke-reduction principle 1-11, 1-12

superchargers 2-7. *See also positive-displacement cabin compressors (superchargers)*

switch 4-74

  high current rush circuit 4-75

  inductive circuit 4-75

  motor circuit 4-75

  types of 4-77

    battery 4-82

    double-pole double-throw 4-77

    double-pole single-throw 4-76

    generator 4-82

    master 4-82

    micro 4-77

    proximity 4-77

    rocker 4-77

    rotary 4-77

    single-pole double-throw 4-76

    single-pole single-throw 4-76

    toggle 4-77

synchro systems 5-9

  types of

    AC 5-9

    DC 5-11

T

tachometer 5-59

  system maintenance 5-62

  types of

    high pressure (HP) 5-59

    low pressure (LP) 5-59

tachometer generators 5-59

taxi lights 7-6

temperature anticipator 2-23

temperature bulb 5-55

temperature compensation 3-53

temperature control 2-19

  temperature controller 2-20

  temperature limiting 2-21

  thermistor 2-20

  wheatstone bridge 2-20

temperature controller 2-20. *See also temperature control*

temperature limiting 2-21. *See also temperature control*

temperature limiting sensor 2-21. *See also temperature control*

temperature sensors 2-23

tetraethyl lead 9-2

thermal expansion 1-5, 1-22–1-23

thermal switch system. *See fire detection systems*

thermistor 2-20. *See also temperature control*

thermocouple 5-69

thermocouple system. *See fire detection systems*

thermosphere 2-4

thread sealer 2-45

three point landing 3-2

three-way control valve 1-15

tire cuts 3-35

tires, aircraft 3-27

  bias tires 3-27

  construction of 3-27

    bead 3-27

    breaker 3-28

chafing strip 3-27
chine 3-28
cord body 3-27
ply rating 3-28
sidewall 3-27
tread 3-27
tread patterns 3-28
dual installations 3-36
inflating 3-33
dual tire 3-36
inspection 3-30
dismounted 3-31
mounted 3-30
maintenance 3-31
dismounting 3-32
divided (split) wheels 3-32
remountable flange wheels. 3-32
mounting 3-32
divided (split) wheels 3-33
remountable flange wheels 3-33
nonrepairable 3-40
preventative maintenance 3-35
dual Installations 3-36
matching criteria 3-36
nylon flat spotting 3-36
uneven tread wear 3-35
radial tires 3-27
rebuilding/retreading 3-38
for tires operated above 120 m.p.h. 3-38
for tires operated below 120 m.p.h. 3-38
size designation 3-28
standard identification markings. 3-28
storage of 3-30
vent markings 3-29
tires and tubes, aircraft 3-27
TKS de-ice and anti-ice **10-8**
torquemeter indicator 5-71
total air temperature probe (TAT) 5-33
total air temperature (TAT) indicator 5-33
totalizer 9-17
total temperature (Tt) 5-33
traffic alert and collision avoidance system (TCAS) 7-10
TCAS I 7-10
TCAS II 7-11
transmitter 5-12, 6-4
transponder 6-16
inspection 6-17
transport category turbine aircraft fuel system 9-21
tread 3-27
patterns 3-28
troposphere 2-3
truck 3-6
true altitude 5-18

true power 4-8
tubes, aircraft 3-37
identification 3-37
inspection 3-37
nonserviceable 3-37
serviceable 3-37
storage of 3-37
tumble 5-42
turbine engine ground fire protection. *See fire prevention and protection*
turbine engine instruments 5-71
engine indicating and crew alerting system (EICAS) 5-72
engine pressure ratio (EPR) 5-71
engine top temperature control 5-72
torquemeter indicator 5-71
turbine vibration indicator (TVI) 5-72
turbine fuel 9-2
turbine gas temperature (TGT) 5-72
turbine inlet temperature (TIT) 5-69
turbine vibration indicator (TVI) 5-72
turbocompressors 2-9. *See also positive-displacement cabin compressors (superchargers)*
turn-and-bank indicator 5-47
ball portion 5-48
turn pointer 5-48
turn coordinator 5-49
turn pointer 5-48
two-stage regulator **10-7**. *See also pneumatic surface de-ice*
types of
typical pump-driven vacuum system 5-41
air filter 5-41
relief valve 5-41
selector valve 5-41
vacuum/suction gage 5-41
typical thermocouple thermometer system 5-69

U

underinflation 3-35
undocking 11-13
uneven tread wear 3-35
causes of 3-35
balance 3-35
misalignment 3-35
overinflation 3-35
tire cuts 3-35
underinflation 3-35
weather checking 3-36
upstream pressure 1-41, 1-60
utility hydraulic system 1-52
pump switch 1-53

V

vacuum-driven attitude gyro 5-42
valve-in-transit indicator lights 9-20
valves, fuel system 9-11
vane pumps 9-13
vane type fuel flow transmitter 5-63
vane-type pump 1-57
dry 1-57
pressure and vacuum 1-60
wet 1-57
vapor-cycle air conditioning system 2-31
recharging 2-39
refrigeration cycle 2-32
vapor lock 9-3
vapor separator 1-57
variable-delivery pump 1-11. *See also hydraulic pump*
demand principle 1-11
stroke-reduction principle 1-11
variable resistance method 4-19
variable resistors 5-7
V-command bars 6-32
vendor service publications (VPs) 11-3
ventilation air 2-15
vent markings 3-29
vertical gyro indicator (VGI) 5-37
vertical scale instruments 5-5
vertical speed indicator (VSI) 5-22
Very High Frequency (VHF) 6-4
Very High Frequency (VHF) system 6-6. *See also communication systems*
communications 6-7
operation (ops) check 6-7
visual/sight gauges 5-4
visual smoke detectors. *See smoke detection systems*
volatility of fuel 9-2
voltage drop (IR drop) 4-36
VOR test facility (VOT) 6-11

W

wafer screen filter 9-11
warning horn 3-10
warning messages 7-15
warning system, fuel pressure 9-19
warning systems 7-11
configuration warning systems 7-12
engine indicting and crew advisory system (EICAS) 7-14
stall warning systems 7-12
air data 7-13
electrical input 7-13
pressure input 7-12

water and toilet drain heaters **10-13**

water separator 2-19

weeping membrane **10-8**. *See also **TKS de-ice and anti-ice***

wheatstone bridge 2-20

wheatstone bridge circuit 5-65

wheels, aircraft 3-40

   assembly 3-41

   cleaning 3-42

   corrosion 3-44

   drop center 3-41

   inspection 3-43

      bearing maintenance 3-44

      corrosion 3-44

      physical damage 3-44

   maintenance of 3-42

      cleaning 3-42

      installation 3-42

      lubrication 3-42

      safety 3-43

   overhaul 3-43

   remountable flange 3-41

   split (divided) wheel 3-41

wheel well 3-4

windings

   burned 4-12

window defrost system **10-11**

windshield alcohol deicing systems **10-11**

windshield icing control systems **10-10**

   maintenance of **10-10**

wing inspection lights 7-6

wiper 1-15, 1-27, 1-40, **10-11**

wire 4-35

   aluminum conductor 4-44

   bundling 4-57

   chafing and abrasion 4-50

   determining current carrying capacity of 4-42

   dielectric strength 4-38

   harness at altitude 4-43

   identifying 4-47

   in a harness 4-43

   insulation 4-38

      effects of heat aging 4-43

   lacing 4-57

   maximum operating temperature 4-43

   mechanical strength of 4-38

   resistance 4-39

   single wire in free air 4-43

   size 4-36

   splice spacing 4-49

   twisted 4-49

   tying 4-57

   voltage drop in 4-38

wire bundle 4-49, 4-57

   radius of bends 4-49

   slack (sag) 4-50

wire group 4-49, 4-57

   radius of bends 4-49

wiring 4-35

   bonding and grounding 4-54

   conduit 4-53

   electrically protected 4-49

   electrically unprotected 4-49

   high temperature environments 4-51

   installation 4-48

   miscellaneous routing precautions 4-52

   protection 4-51

   routing 4-48

   shielding 4-54

   wheel well area 4-51

wiring diagrams 4-69

   pictorial 4-74

   schematic 4-74

World Geodetic System (WGD) 6-19. *See also **navigation systems***

Y

yaw damper 5-81

Z

zone final 11-13

Corrections, Suggestions for Improvement, Request for Additional Information

It is Avotek's goal to provide quality aviation maintenance resources to help you succeed in your career, and we appreciate your assistance in helping.

Please complete the following information to report a correction, suggestion for improvement, or to request additional information.

REFERENCE NUMBER (*To be assigned by Avotek*)		
CONTACT INFORMATION*		
Date		
Name		
Email		
Daytime Phone		
BOOK INFORMATION		
Title		
Edition		
Page number		
Figure/Table Number		
Discrepancy/Correction (*You may also attach a copy of the discrepancy/correction*)		
Suggestion(s) for Improvement (*Attach additional documentation as needed*)		
Request for Additional Information		
FOR AVOTEK USE ONLY	Date Received	
	Reference Number Issued By	
	Receipt Notification Sent	
	Action Taken/By	
	Completed Notification Sent	

Contact information will only be used to provide updates to your submission or if there is a question regarding your submission.

Send your corrections to:

Email: comments@avotek.com

Fax: 1-540-234-9399

Mail: Corrections: Avotek Information Resources
P.O. Box 219
Weyers Cave, VA 24486 USA